ASTROPHYSICS FROM SPACELAB

ASTROPHYSICS AND
SPACE SCIENCE LIBRARY

A SERIES OF BOOKS ON THE RECENT DEVELOPMENTS
OF SPACE SCIENCE AND OF GENERAL GEOPHYSICS AND ASTROPHYSICS
PUBLISHED IN CONNECTION WITH THE JOURNAL
SPACE SCIENCE REVIEWS

VOLUME 81

ASTROPHYSICS FROM SPACELAB

Edited by

PIER LUIGI BERNACCA

Osservatorio di Asiago, Università di Padova, Italy

and

REMO RUFFINI

Instituto di Fisica, Università di Roma, Italy

D. REIDEL PUBLISHING COMPANY

DORDRECHT : HOLLAND / BOSTON : U.S.A.

LONDON : ENGLAND

Library of Congress Cataloging in Publication Data

Astrophysics from Spacelab.
   (Astrophysics and space science library: v. 81)
   Based on a meeting held at the international centre for
theoretical physics, Trieste, in fall 1976.
   Includes bibliographical references and index.
   1. Astrophysics—Congresses. 2. Spacelab project—
Congresses. I. Bernacca, Pier Luigi. II. Ruffini, Remo.
III. Series.
QB460.A88   523.01   80-13036
ISBN 90-277-1064-3

Published by D. Reidel Publishing Company,
P.O. Box 17, 3300 AA Dordrecht, Holland.

Sold and distributed in the U.S.A. and Canada
by Kluwer Boston Inc., Lincoln Building,
160 Old Derby Street, Hingham, MA 02043, U.S.A.

In all other countries, sold and distributed
by Kluwer Academic Publishers Group,
P.O. Box 322, 3300 AH Dordrecht, Holland.

D. Reidel Publishing Company is a member of the Kluwer Group.

Printed in The Netherlands

TABLE OF CONTENTS

PREFACE

A meeting on "Astrophysics from Spacelab" was held at the International Centre for Theoretical Physics, Trieste, in the Autumn of 1976. Scope of
the meeting was to bring to the attention of an increasing number of physi-
cists and astrophysicists, including scientists from developing countries ,
the new facilities made available by the combination of the Shuttle and the
Spacelab programmes.

This book starts from that meeting and includes, together with reports
presented in Trieste, duly updated, a few additional reviews on selected topics.

In the first part, D.J.Shapland and G.Giampalmo ("The Shuttle and the
Spacelab") present the design and the programmatic data of these advanced
transportation systems and orbital laboratories. Vittorio Manno introduces
the scientific programmes coordinated and led to execution by the European
Space Agency ("ESA Programmes in Astronomy and Astrophysics"). J.D.Rosendhal
("The NASA Programmes in Astronomy and Astrophysics") summarizes the activities in solar physics, high-energy astrophysics and astronomy planned in the
United States of America by the National Aeronautics and Space Administration
as well as the expected use of the space shuttle and spacelab in their first
year of operation.

In the second part, reviews are presented on the significance of performing astrophysical observations from space in regions of the electromagne
tic spectrum not attainable from ground observations due to atmospheric scattering and absorption. C.Fichtel,D.Arnett,J.Grindlay and J.Trombka
("Gamma-ray Astrophysics") review the status of gamma-ray astronomy in the
light of observations from OSO-3, SAS-2 and COS-B satellites, and discuss
the progress which could be attained by detectors to be flown from spacelab
or from free flyers launched from the shuttle. They emphasize how the observations of both the gamma-ray diffuse background and the discrete gamma-ray
sources can give basic information on problems ranging from nuclear reactions occurring in supernovae and early cosmology, to the magnetosphere of pul
sars, to the origin of cosmic rays. In recent months, after this review was
completed, great interest has been aroused in the field of gamma-ray astro-

P. L. Bernacca and R. Ruffini (eds.), Astrophysics from Spacelab, vii–xi.
Copyright © 1980 by D. Reidel Publishing Company.

nomy by simultaneous observations in radio, optical, x-ray and gamma-ray of discrete sources (1). The very high energy flux in gamma-rays discovered in some pulsars and quasars and the comparison with simultaneous observations at lower energies promise to be of the greatest relevance for the final understanding of the physical mechanisms leading to the emission of radiation in both these systems.

Francesco Paresce ("Astrophysics in the Extreme Ultraviolet") reviews the current status of observations in the extreme ultraviolet and the implications that this new field of research can have in the understanding of the structure and evolution of stars and of the interstellar medium in the galaxy. Still in the ultraviolet, T.P.Snow and J.L.Linsky (" Ultraviolet Spectro scopy of the Outer Layers of Stars") review the great progress achieved by ultraviolet observations in the study of the mass loss from stars and of the physical processes occurring in their outer layers. Both these reports also consider the improvements to be expected by experimental facilities flown on spacelab.

M.Capaccioli and S.D'Odorico ("Nebular and Extragalactic Astronomy from Space in the Optical Region") review the optical observations already performed from space in the fields of nebular and extragalactic research . After an analysis of the technical reasons why large qualitative and quantitavive improvements in optical observations should be expected by using space telescopes, as contrasted with ground based observatories, they extensively deal with those domains of astrophysics in which important advances are expected to occur in the 1980s. Among these are galactic nuclei, Seyfert gala xies and quasars and the stellar content, structure and evolution of galaxies. Their emphasis is also directed towards a renewed observational approach to the classical aspects of cosmology: the determination of the extragalactic distance scale, the problem of the missing mass, the measurement of the dece leration parameter in the rate of expansion of the Universe. In their words, in this new era of optical observations from space "we will witness a jump of our astronomical knowledge comparable to that caused by the introduction in astronomy of the photographic technique". Finally, E.Bussoletti, R.Fabbri and F.Melchiorri ("Infrared Astronomy") have reviewed the status of infrared astronomy in the range 1-1000 microns. They deal mainly with the significance of infrared observations for cosmological observations of the microwave background and with future developments in this field to be expected with space-borne instrumentation. They also compare and contrast the performance of existing infrared detectors.

We have not reviewed in this book either x-ray or radio observations
from space. Since the launch in 1971 of the UHURU satellite of R. Giacconi
and his group, it has become patently clear that x-ray observations give new
and basic information on gravitationally collapsed objects like neutron stars
and black holes. A vast literature already exists (2) on this most successful branch of astronomy and astrophysics including further developments to
be expected by the use of the shuttle and the spacelab. The latest news in
this field were reported to the Trieste meeting by W.Lewin on x-ray bursters
and that review has already appeared in the literature (3). Different in several respects from x-ray astronomy is the case of radio astronomy already
well performed from the ground . Radio observations from space become essential when very long baselines, larger than the radius of the Earth, are needed in interferometric techniques. Using geostationary satellites could provide in the future an angular resolution down to the millisecond of arc, e-
nabling a complete new set of observations ranging from the analysis of the
radio source cores, to the study of proper motions, galactic structure and
rotation, and to geophysical observations such as earth planimetry and rotation (4).

The third part of the book deals with selected topics on solar system
sciences and spacelab utilisation for application programmes. In "Recent
Planetary Physics and Chemistry" , R. Smoluchowsky reviews the considerable
progress made in the last few years in the understanding of the origin of the
planets and of the structure of their interiors and atmospheres in light of
the Venerea and Viking results. It goes without saying that the physics of
the Earth atmosphere, ionosphere and magnetosphere and of our own environment
is one of the major goals in the study of planets from spacelab. We have not
treated this topic in these proceedings since an extensive publication already exists in the literature (5).

Still in this third part, J.Plevin ("Remote Sensing of Earth Resources
and Environment from Spacelab") discusses the use of spacelab as a remote sensing platform along with more general remote sensing programmes for a variety of investigations such as earth resources, global atmospheric physics,
climatology, oceanography. Finally, G.Seibert ("Material Sciences in Space")
gives a summary of the analyses already performed by the Apollo, Skylab and
Apollo-Soyuz experiments and emphasizes the specific goals of the researches
on materials in space. This program tends to develop new materials with exceptional properties by exploiting the results of physical and chemical analyses of known materials under high vacuum conditions and in the absence of

gravitational fields. We have not reviewed in this section the extensive work
planned from spacelab and the solar probe for the study of the solar wind ,
of the sun's photosphere, corona, chromosphere, and more generally of the so
lar activity. Clearly simultaneous observations in different spectral bands
with high spatial and temporal resolution will allow a detailed analysis
of the processes of energy release in active regions of the sun as well as
of their location, extension, and their development in time. Strictly related to solar energy and to its utilization on the earth are spacelab - born
experiments dealing with earth climatology, the energy transfer in the earth
atmosphere and the coupling of the thermosphere to the lower atmospheric layers. For an introduction to these topics we refer to a study performed by
ESA (6).

At the farthest reaches of the solar system begins the interstellar
medium from which astrophysical galactic bodies were formed. It is also a
screen through which the galactic and extragalactic Universe is seen from
the Earth. The fourth and last section of the book deals with the improved
knowledge of this medium, following the observations from the Copernicus
satellite and examines future developments made possible by observations car
ried out from spacelab. D.G.York ("Abundance determinations for Interstellar Gas") reviews the general abundance studies and discusses the deple
tion of some heavy elements near the solar system, while the current knowledge of interstellar grains and molecules is reviewed by T.P.Snow in "Ultra
violet Observations of Interstellar Molecules and Grains from Spacelab".

It is a pleasure to acknowledge the International Center for Theoretical Physics, its director Prof. Abdus Salam, and the International   Atomic
Energy Agency for sponsoring the meeting. The technical assistance of the
entire staff of the Center has greatly contributed to the success of the meeting. We would like to express our gratitude to Dr. Gallieno Denardo and to
Prof. Margherita Hack for their help and contributions in the organization
of the meeting. Finally the financial contributions by the European Space
Agency and by the Italian Astronomical Society are gratefully acknowledged.

Roma, 25th of September 1979

P.L.Bernacca and R.Ruffini

REFERENCES

1. See e.g. R.N.Manchester and J.H.Taylor: "Pulsars", W.H.Freeman and Co.,
        S.Francisco,1977.

    N.D'Amico and L.Scarsi: "Gamma-Ray Astronomy" in the Proceedings of the
        1979 Austral Summer School, Ed. C.Edwards, Springer,1979.

2. See e.g. H.Gursky and R.Ruffini (Eds.): "Neutron Stars, Black Holes, and
        Binary X-Ray Sources", D.Reidel Pub. Co.,Dordrecht, 1975.

    See also R.Giacconi and R.Ruffini (Eds.): "Physics and Astrophysics of
        Neutron Stars and Black Holes", North Holland Pub. Co., Amster-
        dam, 1978.

3. W.Lewin: "X-Ray Bursters", Proceedings of the 8th Texas Symposium on Re-
        lativistic Astrophysics, Ann. N.Y. Acad. Sci., 302, 210, 1977.

4. See e.g. "Very Long Baseline Radio Interferometry Using A Geostationary
        Satellite", ESA DP/PS(78)15.

5. G.Fiocco (Ed.): "On Mesospheric Models and Related Experiments", Esrin-
        Eslab Symposium, D.Reidel Pub. Co., Dordrecht, 1971.

    See also " Spacelab Sub-Satellites for Atmosphere, Magnetosphere and
        Plasma in Space", ESA DP/PS(76)7, 1976.

    See also " Spacelab Borne Atmospheric Passive Experiments", ESA DP/PS
        (77)13, 1977.

6. "Sun-Earth Observatory and Climatology Satellite Seocs", ESA DP/PS(78)
        10, 1978.

THE SHUTTLE AND THE SPACELAB

D.J. Shapland and G. Giampalmo
European Space Agency, Paris

ABSTRACT

The combination of Space Shuttle and Spacelab represents an important tool for space research in a variety of disciplines. This international orbital facility will be available to all levels of experimenters - be they from small laboratories or large organisations - through the 1980s. A summary description of the Space Shuttle and Spacelab programmes is provided, to help the future user in the fields covered by this book, to better appreciate the potential of the system. The material presented includes design and programmatic data and introduces the potential user to the operations associated with this advanced transportation system and orbital laboratory. The latter includes payload integration activities foreseen for Europe and the US, as well as the launch site operations necessary to put the Orbiter-Spacelab system into orbit. In addition, the general aspects of possible Spacelab utilisation are discussed, stressing the concept of reusability. Mission scenarios and tentative utilisation models are presented, the principal elements of the joint ESA-NASA payload for the first Spacelab flight are described and the proposed means of effecting a ready access for experimenters to space are outlined.

INTRODUCTION

The 1980s will see the introduction of a space transportation system (STS) which represents a revolutionary approach to the performance of space science experiments. Two major elements of the STS are the Space Shuttle and Spacelab. They represent a major advance in space activities since they introduce the concept of reusability of hardware leading, in turn, to reduced transportation costs, relaxed experiment design constraints and increased opportunities for space experimentation.

The Space Shuttle is being produced in the US under the auspices of NASA. Spacelab is being developed in Europe, with European funds,

1

P. L. Bernacca and R. Ruffini (eds.), Astrophysics from Spacelab, 1-50.

under the management of ESA. In operation, Spacelab will be transported
to and from orbit in the cargo bay of the Shuttle Orbiter, and remain
there throughout the flight. Spacelab, therefore, extends the Shuttle
capability and the Orbiter/Spacelab combination represents a short-stay
space station which can remain in orbit for periods of up to 30 days.
This "space station" will provide opportunities for experimenters to
live and work in space under conditions which are hardly more exacting
than those experienced on Earth. Further, the space laboratory can be
re-configured after each flight to satisfy the needs of the next
mission.

The theme of this book is how best to use Spacelab in the fields
of physics and astrophysics. However, because of the close inter-
dependence of Shuttle and Spacelab, one must describe both systems in
order to present the capability of the combination. It is the intent
of this chapter to provide a general summary description of both programmes, that can act as a back-cloth against which subsequent
specialist discussions can be carried out. Further, some general
remarks are provided as to the way Spacelab might be utilised.

THE SPACE SHUTTLE PROGRAMME

    It has been authoritatively stated that the Space Shuttle pro-
gramme was designed "... to help transform the space frontier of the
1970's into familiar territory, easily accessible for human endeavour
in the 1980's and 1990's..." This lofty goal is possible since the
Space Shuttle programme exhibits the following characteristics:

- Provides low cost transportation to and from low-earth orbit;
- Utilises reusability;
- Ensures short turn-around time;
- Expands practical application of space technology;
- Provides a versatile and flexible launch system with relaxed
  constraints;
- Reduces total programme costs;
- Internationalises space.

The Space Shuttle System consists of the Orbiter, a large external
tank containing ascent propellants (liquid oxygen and liquid hydrogen)
to be used by the Orbiter main engines, and two solid rocket boosters.
The only fully expendable element is the external tank. An artist's
impression of the Space Shuttle on the launch pad at Kennedy Space
Center (KSC) in Florida is given in Fig. 1. The total system weighs
2×10^6 kg at launch and measures about 56m overall. Some dimensions
relative to the Space Shuttle vehicle are provided in Fig. 2 (1ft =
0.3048m, 1lb = 0.4536kg). The cargo bay which accommodates Spacelab
has the dimensions 18.3m x 4.6m dia. The Orbiter is about the size of
a DC-9 aeroplane and the total Shuttle system size is compared in
Fig. 3 with some other well known transportation systems.

Figure 1.   Space Shuttle on Launch Pad

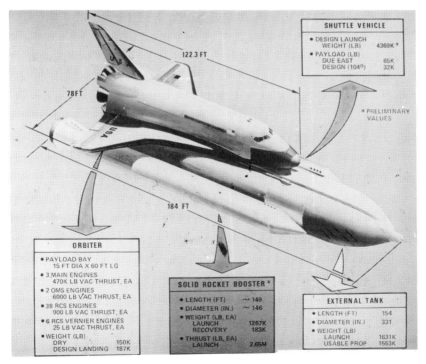

Figure 2.   Space Shuttle Vehicle Characteristics

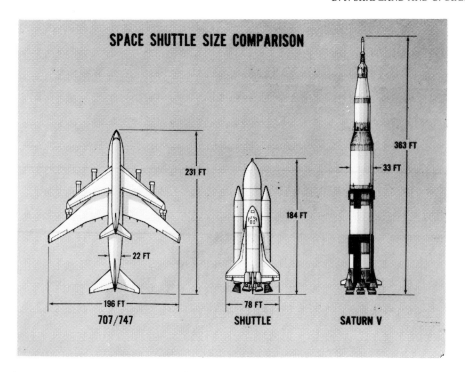

Figure 3. Space Shuttle Size Comparison

Space Shuttle Flight Profile

    A discussion of the complete flight profile of the Space Shuttle (Fig. 4) is helpful in understanding the operating mode of the new transportation system.

    At lift-off (Fig. 5) the solid rocket boosters (SRB's) and Orbiter main engines fire in parallel developing a total thrust of 30×10^6 newtons. After burn-out (at about 125 sec into the flight), the two SRB's are jettisoned over the sea as shown in Fig. 6. The spent rockets descend by parachute and are recovered from the sea using a special retrieval ship and the procedure illustrated in Fig. 7. The Orbiter continues its ascent on its main engines and, after engine shut-down, the large external tank is separated (Fig. 8) just before orbit insertion which is accomplished by means of the orbital manoeuvring system (OMS) and depicted in Fig. 9. The external tank (ET) re-enters the atmosphere and its remnants impact the sea almost 10,000 km down range. The OMS develops a thrust of 53×10^3 newtons burning MMH/N_2O_4 propellants and is used for orbit adjustments and any subsequent manoeuvring.

    On orbit, the Orbiter's cargo bay doors are opened and cargo operations commence. This phase of the flight is illustrated in Fig. 10 which shows a satellite attached to a small propulsion stage being readied for stage ignition with the aid of the remotely controlled

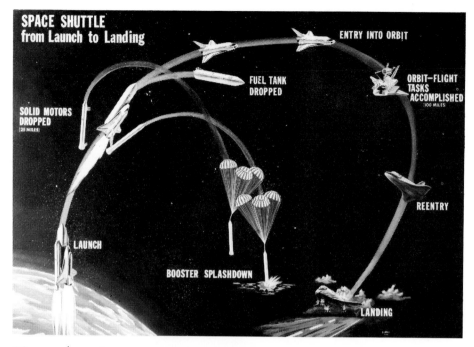

Figure 4.   Space Shuttle Flight Profile

Figure 5.   Space Shuttle Lift-off

Figure 6. Solid Motor Jettison

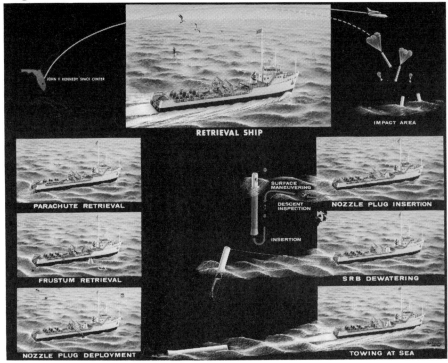

Figure 7. Retrieval of Solid Rocket Boosters

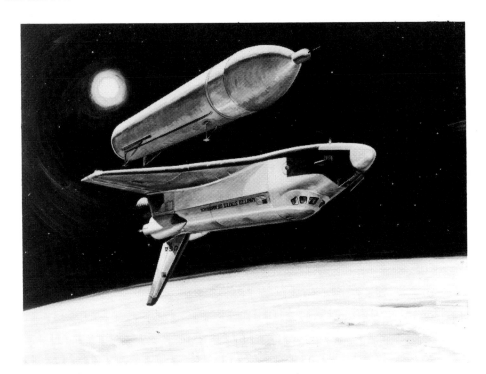

Figure 8.   Large External Tank Jettison

Figure 9.   Orbit Insertion

Figure 10.   Orbital Deployment Operation

manipulator arm. It is during this portion of the flight that, when
carried, Spacelab becomes operational. After the orbital tasks are
completed - on orbit time depends on the tasks performed but, for
Spacelab operation, it can be up to 30 days although the nominal
mission time is 7 days - the cargo bay doors are closed, a de-orbiting
manoeuvre is initiated, and the Orbiter re-enters.

     Re-entry into the Earth's atmosphere (Fig. 11) is made at a high
angle of attack. High temperatures - approximately 1600°C at the
stagnation points and about 300°C on the leeward surfaces - are
generated during this phase of the flight and special radiative cooling
surfaces, composed of silica tiles, are employed over most of the
Orbiter and carbon-carbon inserts are used at the nose and leading
edges. However, the combination of radiative cooling and insulation
permits the Orbiter to be constructed of conventional aluminium
material. At lower speed the angle of attack is reduced and the Orbiter
lands in a gliding, unpowered mode as illustrated in Figs. 12 and 13.
Due to its favourable hypersonic lifting characteristics, the Orbiter
can reach landing sites as far as 2000 km on either side of its initial
entry flight path. The runway, a conventional 4.5 km concrete strip
which can be seen in Fig. 12, already exists at KSC which will be the
initial launch and landing site for the shuttle. Later, Vandenberg Air
Force Base (VAFB) in California will also be utilised. The allowable
launch azimuths and corresponding orbit inclinations attainable from
these two sites are shown in Fig. 14. Using both sites, global coverage
can be achieved.

Figure 11. Re-entry of Orbiter

Figure 12. Orbiter Landing Approach

Figure 13. Orbiter Landing

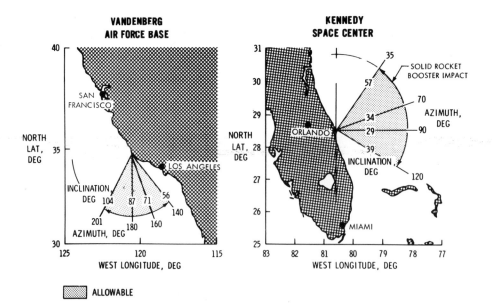

Figure 14. Practical Launch Azimuth and Inclination
          Limits from VAFB and KSC

Space Shuttle Capabilities

   Some capabilities of the Space Shuttle that are of particular
interest to potential users are summarised in Fig. 15. The Shuttle as
a general purpose space transportation system has many applications
as illustrated in Fig. 16. However, the particular concern here is
with its use as a carrier of Spacelab. In this respect, the Shuttle
provides considerable mission flexibility by means of the wide range of
performance parameters which are available.

PAYLOAD WT	29,500 Kg (28½ deg., 400 Km) 17,000 Kg (90 deg., 400 Km) LANDED PAYLOAD LIMIT 14,500 Kg
PAYLOAD VOLUME	CARGO BAY 18.3 m X 4.6 m dia
ORBIT INCLINATION	FROM 28°. 5 UP TO 104° (ETR and WTR) GLOBAL COVERAGE
ORBIT ALTITUDE	CIRCULAR OF ELLIPTICAL ORBITS WITH USE OF OMS CIRCULAR 200 Km to 1,100 Km (ETR)
LAUNCH TIME/WINDOW	NO PARTICULAR CONSTRAINTS
FLIGHT DURATION	NOMINAL 7 DAYS, LATER UP TO 30 DAYS
ORIENTATION	ANY, ACCURACY ± ½ deg (GN&C) ± 0.1 deg (FCS) STABILITY ± 0,01 deg/sec/axis
COMMUNICATION LINK	TDRSS (up to 50 mbps) STDN (4 MHz or 5 mbps)
ENVIRONMENT	GENERALLY BENIGN MAX g ~ 3 ON-ORBIT $< 10^{-5}$ g

        Figure 15. Space Shuttle Capabilities

   In particular, the Orbiter cabin provides a living and working
accommodation for the crew. Fig. 17 shows the Orbiter crew working in
the upper, aft flight deck, from where the manipulator arm, Spacelab
and Spacelab payloads can be monitored and controlled as required. The
standard Orbiter crew is 4 persons (commander, pilot, mission specialist
and payload specialist) but an additional three payload specialists can
be accommodated. Spacelab represents a natural extension of the somewhat
confined working area of the Orbiter. If extra vehicular activity (EVA)
is required for experiment operation, this can be provided by the Orbiter
crew.

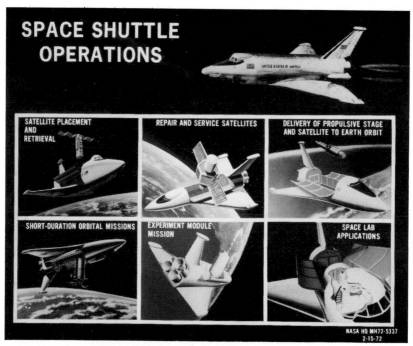

Figure 16. Space Shuttle Applications

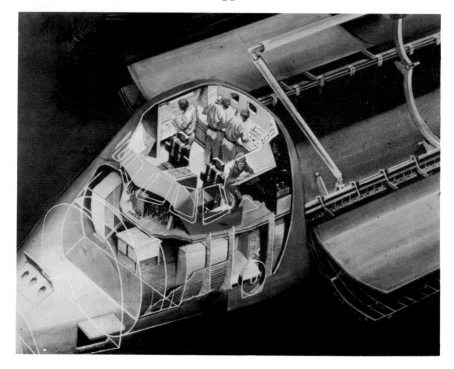

Figure 17. Orbiter Crew Accommodation

Space Shuttle Programme Milestones

    Some principal milestones of the Shuttle programme are given in
Fig. 18. Orbiter 101 will be used for approach and landing tests during
late 1977. For these tests, the Orbiter vehicle will be carried aloft
by, and launched from, a specially converted Boeing 747. Fig. 19
illustrates the proposed configuration. This mode will also be used for
ferrying the orbiter between distant landing and launch sites.

CRITICAL DESIGN REVIEW   AUGUST 1977

ORBITER 101

   . ROLL OUT   SEPTEMBER 1976
   . FIRST CAPTIVE FLIGHT (747 FERRY)   MARCH 1977
   . APPROACH AND LANDING TESTS   JUNE 1977 - MARCH 1978

ORBITER 102

   . TO KSC   AUGUST 1978
   . ORBITAL FLIGHT TESTS (6)   MARCH 1979 - MAY 1980
   . FIRST OPERATIONAL FLIGHT   JULY 1980

Figure 18.   Space Shuttle Programme Milestones

    Orbiter 102 will be used for vertically-launched flights and six
orbital flight tests are planned. The first operational flight of the
Space Shuttle is scheduled for July 1980 and it is on this flight that
Spacelab will receive its orbital baptism carrying a joint ESA-NASA payload composed of various multi-discipline elements - science, applications, technology and verification test equipment.

    The total cost of the Space Shuttle development programme is
approximately 5.5 billion dollars (1972 prices).

THE SPACELAB PROGRAMME

    Spacelab is being designed to capitalise on the advantageous features
provided by the Space Shuttle, and more specifically, it was conceived
to satisfy the objectives listed below:

  • To provide to a large multi-disciplinary user community a
    versatile laboratory and observatory facility;

- To reduce significantly both the time and cost required for space experimentation;
- To make direct space research possible for qualified scientists and engineers without the need of full astronaut training.

Figure 19. Orbiter-Boeing 747 Piggyback Concept

The principal programme requirements can be summarised as follows:

- Pre-determined funding ceiling;
- Delivery of flight unit one year-, engineering model two years-before launch together with associated ground support equipment;
- Low operations costs to be ensured;
- User flexibility to be preserved;
- Experiment payload weight 5000 to 9000kg;
- Provision for follow-on production;
- Flight duration 7 to 30 days;
- Design life 50 reuses or 10 year lifetime;
- Spacelab crew of 1 to 4 payload specialists;
- Compatibility with Space Shuttle.

The resulting concept is fully compliant with these requirements. The artist's impression in Fig. 20 shows Spacelab in a module + pallet configuration accommodated in the Orbiter cargo bay. The illustrated payload is one proposed for an atmospheric and space physics mission.

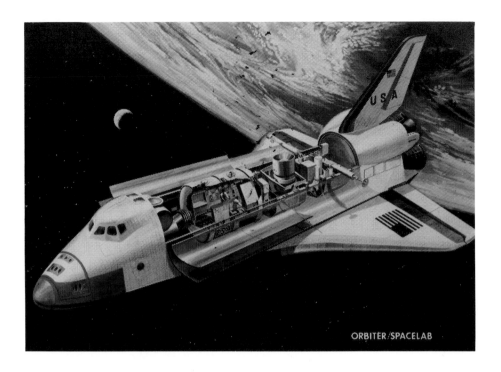

ORBITER/SPACELAB

Spacelab Design Considerations

   Spacelab consists basically of two parts: a pressurised module
that provides a "shirt-sleeve" environment laboratory and an unpress-
urised pallet structure which acts as an observing platform and permits
the direct exposure of instruments to the space environment. The main
external design features are shown in Fig. 21.

   The pressurised module is composed of two identical cylindrical
shells (2.7m length, 4.1m dia.) enclosed by two-end cones. The use of
one or two shells leads to a small- or large-module respectively. The
core segment of the module contains the basic Spacelab subsystems but
also has volume (approx. 7.5m^3) set aside for experiments. The experiment segment provides an additional experiment volume of 15m^3. The subsystem and experiment equipment is housed in 19" standard racks so that
laboratory-type equipment can be used.

   The pallet is also modular in construction and comprises up to five
segments each of 2.9m length. They may be attached individually to the
Orbiter or in a series of up to three (this configuration is referred
to as a pallet train). Electrical and cold plate cooling services are
available on each pallet segment. In addition to the basic Spacelab
facilities, other common payload support equipment required for particular experiments is available, e.g. an optical window of given
spectral performance, view ports for observing the outside, an airlock

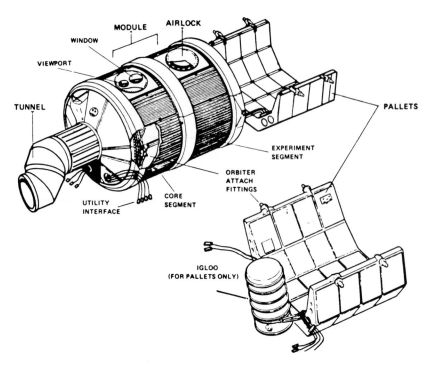

Figure 21. Spacelab External Design Features

for exposure of experiments to vacuum without resorting to pallet-type operations and a film vault for the storage and protection of sensitive film. Ample stowage volume is provided by over-head and under-the-floor containers. The locations of some of these facilities within the module are shown in Fig. 22.

The modularity of Spacelab is further illustrated in Fig. 23 and this principle permits the mission planner to select a configuration which is compatible with the mission objectives. Three basic configurations are immediately obvious - module only, module + pallet and pallet only - but many variations of the basic theme are possible. Fig. 24 shows module and pallet elements just prior to integration of a Spacelab "hard mock-up" and Fig. 25 illustrates the interior of the module laborabory.

In addition to the common payload support equipment mentioned above, only certain mission dependent elements of the subsystems (racks, cold plates, data recorders, etc.) need be flown. The net result is that the payload capability of Spacelab can be maximised for a certain mission.

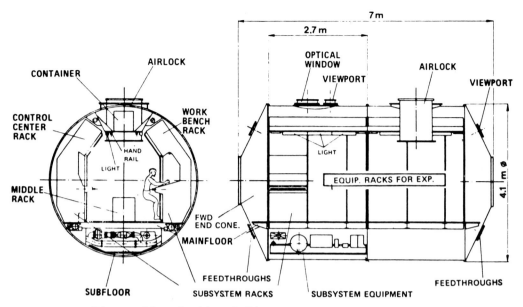

Figure 22.   Sectional Views of Spacelab Module

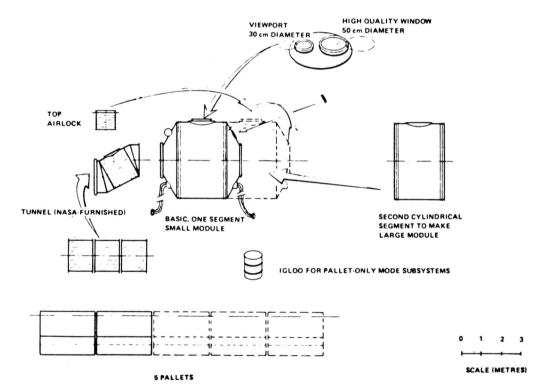

Figure 23.   Spacelab Modularity Approach

Figure 24.   Elements of Spacelab Hard Mock-up

Figure 25.   Spacelab Module Interior

The tunnel is also modular and provides access between Orbiter cabin and Spacelab, while the airlock mounted on the tunnel adapter permits EVA activities without interfering with the work schedule of the laboratory. The igloo is used when a pallet-only mode is employed. It ensures a pressurised and thermally controlled environment for the subsystem elements required for this mode. Experiments located on the pallet may be controlled from the module, the Orbiter, or from the ground, as required.

Spacelab is being designed for at least 50 seven-day flights or for a 10-year life, whichever is obtained first. A crew of up to four "payload specialists" (men or women) can be accommodated for performing the experiments. Normally they would work on a shift basis, with the Orbiter cabin being used for off-duty periods.

Spacelab-Orbiter Interfaces

The close interfaces that exist between Spacelab and Shuttle result in certain advantages and disadvantages for Spacelab. It can draw on the orbit-performance capabilities of the Shuttle, mentioned previously, and on the many resources available in the Orbiter as explained later. However, it also has to conform to certain constraints imposed by the Orbiter. These include the following major constraining factors:

- Dimensional limits of Orbiter cargo bay (18.3m x 4.6m dia.);
- Orbiter landing limit on cargo weight (14,500kg);
- Strict CG control for Spacelab and its payload;
- Functional and physical interfaces.

The landing weight limitation of 14,500kg compared to the actual launch capability for various orbit altitudes is shown in Fig. 26. The advantage of sharing a Spacelab launch with space hardware to be left in orbit arises from this consideration. As a consequence of the CG constraint, Spacelab must be placed as far aft in the cargo bay as possible. This is effected by means of a modular tunnel and the forward and aft extremes of location of Spacelab in the Orbiter cargo bay are given by Fig. 27. Even so, careful loading of Spacelab is essential and the general rule, that heavy elements of the payload must be located as rearward as practical, must be followed.

Services for Spacelab Users

The Spacelab subsystems provide basic services for running Spacelab itself and for its payload. A mix of Orbiter-provided and Spacelab-furnished resources are used. Orbiter support to Spacelab is summarised in Fig. 28. It must be borne in mind that even though a raw resource may be provided by the Orbiter, the Spacelab subsystem equipment ensures that the resource is conditioned so that it is available in a form that is compatible with the needs of the experiment equipment.

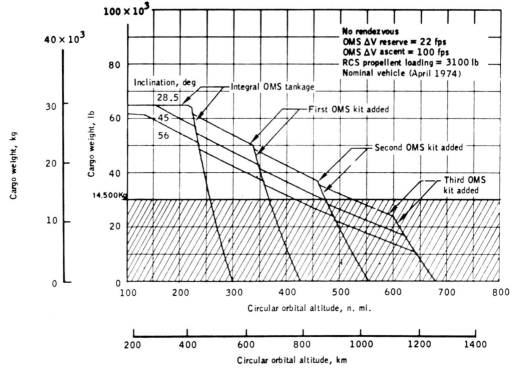

Figure 26.   Shuttle Orbital Capability (for launch from KSC)
             and Orbiter Landing Constraints

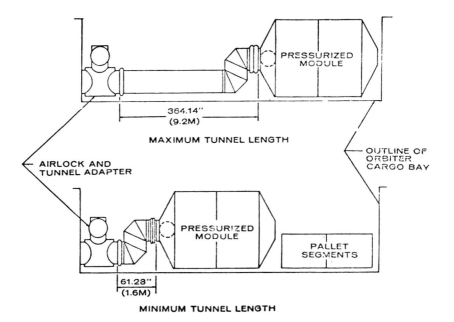

Figure 27.   Tunnel Configurations for Forward and Aft
             Locations of Spacelab

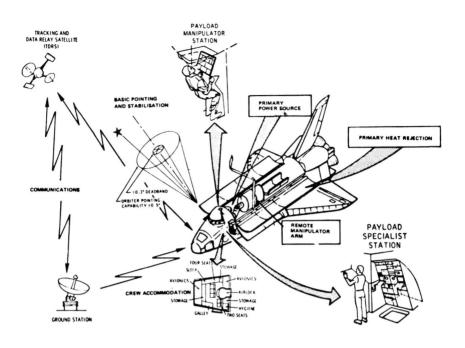

Figure 28.   Space Shuttle Provisions for Spacelab User

As a result, the services shown in Fig. 29, for four typical configurations, are provided in Spacelab with standard interfaces being
maintained between the Spacelab subsystems and the experiment equipment.

Particular mention must be made of the Instrument Pointing Subsystem (IPS), illustrated in Fig. 30.   The IPS will be delivered as part
of the Spacelab programme hardware and hence is available for users with
special pointing needs.   It builds on the Orbiter coarse pointing capability provided by its reaction control subsystem (RCS) and will provide
arc sec pointing capability in 3 dimensions.   Because of its so-called
"inside-out" design there are no strict limits on the geometry of the
instrument to be pointed.   Typically, instruments weighing up to 3000kg
and of diameter 3m can be accommodated by the IPS which incorporates a
mounting ring which presents a clean interface with the payload, a soft
mount, a clamp assembly and an optical-electrical system that permits
on-orbit alignment.   The IPS operates in a closed loop manner with the
Control and Data Management Subsystems (CDMS) which performs the
necessary high speed data computation.

Spacelab Programme Funding and Schedule

The Member States of ESA (with the exception of Sweden) have agreed

SPACELAB CONFIGURATION AVAILABLE TO USERS	SHORT MODULE + 9 METER PALLET	LONG MODULE	15 METER PALLET	INDEPENDENTLY SUSPENDENT PALLET
• PAYLOAD WEIGHT (KG)	5500	5500	8000	9100
• VOLUME FOR EXPERIMENT EQUIPMENT				
PRESSURIZED (M³)	7.6	22.2	–	–
NON-PRESSURIZED (M³) (NO OVERHANG)	99.8	–	167.4	97.1
• PALLET MOUNTING AREA (M²)	51.3	–	85.7	53.6
• ELECTRICAL POWER (28 VOLTS DC/ 115/200 VOLTS AT 400 HZ AC)				
AVERAGE (KW)	3 – 4	3 – 4	4 – 5	4 – 5
PEAK (KW)	8.8	9.0	10.0	10.0
ENERGY (KWH)	∿ 400	∿ 400	∿ 600	∿ 600
• EXP. SUPPORT COMPUTER WITH CENTRAL PROCESSING UNIT	◄———— 64 K CORE MEMORY OF 16 – BIT WORDS ————► ◄———— 320 000 OPERATIONS PER SEC. ————►			
• DATA HANDLING TRANSMISSION THROUGH ORBITER STORAGE DIGITAL DATA	◄———— UP TO 50 MBPS ————► ◄———— UP TO 30 MBPS ————►			

Figure 29.   Spacelab Services for Users

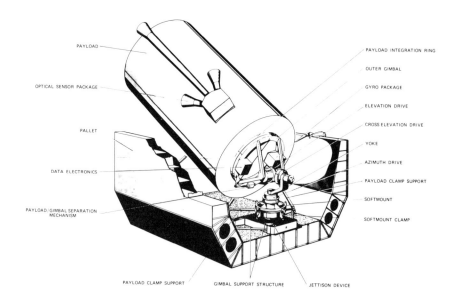

Figure 30.   Instrument Pointing Subsystem

on a fixed funding envelope of 396 million Accounting Units
(1 AU ≃ 1.3 US$), reckoned in mid-1975 prices, for the design, develop-
ment and manufacturing activities associated with Spacelab. Fig. 31
gives a rough breakdown of the budget allocation and shows the percentage distribution of the funding among the ESA Member States. Although
not a Member State, Austria has also agreed to partake in the Spacelab
programme.

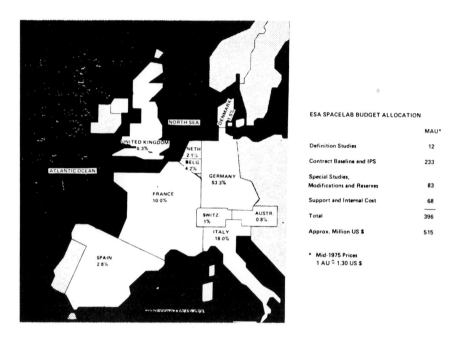

Figure 31. Spacelab Programme Funding

     The programme schedule presented in Fig. 32 identifies some impor-
tant programme milestones. The flight unit of Spacelab will be delivered
to NASA in mid-1979 in preparation for its first flight in mid-1980.
Although this first unit is provided "free of charge" to NASA, additional
Spacelabs can be procured in Europe by any organisation which might
require them to conduct research and technology investigations for
peaceful uses.

SHUTTLE/SPACELAB OPERATIONS

     This discussion is presented to help a potential user to visualise
the ground activities associated with a typical Shuttle/Spacelab mission.
A total operational cycle for the Shuttle/Spacelab system is depicted in
Fig. 33. This profile is repeated from flight to flight but with a
different Spacelab payload complement. The ground operations - whose

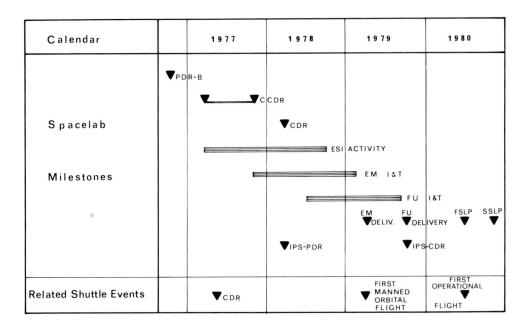

Figure 32. Spacelab Programme Milestones

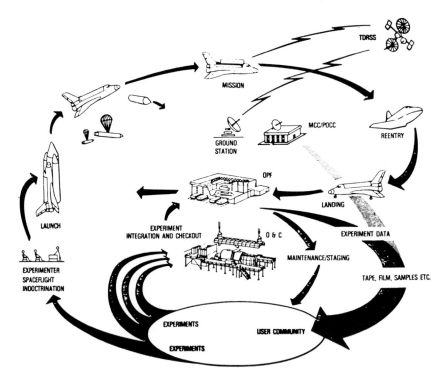

Figure 33. Shuttle/Spacelab Operational Profile

time may be reckoned in months rather than the years associated with
current payload launchings - include activities related to the experiments to be flown and the transportation system.

Experiment Operations

The integration of individual experiments into a payload and check-
out at system level represents an important part of the total Shuttle/
Spacelab operational profile. The type of activities involved in a
joint ESA/NASA mission can be visualised as those illustrated in Fig. 34.

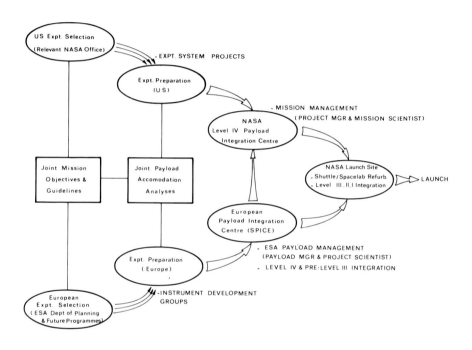

Fig. 34. Spacelab Experiment/Payload Operations

Direct delivery of experiment equipment to the launch site is permitted
if the hardware is considered to be self-contained and fully checked out.
For NASA-only or ESA-only missions, similar flows can be foreseen but,
of course, there would be no interaction at payload level. However,
launch site operations are always performed at the NASA launch site. In
the case of an ESA-only mission the pre-launch operations at the launch
site may be considerably reduced. The geographical locations of the
principal NASA and ESA centres relating to the first Spacelab payload
are shown in Fig. 35. Mission (in flight) operations are controlled from
JSC, Houston. The Mission Control Centre (MCC) and the Payload Operations
Control Centre (POCC) relate to the STS system and payload respectively.

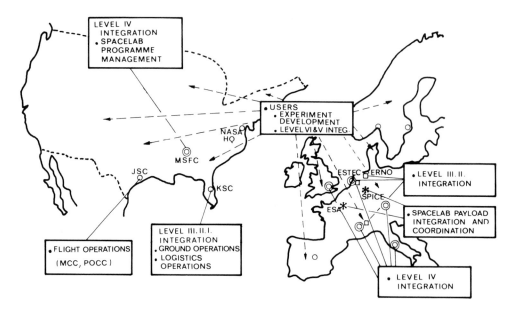

Figure 35. Spacelab Missions-Implementation

It is usual to talk about levels of integration - each level per-
taining to a particular payload-related function. Integration levels,
as currently planned, are defined in Fig. 36. Experiment integration
normally starts at the user's home laboratory or establishment where
the various parts of the experiment are assembled and intra-experiment
interfaces verified. This pre-level IV integration ensures that the
experiment equipment is functioning properly. The remaining levels of
integration (IV through I) are performed at central sites in Europe or
the US. The European integration activities will be performed under the
auspices of SPICE (acronym for Spacelab Payload Integration and Co-
ordination in Europe, see page 45), based at Porz-Wahn, Germany.
Various sites are visualised for integration in the US, but for the first
Spacelab payload, Marshall Space Flight Center at Huntsville is responsible for the level IV work and the launch site activities will take
place at the Kennedy Space Center in Florida.

The pre-level III integration is foreseen for Europe only. It
involves the integration and testing of all the experiments that
comprise the European payload complement. This activity will ensure a
high confidence in the successful operation of the European payload
elements and can result in a relatively short time for integration with
the US complement.

DEVELOPMENT PHASE	No Spacelab Flight Hardware Involved	Pre-Level IV	Integration and "Acceptance" Testing of one or more Spacelab Experiments or Experiment Element(s) using Flight Type GSE. (Early Interface Verification)
INTEGRATION PHASE	Spacelab Flight Hardware Involved	Level IV	Integration and check-out of experiment equipment with individual experiment mounting elements
		Pre-Level III (Europe only)	Combination, Integration and check-out of all European Experiment mounting elements with European Experiment Equipment already installed. Integration of European part of Flight Application and check-out software
		Level III	Combination, Integration and check-out of all experiment mounting elements with experiments equipment already installed and of experiment and Spacelab software
		Level II	Integration and check-out of the combined experiment equipment and experiment mounting elements with the flight subsystem support elements and extension modules, when applicable
		Level I	Integration and check-out of Spacelab and its payload with the shuttle orbiter

Figure 36. Integration Levels for Spacelab Payload Items

Space Transportation System Integration

The STS integration will take place at the launch site and this section deals with some of the activities and facilities planned for this stage of the ground operations. Reference should be made to Fig. 37 throughout this discussion. An aerial view of the KSC Shuttle launch site is given in Fig. 38, but at this time, many of the required facilities are still under construction.

The activities start with the delivery of the payload to KSC. It is then integrated into the refurbished Spacelab in the Operations and Check-out (O&C) building. Typical activities in the O&C building are illustrated in Fig. 39. The Spacelab and its payload are then taken to the Orbiter Processing Facility (OPF) where the Orbiter is normally refurbished and Spacelab is installed in the cargo bay in the horizontal position (Fig. 40). The Orbiter, with Spacelab and its payload in the cargo bay, is then towed to the Vehicle Assembly Building (VAB) where it meets other Shuttle components (ET and SRB's, see flow of Fig. 37). The total Shuttle system is then assembled in a vertical position as shown in Fig. 41.

The total integrated Shuttle/Spacelab system is then taken to the launch pad by means of the crawler vehicle depicted in Fig. 42. The actual pre-launch check-out and the launch itself is controlled from the Launch Control Centre (Fig. 43). The front room contains the consoles of the Launch Processing System (LPS) which are used to check-out various elements of the system. It is likely that one such console and its associated computer system will be assigned to Spacelab. The total system just prior to launch is illustrated in Fig. 44 showing

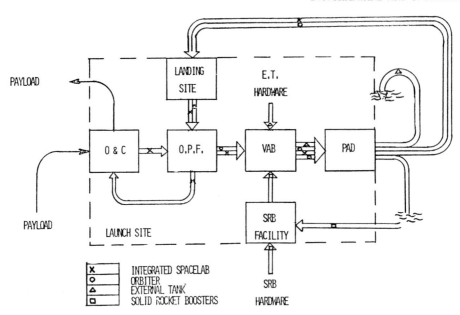

Figure 37.   Shuttle-Spacelab Turnaround Flow

Figure 38.   General View of KSC Launch Site

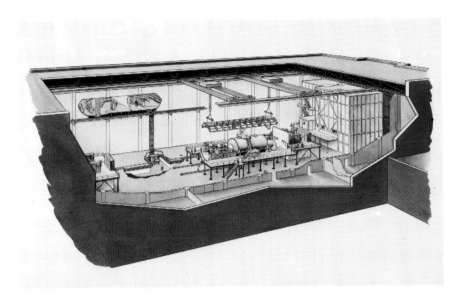

Figure 39. Activities in O&C Building

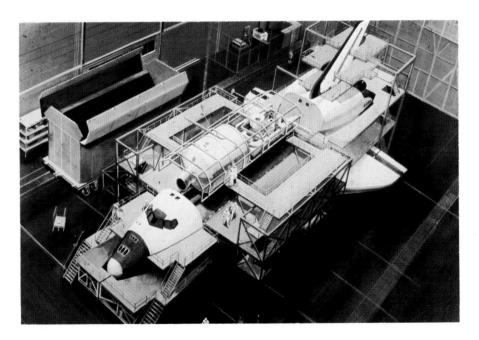

Figure 40. Orbiter in OPF

Figure 41.   Space Shuttle Assembly in VAB

Figure 42.   Assembled Shuttle on Crawler

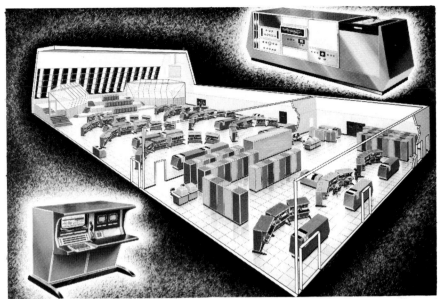

Figure 43. Launch Operations Control Centre

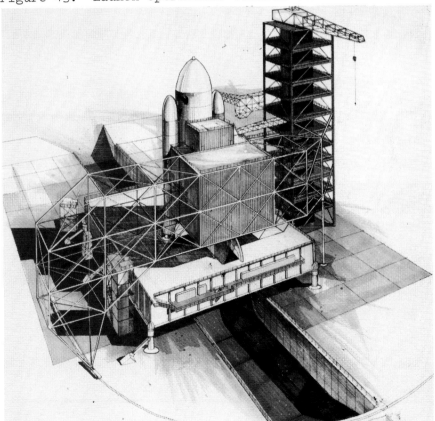

Figure 44. Shuttle Vehicle on Launch Pad Showing Payload
            Closeout Room

the launch pad service and access tower and the payload changeout room
which can be used for last minute operations with the Shuttle cargo.
The blast-off depicted in Fig. 5 serves to complete the KSC operations.

After landing at the conclusion of the flight, the Orbiter is
taken to the OPF where Spacelab is removed and returned to the O&C
building. The payload equipment, stored data, specimens, etc. are
removed from Spacelab in the O&C building and despatched to the user.
The Orbiter and Spacelab are then refurbished in the OPF and O&C
buildings respectively. A new payload arrives at KSC and the whole
cycle begins again. The total turnaround time for the Shuttle and
Spacelab is two weeks, but payload integration may be performed off-line
at a more leisurely pace.

SPACELAB UTILISATION

Both NASA and ESA have been actively engaged for many years in the
study of the opportunities provided by Spacelab in the different fields
of scientific research and application disciplines, including the new
fields which are opened up by the availability of a manned space
laboratory.

NASA has elaborated a number of mission scenarios out of which
selected missions have been incorporated in the early STS mission
planning and is already preparing the first three Spacelab missions.

ESA has undertaken the promotion and the establishment of a
European Spacelab utilisation programme, having as its nearest target
the participation of European experiments in the joint NASA/ESA first
Spacelab payload (FSLP) and is defining further missions following
FSLP.

The purpose of this section is to give an overview of the main
results of these activities by summarising the main features of
Spacelab from the users' point of view, by outlining its possible
utilisation in the different disciplines and by providing some information on how this activity will be organised.

Spacelab Features for the Users

The characteristics of Spacelab when compared with the pre-
Spacelab era are almost revolutionary and will have a profound influence
on the design, construction and implementation of the experiment conceived to make use of this new tool, resulting in a low cost and easy
access to space for an increasing number of users.

The services provided by Spacelab to the users have been described
in a previous section (see page 19).

The main advantages offered by Spacelab can be summarised as
follows:

- The volume and weight capabilities of the system are extremely
  high (up to 9,350kg). Design constraints can therefore be
  relaxed and in particular there is no need for miniaturisation.

- A pressurised volume is available, allowing laboratory-type
  hardware to be used, with minor modifications.

- Increased mission opportunities are provided which, by the
  reuse of the equipment and the early recovery of data, allow
  simplified experiment life cycles.

- Short access time and the possibility of changing objectives
  during the mission provide a flexibility which the experimenter
  did not have before.

- Man is directly involved. This is the main characteristic of
  Spacelab, since Payload Specialists, i.e. experimenters
  and scientists who have undergone minimum training for space
  flight, will be on board during each mission. This will
  permit technical intervention to repair failures, to control
  certain sequences of the mission (e.g. target acquisition), and
  to calibrate optical systems, and will also allow active
  involvement in the experiment cycle for tasks such as the
  manipulation of biological samples, the evaluation of data
  during the flight, the performance and direct observation of
  particularly tricky experiments and even the modification of
  the programme in the event of unexpected occurrences.

The presence of men on board implies, of course, some limitations
which are outweighed by the advantages, but which are to be mentioned
for the completeness of the picture.

These limitations are:

- Relatively short duration of the mission (one week to one
  month), not suitable for survey-type experiments.

- Weight, power and energy penalty to provide adequate living
  conditions for crews who have not had intensive astronaut-
  type training (clear volume for habitability, light, atmosphere
  regeneration).

- Special safety requirements for experimenters' equipment
  to prevent life hazards (restriction on choice of material,
  particular tests - flammability, toxicity, etc.)

- Possible perturbation of the attitude due to crew motion.

- Contamination of the close outside environment.

A synthesis of the main characteristics and of the favourable and
unfavourable features of Spacelab, as regards the user, is given in
Fig. 45.

Programme Characteristics	Unfavourable Features	Favourable Features
1. Mission Duration one week – one month	No survey type of experiment possible	Relaxation on reliability of equipment
2. Volume and weight capabilities of the system are extremely high (up to 9,100kg)		No need for miniaturisation
3. Pressurised volume available	Possible contamination of Spacelab environment	Use of ground type of equipment with little modification
4. Manned system	- Resources penalty to provide living conditions - Safety requirements on payloads	- Technical intervention for monitoring and repair - Active involvement in the experiment cycle
5. System comes back to Earth		- Immediate recovery of data - Same equipment can be reused - Failure of equipment not catastrophic (since recovered)
6. Modular system		- Easiness of integration (even on user's site) - Quick access - Short turn around time - Greater number of flights

Figure 45. Main Features of Spacelab of Interest to Users

Spacelab Utilisation for Science

The scientific and technical community has been involved for many years in the study of the possibilities offered by Spacelab in the different disciplines. Conceptual mission studies performed both by NASA and ESA in order to orient the experimenters in the use of Spacelab have resulted in the elaboration of a number of possible mission scenarios. The following is an outline of what has been postulated, in the scientific area, for Spacelab missions.

All the different disciplines under the very general heading of Astronomy and Astrophysics will greatly benefit from the availability of Spacelab.

In Solar Physics, the Spacelab will enable measurements to be made for a period of several days and will be able to carry enough apparatus to cover all wavelengths. A set of major telescopes optimised for different wavelengths and serving to feed interchangeable instruments could become the major facility of the solar physics community making it possible to attack specific problems such as the structure of chromosphere and corona, and through the short lead time of Spacelab to study transient phenomena in all layers of the solar atmosphere whenever interesting events show up.

In Ultra-Violet and Optical Astronomy, the cornerstone of the overall programme is identified in the Space Telescope (ST). Nevertheless, a number of instruments can be identified which, flown on Spacelab, would both support the ST programme and be of unique scientific interest themselves. Small telescopes could be flown in multidisciplinary payloads to study simultaneously transient objects such as comets, novae and supernovae.

High resolution imagery could be obtained by means of specialised cameras operated through an airlock in the main cabin.

Overall sky surveys with all-reflecting Schmidt cameras will become feasible when the mission length is extended to more than seven days, thus making it possible to direct the ST towards new sources requiring more detailed study.

In Infrared Astronomy, due to technological difficulties associated in particular with cryogenics, no automated satellites have been flown or are at present planned. These difficulties are by far less critical on Spacelab. ESA has, in fact, studied an infrared telescope for Spacelab (the LIRTS) of 2.8 meter diameter.

As far as High Energy Astrophysics is concerned, the sensitivities required and the overriding aim of increasing statistics, particularly in the gamma and cosmic ray field, will no longer be achievable with sounding rockets and balloons. The typical instrumentation used on those carriers, qualitatively improved, will still be used in the

1980's, and since it is not likely to place heavy demands on Spacelab, could be flown in a piggy-back manner in such a way that requirements on rapid turnaround and multiple flight opportunities can be met.

A comprehensive example of typical payload elements in the disciplines outlined above can be drawn from the relevant mission scenarios elaborated by NASA: Fig. 46 for astrophysics payloads (APP); Fig. 47 for solar physics payloads (SPP).

Instrument	Study	
	Origin and future of matter studies	Stellar evolution studies
Ultraviolet photometer/ polarimeter	X	X
High-resolution ultraviolet spectrograph	X	X
Narrow field imaging telescope and objective spectrograph	X	
X-ray detector/counter	X	X
High-resolution gamma ray spectrometer	X	
Extreme ultraviolet imaging telescope		X
Small infrared cryogenic telescope		X
Medium-energy gamma ray detector		X
Cosmic radiation electron detector/spectrometer		X

Figure 46. Typical Astrophysics Instrumentation

It may be concluded that there is a great number of research themes, in Astronomy and Astrophysics, which will find in Spacelab unique opportunities for new experimental approaches. Clearly, Spacelab will not replace automatic spacecraft. Free-flying satellites with very accurate pointing, staying in orbit for long periods, and free of contaminants will be necessary to make progress in high resolution astronomy.

But even in this case Spacelab will provide low-cost means to develop and test equipment and experiment methodologies to be used later in expensive automatic spacecraft.

Other disciplines find in Spacelab a new experimental dimension.

Instrument	Study	
	Solar wind/Sun	Acceleration processes
X-ray telescope	X	X
White light coronograph	X	X
Ultraviolet scanning spectrometer	X	
X-ray spectrometer/ spectrograph		X
Extreme ultraviolet and X-ray spectrometer	X	
Extreme ultraviolet monitor		X
Extreme ultraviolet spectroheliograph	X	X
Hard X-ray collimator		X
Helioscope		X
Hard X-ray spectrometer		X
Solar gamma ray telescope		X

Figure 47. Typical Solar Physics Instrumentation

Meteorite Physics will benefit from the possibility of deploying large collectors or of powering direct detection devices.

In Atmosphere Science, the interplay between the chemistry, dynamics and energy transport below about 120km will be examined by means of passive remote sensing techniques and by means of active probing with powerful laser beams.

In the study of the Earth's Plasma, and of the natural processes which dominate it, instruments on Spacelab will be used to inject energised particles, electromagnetic waves of various kinds, and cold plasma. A wide range of detectors placed either on board or on sub-satellites, away from the perturbations of the Shuttle, will examine the perturbations thus generated.

The Pure Plasma scientist will find in Spacelab a unique laboratory which provides access to plasma conditions that are unobtainable in ground-based laboratories.

Fig. 48 lists the typical instrumentation of the atmosphere, magneto-sphere and plasma-in-space (AMPS) payload.

Finally, there are completely new fields for space research which will be opened up by Spacelab, such as Material and Life Sciences.

This last general heading covers both medical and biological disciplines. The Spacelab will provide an ideal tool for conducting a variety of observations and experiments on cells, plants and on man himself.

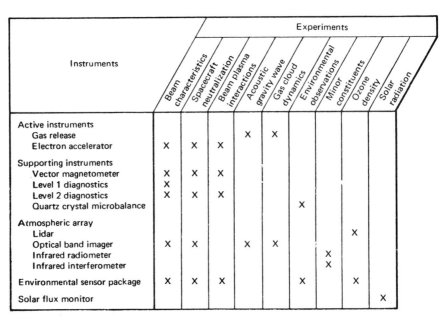

Instruments	Beam characteristics	Spacecraft neutralization	Beam plasma interactions	Acoustic gravity wave	Gas cloud dynamics	Environmental observations	Minor constituents	Ozone density	Solar radiation
Active instruments									
Gas release				X	X				
Electron accelerator	X	X	X						
Supporting instruments									
Vector magnetometer	X	X	X						
Level 1 diagnostics	X								
Level 2 diagnostics	X	X	X						
Quartz crystal microbalance						X			
Atmospheric array									
Lidar								X	
Optical band imager	X	X		X	X				
Infrared radiometer							X		
Infrared interferometer							X		
Environmental sensor package	X	X	X			X		X	
Solar flux monitor									X

Figure 48.   Representative Experiment Instrument Complement (AMPS)

The long module configuration makes available a large pressurised volume where researchers could embark various basic instruments which they normally use on the ground. A centrifuge could be installed providing the capability to study the effects of gravity from 0 up to several g. A specific facility (the SLED, developed by ESA) is already foreseen for the first Spacelab flight for research on the mechanisms of vestibular adaptation to weightlessness.

In a similar vein, NASA is developing its Minilab, which is conceived for the investigation of biological activities in space.

With Material Sciences we come to the borderline with application. There is still a great deal of basic and applied research to be performed before new products can be currently produced.

The 0-g environment will allow the study of the laws and the effects of the extremely weak forces that govern the formation of composite materials, the investigation of the growth of homogeneous single crystals, the behaviour of fluids in a weightless environment with the aim of understanding the basic processes in a liquid melt.

These are in fact the main research themes which have guided the conception of the European Material Sciences Facility, which is one of the main components of the First Spacelab Payload, and consists of a double Spacelab rack equipped with four furnaces of different characteristics, a fluid physics module and other specialised equipment for the study, among others, of crystal growth, capillarity and adhesion.

Spacelab Utilisation for Applications

Spacelab will play a role in all phases of an application programme, beyond the research and development phase where, as already
mentioned for Material Sciences, it can be used for developing
operational methodologies and for evaluating fundamental principles
associated with various techniques, as well as for development testing
of new equipment to be used in a later operational phase.

In the experimental or transition phase of an application programme, Spacelab will be used as a flying test bed and for limited
operations to help bridge the transition from the experimental to
operational activities. The utilisation of Spacelab for this will prove
to be very economical when compared with existing experimental satellite
programmes.

In the operational phase, due to its relatively short missions,
Spacelab will be used in a supporting role for activities such as calibration of sensors and for specialised techniques such as high resolution
imagery and the verification of specific data supplied by the automatic
systems.

Looking now at specific areas, the picture of the Material Sciences
programme can be completed by saying that its specific goal is the
development and ultimately the commercialisation of new materials and
products to be manufactured or refined in space for use on Earth. Also
new ways and methods of preparation and shaping of materials are expected
to be applied in space. Another field of application is the electrophoretic separation method, which is disturbed by gravity segregation.
Sera, vaccines and other pharmaceutical products are expected to be
separated with higher resolution, i.e. higher purity.

In Remote Earth Sensing, techniques will be refined and upgraded
for cataloguing and monitoring earth and ocean resources, for controlling
air quality, for land mapping and land-use planning for urban development,
agriculture and forestry, and inland water resources. These are some of
the uses of the data which will be gathered by instruments such as radiometers, side-looking radars, multispectral scanners, spectrometers, and
microwave scatterometers, all of which will make their first space
flight on board Spacelab.

Specialists in Meteorology, Geodesy, Communications and Navigation
are also preparing experiments for Spacelab to test new systems and to
assess the validity of new operational concepts.

Both NASA (with the Earth viewing applications laboratory, EVAL)
and ESA (with the Earth observation payload, EOP) have analysed the
possible mission scenarios taking into account the prospected needs of
the users from the respective countries. Fig. 49 lists specific EVAL
missions, whereas Fig. 50 shows a typical EOP configuration.

Earth resources	• World crop survey • Vegetation stress • Urban planning • Timber inventory • Mineral survey • Water inventory
Weather and climate	• Cloud climatology • Ozone mapping • Solar energy monitoring • Atmospheric X-ray emission • Weather modification
Earth and ocean dynamics	• Crustal motions • Geomagnetic fields • Sea surface temperature • Ocean currents • Ocean waves • Geoid measurements
Environmental quality/ pollution	• Stratospheric pollution • Tropospheric trace constituents • Radiative flux changes • Thermal balance • Water pollution
Communication and navigation	• Electromagnetic mapping • Data collection • Search and rescue • Propagation effects • Laser communications

Figure 49.   Specific EVAL Missions Being Considered

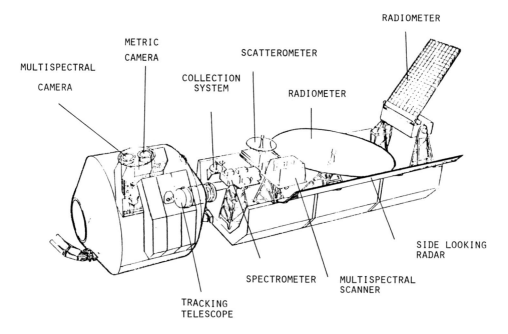

Figure 50.   Typical Earth Observation Payload Configuration

This engagement in all fields implies the development of advanced technologies. An increasing effort is in fact being dedicated to Technological Research on Spacelab, in fields such as cryogenics, thermal control, antennas and large structures. Whole missions are already planned to be entirely dedicated to the development of space-borne systems, as in the advanced technology laboratory (ATL), with which NASA intends to provide an organised and systematic approach for carrying on and extending technological research.

Spacelab Utilisation Planning

This activity of conceptual analysis and long term planning has its milestone and a turning point in the first flight of Spacelab (second half of 1980), when a joint NASA/ESA payload representing all disciplines quoted above will demonstrate the prospected possibilities of this new approach to space research.

Figs. 51 and 52 list the selected FSLP experiments of ESA and NASA respectively.

DISCIPLINE	TITLE	PI COUNTRY	CURRENT WT
ATMOSPHERE	GRILLE SPECTROMETER	FRANCE	65 KG
	WAVES IN THE OH EMISSIVE LAYER	FRANCE	8
	TEMPERATURE, WIND IN MESOPHERE, THERMOSPHERE	FRANCE	22
	SOLAR SPECTRUM FROM 1900 Å 4 MICRON	FRANCE	10
	H AND D LYMAN ALPHA	FRANCE	7
PLASMA PHYSICS	MAGNETIC FIELD MEASUREMENT	GERMANY	11
	LOW ENERGY ELECTRONS	GERMANY	16
	PHENOMENA INDUCED BY CHARGED PARTICLE BEAMS	FRANCE	35
SOLAR PHYSICS	SOLAR CONSTANT	BELGIUM	6
ASTRONOMY	VERY WIDE FIELD CAMERA	FRANCE	85
	X-RAY SPECTROSCOPY (GAS SCINTCOUNTER)	UK-ESA	15
	HEAVY COSMIC RAY ISOTOPES	GERMANY	20
LIFE SCIENCES	SLED (FACILITY)	ESA	150
	SLED EXPERIMENTS	GERMANY	25
	LYMPHOCYTE PROFIFERATION IN WEIGHTLESS-NESS	SWITZERLAND	
	MASS DISCRIMINATION	UNITED KINGDOM	
	MEASUREMENT OF INTRATHORAXIC BLOOD PRESSURE	GERMANY	
	ADVANCED BIOSTACK	GERMANY	30
	THREE-DIMENSIONAL BALLISTOCARDIOGRAPHY	ITALY	
	EFFECT OF RADIATION	GERMANY	
	ELECTROPHYSIOLOGICAL TAPE RECORDER	UNITED KINGDOM	
	COLLECTION OF BLOOD SAMPLES	GERMANY	
MATERIAL SCIENCES	MATERIAL SCIENCE FACILITY (4 FURNACES: ISOTHERMAL, LOW TEMPERATURE GRADIENT, HIGH TEMP. GRADIENT, MIRROR; FLUID PHYSICS MODULE; SPECIAL EQUIPMENT FOR CRYSTAL GROWTH, ELECTROLYSIS, CAPILLARITY, ADHESION OF METALS).	EXPERIMENTS FROM ALL ESA MEMBER STATES	460
EARTH OBSERVATION	METRIC CAMERA	AO FOR USE OF DATA FROM THESE FACILITIES TO BE ISSUED IN 1978 (OPEN TO USERS ON WORLD WIDE BASIS)	219
	MICROWAVE SCATTEROMETER (FACILITY) RADIOMETER		150
TOTAL			1334

Figure 51. First Spacelab Payload – Selected ESA Experiments

DISCIPLINE	TITLE	PI COUNTRY	CURRENT WT
ATMOSPHERE	AN IMAGING SPECTROMETRIC OBSERVATORY	USA	220 KG
	ATMOSPHERIC EMISSION PHOTOMETRIC IMAGING (LLLTV)	USA	149
	ATMOSPHERIC TRACE MOLECULES OBSERVED BY SPECTROSCOPY	USA	170
PLASMA PHYSICS	SPACE EXPERIMENTS WITH PARTICLE ACCELERATORS	JAPAN	300
SOLAR PHYSICS	ACTIVE CAVITY RADIOMETER SOLAR IRRADIANCE MONITOR	USA	170
ASTRONOMY	FAR UV OBSERVATIONS USING THE FAUST INSTR.	USA	65
ASTROPHYSICS	STUDIES OF THE IONISATION STATUS OF SOLAR AND GALACTIC COSMIC RAY HEAVY NUCLEI	INDIA	13
	GEOPHYSICAL FLUID FLOW	USA	300
LIFE SCIENCES	HZE PARTICLE DOSIMETRY	USA	3.0
	CHARACTERISATION OF PERSISTING CIRCADIAN RHYTHMS	USA	10
	L.S. MINI LAB (FACILITY)		197
	NUTATION OF HELIANTHUS ANNUS	USA	(132)
	VESTIBULAR EXPERIMENTS	USA	(26)
	INFLUENCE OF SPACE FLIGHT ON ERYTHROKINETICS IN MAN	USA	(14)
	VESTIBULO-SPINAL REFLEX MECHANISMS	USA	(24)
	EFFECTS OF PROLONGED WEIGHTLESSNESS	USA	(1)
TECHNOLOGY	WETTING, SPREADING, AND OPERATING CHARACTERISTICS OF BEARING LUBRICANTS IN A ZERO GRAVITY ENVIRONMENT	USA	
	TRIBOLOGICAL STUDIES OF FLUID LUBRICATED JOURNAL	USA	63
TOTAL			1290

Figure 52. First Spacelab Payload - Selected NASA Experiments

As can be seen, despite the strict limitations in resources and flight attitude with respect to operational missions, due to the verification nature of the flight, a large number of experiments has been accommodated, most of which have been conceived as forerunners of more complex and ambitious Spacelab missions.

The payload complement in the various disciplines will probably constitute the core of future discipline-dedicated missions.

As a consequence, reuse of the experiment equipment after refurbishment/reconfiguration/upgrading phases on the ground is largely foreseen in the various Spacelab utilisation models which have been developed by both Agencies.

These models are planning tools, enabling to assess the needs in terms of investments, manpower, facilities related to different levels of activity and to tailor the means to the users demands, with the overall guideline to keep Shuttle/Spacelab usage costs as low as possible.

Figs. 53 and 54 are examples of NASA and ESA Spacelab utilisation models respectively.

Spacelab Utilisation Costs

As regards costs for utilising the Shuttle/Spacelab system, it is judged appropriate to give some information on the matter in order to complete the picture, although their amount and their structure has not been fully defined by NASA at the time of writing.

SHUTTLE FLIGHT No	SL FLIGHT No	1980–1982 Schedule	RESPONS.	CONFIG.
8	1	FSLP	NASA/ESA	LM + 1P
10	2	SSLP	NASA	5P
12	3	MULTI-USER 81-1	NASA	LM + 1P
14	4	LIFE SCIENCE (Mod1)	NASA	LM
17	5	MULTI-USER 81-3	NASA/ESA	M + P
19	6	ATL (NO.1)	NASA	M + 2P
21	7	MULTI-USER 81-2	NASA/ESA	SM + 3P
23	8	LIFE SCIENCE (Mod. 1)	NASA	LM
25	9	COMBINED ASTRONOMY	NASA/ESA	5P
27	10	MULTI-USER 82-1	NASA	LM + 1P
30	11	MULTI-USER 82-2	NASA/ESA	SM + 3P
34	12	LIFE SCIENCE (Mod.2)	NASA	LM
36	13	AMPS	NASA/ESA	SM + 3P
38	14	MULTI-USER 82-4	NASA/ESA	LM + 1P
40	15	ATL	NASA	
42	16	EVAL	NASA	
44	17	MULTI-USER 82-3	NASA/ESA	LM + 1P
46	18	LIFE SCIENCE (MOD.2)	NASA	LM
48	19	ASTRO / HIGH ENERGY	NASA	5P
MFSC June 22, 76		1980 1981 1982	Early SL-Missions	

Figure 53. Example of NASA Spacelab Utilisation Model

A large share of these costs (which, in the pre-STS era, corresponded to the procurement of an expendable launcher and related launch services) will be due to the use of the Shuttle itself. However, the operation of a manned system like Spacelab also implies other tasks, as described in a previous section, which will amount to a sizeable part of the final bill.

The former is referred to as Shuttle standard charge, accounts for a standard one-day mission with given orbital parameters and pertinent basic operations support and has been officially published in January 1977. It amounts, for non-commercial users belonging

to the United States and to other countries who contributed significantly to the STS (ESA and its Member States and Canada) to 16.1
million $ (1975).

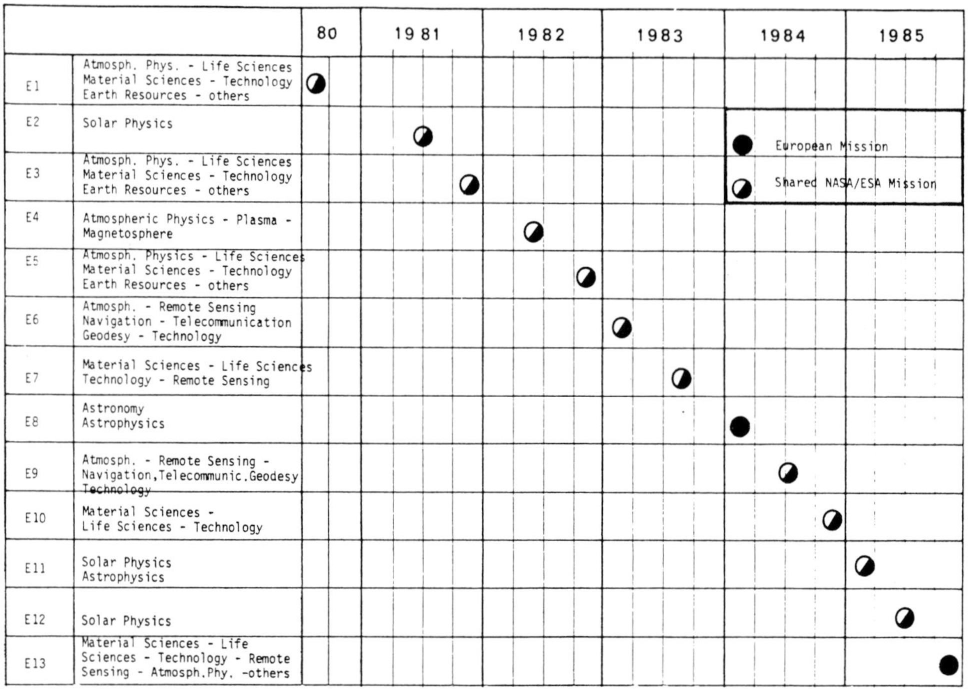

		80	1981	1982	1983	1984	1985
E1	Atmosph. Phys. - Life Sciences / Material Sciences - Technology / Earth Resources - others	◓					
E2	Solar Physics		◓			● European Mission	
E3	Atmosph. Phys. - Life Sciences / Material Sciences - Technology / Earth Resources - others			◓		◓ Shared NASA/ESA Mission	
E4	Atmospheric Physics - Plasma - Magnetosphere			◓			
E5	Atmosph. Physics - Life Sciences / Material Sciences - Technology / Earth Resources - others				◓		
E6	Atmosph. - Remote Sensing / Navigation - Telecommunication / Geodesy - Technology				◓		
E7	Material Sciences - Life Sciences / Technology - Remote Sensing				◓		
E8	Astronomy / Astrophysics					●	
E9	Atmosph. - Remote Sensing - / Navigation,Telecommunic.Geodesy / Technology					◓	
E10	Material Sciences - / Life Sciences - Technology					◓	
E11	Solar Physics / Astrophysics						◓
E12	Solar Physics						◓
E13	Material Sciences - Life Sciences - Technology - Remote Sensing - Atmosph.Phy. -others						●

Figure 54. Example of ESA Spacelab Utilisation Model

The special mission surcharge, due for payload integration and
check-out, for special Shuttle provisions to fly a Spacelab mission
(consumables for additional days, power kits, non-standard orbit
requirements, etc.), for flight support operations and data transmission,
for post flight activities and for crew training is not finalised
yet. The authors estimate it at the order of 5 million $ (1975),
thus bringing the total charge for a typical 7 days Spacelab mission
at the order of 21 million $ (1975).

This charge will, of course, be shared among the users according
essentially to the resources and the mission time allocated to the
various experiments.

System Access and User Support

The co-operation between NASA and ESA in the development of Spacelab
is expected to continue for its utilisation.

Due to its overall responsibility for Shuttle operations, NASA is ultimately responsible for Spacelab operations as well.

ESA, however, is called on to play a major role in programming the European utilisation of Spacelab and supporting its execution. To this end the SPICE has been created with the major function of the overall planning and co-ordination of the preparation, execution and exploitation of European missions on Spacelab.

This new centre, together with the NASA centers will play an essential role in assisting the user throughout the whole process from the conception of an experiment for Spacelab to its execution (e.g. supply of GSE from a central pool, design guidance, mission analysis).

The two Agencies, which have established in the frame of the close co-operation for the FSLP a number of channels for planning co-ordination, exchange of information, management, have developed minimum paperwork procedures so that modest documentation is required to enter the system.

Both carry on various promotional activities intended to bring new users to the Spacelab and are publishing various types of literature, addressed to potential users, such as Handbooks, Manuals, Guides, etc. which can be requested if desired. (NASA: Space Transportation Systems Operations Office, Mail Code MO, National Aeronautics and Space Administration, WASHINGTON, DC 20546. ESA: European Space Agency, 8-10 rue Mario Nikis, 75738 PARIS Cedex 15, France).

Note by Authors (added November 1979)

    Since this chapter was written (mid-1977), there have been some
significant developments in the Shuttle and Spacelab programmes.
Although the descriptive and performance data provided here are still
applicable, the natural evolution of both programmes has led to some
reorientation. For completeness, the principal events and some recent
planning trends are summarised here.

    The roll-out of Orbiter 101 (now called Enterprise) occurred in
September 1976 and the approach and landing tests, which proved its
subsonic flying characteristics, were successfully completed by
October 1977. This Orbiter, together with other Shuttle elements then
underwent structural tests at NASA's Marshall Space Flight Center.
It was then shipped to the Kennedy Space Center where the whole space
shuttle integration and on-the-pad tests were carried out in April/May
of this year, to ensure that all geometric and functional interfaces were
compatible. Fig. 55 shows the inert Space Shuttle assembly on the
launch pad for its facility verification tests after its 6km journey on
the crawler vehicle from the VAB to Pad 39A. All the integration facilities required for Shuttle/Spacelab ground operations at KSC have been
completed.

Figure 55.   Space Shuttle on Launch Pad for Facility
             Verification Tests

The major concern of the Space Shuttle programme is the Orbiter
main engine certification. A number of test failures have been
experienced and the planned launch schedule has had to be revised. It
now seems likely that the first launch will occur in late-1980 and the
first Spacelab flight is foreseen for mid-1982. An additional technical problem has been experienced with the application of thermal
tiles to the Orbiter Colombia. This Orbiter, the first to actually
fly in space, is now in the OPF and the associated activities will be
completed in early-1980. The combined technical problems and schedule
slips have led to cost increases in the programme.

Spacelab too has passed some significant milestones and has also
encountered schedule and cost problems. The important Critical Design
Review which resulted in design acceptance and cleared the way for the
approval of flight hardware fabrication was successfully held in early-
1978. Models in the form of a hard mock-up, an electrical systems
integration model and a development fixture have been manufactured for
such tasks as - verification of subsystems, equipment-interfaces,
routing of cables and fluid lines and verification of fabrication
techniques.

Both Engineering Model (EM) and Flight Unit (FU) have reached an
advanced stage of integration and test. Fig. 56 shows the EM (left)

Figure 56.   Engineering Model and Flight Unit of Spacelab under-
             going Tests in the Integration Hall at Bremen

and FU (right) in the Integration Hall at Bremen. The EM is covered
with thermal protection material. Its delivery to NASA is scheduled
for late 1979 and some of the pallets will be used in the orbital
test flights of the Shuttle. The FU will be provided in two deliveries -
long module + 1 pallet in early-1981 and the remaining igloo and
pallets by mid-1981. The EM module being integrated with three pallets
is shown in Fig. 57.

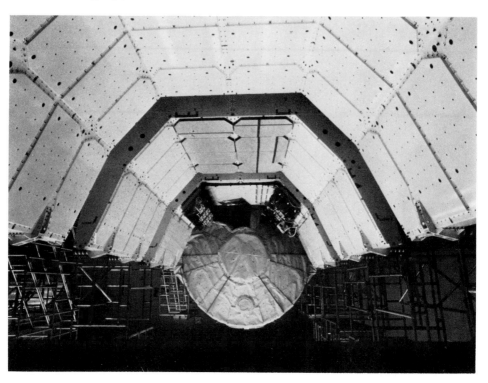

Figure 57.   Integration of Spacelab Engineering Model
Module and Pallets

Although the schedule slips of the Spacelab are significant the
delivery dates are still compatible with its flight in the early
operational programme of the Space Shuttle. The most serious problem
associated with the Spacelab programme is cost. The current cost-to-
completion estimate is 680 MAU (1978 price levels) i.e. equivalent to
140% of the original estimate.

As far as the utilisation of Spacelab by Europeans is concerned, ESA
is in the course of implementing a Spacelab Utilisation Programme. The
principal aims of this programme are to provide easy access to space
for European experimenters, to establish optimum relationships with
user organisations and to encourage and guide European participation in
future manned space activities. Two particularly important aspects of
the programme are the setting up of an instrument pool and the partici-

pation in early Spacelab missions. The pool will be used to provide experimenters with common ground- and flight-equipment when and if required. It should ensure the avoidance of duplication and save money for the users. The early Spacelab flights will illustrate the use of Spacelab in science and applications fields. To this end, a joint ESA-NASA payload will be provided for the first Spacelab flight and later participation in NASA missions is envisaged by the provision of payload elements to the total Spacelab payload. These elements would be at full rack or pallet level so that integration in Europe is still desirable. Participation in three missions is currently envisaged through 1985 viz microgravity (with a rack), Earth observation (with a pallet) and solar/astrophysics (with a pallet). Beyond 1985, increased usage of Spacelab can be envisaged, leading eventually, to dedicated-ESA missions.

The definition of Spacelab utilisation costs is also evolving and, unfortunately, recent estimates imply somewhat increased usage costs. The Shuttle standard charge has now been fixed by NASA at 18 million $ (1975) for non-commercial users. Commercial users (that is to say, where payload has near-term commercial implications) would have to pay an additional user charge of 4.3 million $ (not subject to inflation).

The Spacelab mission surcharge, comprising a set of standard services, is estimated at 6.3 million $ (1975). Thus the total NASA standard charge for a non-commercial Spacelab mission of seven days can be set at 24.3 million $ (1975). Inflation should be taken into account to obtain current rates.

To these charges, the costs of the activities to be performed in Europe should be added, in order to complete the picture. These costs are strongly dependent upon the complexity of the payload, but they may amount to 20 million $ (1975) for a full Spacelab mission.

As regards Spacelab operations, two changes are worthy of note. The Orbiter crew will now include two mission specialists, who have had scientific training, so that the standard crew consists of commander, pilot and two mission specialists. Up to three payload specialists will also be carried.

The integration activities foreseen for Europe are no longer termed pre-level III but conform to the NASA level IV standards. The integrated payload or partial payload will now be shipped directly to the launch site. Thus the European first Spacelab payload complement will be delivered to KSC after integration which will be carried out in the Integration and Test Hall at ERNO/Bremen.

References

Space Shuttle System Payload Accommodation Handbook, Volume XIV, Revision D of Level II Program Definition and Requirements, NASA/JSC/07700, November 1975.

Space Transportation System User Handbook, NASA HQ, June 1977.

Spacelab Payload Accommodation Handbook (SPAH), ESA Report, SLP/2104, Issue 1, July 1978.

Report on Long-Term Planning, Astrophysics Working Group, ESA/ASTRO(76)13, August 1976.

An Approach to Long-Range Planning, Solar System Working Group, 1980-1990, ESA/SOL(76)9, September 1976.

Spacelab Utilisation Programme - Conception, Organisation and Funding, ESA/EXEC(77)4, April 1977.

NASA Early Space Transportation System Mission Plan, Marshall Space Flight Center, June 1976.

Spacelab Users Manual, ESA/DP/ST(79)3, July 1979.

ESA's PROGRAMMES IN ASTRONOMY AND ASTROPHYSICS

V. Manno
European Space Agency

1. INTRODUCTION

    The European Space Agency (ESA) has the task of elaborating and
executing a scientific programme with space systems and of coordinating
within Europe the individual national space programmes.

    To fulfil these tasks, ESA undertakes a number of activities
which lead to the execution, under its own guidance and funding, of
scientific projects of general interest to the European community and
to the identification of a number of missions and instruments which
could be developed and funded within a national context. In some cases,
the magnitude of the effort and/or the global interest of a particular
mission lead ESA to establish cooperative arrangements with other Space
Agencies. In the programme of Astronomy and Astrophysics, all the
features outlined above are found. The programme itself contains a
number of missions in the development phase either solely European or
in cooperation with NASA, and a number of missions still in the study
phase which could either be developed in the future under ESA or national sponsorship. In addition, a number of missions are currently being
developed within the national programmes, which, although strictly
speaking, they cannot be considered within the ESA framework, they,
nevertheless, find a proper place in the overall planning and views of
the Agency on the development of Space Astronomy and Astrophysics in
Europe.

2. THE OVERALL PLANNING

    A first attempt to elaborate a consistent and homogeneous picture
of the developments and requirements of Space Astronomy and Astrophysics
in Europe was made by the Astronomy Working Group of the Agency in 1975,
and has led to an overall view which testifies to the fertile imagination and vitality of European space science (cf. ESA Astronomy Working
Group Report on Long-Term Planning - 1980-1990, ASTRO(76)13).

P. L. Bernacca and R. Ruffini (eds.), Astrophysics from Spacelab, 51-109.

The necessity for such an exercise became particularly pressing when, in 1973, the initiation of the Spacelab programme and the rather definite plans in NASA for a large space telescope appeared to open a bewildering range of research possibilities. It was, thus, that the Astronomy Working Group of ESA (an advisory group of some 15 European scientists, appointed on a rotation basis of three years) began addressing the problem. As a first step, it identified five main domains of research, viz.: (i) the Interstellar Medium, (ii) Solar and Stellar Physics, (iii) the Structure of Galaxies, (iv) Extragalactic Astronomy and Cosmology, and (v) General Physics, and, with the help of a few invited experts, discussed the likely developments in each domain and the space projects that could help to bring them about. The discussions implied a continuous balancing of likely technical developments, with re-requirements based on purely scientific considerations. A number of conclusions were reached and certain trends identified in the disciplines of Astrophysics and Astronomy.

In Infrared Astronomy, it was concluded that an overall sky survey was an essential first step. Subsequent to it, telescope facilities could be flown on the Shuttle/Spacelab system for studies with high spatial and spectral resolution of the identified sources. Developments beyond that phase would depend on the scientific findings and further technical progress.

In Ultraviolet and Visible Astronomy, it was concluded that, after the initial survey work and the more detailed source studies currently being conducted or planned for the near future, the Large Space Telescope would open up the field completely and produce unparalleled and also unexpected results. However, it was necessary to complement the space telescope with a number of instruments on Spacelab especially suited for very high spectral and temporal resolution work and surveys in larger fields. Also, the new field of EUV should be opened up with an unbiased sky survey, to be followed, if successful, by a spectrometer on Spacelab for detailed studies of the identified sources.

In X-Ray Astronomy, it was concluded that the future emphasis should be on higher spatial and spectral resolution, but the emphasis would depend on the results of the forthcoming European and American satellites.

In Gamma-Ray Astronomy, it was felt that it was too early to judge with confidence, but one would expect the field to develop into one of the most exciting areas in the 1980s. In the meantime, major technical developments would have to take place, directed at higher spectral and spatial resolution.

In Cosmic Ray Astronomy, it was felt that after the already planned future missions, which would allow for a statistical determination of the mean mass of cosmic ray particles and investigation of heavy primaries, a superconducting magnet facility should be developed on Spacelab, to add a new dimension – separation by mass – to our present knowledge of the cosmic ray particle spectrum, and to make critical searches for antimatter.

In the field of Solar Physics, it was concluded that, in future, simultaneous correlated measurements in a wide spectral band of the quiet and active Sun were essential, and that Spacelab was an ideal carrier for such instruments.

In Gravitational Physics, it was recognised that space science, by providing access to regions of differing gravitational potential, offered the possibility of exploring the relationship between gravity and other branches of physics. This area required a definite effort in the field of advanced technology.

In general, one could discern an evolution common to all disciplines:

i)   the survey phase, using conventional satellites;

ii)   the development phase, where the emphasis would be on the detailed study of a number of astronomical objects to define more precisely technological and astrophysical objectives. The use of Spacelab could be very effective in this phase also to test new instruments;

iii)   the space observatories' phase, dominated by the use of large free-flying facilities (possibly Shuttle serviced), to be operated as supranational instruments. The characteristics of these will, however, depend on the scientific discoveries of the previous period and on the technological developments.

The missions and experiments using conventional launchers or being Spacelab-borne, identified in this first phase of the discussions, are listed in Table 1.

TABLE 1 : EXPERIMENT/MISSIONS CONSIDERED NECESSARY IN THE
OVERALL DEVELOPMENT OF ASTRONOMY/ASTROPHYSICS

FREE-FLYERS		SPACELAB	
Satellites	Facility satellites	Spacelab experiments	Spacelab facility
IR Diffuse flux	1m X-ray Wolter Telescope	Wide field camera	3m IR Telescope
IR Sky Survey	$3m^2$ Focussing X-ray Collector	EUV/XUV Tel.	IR/UV Spatial Interferometer
Astrometry	10m IR Collector	Gamma-ray Compton Tel.	1m UV Telescope
EUV/XUV Survey	H.E. Gamma-ray Observatory	Astrometry	$1m^2$ X-ray Coll. for spectro/pol.
Very H.E. Cosmic Ray	Cosmic Ray fac.	40cm-1m UV Photometer	Cosmic Ray fac.
Out-of-ecliptic		Solar Const. Telescope	Solar Grazing Inc. Telescope
Solar Max.Miss.	L.E. Gamma-ray Observatory		Solar UV Tel.
Solar Probe	UV/Optical Space Tel.		Solar IR Tel.
Mercury Orbiter			
Solar Orbiter			
Interplanetary Satellite			

In a second step, the Astronomy Working Group reviewed the various projects known to be at present developed in Europe and in the world in the various disciplines of Astronomy and Astrophysics. These are shown in Figure 1, with the planned launch dates and operation periods, and described briefly as follows:

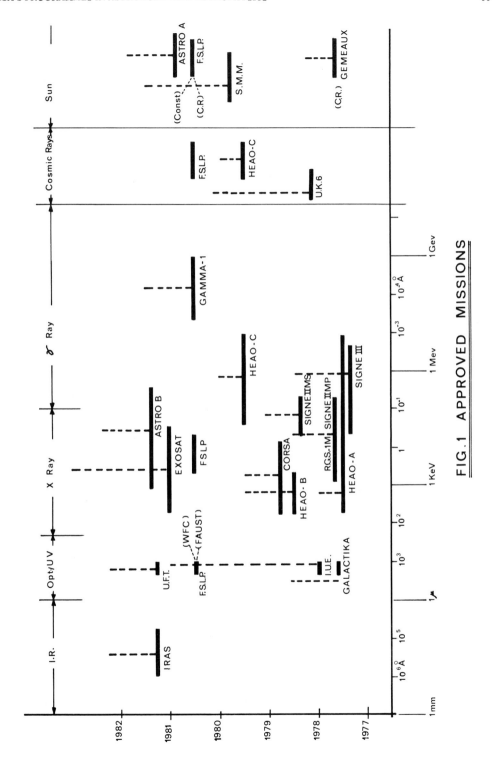

FIG. 1 APPROVED MISSIONS

IR Astronomy

IRAS : An NL/NASA/UK satellite for an all-sky survey of IR sources, in the 8 - 120 μ (and possibly up to 300 μ) region. 60 cm telescope, f/10 Cassegrain system and cryogenically cooled.

UV/Visible Astronomy

GALACTIKA : A French spectrometer for detection of galactic radiation on the Soviet Prognoz.

IUE : ESA/NASA/UK International Ultraviolet Explorer. A 45 cm telescope for spectroscopy in the 1120 - 3250 A region.

FSLP : First Spacelab Payload - UV packages:

a) Wide Field Camera : 60° f.o.v., direct photography of large-scale features in 1300 - 3000 A.

b) Faust Telescope : a Wynne camera with 7.5° f.o.v, for observation of faint astronomical sources in the 1200 - 3000 A region.

UFT : UV Telescope of 0.8 metres. Joint undertaking of France and USSR. About as Copernicus (OAO-C) low resolution mode.

X-Ray Astronomy

CORSA : A Japanese satellite to survey the sky in the 0.1 - 20 keV region. In fact the reflight of the ill-fated 1976 mission, which would have given, if successful, the first sky survey in the low energy range.

HEAO-A : NASA observatory to survey the sky at high sensitivity for X-ray and low energy gamma-ray sources.

HEAO-B : NASA observatory, with a four-element nested array, Wolter type I grazing incidence telescope to locate accurately X-ray sources, study their structure, spectra and variability in the 0.2 - 4 keV region.

RGS-IM : A USSR experiment on PROGNOZ for study of sources and bursts in the 2 - 200 keV region.

FSLP : First Spacelab Payload - A gas scintillation proportional counter for low resolution spectral observation of X-ray sources in the 2 - 20 keV region.

EXOSAT : An ESA X-ray observatory to measure with high accuracy the position, structural features and spectral and temporal characteristics of X-ray sources in the range from approximately 0.1 keV to 50 keV.

ASTRO-B : A Japanese satellite for observations of X-ray sources, soft X-ray nebulae, gamma-ray bursts.

Gamma-Ray Astronomy

HEAO-A : see X-ray Astronomy

SIGNE III : A French satellite launched with a USSR rocket, for a study of cosmic background sources and bursts in the 0.02 - 2 MeV region.

SIGNE II MP : A French experiment on the USSR PROGNOZ for study of gamma bursts $>$ 80 keV in correlation with SIGNE III and Helios-B.

SIGNE II MS : Two French experiments on the 2 VENERA USSR probes for study of gamma burst $>$ 80 keV in correlation with SIGNE III.

GAMMA 1 : A French/USSR mission for high resolution study of gamma-ray sources in the 50 to 1000 MeV range.

HEAO-C : A NASA spacecraft for cosmic ray and gamma-ray studies. A gamma-ray line spectrometer will search for gamma-ray line emission from a variety of source phenomena in the 0.06 to 10 MeV region.

Cosmic Ray Physics

UK-6 : A UK satellite for measurement of ultra-heavy primaries ($Z \geqslant 26$) in the 0.5 - 6 Gev/n energy range.

HEAO-C : A NASA observatory with experiments to measure the charge spectrum of nuclei with $17 \leq Z \leq 120$ in the energy interval 0.3 to 10 GeV/n, and to measure the isotopic composition of cosmic rays, with Z from 4 to 26 and the elemental abundance up to Z = 50.

FSLP : First Spacelab Payload - Experiments to study low energy galactic ions and heavy cosmic rays with $Z > 3$.

Solar Physics

GEMEAUX S2 : A French experiment on Prognoz for the study of solar cosmic rays.

SMM : The NASA Solar Maximum Mission, for simultaneous high resolution measurements in a wide spectral band from the UV to gamma regimes of the active Sun, and to study the variations in the solar constant.

FSLP : First Spacelab Payload - Instruments to study the ionization states of heavy elements from 0 to Fe in energetic solar particles, and to study the variations of the solar constant.

ASTRO-A : A Japanese satellite for observation of solar hard X-ray flares, X-ray bursts and solar particles.

In a third step, the Astronomy Working Group, basing itself on the expected progress that the approved missions will bring about in each discipline and having confronted the approved programme (Figure 1) with all missions identified in Table 1, narrowed down its choice of missions and formulated a number of alternative models for the future scientific planning, each one of them containing a different number of preferred missions and having a different scientific content.

The ground rules to be satisfied by the mission models were selected to be the following:

a)   the scientific needs of an ever-growing European space community should be satisfied;

b)   the space programme should profit the astronomical community at large and enhance the interaction with ground-based observatories;

c)   a balance should be maintained among the different disciplines, not only to satisfy as large a community as possible, but because the advances of the whole field of Astrophysics depends on a balanced advance of all its disciplines;

d)   the missions and experiments foreseen should offer the best possibilities for international cooperation also with other agencies;

e)   the best and most efficient use of the Agency's means should be made;

f)   the scientific return for the invested capital should be optimized.

The basic financial constraint was set by the opportunity to realistically limit the prospections in the future to the next decade, and it was then imposed that no mission model should exceed financially the present allocation to the scientific programme of ESA over the next decade. This was called the "baseline" programme, and the alternative mission models in it are shown in Table 2. However, the real world is such that the baseline programme, meagre as it is, is in reality optimistic, since it presupposes that Astronomy would be allotted all of the scientific budget. The Astronomy Working Group, therefore, further considered the case where the present budget would not change in the future and where today's rough equipartition of funds between Astronomy and other Space Sciences would be maintained. The mission models would, therefore, shrink financially to 50% of the above, and would constitute the alternative options of the "barebones" programme, also shown in Table 2.

A common feature of all models proposed is that they include a significant number of Spacelab and satellite facilities, thereby meeting at least partially, most of the criteria mentioned above.

TABLE 2 : MISSION MODELS AND BASIC PROGRAMMES

Instrument/Satellite Mission	Category	Barebones programme								Baseline programme						
		A	B	C	D	E	F	G	H	T	U	V	W	X	Y	Z
~3 m IR Telescope	S.L.	x	x	x	x		x		x	x	x	x	x	x	x	x
~1 m² X-Ray Collector for Spectro-Polarimetry	S.L.	x	x		x	x	x	x		x	x	x	x	x	x	x
Gamma-Ray Compton Tel.	S.L.	x	x		x	x	x	x		x	x	x		x	x	
Cosmic Ray Magnet	S.L.		x			x	x			x	x	x	x			
1 m UV Telescope	S.L.								x	x	x	x		x		x
40 cm - 1 m UV Photometer	S.L.	x					x							x		
Solar IR Telescope	S.L.	x					x					x		x		
Solar UV Telescope	S.L.	x					x					x		x		
Solar Grazing Incidence Telescope	S.L.											x		x		x
IR/UV Interferometer	S.L.			x					x			x				x
Contribution to the 2.4 m Space Telescope	F.F.	x	x	x	x	x	x	x		x	x	x	x	x	x	
Astrometry	F.F.		x	x		x	x	x		x		x	x	x	x	x
IR Diffuse Flux	F.F.					x				x		x	x			x
Solar Probe	F.F.									x	x					
VHE Cosmic Rays	F.F.	x								x	x					
Low-energy Gamma-Ray Observatory	F.F.											x	x			x
X-Ray Facility	F.F.					x					x			x	x	
EUV/XUV Survey	F.F.		x				x				x				x	x
Out-of-Ecliptic	F.F.											x				x

S.L. = Spacelab payload

F.F. = Free-flyer

In addition, it should be noted that practically all alternative mission models in both programmes include the focal plane instrument of the Space Telescope, the Large Infrared Telescope on Spacelab, the Large X-Ray Spectropolarimeter on Spacelab and the Astrometry Mission, all of which are to be found either in the approved or in the study programme of ESA, and described in more detail in the next Section.

The maximum achievable observation time per each Spacelab payload instrument in the various alternative options is shown in Figure 2. Figure 3 shows the contents of each alternative mission model in terms of free-flying satellites and of Spacelab equivalent payloads (SLEP). SLEP is equal to unity when the ensemble of instruments on Spacelab draws 100% of the Spacelab resources in each particular mission.

It can be seen that in the "barebones" programme an average re-quirement exists for about 1.5 SLEP in 10 years. In the "baseline" programme, while the SLEP requirements increase by about 40%, the free-flyers increase by a factor of 2, thereby reflecting the general trend discussed above of the Astronomy and Astrophysics disciplines towards long duration free-flying facilities.

The analysis of the various mission models clearly outlines the difficulty which a balanced progress of Astronomy and Astrophysics will encounter if constrained within the "barebones" programme. The negative feature of this programme is its rigidity and severe lack of flexibility, i.e. of the possibility to adapt it to a situation which is continuously changing either because of scientific discoveries or because of other agencies' undertakings in this field. The "barebones" programme is quickly filled with "today's" instruments, and there is no room for future complex instruments opening a window onto the next decade.

As opposed to this, the "baseline" programme is obviously flex-ible and allows for internal modifications to adapt itself to the changing world. It also introduces at the end of the decade, large facilities to look into the future.

The Astronomy Working Group, in a final analysis, ventured into expressing a somewhat higher preference for the Model D of the "bare-bones" programme, and for the Models U and X of the "baseline" pro-gramme, the one emphasizing the Spacelab observing time, the other the satellite programme. It, nevertheless, considered the "barebones" programme unsatisfactory, and strongly requested that the "baseline" programme of Astronomy and Astrophysics be pursued.

3.   THE ESA PROGRAMME

The programme of ESA in Astronomy and Astrophysics includes missions at various stages on the long line which begins with the in-ception of the idea and proceeds through a number of studies of in-

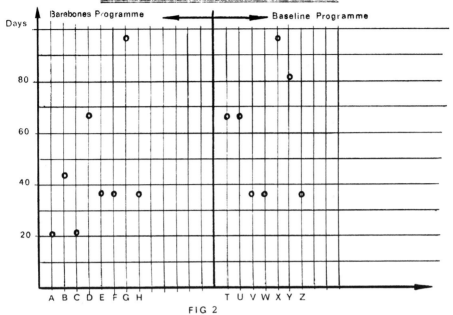

MAXIMUM ACHIEVABLE OBSERVING TIME
PER EACH SPACELAB PAYLOAD ELEMENT

FIG 2

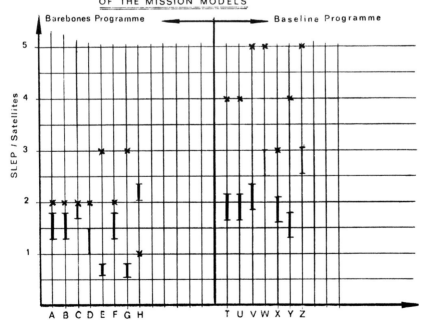

NUMBER OF FREE-FLYERS (×) AND SLEPs(I)
OF THE MISSION MODELS

FIG 3

creasing depth before reaching the development phase and, ultimately, the final goal of operations in space.

It is perhaps appropriate to describe the projects in the future Astronomy and Astrophysics programme of ESA rearward along the line, and subdivide them in classes as follows:

Class 1 :   Approved projects in the development phase or in the detailed system design study phase B (see Figure 1).

Class 2 :   Projects (not yet approved) ready to enter the phase B.

Class 3 :   Projects whose scientific definition is completed, but which have to be subjected to a preliminary system design to ascertain the feasibility with regard to engineering, schedule and cost considerations (phase A).

Class 4 :   Projects in the mission definition study phase, during which the scientific objectives and aims and the typical scientific payload are defined, without much consideration of the engineering constraints.

Only the projects in Class 1 are approved and can be considered as part of the Agency's programme, whereas those in the other classes belong to the study programme and are subject to competition and selection before acceeding to the "approved" status.

The projects in the different classes are briefly discussed below, with the main emphasis on their principal characteristics, without any pretention to exhaustiveness.

CLASS 1

Projects in the development phase (IUE + FSLP); projects in the detailed system design study phase B (EXOSAT + Faint Object Camera on the Space Telescope). The projects of this class are approved and are shown in Figure 1.

1. The International Ultraviolet Explorer (IUE)

The International Ultraviolet Explorer (IUE) is a joint project between ESA, NASA and the Science Research Council (SRC) of the United Kingdom. Under the agreed Arrangements, ESA will contribute the deployable solar array and design, build and operate a ground station in Spain.

The main scientific objective of this mission is to advance the study of the physical characteristics of stars of all spectral types, with improved spectral resolution and to a fainter magnitude, by comparison with the pioneering missions, such as TD-1, OAO-2 and Copernicus.

The observatory will be launched into a geosynchronous orbit in early 1978, and will thus be in continuous contact with, and under the control of, the U.S. and Europe-based ground stations. The originality of the mission lies in this continuous interaction between the astronomer on the ground and the telescope in orbit, thus reproducing the familiar conditions existing at all ground-based observatories.

The scientific payload

The IUE will contain a 45 cm Cassegrain telescope which will be used exclusively for spectroscopy. The general layout of the telescope/spectrograph system is shown in Figure 4. The telescope field of view can be recorded with a television camera and displayed on the ground so that the observer may identify his target star. The scientific aims of IUE demand a capability of obtaining both high-resolution (0.1 Å) spectra of bright objects and low resolution spectra of fainter objects to a limiting magnitude of at least 12 (a 14th or 15th magnitude limit is highly desirable). An echelle spectrograph was selected to obtain the high resolution spectra desired for brighter objects. With this type of instrument a high dispersion is easily achieved, and an additional advantage is that the format of the spectrum consists of a series of adjacent spectral orders displayed one above another in the raster-like pattern. This format makes efficient use of the sensitive area of the SEC Vidicon television tubes which will be used to

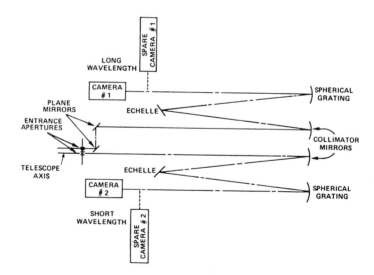

Figure 4. The IUE telescope/spectrograph system.

integrate and record the spectrum. The instrument is easily converted into a low resolution spectrograph by simply inserting a plane mirror in front of the echelle, leaving the low dispersion grating to act alone. In order to achieve the dispersion required in the high resolution mode, the spectral range of operation, 1150 Å to 3200 Å, is

into two ranges. Therefore, two exposures are required to record the
entire spectrum. A summary of the characteristics of the telescope
and spectrograph unit is shown in Table 3.

The IUE can be pointed in any direction in the sky by means of
a stabilization and control subsystem consisting of a set of reaction
wheels which slew the spacecraft or stabilize it while it is pointing
at a star, and a set of gyroscopes to control the slews and pointing.
Fine guidance control during exposures is provided by a star tracker
in the telescope. The satellite can be slewed to any point in the sky
with an accuracy of $\pm$ 2 arc min. Slews are executed about one axis
at a time, at rates of 4 to 5 degrees/minute. When guiding on a stellar target, the spacecraft is expected to maintain $\pm$ 1 arc sec. pointing accuracy indefinitely.

TABLE **3**

CHARACTERISTICS OF THE IUE TELESCOPE
AND SPECTROGRAPH

TELESCOPE

Type	Ritchey Chrétien
Aperture	45 cm
Focal ratio	f / 15
Image quality	1 arc sec
Acquisition field	10 arc min diameter

SPECTROGRAPH

Type	echelle
Entrance apertures	two 3 arc sec circles and
	two 10 × 20 arc sec ellipses

DETECTORS

Type	two SEC Vidicon cameras	
	Camera 1	Camera 2
High-dispersion wavelength range	1192-1924 Å	1893-3031 Å
Resolving power	10^4	1.5×10^4
Low-dispersion wavelength range	1135-2085 Å	1800-3255 Å
Resolution	6 Å	6 Å

Figure 5 shows an artist's conception of the observatory.

2. The First Spacelab Payload in Astronomy and Astrophysics

The First Spacelab Mission is a joint venture between ESA and
NASA, and has a primary objective of testing the Shuttle/Spacelab
system. It is, nevertheless, possible to simultaneously perform investigations in the scientific and applied research fields. The
mission is planned for the mid-1980s, and will have a duration of
only 7 days. The principal emphasis of the mission will be generally
oriented towards the Earth. Nevertheless, some 48 hours will be avail-

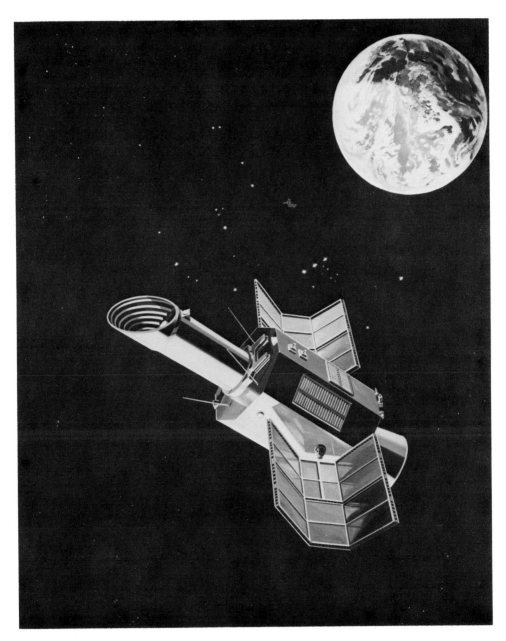

FIG.5   THE INTERNATIONAL ULTRAVIOLET EXPLORER (IUE)

able for deep space observations and about 12 hours for solar observations. A call for proposals for experiments on the First Spacelab
Mission was issued by ESA late in 1976, and as a result of this action,
a payload was selected, approved, and is now in the development phase.

This payload comprises three experiments in Astronomy and Astrophysics which are, strictly speaking, not ESA's development items,
since they are developed, managed and funded under national auspices,
but are ultimately integrated, tested and flown on Spacelab under
ESA guidance and responsibility.

a)   Very wide field camera, developed by the Laboratoire d'Astronomie
     Spatiale, Marseille. The instrument consists of a hyperbolic
     collector, followed by an anastigmatic Schmidt camera with micro-
     channel or electrostatic detector, yielding an observable field
     of 56° in the sky. Using wide band filters, the camera will
     make direct photography in the 1300 - 3000 $\mathring{A}$ region of extended
     sources and of star and star-like objects. In addition, by in-
     troducing an objective grating, the camera will work in a
     spectrographic mode and measure spectra of stars down to 11th
     magnitude and slit spectra of extended sources. The camera will
     be installed in the Spacelab airlock and operate during the night
     part of the orbit.

b)   X-ray astronomy spectroscopy using a gas scintillation proportion-
     al counter, a cooperative experiment between the Laboratorio di
     Fisica Cosmica e Tecnologie Relative, Milan; Mullard Space Science
     Laboratory of the University College London, and the Space Science
     Department of ESA.

     The scientific aim is the measurement of spectral features in
     celestial X-ray sources and the interstellar medium, with a
     significantly better energy resolution than that available at
     present. A new type of detector will be used; the gas scintil-
     lation proportional counter, which has an appreciably better
     energy resolution ($<10\%$ at 6 keV) that the conventional propor-
     tional counters. This improved resolution, combined with high
     photon efficiency and effective area, will permit the search
     for and study of the elusive emission lines and absorption edges
     as well as a more precise determination of the continuum emission
     of X-ray sources. The instrument will be located on the pallet
     and will be built out of four modules. It will have a 5° FWHM
     field of view and windows and gas to give a useful energy band-
     width of 1.5 - 50 keV,

c)   Isotopic stack, developed by the Institut für Reine und Angewandte
     Kernphysik, Kiel. The proposed experiment aims at measuring
     elemental and isotopic abundances of heavy cosmic ray particles
     with charges $Z \geq 3$. Galactic or trapped particles within the
     energy interval 50 MeV/amu $<$ E $<$ 2 GeV/amu can be detected. The
     main objectives are isotopes of cosmic ray elements Al, Cl, Ca,

Ti, V, Cr, Mn, Fe and rare trans-iron elements. An isotopic
separation of 0.4 - 1 amu can be achieved, depending on the charge
of the particle.

The stack is a passive experiment using plastic visual track detectors. Heavy ionizing particles penetrating the plastic layers
produce latent tracks which can be revealed by means of etching.
Further evaluation of the stored track information involves
measuring cone lengths of a track at different residual range
by means of an optical microscope.

The proposed detector consists of several 250 u sheets, the size
and number of which are adjustable to Spacelab facilities (but
minimum size = 30 x 30 x 5 cm^3). The detector sheets should be
housed in a sealed aluminium container to achieve controlled
surrounding conditions.

3. The European X-ray Observatory Satellite (EXOSAT)

EXOSAT was approved by the ESRO Council in 1973, but the start of
the detailed system design study, phase B, was delayed until early 1977
for programmatic and financial reasons. However, in the meantime, a
number of studies were carried out on the system's critical areas and,
in particular, on the scientific payload, a scientific model of which
has already been produced and subjected to scientific evaluation. The
present planning foresees the phase B in 1977 and the start of the
phase C/D (development) early in 1978. Launch of the satellite is foreseen at the end of 1981, and its operational lifetime will be two years.

The scientific mission

The main scientific objective of this mission is to advance the
study of X-ray sources and of energy emitting mechanism through:

a)   the position determination of X-ray sources to some few seconds
       of arc, in order to follow for identification with objects ob-
       served in other wavelength ranges;

b)   the determination of the structure of extended sources and of
       their dimension; and

c)   the determination of the time structure in the emitting mechan-
       ism.

The observatory has two optional modes:

i)   the occultation mode for the precise determination and identi-
       fication of sources by the occultation technique using the Moon
       (primarily) or the Earth as the occulting body;

ii)   the arbitrary pointing mode for the study of the temporal and
       spectral variability of sources over long, uninterrupted time
       intervals (up to 3 days).

The determination of the position and diameter of celestial objects by the observation of the time and speed of their disappearance
behind the shadow of the Moon during periods of occultation is a classical astronomical technique. For observations from the Earth, the
technique is limited to those objects lying along the direction Earth-
Moon. The motions of Moon and Earth relative to the celestial sphere
are such that 8% of the sky is covered over a period of 18 years.

However, an observatory, placed in a highly eccentric orbit with
its apogee at a height of 200,000 km and with the line of apsides perpendicular to the Moon's orbital plane, will be capable of observing
the lunar occultation of sources over 20% of the celestial sphere within a year. The velocity of the satellite for the major parts of its
orbit is small compared to that of the Moon, so the full occultation
time as considered from the ground is conserved. A further part of
the sky will be occultable by the Earth.

EXOSAT employs this technique for the localization and measurement of size and structure of X-ray sources. For lunar occultation,
the limit of accuracy on bright sources ($>10^{-2}$ photons cm^{-2} s^{-1} in
the range greater than 1.5 keV) is about 1 arc second; the limitation
is caused by the inaccuracy of measurement of the position of the
satellite and by the uncertainty in the profile of the lunar limb.
For weaker sources, the accuracy is limited by statistics, i.e. the
total number of X-ray quanta received during the time of the corresponding angular displacement of the Moon. The measurement of sizes
is only limited by statistics, and it is expected that for strong
sources size measurements will be achieved to an accuracy of 10^{-1} to
10^{-2} arc sec (see Figure 6).

The capability of EXOSAT to resolve the spatial features of a
source during occultation depends on the structural features of the
source itself. In the case of the source in Perseus, which has a
reported extent of 40' and appears to be centred on the Seyfert Galaxy
NGC 1275, an angular resolution down to 1' may be expected, assuming
a quasi uniform surface brightness.

The accuracy of Earth occulation is limited by smearing caused
by the Earth's atmosphere, but since the frequency of occultations is
much higher (on the average, 30 per year), numerous independent measurements can be made. For the weak sources, this is particularly advantageous. Angular resolution down to minutes of arc or lower may be
obtained with this method, on the extended supernova remnants Cygnus
Loop (3°) and Puppis A (0.5°).

Rather than leave occultations to the chance lining up of the
source, Moon and observatory, an orbit control system is incorporated

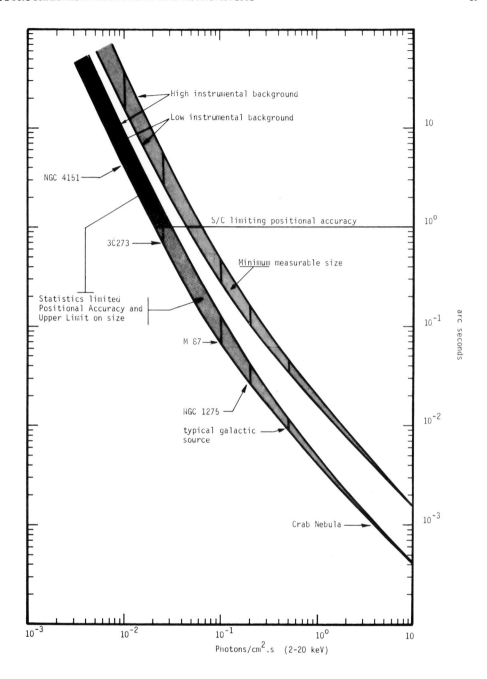

Figure 6 — Performance figures on positional accuracy, upper limit on angular size, minimum measurable size for the 2-20 keV experimental package. Two instrumental backgrounds are considered (4×10^{-2} and 10^{-2} counts/cm^2 s). The relevant position of known X-ray sources is indicated. On the abscissa is the total flux of a source between 2 and 20 keV.

in the satellite so that the orbital period may be modified in order to
maximize the number of occultation possibilities. When not engaged in
occultation observations, the observatory can view any chosen direction
of the sky (except for 60° about the solar direction) for long uninter-
rupted periods of up to the orbital period above the van Allen belts
($\sim$ 80 hours). With accurate time-keeping on board, and this capability
of long, continuous observation, EXOSAT can determine regular and irregular variations of the intensity of X-ray sources with time scales
from tens of microseconds to tens of hours.

The payload

      The EXOSAT payload consists essentially of two main parts: the
low energy system from 0.1 to 2.5 keV and the medium energy system
from 1.5 to 50 keV. A number of European institutes are participating
in the development of the payload which is managed and funded entirely
by ESA. The reason for departing from the traditional procedure whereby payloads are funded by the institutes, with ESA supplying the spacecraft and launcher system, was that a large scientific community, extending well beyond the X-ray community, would wish to use the EXOSAT
data and plan correlated measurements. The mission thus became an
"observatory" for the European astronomy community.

      The low energy system consists of a pair of Wolter type I imaging
systems, each of them with the characteristics given in Table 4.
Directly behind the telescopes may be placed a transmission grating
for spectroscopic observations with a resolution of 4% at 0.5 keV.
Two different detectors, to be selected on command, can be positioned
at the focus of the reflecting optics. These are the position-sensitive proportional counter and the channel multiplier array, the former
to be used for accurate positioning of star-like objects and for high
resolution spectroscopic observations with the transmission grating,
the latter to be used for mapping of extended sources and for studies
of the brightness distribution of the soft X-ray background.

TABLE 4

OPTICS OF IMAGING MIRROR

Wolter 1 imaging telescope

Max. diameter	: 29 cm
Focal length	: 90.2 cm
Geometric aperture	: 110 cm^2
Average grazing angle	: 2 degrees
Coating	: gold
Image quality	: < 10 arc sec over 10 arc min field

The expected performance figures of the low energy system are shown in Table 5 and Figure 7.

TABLE 5

PERFORMANCE FIGURES OF LOW-ENERGY SYSTEM
IN THE ARBITRARY POINTING MODE

		Accuracy	Sensitivity in 10^4 s
Position	PSD	40 arc sec	10^{-6} Sco (PL)[x]
	CMA	10 arc sec	10^{-5} Sco (PL)
Brightness distribution	PSD	40 arc sec	3×10^{-2} Puppis in 40 x 40 arc sec (PL)
			1 x Coma in 40 x 40 arc sec (PL)
	CMA	10 arc sec	1 x Puppis in 25 x 25 arc sec (PL)
			Coma no mapping on arc min scale
Broadband photometer		$\lambda/\Delta\lambda = 1.6$	
Grating spectrometer		$\lambda/\Delta\lambda = 25$ at 25 Å	Coronal lines of sun-type stars at 1 parsec in 3×10^4 s. Emission lines 10^{-4} Puppis in 3×10^4 s
Temporal resolution		10 us	
Soft X-ray background	PSD		In 10^5 s mapping in 1 arc min cells over a field of 1 degree (PL) for a uniform brightness distribution
	CMA		No mapping possible

[x]PL = photon limited observation

The medium energy system consists of an array of argon-filled proportional counters backed by xenon-filled counters with an effective area of some 2000 cm^2 covering the range from 1.2 keV up to some 50 keV. The array is divided into four sections, each of which can

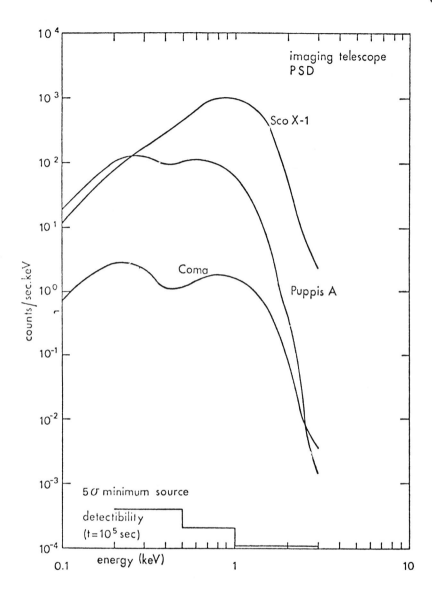

FIG. 7   Minimum source detectability (5σ-level of confidence)
          for low energy system, assuming an integration time
          of 10^5 seconds. For comparison, the spectra of
          Sco X-1, Puppis-A and the Coma cluster, folded with
          the instrument response, are given.

offset from the pointing direction to provide for a 'variable flat top' collimator response. At the smallest, this is approximately 5 arc min for observing point sources with maximum sensitivity. At the largest, this is 45 arc min, the maximum size of the lunar disc desired for lunar occultations of extended objects, such as supernova remnants and extragalactic clusters. This overcomes the basic incompatibility between the needs of the occultation and arbitrary pointing modes, since the flat-top size can be optimized for each observation, ensuring maximum area and minimum source confusion.

The expected performances are shown in Table 6.

TABLE 6

LIMITING POINT-SOURCE SENSITIVITY FOR MEDIUM-ENERGY EXPERIMENT

Energy (keV)	Point-source strength $(cm^2 \ s \ keV)^{-1}$	Fraction of Crab*
1 - 10	2.5×10^{-5}	2.8×10^{-5}
10 - 20	1.25×10^{-5}	2.8×10^{-4}
20 - 40	2.75×10^{-5}	2.4×10^{-3}
40 - 50	1.0×10^{-4}	2.2×10^{-2}

* Assuming differential Crab spectrum as $9 \times E^{-2}$ photons $(cm^2 \ s \ keV)^{-1}$.

An additional instrument completes the medium energy system: a gas scintillation proportional counter of $\sim$200 cm^2 for studies with higher spectral resolution than achieved by the proportional counters, of continuum emission and of emission and absorption lines.

The satellite

The observatory platform will be a three-axis stabilized space-craft with the pointing direction of the experiments held within a cone of better than 1 arc min. The reconstituted attitude is determined by a star sensor system and is expected to be better than 10 arc sec. The corresponding requirements in mechanical alignments of the experi-ment parts ⊤ both together and with respect to the star sensors – seem stringent, but are typical for astronomical satellites of this kind.

The observation programme

The observatory will be placed on a highly eccentric orbit (200,000 km), with apogee at high Northern latitudes. This orbit allows the occultation of 33 sources of the UHURU catalogue and is particularly suited for occultation of sources in the galactic centre region. In order to prepare the scientific utilization of EXOSAT by

the scientific community at large and to have an input early in the
design phase of the payload, a call for proposals for the observation
programme was issued at the end of 1973. Eighty-five replies were
received for this first solicitation, involving ninety scientists, at
thirty-three institutes in six Member States. An advisory group –
the Observation Programme Panel – composed of scientists from among
the proposers, considered the observational aspects of the mission,
evaluated and classified the proposals received and made a study of
the orbit selection. It is noted here that further calls for observation proposals will be issued later in 1977 and one year before the
launch, which is scheduled for late 1980.

4. The European contribution to the Space Telescope Programme

The NASA 2.4 m Space Telescope is the most ambitious space
astronomy project currently planned and will dominate astronomical
research for the rest of the century. It will be able to probe 10
times farther away in the Universe and see objects 100 times fainter
than what is achievable with the largest of the ground-based telescopes.
Its contribution to cosmology, and hence to our knowledge of the
origin of the Universe will be dramatic.

The telescope, in summary, would have a 2.4 m primary mirror
and a number of instruments in the focal plane for high spatial resolution, photometry and spectroscopy of faint sources. The telescope
is designed to operate near its diffraction limit and will concentrate
light from a star in a fraction of a tenth of an arc second. Its
fine guidance system will give a pointing accuracy of about 0.03" and
a pointing stability of 0.007" (rms).

The Space Telescope will be launched by the Space Shuttle into
a nominally circular orbit of 28.8° inclination and with an altitude
of 529 km, in late 1983. The telescope is a long-term enterprise,
with an operational lifetime of at least 15 years, an unusual feature
that can only be realized by development of the Shuttle and reflight
capabilities. It is foreseen that the telescope will, in fact, be
refurbished in orbit via the Shuttle every 2.5 years and that it will
be brought back to Earth for major refurbishment at five-year intervals.

ESA participation in the programme

Clearly, there is not an astronomer who would not wish to avail
himself of the possibility of using such a 'far-reaching' facility
as the Space Telescope. This is all the more true for European astronomers, in view of the recent (or imminent) completion of the large
optical telescope projects by the European Southern Observatory (ESO)
and by individual European countries, as well as the setting up of new
European infrared facilities. By the 1980s, there will be a high level
of European competence and interest in fields where the new telescope
offers unique opportunities.

Although it can be argued that European astronomers could have had access to the data from the Space Telescope via the guest observing programme, there were a number of reasons for ESA's securing a well-defined fraction of the observing time for the European community. Not least, this should stimulate orderly and co-ordinated development of the European astronomy effort, leading ultimately to well-planned and optimal use of the telescope by Europe, as distinct from the more fragmented opportunities offered by the guest observing programme.

It was thus that the Agency undertook to open discussions with NASA concerning direct participation in the Space Telescope programme.

The question then was what form such participation should take. There was a consensus within ESA, following discussions with its advisory bodies - the Astronomy Working Group and the Science Advisory Committee - that European astronomers should be directly involved in the production of the tools by which the science of the Space Telescope would be achieved, and that they should also be afforded the possibility of participating directly in the running of the scientific operations of the telescope and in the planning of its scientific programme. Last but not least, the amount of observing time allocated to Europe should be sufficient to meet the requirements of European astronomers and the related level of participation by ESA in the Space Telescope Programme such as to make it significant in the eyes of both NASA and the United States Congress.

These three principles, formulated with an approved by the European scientific community, guided the discussions with NASA and the American astronomers and led to the present philosophy of ESA's participation.

According to the Memorandum of Understanding, ESA's participation in the Space Telescope programme will comprise:

i)   one scientific instrument, namely, the Faint Object Camera
        and associated Imaging Photon Counting System;

ii)   the Solar Arrays and deployment mechanisms;

iii)   support of the activities of the Scientific Centre which
        will be established to guide and perform the scientific
        programme of the Space Telescope in its operational phase.

In return for this participation, ESA would be guaranteed 15% of the observing time of the Space Telescope during its space operations. This figure does not seem very large at first sight, and, in the context of the overall project, the Agency's role would still be a relatively minor one; but, even 15% of the observing time would be invaluable to Europe: in terms of hours per year, this share would exceed the entire clear dark time on a ground-based telescope, and it would give European astronomers the chance to share in the order-of-magnitude improvement offered by the Space Telescope.

TABLE 7

Distance pc	Distance Modulus	Redshift km/s	Redshift $\Delta\lambda/\lambda_o$	Objects	M_V Limiting Absolute Magnitude (S/N = 3)	Resolution AU	Resolution pc	Measurable Transverse Velocity km/s
10	0			Nearby stars	+29	2		
10^2	5			Hyades	+24	20		
10^3	10			Orion Nebula	+19	200		
10^4	15			Globular Clusters and the Galactic Centre	+14	2000	0.005	5
10^5	20			Magellanic Clouds	+ 9		0.05	50
10^6	25	50	0.0002	Local Group	+ 4		0.5	500
10^7	30	500	0.002	Virgo Cluster	− 1		5	5000
10^8	35	5000	0.02	Radio & Seyfert Galaxies	− 6		50	
10^9	40	50000	0.2	Quasars	−11		500	
10^{10}	45		$\simeq 2$	The Universe	−16		5000	

For the more distant objects, the assumption of Euclidean Space is made for convenience.

The faint object camera (FOC)[x]

The faint object camera on the Space Telescope will have extra-ordinary capabilities. Its limited magnitude (signal-to-noise ratio of three) in a ten-hour exposure will be around $V \sim 29$ and it will re-solve objects with sizes of $\sim 0.1"$. It should be possible to measure the positions of stars to $\pm 0.001"$ (if 10^4 photons are recorded) against the background of faint galaxies. Thus, ten years after the first epoch observatories, it will be possible to obtain proper motions accurate to $\pm 0.0001"$ per year. Table 7 illustrates what these capabilities mean in terms of absolute magnitudes, resolution and transverse velocities at the distances of several interesting objects.

Cursory examination of this table will suggest to all astronomers new and startling observations that will be of relevance to their re-search programmes. The uses to which the faint object camera can be put fall into three main categories:

a)   the study of extremely faint objects - for this the ability to count individual photon events is vital;

b)   the study of faint structures near brighter sources - this involves the requirement for a "coronagraphic" facility;

c)   the study of the structure or positions of bright objects.

A basic system incorporating a camera matched to the estimated Space Telescope performance and using a photon counting image tube will achieve the resolution sought over a field of more than 10^5 pixels and will reach the faint limiting magnitude required. However, the image tube will not be linear for point sources brighter than $V \sim 22$.

For uses of type b), it is necessary to provide a set of occult-ing and attenuating masks (a coronagraphic facility) which cover the bright object but not its surroundings.

A set of interchangeable colour filters should isolate both broad and narrow bands in the visual and ultraviolet spectral ranges. The broad bands would be used to obtain colours and would tie up with pre-existing systems. The narrow bands would primarily be used with emission-line objects at zero and low redshifts.

For c), the provision of a set of attenuating filters to extend the dynamic range of the instrument would be useful. Sufficient atten-

---

[x]cf. The Faint Object Camera, F. Macchetto and R. Laurance, ESA SN-126, March 1977

uation should be provided to permit the imaging of objects in the
solar system.

It would be desirable to incorporate a second camera with a longer focal ratio, which would enable the full resolution of the Space
Telescope to be used in the ultraviolet, a deconvolution process being
employed to combine the information in the speckle pattern. Resolutions
of ~0.03" would then be possible.

The ability to measure polarization, particularly of extended
sources, at high resolution would be very interesting. Polarization
measurements on normal stars and planetary nebulae can also lead to
important results.

The photometer capability of the FOC would give an improvement
over the best ground-based performance in stability (due to the absence
of atmospheric effects) and would permit good time resolution for faint
sources (because of the reduced sky background in each image point) in
both visible and ultraviolet regions. By reducing the number of pixels
scanned to approximately 10^3, a time resolution of less than one millisecond can be achieved. The same procedure will also enable the dynamic range for photometry to be increased.

Optical design

The optical design should preserve the diffraction limited quality
of the Space Telescope image.

The optical quality of the Space Telescope will be such that at
a wavelength of 632.8 nm, the point-spread function (70% encircled
energy contour) will have a radius of 0.10 arc seconds or better.
Working at the f/24 focal plane, this will represent a radius of 28 μm.
It will, therefore, be necessary to magnify the image further to match
the resolution limit of the telescope to the detector pixel size. The
magnification should not be excessive, because field of view will be
reduced and detector thermal noise per star image increased with increasing magnification. The basic NASA requirement states that, in
the prime mode, the pixel size shall be 0.022 x 0.022 arc sec^2. This
corresponds to a focal ratio of f/96 with a 25 um pixel size. However,
because of the electronic zoom capability of the detector, which can
change the pixel size from 20 μm to 50 μm, there is some choice in the
final focal ratio.

The FOC can be operated in two modes; the 'normal imaging' mode
at f/96 and the 'high focal ratio' mode. Its optical layout is shown
in Figure 8. The normal imaging mode relay optics consist of a two-
element reflector (c, d) which corrects the astigmatism of the telescope, together with a flat mirror (e) to fold the beam back on to
either of the two (redundant) detectors. The powered mirrors are two
off-axis parabolas, off-centre with respect to the field centre, in
such a way as to match the astigmatic focal length of the telescope.

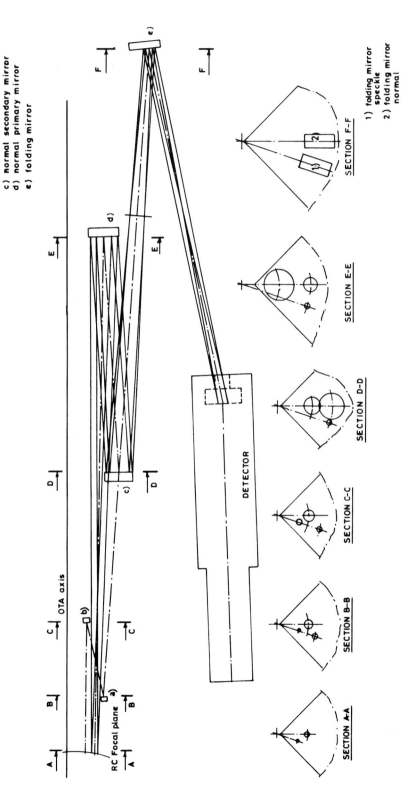

a) speckle secondary mirror
b) speckle primary mirror
c) normal secondary mirror
d) normal primary mirror
e) folding mirror

1) folding mirror speckle
2) folding mirror normal

SECTION A-A

SECTION B-B

SECTION C-C

SECTION D-D

SECTION E-E

SECTION F-F

DETECTOR

OTA axis

RC Focal plane

FIG. 8   OPTICAL LAYOUT OF THE FOC

The two surfaces are surfaces of revolution, and the relay has a plane
of summetry which simplifies the fabrication and adjustment of the
optics. Each parabola works at its optical focus. The magnification
of the relay is given by the ratio of the focal lengths of the two
parabolas.

The wavelength range will be selected by a set of filters. Up to
three filter wheels can be accommodated in the present design, each
with 16 positions. In each filter wheel there may be one clear (i.e.
100% transmission) and 3 neutral density filters. The use of either
one or two wheels in series can thus provide a combination of 8 neutral
density filters ranging from Δ m=0 to Δ m=16 which may then be used
with the colour filter in the third wheel.

In the high focal ratio (f/200) mode, a second set of relay
optics can be introduced by removing the secondary mirror (c in Fig. 8)
of the normal imaging mode. This second set, also consisting of a two-
element reflector (a, b), has a separate entrance aperature in the
Ritchey-Chrétien telescope focal plane (as shown in section drawing AA
of Fig. 8) and is contained in a radial plane different from that of
the prime mode relay. Both relays will have separate shutters and
filter wheels, so designed that no failure in one relay will affect
the operation of the other.

It is possible to include a grating filter in the high focal
ratio mode, to provide a continously variable narrow-band filter in
the ultraviolet. This filter would be an ideal means of permitting
operation in the speckle mode.

If a sufficiently narrow bandwidth is used, the speckle image
resulting from the defects in the mirror surface would be seen in the
wavelength region 120 nm to 350 nm. Unlike the speckle image produced
in ground-based observations, this image will be stationary. With
suitable ground image processing, some special objects, such as stellar
discs and close double stars, may be resolved to an extent approaching
the theoretical resolution limit of the telescope. The field of view
with an f/200 focal ratio in the speckle mode when using 20 um pixels
will be 4 x 4 arc sec^2, but this will increase to 10 x 10 arc sec^2
when the 50 μm pixel is used.

For the study of faint structure near bright sources, a coronagraphic facility can be added to the prime mode with little additional
complexity. A mask is placed at the telescope focal plane which re-
jects the majority of the light from the bright object. A second
apodiser mask is placed at the relayed aperture plane to remove light
scattered and diffracted by the telescope secondary mirror and spider.

Detector design

The main task of the FOC is the imagery of very faint objects
where an exposure time of many orbital periods may be required to

accumulate sufficient photons to obtain a satisfactory signal-to-noise ratio. It is, therefore, important to have a detector system that does not significantly degrade the limiting signal-to-noise ratio in an exposure by the introduction of read-out noise, etc. For this reason, a detector working in the photon counting mode has been selected, referred to as an Image Photon Counting System (IPCS).

The baseline configuration selected consists of a two-stage magnetically focused intensifier lens coupled to a sensitive camera tube call EBSICON. This system should be capable of giving an image of over 1000 x 1000 pixels (50% MTF). Unfortunately, it will not be possible to image this total field at high resolution at one time, because the size of the data store that can be accommodated will limit the picture to 512 x 512 pixels. The resolution performance may be utilized, however, by changing the pixel size and thereby changing the FOC field of view.

Two additional detector configurations are currently being studied, and component and system tests are planned, to enable a selection of one of these systems before entering the development phase.

Figure 9 gives an overall view of the FOC.

Performance

Preliminary calculations have been made of the photon fluxes and signal-to-noise ratios (S/N) to be expected when working with the FOC on the Space Telescope. Table 8 gives the exposure times required to detect stars of various magnitudes, as well as the S/N achievable in 10^3 seconds, and 10 hours. The performance is based on the assumption that 70% of the star's energy will fall within a radius of 0.1 arc seconds, that the telescope surfaces have a reflectivity of 0.9, and that the noise background is 23 m_V (arc sec)$^{-2}$.

The brightest stars that the detector can image require each photon to be sufficiently separated in position and time for detection to occur. Obviously, count corrections can be made at high count rates, but the maximum allowable counting rate which will give a good photometric accuracy will be of the order of 10% of the scan rate. The resulting star magnitudes that can be imaged will depend on the camera frame time and the working f number. Typically, working at f/96 at full format, the limit will be 14 magnitude.

FIG. 9 **THE FAINT OBJECT CAMERA**

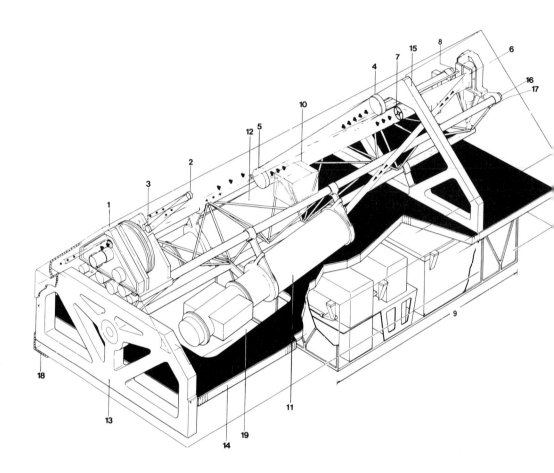

1. Focal Plane
2. Long focus Primary Mirror
3. Long focus Secondary Mirror
4. Normal mode Primary Mirror
5. Normal mode Secondary Mirror
6. Normal mode Fold Mirror
7. Apodizer mask
8. Long focus Fold mirror
9. Electronics bay
10. Calibration Source

11. Detector Head Unit I
12. Detector Head Unit II
13. Primary Structure Forward Frame
14. Primary Structure Sandwich Plate
15. Primary Structure Aft Frame
16. Optical bench
17. Optical bench struts
18. Superinsulation
19. Cut-out for radiators

TABLE 8 FOC Nominal Performance for Point Sources

m_v	Detected Photons/s	Time for S/N = 10	S/N in 10^3 s	S/N in 10 hrs.
21	6.05	17	77	464
22	2.41	43	48	290
23	0.96	113	30	179
24	0.38	314	18	107
25	0.15	983	10.1	60
26	0.06	3700	5.2	31
27	0.02	17100	2.4	14.5
28	0.01	92100	1.04	6.26
29	<0.01		0.43	2.58

Time resolution photometry

With no change to the optical layout of the instrument, it will be possible to carry out time resolution photometry. This can be achieved by modifying the mode of operation of the detector. In this mode, the number of photons detected in each frame is counted, and the count from each frame is stored in sequence in the data store. The frame format is made small (32 x 32 pixels), so that it covers only the object of interest. The inclusion of this facility will have little impact on the reliability of the prime mode, especially if there are redundant detectors. The brightest star that can be measured in this mode without attenuation will be about 13 magnitude. A time resolution of a third of a millisecond will be possible, and with this resolution a continuous photometric time history can be recorded over a period of 87 seconds. It will also be possible to reduce the resolution by adding adjacent scan counts in the detector, and, therefore, to monitor variations over longer time periods.

Status of the project

The ESA participation in the Space Telescope Programme has been approved by the competent authorities, and the detailed design study, phase B, is being performed in 1977. The development phase will be

initiated in mid-1978, leading to the delivery of the FOC to NASA in
1982, for integration on the Space Telescope.

Besides the technical realisation of the technologically advanced
hardware, it is ESA's present concern to promote increased direct involvement of the European astronomy community in all future phases of
the Space Telescope project. It is of paramount importance that
astronomers should be closely associated with the design, development
and testing of the Faint Object Camera and Imaging Photon Counting
System from the earliest stage. This is typically one of the Agency's
responsibilities, and it is in this connection that an ESA Instrument
Science Team, to advise the Agency on the Faint Object Camera project,
has been set up, following the widest solicitation and consultation
within the European astronomy community.

However, of even greater importance ultimately is the preparation
of the European astronomy community for the utilization of this exceptional facility and the definition of the ways and means by which
European astronomers will participate in the scientific preparations,
and the method of allotting the observing time allocated to ESA. Progress here will require in-depth discussions within the European
astronomy community and ultimately the setting up of proper schemes
and a suitable framework. In this respect, ESA has taken the initia-
tive to set up an ad hoc advisory group, the Space Telescope Working
Group, composed of qualified and experienced European astronomers.
This group will be the forum where such vital problems will be addressed.

CLASS 2

Projects (not yet approved) ready to enter the phase B (LIRTS,
EXSPOS, UTEX).

1. The Large Infrared Telescope on Spacelab (LIRTS)[*]

A wide variety of celestial objects emit predominantly in the
infrared region. These include intrinsically cool objects, but there
are also hot objects such as early type stars, whose energy is re-
emitted by the circumstellar clouds at infrared wavelengths. Progress
has been made in the past by ground observatories looking at celestial
objects in the few near-infrared windows of the atmosphere. The region
above 20 μm is inaccessible to ground observatories and can be explored
only by reaching above the atmosphere. Balloon- and aircraft-borne
instruments are still severely hampered by the residual atmosphere
and by the thick forest of molecular lines, from which only instruments
in space will be freed. However, the cryogenic requirements linked
with any infrared observation have until recently made it impracticable

_____

[*]cf. A Large Infrared Telescope on Spacelab (LIRTS), V. Manno,
    ESA Journal, 1977, Vol. 1

to fly such instruments on a satellite. The forthcoming availability of the Shuttle/Spacelab system at last opens the door of space to infrared instruments.

A set of different instruments can be identified as required, to expand infrared research from space. Free-flying satellites can best be used for all sky surveys, while large telescopes can appropriately be embarked on Spacelab for very high sensitivity and high spectral and spatial resolution observations of the infrared-emitting regions. These telescopes will take advantage of the repeatability of the Spacelab flights to change and refurbish the focal plane instrumentation and to optimize its configuration to the specific scientific objectives of the flight.

In the following, a summary description is given of a large uncooled telescope for high sensitivity and high spatial resolution observations in the 30 μ to 1 mm region, studied by ESA as a candidate for repetitive flights on Spacelab.

Scientific mission

The scientific mission objectives are summarized in Table 9.

TABLE 9 : SUMMARY OF LIRTS' OBJECTIVES

Type of object	Astrophysical interest	Type of measurement
Planets	Cloud structure, thermal profiles, chemical composition	High spatial (arc sec) and spectral ($\lambda/\Delta\lambda > 10^3$) resolution observations. $\lambda\lambda = 10$–$300\,\mu m$.
Stars and 'star-like' objects	Circumstellar envelopes. Mass loss. Composition. Novae shells.	Spectrophotometry, high-resolution spectroscopy ($\lambda/\Delta\lambda > 10^3$). High spatial resolution (arc sec). Polarimetry. Time variability.
Molecular clouds, H II and H I regions, Galactic Centre	Nature and distribution of dust. Protostars. Stellar evolution. Development of H II regions.	Multiband photometry ($\lambda/\Delta\lambda \sim 1$–$20$); $\lambda\lambda = 20\,\mu m$ – 1 mm. High spatial resolution mapping (arc sec). Polarimetry.
	Atomic and molecular abundances. Physical conditions. Molecule formation. Cloud dynamics. Masers. Cooling mechanisms.	High ($\lambda/\Delta\lambda > 10^3$) and very high ($\lambda/\Delta\lambda > 10^5$) resolution spectroscopy. $\lambda\lambda = 20\,\mu m$ – 1 mm.
Galaxies	Nature of emission mechanisms. Structure. Evolution.	Multiband photometry $\lambda\lambda = 20\,\mu m$ – 1 mm. High spatial resolution (arc sec). Time variability. Spectroscopy. Polarimetry.

Telescope configuration and its instrumentation

The telescope is intended primarily for measurements in the 30 um to 1 mm region, with very high spatial and spectral resolution. The major design aim was, therefore, to maximize the mirror diameter in compatibility with the Spacelab constraints, and to achieve the lowest noise level compatible with realistic detector performance, by careful control of the temperature of the optics and by optimum optical design.

The performance characteristics of the LIRTS are summarized in Table 10, and its configuration and accommodation on Spacelab is shown in Figure 10.

TABLE 10 : LIRTS PERFORMANCE SUMMARY

Configuration	Cassegrain
Primary mirror diameter	2.8 m
Primary focal ratio	F/2
Overall focal ratio	F/15
Figure	Aspheric between Ritchey Chrétien and tilted aplanatic
Aberration blur	Diffraction limited at $\lambda \gtrsim 10\ \mu$m
Spatial resolution	$5''(\lambda/30)$
Plate scale	$\simeq 5''/$mm
Total field of view	$10'$
Oscillating secondary	Max. throw $\pm 10'$
	Max. frequency 150 Hz (sine)
	20 Hz ($2 \times 40\%$ dwell)
Pointing (IPS)	Stability $1''$ (LOS) Accuracy $\simeq 2''$ Raster field $\simeq 2'$
Thermal background noise	$10\ \mu$m $-$ $100\ \mu$m $< 3 \times 10^{-15}$ W Hz$^{-1/2}$
(Diff. limit $\Delta\lambda = 0.1\lambda$)	$>100\ \mu$m $< 10^{-15}$ W Hz$^{-1/2}$
Temperature fluctuation noise	$< 2 \times 10^{-17}$ W
Variation in imbalance signal	
over 30 min	$< 7 \times 10^{-17}$ W

The optical design of the telescope lies between that of a Ritchey-Chrétien and that of a tilted aplanat. This choice is a compromise in response to the need to have good image quality to 10' off axis, which is typical of the Ritchey-Chrétien, and to maintain a good optical quality at the focus while the secondary mirror is tilted, which is a characteristic of the tilted aplanat. The secondary mirror is slightly elliptical and undersized, to keep the beam always on the primary mirror. Ray tracing and calculation of the energy distribution function show that, indeed, the telescope will be diffraction-limited at 30 um for $\pm$ 1' field of view and $\pm$ 2' beam movement in space. Tilt and decentre controls are included in the design, as well as alignment control of the line of sight to the star-tracker reference. In additon, a focus control sensor function is included. A TV camera displays the 10' x 10' focal plane.

The primary mirror is made from ULE glass (produced by Corning in the USA), and the secondary mirror from 2.35 mm thick beryllium, covered

FIG. 10 : LIRTS CONFIGURATION AND ACCOMMODATION ON SPACELAB

with gold by a replication technique. The secondary mirror's chopping
capability is realised by linking it to a torsion bar about which the
mirror oscillates naturally at 150 Hz. A different frequency can be
chosen by retiming the mirror's release by the two electromagnets that
hold it at each end of its oscillation.

Of particular concern is, of course, the thermal design of the
system. The main concern is not the lowest temperature that the telescope will attain, but rather the temporal variations of the temperature
gradients across the primary mirror at the two extremes of chopping
angle. Different thermal designs have been considered to ensure that
the temperature variations during the chopping cycle and the observation period do not introduce spurious signals. The temperature variation must be less than 10^{-4}K/s and $2-3 \times 10^{-4}$K at the chopping angle
extremes for the primary mirror. The pointing and stability functions

are realised by a group of three star trackers on Spacelab's Instrument Pointing System (IPS). The overall weight budget for LIRTS is shown in Table 11.

TABLE 11 : Φ 2.8 m TELESCOPE WEIGHT BUDGET

Assembly	Weight (kg)	
	Unit	Sub. total
− *Optical telescope assembly*		2843
Structure	1547	
Optics		
− Primary mirror	1090	
− Secondary mirror assembly	25	
Thermal coatings		
−Super insulation	30	
− Black paint	20	
− Heat pipes	8	
Alignment and pointing		
− Internal alignment sensor	5	
− Alignment monitoring instrument (AMI)	30	
− IPS Star Trackers		
boresighted sensor + fixed baffle	25	
roll sensors + baffles	30	
Power distribution		
− Motors and switches for the extendable shield	13	
− Cabling	20	
− *Extendable shield*		242
Structure	218	
Thermal coatings	24	
− *Development weight margin* (10%)		300
		3385
− *FPIP*		400
Michelson interferometer + photopolarimeter	400	
Telescope total weight		3785 kg

Although the focal plane instrumentation awaits final definition and selection, it is expected that three basic instruments will be developed for use with LIRTS. These are a multiband photopolarimeter, a spectrometer and a heterodyne line receiver.

For all instruments mentioned above, cooling of detectors and filters or optics to liquid helium temperatures, and perhaps lower, is required. It is possible to envisage a common dewar for all the proposed focal plane instruments, with a lifetime of 10 to 30 days.

The dewar would weigh approximately 100 kg and have a size of 90 cm x 70 cm diameter. A single chamber containing superfluid He^4 is assumed, venting to space through a porous plug. A possible He^3 stage is also being considered.

Helium dewars for space application have not yet been developed in Europe, and ESA is undertaking the development of such an item within its Development and Technology Department at ESTEC.

It is planned to mount the LIRTS on Spacelab's Instrument Pointing System (IPS), which is in turn mounted on a train of two pallets in the Space Shuttle's cargo bay. Its accommodation is compatible with both the pallet-only and the short module configuration.

Contamination problems

There are three classes of 'contaminant' sources which may affect the observations of LIRTS. Of these, the spurious signals induced in the detectors by high energy particles are the most serious. Both electrons and protons in the Van Allen belts and in the South Atlantic Anomaly cause these unwanted signals, which can be reduced to an acceptable level both by using screening material with high and low Z numbers and by electronics means. In addition, a low altitude, low-inclination orbit is to be preferred.

A second source of potential trouble is molecular emission from the residual atmosphere in the Shuttle environment. Here one can say that LIRTS is definitely less vulnerable than the cooled NASA-projected Spacelab Infrared Telescope Facility (SIRTF) telescope, and can tolerate column densities a hundred times greater than imposed by SIRTF on the Shuttle ($\sim 10^{12} mol/cm^2$ of H_2O).

A related problem is that of condensation of ambient gas on cold optics and detectors. This is particularly true when the telescope is pointed in the direction of the velocity vector. Careful planning of the observation sequence is, therefore, mandatory.

The third source of 'contamination' is due to emission from dust particles released by the Shuttle/Spacelab. A simple analysis in the case of LIRTS shows that, to be detectable, such particles should be larger then 100 um. Cleaning procedures can ensure that no such particles flake off during the mission.

Orbital considerations

The best choice of orbital parameters is dictated by a number of considerations and constraints. The need to reduce contamination from high energy particles calls for a low-altitude, low-inclination orbit, while a higher altitude is preferred from the points of view of condensation of ambient gas on cold surfaces, and the fraction of the celestial sphere accessible during any one mission. The preferred ranges of altitude and inclination, $28^o < i < 50^o$ and 300 km $< h < 400$ km, are consistent with launches from Cape Canaveral.

On any one mission, about 60% of the celestial sphere can be observed with integration times along any one permitted direction of up to 1000 s. By proper selection of orbital parameters and launch dates, it is possible to have access to the total sky with a set of about four 7-day missions, or with two 30-day missions at six-month intervals. This is shown in Figures 11 a and b, which indicate, in celestial coordinates, the accessible sky during missions of 7 days, with launch dates spaced over three-month periods. Figure 11 b shows that complete coverage of the galactic plane can be achieved with particular orbital parameters and launch dates.

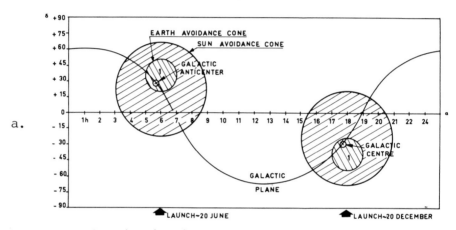

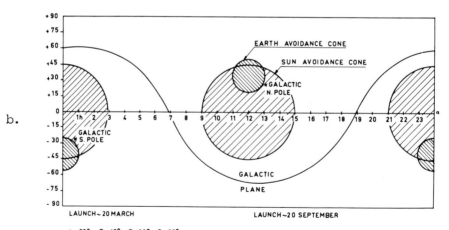

FIG 11 : PARTS OF SKY AVAILABLE FOR OBSERVATION
(7-day sorties)

Status of the project

A supplement of studies will be necessary before entering the phase B. It is, in fact, intended to study the possibility of leaving the telescope in orbit between successive flights of the Shuttle, in order to reduce the otherwise extremely high operational costs related to the bringing up and down with the Shuttle of the whole system. In addition, preliminary design studies of the focal plane instrumentation have been initiated.

2. A Spacelab Facility for Spectroscopy and Polarimetry of X-ray Sources (EXSPOS)[x]

There are, presumably, a number of mechanisms giving rise to the emission of X-ray sources. In many cases, which depend on the physical conditions at the source, the presence of absorption or emission lines, or of polarized light, will be revealed. This field has just been opened up by recent measurements of ionized Iron lines from such diverse X-ray objects as compact binaries, supernovae, and galactic clusters. Also, lines are expected of different width and at different energies, depending on the chemistry of the source and on its dynamic status.

EXSPOS is a homogeneous ensemble of instruments which will come to grips with this problem and simultaneously offer the possibility of detecting and studying line features with very high as well as low resolution in the overall range of interest (0.2 - 20 keV), with some space resolution and with detection limits which are more than one order of magnitude below the presently detected or expected line strength.

The project was studied by ESA as a candidate facility to be flown on Spacelab, thereby taking advantage of its reflight capabilities for a continuous updating of the instrumentation and for reconfiguring the payload to satisfy the evolving demand of research.

Mission objectives

A summary of the scientific objectives and requirements is given in Table 12.

---

[x]An expanded version has been published in ESA SP-132, "Physics and Astrophysics from Spacelab", ESA contributions to the Trieste Symposium, 9-12 September 1976.

TABLE 12 : SCIENTIFIC REQUIREMENTS, SUMMARY TABLE

Scientific objectives	Energy range (keV)	Spectral resolution ($\Delta E/E$)	Spatial resolution (arc min)	Time resolution (s)	Polari-metry	Remarks
Compact objects & transients						
(a) Emission lines	0.5–10	1–4×10^{-2}	–	$\sim 10^{-3}$	–	e.g. Sco X–1; $\sim 10^{-2}$ photon/cm²s in Fe XXV line core
(b) Absorption edges	0.5–2	10^{-2}–10^{-3}	–	–	–	e.g. Cyg X–3
(c) Continuum	0.5–10	$<10^{-1}$	–	10^{-3}	$<1\%$	e.g. Her X–1
Supernova remnants						
(a) Emission lines (old)	0.5–2	10^{-2}–10^{-3}	1–10	–	–	e.g. Cyg Loop; $\sim 10^{-4}$ photon/cm² arc min² in O VIII lines
(b) Emission lines (young)	0.5–10	$<10^{-3}$	1	–	–	e.g. Cas A; $\sim 10^{-2}$ photon/cm²s in Si XIII lines
(c) Non-thermal component	2–10	$<10^{-1}$	1	10^{-2}	$<1\%$	search for central objects
Clusters of galaxies	0.5–10	10^{-2}–10^{-3}	1–10	–	–	Perseus Cluster; $\sim 10^{-3}$ photon/cm²s in Fe XXV lines
Extragalactic objects	0.5–10	10^{-2}–10^{-3}	1–10	–	$<1\%$	Cen A, Perseus Cluster, 3C273
Active stars and coronae	0.2–10	10^{-2}–10^{-3}	–	10	$<1\%$	U Gem Systems, and nearby stars
Interstellar medium	0.3–2	10^{-1}–10^{-3}	–	–	–	Sources at known distance, e.g. Crab Nebula
Low-energy background	0.5–2	$<10^{-1}$	~ 60	–	–	North Polar Spur

The payload elements

The payload consists of four different systems which together cover the 0.5 - 8 keV range, with a spectral resolution (Δ E/E) spanning the 10^{-4} to 1 range. The four different systems will now be considered in turn.

i) Bragg crystal spectrometer and polarimeter

This system is composed of a number of free-standing, large area crystals disposed in a flower petal arrangement of 12 sectors.

The crystals selected in the present design are given in Table 13.

TABLE 13 : FREE-STANDING SPECTROMETER CRYSTALS

Panel no.[x]	Crystal	Angular range		Ion
		B^{o}	L^{o}	
1, 2, 3	Graphite[xx]	44.0	46.1	S XVI
4, 5, 6	Graphite[xx]	48.3	50.3	S XV
7, 8	PET[xx]	49.3	50.8	Si XIII
9, 10	RAP[xxx]	55.0	60.0	{ 0 VII { 0 VIII
11, 12	Lithium fluoride[xxx]	40.0	41.5	{ Fe XXV { Fe XXVI

[x]Each panel has a projected area of approx. 2 500 cm^2

[xx]Doubly curved

[xxx]Conical focus

A fixed honeycomb collimator with a field of view of 2° will be mounted above all crystals.

The polarimetric capability of this system is achieved by making use of the property that a crystal operating at 45° acts as a perfect polarization analyser over an energy range corresponding to the crystal rocking curve width. As can be be seen in Table 13, the selected Bragg angles for all panels (excluding the RAP panel) are close enough to 45° to act as polarizers. The complete crystal arrangement (and associated detector) is rotated as a unit at an angular rate of 90° per minute, and any polarized emission will be revealed by the signal modula-

tion at twice the rotation frequency. The large area crystals selected
in the present design are all curved in the plane perpendicular to the
plane of dispersion. This is done in order to obtain a focussing
effect of the radiation onto a line detector, thereby reducing the instrumental background and, hence increasing the sensitivity of the instrument.

The crystals in Table 13, which present this single curvature,
can be scanned in the plane of dispersion to give a monochromatic,
one-dimensional map of an extended source (provided there is no confusion with neighbouring lines), with the angular resolution given by
the crystal rocking curve (12' for RAP). A two-dimensional map can
be obtained by rotating the entire spectrometer assembly. The graphite and PET crystals present, in addition, a convex curvature in the
plane of dispersion (see Figure 12), so as to diffract a small range
of Bragg angles (typically 0.5^{o} to 5°) and to display an especially
interesting spectral region. Diffracted photons are registered by a
one-dimensional position-sensitive detector. Each panel is served by
a detector, with suitable window to the energy of the expected line.
No problems are foreseen for the detectors, which are either already
developed and flown or in a state of advanced development, for HEAO
and EXOSAT, among others. However, the feasibility of properly bending the large area crystals has not yet been demonstrated, and requires
some development.

ii) The transmission grating spectroscopy system

High resolution spectroscopy in the soft X-ray region can be
profitably performed with a transmission grating spectrometer
whose dispersing element is place immediately behind an imaging
telescope. The line-shaped spectrum of a point source is then
focused onto a position-sensitive detector. For extended
sources, acceptable spectral resolution is limited to structure
of less than 1 arc minute.

A 66 cm Wolter type I is considered with a geometric area of
400 cm^2 and a field of view of 1°. Two collectors can be in-
serted at the focus behind the grating, either a channel multiplier array (CMA) or a position sensitive proportional counter
(PSP).

iii) The non-dispersive spectrometer

This system meets the requirements for low spectral resolution
in the 0.2 to 2 keV region over a field of a few minutes of arc.
The system consists of a 3 m focal length (parabolic mirror)
telescope with two detectors mounted on a turret to be alternatively brought to the focus, a cooled solid state detector (SSD)
and a gas scintillation proportional counter (GSPC). A nested
set of staggered parabolic aluminium mirrors is assumed, with
a maximum frontal diameter of 58 cm, yielding a total area of
1415 cm^2.

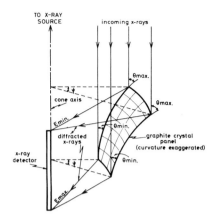

FIG. 12   DOUBLE CURVATURE CRYSTALS

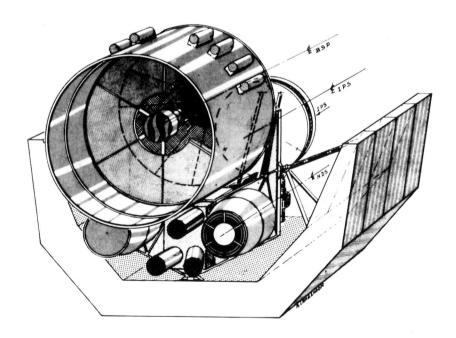

FIG. 13   EXSPOS – OVERALL VIEW OF PAYLOAD

iv) The large area non-dispersive spectrometer

> This system meets the requirement for low spectral resolution
> in the medium energy range (>2 keV). The system consists of
> eight independent gas scintillation counters, adding up to about
> 2000 cm^2 effective area. The energy resolution is about 10% at
> 6 keV. This type of detector is being developed in European
> laboratories.

The overall payload

An overall view of the payload is shown in Figure 13. The payload is mounted on the IPS on a single pallet element of Spacelab. All
four experiment systems are modularly mounted on a common structure,
which provides a strong frame necessary to maintain the total system
alignment throughout the mission. The modular mounting concept allows
the different spectrometers to be flown separately, if desired. This
permits great flexibility and optimization of the flight configurations,
and also a staggering in the development schedule of the overall payload. A star tracker system is mounted on the structure for pointing
and stability. The overall payload weight of about 1200 kg is not un-
duly high for the IPS and Spacelab capabilities.

Expected performance

Detailed calculations have been made of the sensitivity of the
various payload elements for the detection of lines and the study of
polarized emission over integration periods of 10^4 seconds. The results
are summarized in Figure 4.

The curves for the grating spectrometer and the solid dots for
the crystal spectrometer refer to the photon-limited and background-
limited gases, respectively. It is seen that the continued dispersive
systems provide a limiting sensitivity of about 2×10^{-4} photons/cm^2 s.
This is about 1 to 2 orders of magnitude below known and expected line
strengths from various cosmic X-ray sources. For comparison, expected
strengths for a few lines in the spectra of two optically thin, but
relatively weak, supernova remnants are shown. The non-dispersive
systems provide a good sensitivity at the expense of spectral resolution. On the basis of these performance characteristics, EXSPOS is
expected to be able to measure line strengths in a great number of
cosmic X-ray sources and to be an adequate facility for progress in
this important new field.

Status of the project

A supplement of studies is to be undertaken, to demonstrate the
feasibility of bending the large area crystals to the required curvature.

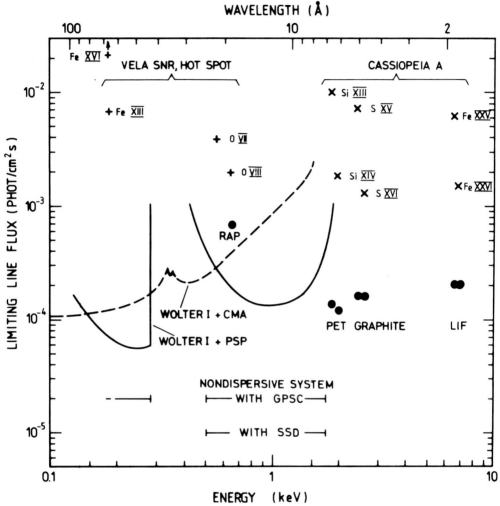

FIG. 14

3. The U.V. Facility on Spacelab (UTEX)

There is a considerable scope in the epoch dominated by the Space Telescope of NASA for less ambitious facilities to carry out astronomical research in those areas in which ST will not be employed, and hence independently of it, or to support ST itself.

One such facility (UTEX) was studied by ESA in the frame of the accommodation sutdies on Spacelab. The performance of this facility is important both for the particular goals it realises and also in support of ST in particular through its wide field imagery capability.

Mission and Facility

UTEX has been conceived as a Facility telescope capable of being operated, during a same mission, in the following basic modes:

- High time resolution correlated photometry in the optical and X-ray bands.

- Moderate field imagery.

- Low resolution spectroscopy of both point and extended sources.

- Objective grating spectroscopy over a field of moderate size.

These modes will allow observations of:

- Optical counterparts of galactic X-ray sources and fast variables.

- Selected galactic and extragalactic stellar-like fields.

- Galactic and extragalactic extended sources.

- Least spectroscopically studied classes of stars to fainter limiting magnitudes than previously reached by other UV experiments.

A telescope aperature of 60 cm was selected to limit weight and dimensions in order to take advantage of every flight opportunity on Spacelab. Such a diameter is fully compatible with the scientific objectives and it is sufficient to provide the necessary calibrations for ST observations of fainter sources.

The chosen optical design relies on the functions that the telescope has to perform.

i)   Light collector for narrow field instruments, such as photometers and spectrographs. This function requires a moderately high f-number and a small field of view is sufficient.

ii)   Wide field imagery. This function calls for a wide field of
       view and a low f-number for the observation of faint extended
       sources.

       In order to fulfil the above requirements during the same mission,
the telescope must have the possibility to work with two different
focal lengths. An optical system with f/8 and f/2.4 and providing a
corrected field of 1.65⁰ has therefore been chosen. The overall system
is composed of a Ritchey-Chretien telescope and f Focal Reducer. The
characteristics of such system are shown in Table 14.

TABLE 14 : THE TELESCOPE

Configuration: Ritchey-Chretien

Aperture	600 mm
Focal length	4800 mm
Relative aperture	f/8
Primary relative aperture	f/2
Diameter of secondary mirror	200 mm
Secondary magnification	- 4
Distance between mirrors	880 mm
Back focal distance	400 mm
Mirrors material	Zerodur
Mirrors weight	31 Kg
Coating	Al + MgF_2
Diameter observable flat field	20' = 28 mm
Diameter full field (to be used by the Focal Reducer)	1°65 = 138 mm
Plate scale	42".9/mm

       The focal reducer consists of an all reflecting Schmidt camera
with an intensifier and films as detector. In the instrument bay in
the focal plane of the telescope a number of instruments can be mounted
in varied configurations. This allows for the optimum use of any
flight opportunity. In addition, an X-ray detector sensitive to X-rays
in the range of 0.1 to 150 keV is mounted externally on the telescope
for simultaneous correlated measurements with the U.V. instruments.
The basic instruments and configurations are indicated in Table 15.

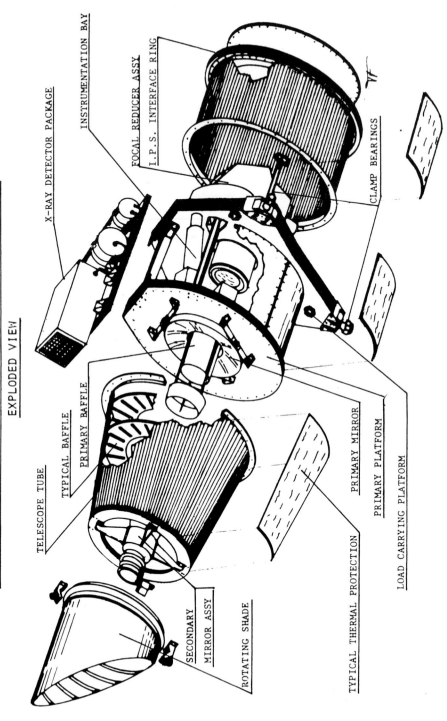

UTEX - BASELINE CONFIGURATION

EXPLODED VIEW

INSTRUMENTATION BAY

X-RAY DETECTOR PACKAGE

FOCAL REDUCER ASSY

I.P.S. INTERFACE RING

CLAMP BEARINGS

TELESCOPE TUBE

TYPICAL BAFFLE

PRIMARY BAFFLE

PRIMARY MIRROR

PRIMARY PLATFORM

LOAD CARRYING PLATFORM

TYPICAL THERMAL PROTECTION

SECONDARY MIRROR ASSY

ROTATING SHADE

FIG. 15

The whole Assembly is mounted on the ESA supplied Instrument Pointing System, whose performance is fully satisfactory to achieve the requested aims. An ad hoc pointing and stabilisation system has also been considered in case of IPS unavailability. This latter system could require active crew intervention to achieve ultimate accuracy of a few arc seconds.

An overall representation of the UTEX is shown in Figure 15. The overall weight of the facility, excluding the IPS, is around 1 ton.

The performance of the UTEX in different modes is expected to achieve the following limits:

Medium resolution Spectroscopy	Range	1100-2900 (912-3200) $\overset{\circ}{\text{A}}$
	Resolution	$10^3 - 5.10^3$
	Limiting mag.	15
Fast Photometry X-ray Correlated	Spectral Range	1500-3000 (3000-8000)$\overset{\circ}{\text{A}}$
	X-ray Range	0.1 - 150 keV
	Time Resolution	10 us
Wide Field Imagery	Focal Ratio	f/2.5
	Spectral Range	1200-3000 (3000-8000)$\overset{\circ}{\text{A}}$
	Bandpass	50-500 $\overset{\circ}{\text{A}}$
	Field Diameter	1$\overset{\circ}{.}$65
	Spatial Resol.	3"
	Limiting mag.	25/arcsec^2
Objective Grating Spectroscopy	Spectral Range	1500-3000 $\overset{\circ}{\text{A}}$
	Resolution	$\sim$ 50
	Field Diameter	1$\overset{\circ}{.}$5
	Limiting mag.	19
Long Slit Spectroscopy	Range	1100-2900 $\overset{\circ}{\text{A}}$
	Resolution	10^2
	Slit Length	10'
	Limiting mag.	16

TABLE 15 : INSTRUMENT FLIGHT CONFIGURATIONS

Configuration	Description	Operation modes	Comments
FP1	-Echelle Spectrograph for 10^3 and 5×10^3 resolution	-Spectroscopy of point sources	This is thought to be the base-line flight configuration.
	-Focal Reducer Camera with replaceable objective grating	-Imaging and Spectroscopy over $1^{\circ}65$	
	-Photometer (2 photomultipliers)	-UV and X-ray correlated photometry	
FP2	-Long-Slit Spectrograph with $\sim 10^2$ resolution	-Long-Slit Spectroscopy	Mission with emphasis on extragalactic and extended sources, Spectroscopy is performed with long-slit (configuration for missions with less stringent requirements for pointing and stabilization).
	-Focal Reducer Camera with replaceable objective grating	-Imaging and Spectroscopy over $1^{\circ}65$	
	-Photometer (2 photomultipliers)	-UV and X-ray correlated photometry	
FP3	-Concave grating Spectrograph (Rowland mounting) on axis	-Spectroscopy down to 912 Å	Spectroscopy below 1200 Å on sources of limited angular extent. Suitable coating of the telescope mirrors is needed.
	-Photometer (2 photomultipliers)	-UV and X-ray correlated photometry	
	-Direct Camera at f/8 (DIC) or other	-(Imaging over $0^{\circ}5$ with 1" resolution)	

Status of the project

The project is likely to be considered in the context of the recommendations on the future scientific missions of the Agency.

CLASS 3

Projects in the feasibility study phase A (EXUV satellite, Astrometry satellite, GRIST).

1. Extreme U.V. and soft X-ray survey satellite (EXUV)

The pioneering experiment of the Berkeley group on the Apollo-Soyuz mission led to the discovery of the first EUV emitting star. This somewhat unexpected result, if one believed the current ideas on the interstellar atomic hydrogen distribution, opens up an unexplored region, with the prospect of new exciting discoveries.

The mission definition study performed in 1976 on this mission, identified an unbiased all-sky survey in the EUV as the principal scientific aim. A sensitivity of 1/100 HZ 43 (the EUV star discovered by the Berkeley Group) is being aimed at, and a space resolution of a few minutes of arc. A complementary objective will be the brightness distribution in the XUV range, i.e. longwards of about 40 Å, with a spatial distribution of about 1 arc min., thereby well improving the present biased data.

The payload considered consists of two Wolter type I nested tele-scopes of 900 cm^2 for both the EUV and XUV regions, with microchannel plate arrays and proportional counters in the focus. The character-istics of the two telescopes are given in Table 16.

TABLE 16 : CHARACTERISTICS OF THE TELESCOPES

	EUV	XUV
Energy range	100 – 912 Å	30 – 250 Å
Front diameter (cm)	80	80
Focal length (cm)	97	163
f.o.v.	5°	2.1°
(at f.o.v. edge)	11'	3'

The spacecraft is designed for launch on an Ariane launcher into a highly eccentric orbit (apogee 100.000 km). The unique advantage of this orbit consists of the fact that it removes the geocoronal EUV flux which is otherwise an almost unsurmountable source of background noise. The project is at present being subjected to a detailed re-vision of the scientific objectives and payload. In the course of 1977, an industrial study will perform an analysis of the system and

and of its feasibility, as well as of its cost. The project will be considered in 1978 as one of the candidates for future inclusion in the scientific programme of the Agency.

2. Astrometry satellite

Observing the position of stars and planets as a function of time has been a cornerstone of astronomy throughout the centuries. Ground-based astrometry has produced impressive results (e.g. the Copernican system), however, the increase in accuracy has levelled off in the last half century and improvements of more than maximally a factor of 3 are not envisaged for the next two decades. Space astrometry, by moving the observing equipment aboard a platform above the atmosphere, will remove most of the effects plaguing the ground-based observations. Compared to ground-based astronometry, it is expected that a space project will improve by one order of magnitude at least position and parallaxes and observe stars four magnitudes fainter and a hundred times more numerous, in a homogeneous way with no systematic errors. If these predictions could be attained, the impact of this mission would be tremendous on such all important aspects of astrophysics as galactic structure, stellar evolution, stellar masses, extragalactic distance scale, existence of planetary systems, etc.

In order to construct an accurate celestial sphere, it is necessary to measure with high precision the angular distance between stars separated by a large angle, of the order of 90°. Such accurate measurements are obtained in the proposed projects by a differential method, comparing at the focal plane of a single telescope the positions of two stars imaged from two different fields of view. The two fields of view are superposed by means of a mirror with two reflective plane surfaces, inclined at a constant angle. The mirror, as well as the telescope itself, should be thermally very stable, and the design should minimise thermal variations.

Two basic options are at present being discussed:

Option A – characterised by the possibility of establishing a preselection programme of observed stars, including an active attitude control. The stars can be observed up to magnitude 14.

Option B – characterised by a systematic scan of the sky and observations of all stars of magnitude smaller than 10.5 as they enter the fields of view. The attitude control can preferably be passive (gravity-gradient stabilisation), but if necessary, may also include active sub-systems.

The performance of the two options, as well as comparative data of ground-based instruments, are shown in Table 17.

TABLE 17 : OPTICAL ASTROMETRY PROJECTS

Project	FK4		HMC		Option A		Option B	
Instrument	Visual Meridian Circle		Automatic Meridian Circle		Spacecraft		Spacecraft	
Observing time	70 years		2 years		3 years		2.4 years	
Position — Mag. m	N	σ arc sec	N	σ arc sec	N	σ arc sec	N	σ arc sec
6	1.5 10³	0.04						
9			10⁴	0.015	3.5 10⁴	0.002	6.10⁴	0.002
10							10⁵	0.003
10.5							5.10⁴	0.004
11			4.10⁴	0.05	6.10⁴	0.005		
14			10⁴	0.05	10⁴	0.012		
Proper motions (arcsec/year)	1.5 10³	0.0016	–	–	10⁵	σ	2.10⁵	1.4 σ
Parallaxes	6.10³	0.013	–	–	10⁵	1.6 σ	2.10⁵	1.2 σ
Minor planets	–		yes		(yes)		(yes)	
Double stars	–		no		(yes)		(yes)	
Present status	Existing		Proposed		Proposed		Proposed	
Astrometric index	1		10²		more than 10⁴		more than 10⁴	

The astrometric index, giving comparative evaluation of the various projects, is approximately equal to the number of observed stars N times the inverse squares of the mean error σ^{-2} (σ in units of 0.001 arc sec).

The project is being subject, in 1977, to a detailed feasibility study. Two studies will be conducted simultaneously: 1) A "software" study to verify through proper simultations of the system and its be- haviour, the feasibility to achieve the predicted astrometric perform- ance, and 2) A "hardware" study to proceed with the analysis of the spacecraft project, of the optical system and to assess the feasibility and cost implications of such mission.

The project will be considered in 1978 among the possible candi- dates for inclusion in the Agency's scientific programme.

3. Grazing Incidence Solar Telescope (GRIST)

A large facility-class grazing-incidence telescope has been identi- fied as a system which will enable a wide range of new solar problems to be studied. Such a telescope, which has a nominal wavelength range 90 Å to 1700 Å, covers a very wide range of solar phenomena including line spectra from temperatures 10^4K to 2.10^7K & continuum down to the solar minimum temperature. In order to maximise this temperature range, the wavelength range should also be as wide as possible. In practice, the short wavelength limit will be enforced by technical difficulties and the chosen value will result from a trade-off between such difficulties and the scientific benefits. There is no rigid upper wavelength limit imposed by the telescope, other than the principle that one should not extend it too far beyond the point at which it is more efficient to use a normal-incidence telescope.

The following general problem areas can be covered by such tele- scope:

  i)   heating and energy balance of the quiet solar atmosphere;
 ii)   mass flow and balance;
iii)   formation, evolution and heating of active regions;
 iv)   coronal structure, holes, streams, bright points.

To achieve these objectives, the summary requirements are as follows:

Wavelength range	90 - 1700 Å
Wavelength resolution	10^5 at 1000 Å
Field of view	5' x 5' and 50' x 50'
Spatial resolution	1 arc - sec
Time resolution	10 mS

The proposed payload consists of a Wolter type II sector grazing incidence telescope with paraboloid and hyperboloid surfaces. The focal length of the telescope has been preliminarily defined at 412 cm.

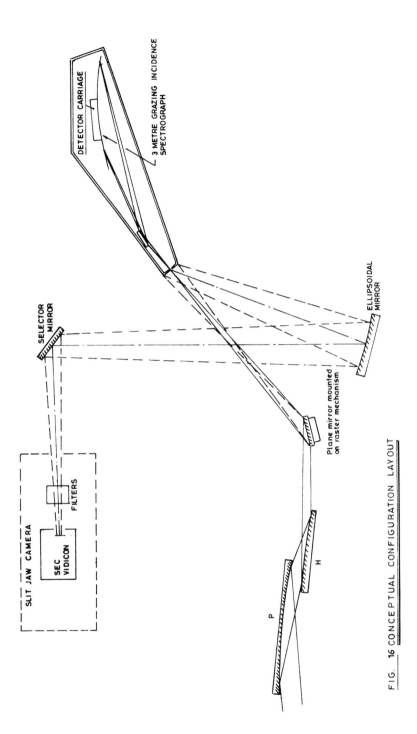

DETECTOR CARRIAGE

3 METRE GRAZING INCIDENCE
SPECTROGRAPH

SELECTOR
MIRROR

ELLIPSOIDAL
MIRROR

Plane mirror mounted
on raster mechanism

SLIT JAW CAMERA

FILTERS

SEC
VIDICON

H

P

FIG. 16 CONCEPTUAL CONFIGURATION LAYOUT

At the focal position of the telescope, a number of focal plane instruments are installed, which are modular and interchangeable in character in order for the facility to accommodate a large range of possible instruments and to optimise for the specific mission. Typically three FPIs will be flown at any one time. Typical instruments considered are : objective grating for slitless spectroscopy; photo-polarimeters; grazing-incidence spectrometers and normal incidence-spectrometers.

The model payload also contains one or more field cameras to record the solar image.

A schematic view of the configuration with a grazing incidence spectrograph is given in figure 16.

The overall system will be mounted on the ESA supplied IPS and flown repeatedly on Spacelab.

The project is subjected, in 1977, to an industrial system study and to an analysis of its feasibility and cost implications.

The project will be considered, in 1978, as a possible candidate for inclusion in the Agency's scientific programme.

CLASS 4

Projects in the Mission Definition Study Phase.

1. Superconducting Magnet Facility for Cosmic Ray Research on Spacelab

A preliminary study was carried out in 1976 to assess the feasibility of a magnetic spectrometer to be flown on Spacelab in the 1980s for cosmic ray research. It is considered, in fact, that the next major step in this branch of physics will require such facility to be carried out.

The major scientific aims in cosmic ray astrophysics have been identified to be (a) the knowledge of the isotopic composition of cosmic rays with charge between 1 and 30 and energy in the range up to 20 GeV/nucleon, (b) the charge composition spectra at least up to 100 GeV/nucleon, (c) the electron and positron spectra up to 1000 GeV, and (d) the amount of antiproton and heavier antinuclei in cosmic rays.

These aims lead to the following requirements for a spectrometer in a Spacelab flight:

1)   a maximum magnetic bending power of about 3 Tesla.m;

2)   an exposure factor of a few m^2 ster.days; and

3)   a track definition system with high resolution, possibly down to 10 microns, and "massless" detectors.

Several configurations of superconducting magnets that would approximately provide the required performance have been studied numerically, and the results indicate that the instrument is feasible within the Spacelab constraints; but a more specific study of the superconducting magnet, including the problems related to the behaviour of the cryogenics in space, is necessary in future.

With respect to the track definition systems, the study has shown that the instruments closest to the required performance are the multiwire proportional chambers used for elementary particle physics in the accelerators.

At the present state of the art, it is possible to achieve 30 μm space resolution, but continued study in the laboratory is necessary, in particular to make sure that the same performance can be obtained in the presence of a strong magnetic field.

A total number of three flights on Spacelab, of a duration of between 10 and 20 days, are envisaged. Ultimately, the facility should be made a free-flyer.

This project will be a candidate for an industrial feasibility study in the course of 1978.

4. Conclusions

The ESA projects in astrophysics currently in development or under study and the projects identified in the Long-Term Planning effort of the Astronomy Working Group, clearly testify of the richness of ideas and of the imagination of the European Community of Astrophysics. However, the financial constraints within which the Astrophysics programme is now to be planned will impose painful cuts therein.

In terms of conclusion, the best that can be done is to reproduce the conclusions and recommendations arrived at by the Astronomy Working Group in its Long Term Planning Report, 1980-1990.

"It is the considered recommendation of the Astronomy Working Group that serious consideration be given to the possibility of increasing the scientific budget of the Agency and to allow the planning of Astronomy and Astrophysics in the next decade within the limits of the baseline programme, ..." (as defined in Chapter 2).

5. Acknowledgements

The substance of this article derives from the efforts of the Astronomy Working Group of ESA, of the scientific consultants and of a number of ESA staff who were involved in the elaboration of the overall planning and detailed definiton of the scientific missions in Astronomy/Astrophysics.

THE NASA PROGRAMS IN ASTROPHYSICS

Jeffrey D. Rosendhal
National Aeronautics and Space Administration
Washington, D.C. 20546

ABSTRACT

   A summary is given of the NASA programs in Solar Physics, High
Energy Astrophysics and Astronomy. Descriptions are given of future
missions which are in the currently approved program (The Solar Maximum
Mission, the High Energy Astrophysical Observatories, the International
Ultraviolet Explorer and the Space Telescope) as well as of missions
which either have been studied or are currently in the planning stage.
Some of the major scientific problems to be addressed by these missions
are outlined. Plans for the use of the Space Shuttle and Spacelab during
the first few years of their operation are discussed.

I.   Introduction

   The purpose of this paper is to present a short summary overview
of the NASA program in Astrophysics. Emphasis will be placed on future
missions including both the currently approved program and potential
missions which are now in the planning stage and are serious candidates
for implementation having been subjected to detailed study during the
past few years. An attempt will be made to indicate the specific scientific contribution of each mission. However, of necessity, the scientific discussion will be telegraphic in nature and limited to broad
statements of general goals. The reader is referred to other papers in
this book for detailed reviews of specific scientific results which have
been obtained from a variety of satellite experiments and which constitute the scientific foundation upon which the future program will be
built.
   For ease of presentation, following some general remarks in Section
II, astrophysics will be subdivided into the disciplines of Solar Physics
(Section III), High Energy Astrophysics (Section IV) and Astronomy
(Section V). Some additional missions which are in a less advanced stage
of definition as well as some possible new directions for research in
Space Astronomy during the 1980's will be discussed very briefly in
Section VII.
   As is evident from the title of this volume, one of the major new

P. L. Bernacca and R. Ruffini (eds.), Astrophysics from Spacelab, 111-143.
Copyright © 1980 by D. Reidel Publishing Company.

tools which will be available for space research during the decade of
the 1980's will be the Space Shuttle. An outline of the unique features
of the Space Shuttle for conducting scientific research and a discussion
of current thinking concerning the use of the Shuttle during the first
few years of its operation are presented in Section VI.

II.   General Considerations

Before proceeding to a discussion of individual disciplines and
specific missions, it will be useful to make a few general remarks concerning the philosophy of development of a program in space science and
the relationship of various missions to the major problems in contemporary astrophysics.

First, while the details will differ from discipline to discipline
and different disciplines differ in their degree of development, it is
possible to identify several more-or-less distinct chronological stages
in the development of a space research program. These are:
   (1) Preliminary surveys - detection of gross features
   (2) Detailed surveys - gross statistics; approximate location of
sources
   (3) Detailed studies of individual objects; deep restricted
surveys
   (4) Special studies
The discipline of X-ray astronomy provides an excellent illustration of
this development. The first detections of celestial X-ray sources with
relatively crude spatial and positional information were made using
sounding rockets. More sophisticated instrumentation flown on later
generations of sounding rockets was used to make more detailed studies
of individual strong sources. The first X-ray sky survey was made using
the Uhuru satellite (SAS-A). Positional accuracy was sufficiently high
to permit the optical identification of a significant number of sources.
Other satellites (eg. Copernicus, SAS-C) were then used to make detailed
studies of individual sources discovered by Uhuru. Current and future
missions will both greatly extend the sensitivity of the Uhuru sky
survey (HEAO-1) and permit the first detailed spatial mapping of a large
number of individual extended X-ray sources using a true imaging telescope (HEAO-B). Additional generations of instruments are now being
studied.

Other disciplines can be characterized in a similar fashion. Extremes of development range from EUV Astronomy where the first sources
have recently been detected to Solar Physics for which the next approved
mission will involve the dedication of an entire payload to the definitive
study of a single problem - the understanding of solar flares. In each
case, however, potential new missions can be characterized as either
building on and extending the results of previous missions or as opening
up entirely new areas of investigation.

Second, the enormous number of observational tools employed in
contemporary astrophysics (ground-based optical telescopes, millimeter-
waves and radio telescopes, infrared telescopes on airplanes and balloons,
sounding rockets, and satellites covering the spectrum from gamma rays
through infrared and the long wave radio region) are, in the final ana-

TABLE 1

SOME MAJOR THEMES AND QUESTIONS ADDRESSED BY CONTEMPORARY ASTROPHYSICS

. THE NATURE OF THE UNIVERSE
  - HOW DID THE UNIVERSE BEGIN?
  - HOW DO GALAXIES FORM AND EVOLVE?
  - WHAT ARE QUASARS?
  - WILL THE UNIVERSE EXPAND FOREVER?
  - WHAT IS THE NATURE OF GRAVITY?

. THE ORIGINS AND FATE OF MATTER
  - WHAT IS THE NATURE OF STELLAR EXPLOSIONS?
  - WHAT IS THE NATURE OF BLACK HOLES?
  - WHERE AND HOW WERE THE ELEMENTS FORMED?
  - WHAT IS THE NATURE OF COSMIC RAYS?

. THE LIFE CYCLE OF THE SUN AND STARS
  - WHAT ARE THE COMPOSITION AND DYNAMICS OF INTERSTELLAR MATTER?
  - WHY AND HOW DOES INTERSTELLAR DUST CONDENSE INTO STARS AND PLANETS?
  - WHAT ARE THE NATURE AND CAUSE OF SOLAR ACTIVITY?
  - WHAT IS THE INTERACTION BETWEEN STARS AND THEIR ENVIRONMENT?
  - WHAT IS THE ULTIMATE FATE OF THE SUN?

TABLE 2

REPRESENTATIVE SCIENTIFIC PROBLEMS IN SOLAR PHYSICS DURING THE NEXT DECADE

. PHYSICS OF SOLAR FLARES AND ENERGETIC PARTICLE ACCELERATION
. SOLAR ACTIVE REGIONS
  - ACTIVE REGIONS AT VISIBLE, EUV, RADIO AND SOFT X-RAY WAVELENGTHS
  - MASS AND ENERGY TRANSPORT IN ACTIVE REGIONS
  - STRUCTURE AND DYNAMICS OF CORONAL PLASMA LOOPS AND ARCHES
  - MASS AND ENERGY TRANSPORT IN PROMINENCES AND FILAMENTS
  - SUNSPOTS AND THE MAGNETIC CYCLE
. STRUCTURE AND ENERGETICS OF THE QUIET SOLAR ATMOSPHERE
. STRUCTURE OF THE CORONA
. INTERACTION BETWEEN THE SOLAR CORONA AND THE SOLAR WIND
. SOLAR - TERRESTRIAL INTERACTIONS
  - CLIMATE SENSITIVITY TO SOLAR CONSTANT CHANGE
  - CLIMATE SENSITIVITY TO OZONE CHANGES (UV FLUX)
  - INFLUENCE OF SOLAR PARTICLES ON THE TERRESTRIAL ENVIRONMENT
  - FLARES AND SHORT TERM VARIABILITY

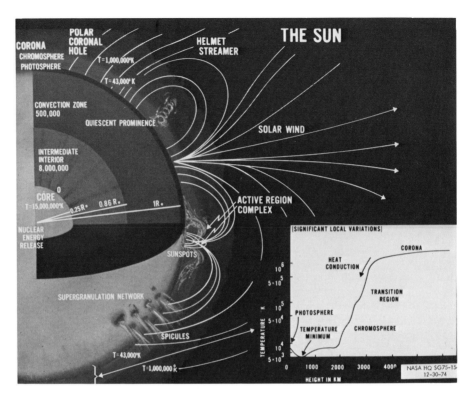

Fig. 1. A schematic cross section of the Sun showing the wide variety of
phenomena associated with the solar atmosphere.

lysis, all being used to try to obtain answers to a relatively small
number of very general questions concerning the origin and evolution of
the Universe. Some of the major themes and questions of contemporary
astrophysics are listed in Table 1 which has been taken from the recent
NASA study Outlook for Space (NASA SP-386, 1976). Obviously a much more
detailed set of questions or problems can be associated with each of
these general themes. Some of these more detailed questions as well as
the relationship between the information obtained using different techniques are outlined in the next three sections. All these tools are
needed because different wavelength ranges provide information on different types of objects or on different physical processes which may be
taking place in a given object. For example, infrared radiation reveals
the presence of thermal emission from cool objects while ultraviolet or
extreme ultraviolet radiation may result from thermal radiation from
hot objects or from synchrotron or bremsstrahlung radiation. Various
types of violent events may lead to the emission of high energy photons
or particles. Access to all wavelengths is needed to have access to all
types of objects which may be present in the Universe. At the same time,
a given object such as the Crab Nebula has been observed at all wave-

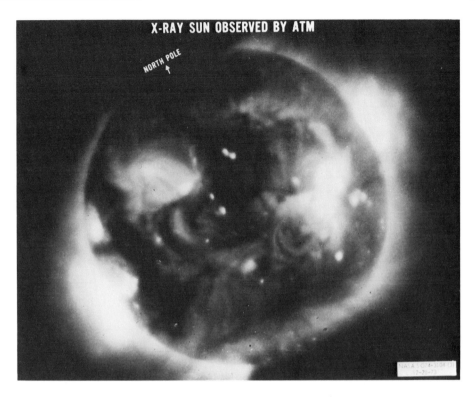

Fig. 2. The X-ray Sun as observed by the Apollo Telescope Mount aboard
        Skylab.

lengths from the radio to the γ-ray region of the spectrum and may be a
source of cosmic rays as well. Access to all wavelengths is obviously
needed to obtain a complete physical understanding of such an object.

III. Solar Physics

        The Sun, as the only star which can be observed in detail, dis-
plays a bewildering variety of phenomena (Figure 1). While the general
character of many solar features can be discerned from the ground,
there are a number of important features which are, in fact, only observable at ultraviolet, extreme ultraviolet or X-ray wavelengths. With
the advent of the space age, a series of experiments starting with
sounding rockets, proceeding through several Orbiting Solar Observatories,
and culminating in Skylab have revealed a new picture of the Sun (eg.,
Figure 2). These measurements, together with high resolution ground-
based studies, have revealed for the first time the full complexity of
the solar atmosphere and solar activity.
        In terms of the stages of development outlined in Section II,
Solar Physics has, in many respects, now passed beyond the initial exploratory phases with the next step being detailed studies whose purpose

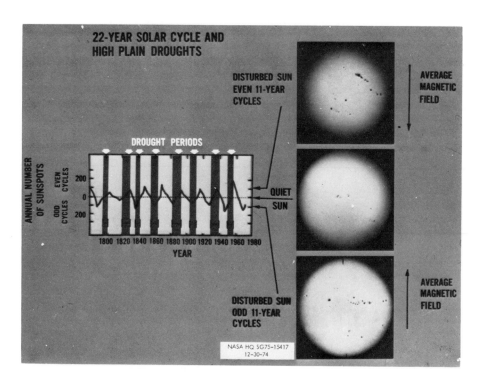

Fig. 3. The possible correlation between the occurrence of droughts in
the High Plains of the United States and the Solar Cycle. The
significance and even the reality of correlations such as these
are still the subject of considerable controversy.

will be to determine the precise physical nature of each phenomenon.
Some representative specific problems which are likely to be the subject
of such detailed studies during the next decade are listed in Table 2.
This list has been abstracted from the report of the Solar Physics Study
of the Space Science Board (Washington: National Academy of Sciences –
1976).

In broad terms, two categories of problem are present on this list.
The first deals with the detailed physics of the solar atmosphere. The
results from such studies can then be applied to the study of stellar
atmospheres and other astrophysical objects. For example, the solar wind
is the only example of a stellar wind subject to comprehensive examination and theoretical modeling. The solar flare is the only example of
the stellar flare phenomenon subject to close scrutiny and analysis.
Possibly similar violent ejections of relativistic particles and hot
plasma are seen in active galaxies. The second category of problem
deals with the interaction between the Earth and the Sun. Recent studies
have suggested that there may be both a long- and short-term relationship between variations in the Earth's weather and climate and variations
in solar activity (cf., Figure 3). The verification of these suggestions

TABLE 3

SOLAR MAXIMUM MISSION - PAYLOAD

EXPERIMENT	P.I.	ORGANIZATION	SPECTRAL RANGE	SPECTRAL RESOLUTION	SPATIAL RESOLUTION
GAMMA RAY SPECTROMETER	E.L. CHUPP	U. NEW HAMPSHIRE (MAX PLANCK INSTITUTE, GARCHING, & NRL)	0.3-17MeV 10-160MeV	≤ 7.5% (AT 0.66 MeV)	FULL SUN
HARD X-RAY SPECTROMETER	K. FROST	GODDARD SPACE FLIGHT CENTER	20-300keV	16 channel (0.1 sec time)	FULL SUN
HARD X-RAY IMAGING SPECTROMETER	C. DE JAGER	UTRECHT (BIRMINGHAM)	3.5-30keV	6 CHANNEL	8" x 8"
SOFT X-RAY POLYCHROMATOR	L. ACTON J.L. CULHANE A. GABRIEL	LOCKHEED MULLARD SPACE SCIENCE LAB. APPLETON LAB.	1.4-22.4Å	25000 at 2Å	10" x 10"
XUV SPECTRO-HELIOMETER	E.M. REEVES	HARVARD COLLEGE OBSERVATORY	20-716Å and 929-1336Å	0.1Å	4" x 4"
UV SPECTRO-METER & POLARIMETER	E. TANDBERG-HANSSEN	MARSHALL SPACE FLIGHT CENTER/ GODDARD SPACE FLIGHT CENTER	1100-3000Å	0.02Å	4" x 4" VARIABLE
CORONAGRAPH/ POLARIMETER	R. MACQUEEN	HIGH ALTITUDE	4435-6583Å	200Å +20Å, ± 2Å	7.5" x 7.5"
SOLAR CONSTANT MONITORING PACKAGE	R.C. WILLSON	JPL	UV-IR		FULL SUN

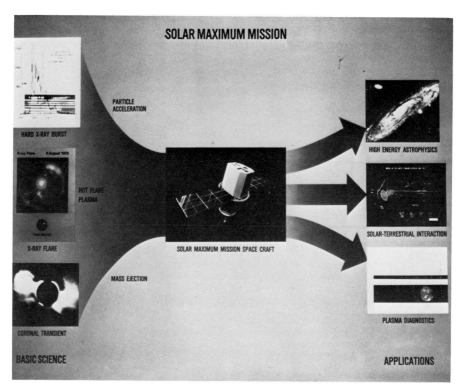

Fig. 4. Flare phenomena to be observed by the Solar Maximum Mission. The
relationship between the study of flares and other astrophysical
problems is also illustrated.

and the achievement of a detailed understanding of cause and effect
relationships clearly would be of enormous practical significance.
     The next approved mission in Solar Physics is the Solar Maximum
Mission (Figure 4) which is scheduled to be launched near the peak of
solar activity in 1979. Its purpose will be to achieve a detailed understanding of the physics of solar flares. The payload (Table 3) contains
a comprehensive set of instruments for performing a coordinated set of
measurements to study the entire flare event from preflare energy storage
to the injection of high energy material into the corona. Key questions
which will be addressed include the relationship between precursor signals
and the beginning of the flare instability, the nature of the mechanism
for the conversion of magnetic energy into flare energy, and the location
and nature of the primary and secondary particle acceleration processes.
Although most of the flare phenomena have been studied separately in
past missions (Figure 5), the physical mechanisms involved in the different phases and their causal relationships can only be determined by
determining the spatial and temporal relationships among the various
forms of flare emission. The Solar Maximum Mission will provide this
capability. Observations from space will be supplemented by ground-based

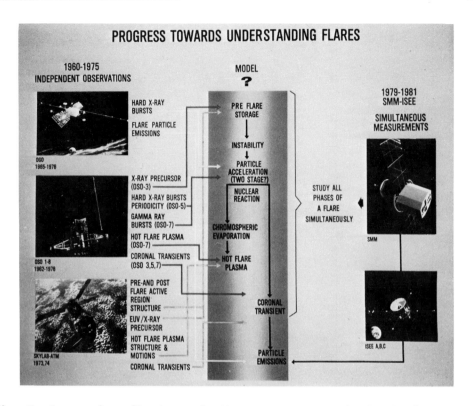

Fig. 5. Some major milestones in the progress towards developing an
understanding of solar flares.

radio and optical observations and by interplanetary particle measurements. The solar flare problem provides a particularly good example of
the importance of obtaining observations over a broad wavelength interval.

IV.   High Energy Astrophysics

As was discussed in Section II, during the past several years, a
set of increasingly sophisticated rocket and satellite experiments have
provided us with a new high energy view of the Universe. The first pictures of the γ-ray sky have been obtained (Figure 6). The crude map
illustrated has, of course, been greatly refined by the measurements
obtained with the SAS-B satellite. The combination of X-ray measurements
and optical observations (Figure 7) have shown that X-ray emission is
associated with many familiar types of astronomical objects. It has also
shown that many seemingly familiar astronomical objects such as binary
star systems are, in fact, quite exotic and extremely interesting when
observations are obtained over an extended wavelength range. Totally new
types of objects have also been revealed (see, for example, the paper
by Lewin in this volume).

TABLE 4

REPRESENTATIVE SCIENTIFIC OBJECTIVES OF HIGH ENERGY ASTROPHYSICS IN THE SHUTTLE ERA

SUB-DISCIPLINE	SCIENTIFIC PROBLEM						
COSMIC RAYS	ORIGIN OF COSMIC RAYS; ABUNDANCE OF ELEMENTS; NUCLEOSYNTHESIS	ACCELERATION MECHANISM AND PROPAGATION OF COSMIC RAYS; SPECTRUM OF COSMIC RAYS; SPECTRUM OF ELECTRONS + POSITRONS	ISOTOPIC RATIOS OF ELEMENTS PRODUCED IN SUPERNOVAE	PHYSICS OF SUPERNOVAE	NATURE OF COLLAPSED OBJECTS	STRUCT. OF OUR GALAXY	COSMOLOGY
GAMMA RAYS	DIRECTION OF SOURCE AND TIME STRUCTURE OF EVENTS PRODUCING THE COSMIC RAYS	GAMMA RAY SPECTRUM REFLECTS SPECTRUM OF C.R. MAGNETIC FIELDS ETC. AT SOURCE		CORRELATION OF DIRECTION AND TIME STRUCTURE WITH EVENTS		STUDY OF THE NATURE OF γ-RAY BURSTS(?) MAPPING OF GALACTIC ARMS	STUDY OF DIFFUSE GAMMA-RAY BACKGROUND
X-RAYS				CORRELATION OF DIRECTION AND TIME STRUCTURE WITH EVENTS	DETERMINATION OF CHARACTER OF BINARY STAR SYSTEMS WHICH CONTAIN COLLAPSED OBJECTS	MAPPING SOFT X-RAYS FROM INTERSTELLAR MEDIUM	MAPPING OF DIFFUSE BACKGROUND; SPECTRAL MEASUREMENT OF SOURCES IN CLUSTERS OF GALAXIES

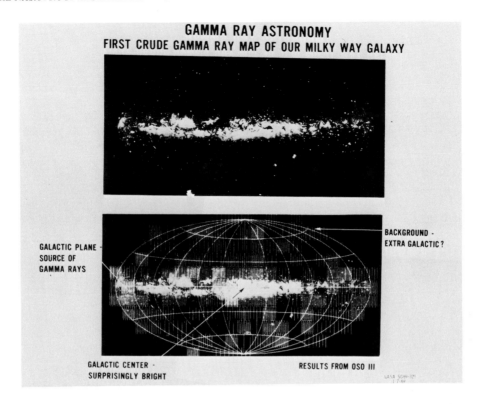

Fig. 6. A map of the X-ray sky as observed by Uhuru. Several particularly
        interesting examples of optical objects associated with X-ray
        sources are also shown.

TABLE 5

A SUMMARY OF THE HIGH ENERGY ASTRONOMY OBSERVATORY (HEAO) MISSIONS

MISSION A

OBJECTIVE:	TO SURVEY THE SKY AT HIGH SENSITIVITY FOR X-RAY AND LOW ENERGY GAMMA RAY SOURCES.
INTRUMENTATION INVESTIGATION:	FRIEDMAN (NRL): LARGE AREA X-RAY SURVEY BOLDT (GSFC)/GARMIRE (CIT): DIFFUSE X-RAY BACKGROUND INVESTIGATION GURSKY (SAO)/BRADT (MIT): SCANNING MODULATION COLLIMATOR PETERSON (UCSD)/LEWIN (MIT): HARD X-RAY, LOW ENERGY GAMMA RAY SURVEY

(CONTINUED)

TABLE 5

A SUMMARY OF THE HEAO MISSIONS (CONTINUED)

MISSION B

OBJECTIVE: TO LOCATE ACCURATELY X-RAY SOURCES, STUDY THEIR
                 STRUCTURE, SPECTRA AND VARIABILITY IN ORDER TO BETTER
                 UNDERSTAND THE PHYSICAL BEHAVIOR AND DISTRIBUTION OF
                 X-RAY EMITTERS.

INSTRUMENTATION: FOUR ELEMENT NESTED ARRAY, WOLTER TYPE I GRAZING IN-
                 CIDENCE TELESCOPE EFFECTIVE AREA 396 cm^2 at 0.4 keV
                                                   47 cm^2 at 4.13 keV

                 FOCAL PLANE INSTRUMENTATION:
                   HIGH RESOLUTION IMAGING DETECTOR
                   IMAGING PROPORTIONAL COUNTER
                   SOLID STATE SPECTROMETER
                   CURVED CRYSTAL SPECTROMETER
                 ASPECT: 0.1°, ARC SEC RESOLUTION USING GROUND ANALYSIS
                 PRINCIPAL PARTICIPANTS: R. GIACCONI (SAO), E. BOLDT
                   (GSFC), G. CLARK (MIT), H. GURSKY (SAO), R. NOVICK
                   (COLUMBIA).

                 HEAO-B WILL BE OPERATED WITH EXTENSIVE GUEST INVESTI-
                 GATOR PARTICIPATION-GUEST INVESTIGATORS WILL BE SELECTED
                 BY A. O.

MISSION C

OBJECTIVE: TO MEASURE THE CHARGE COMPOSITION OF VERY HEAVY COSMIC
                 RAYS, THE COMPOSITION OF COSMIC RAY ISOTOPES AND TO
                 SEARCH FOR CELESTIAL GAMMA RAY LINES.

INSTRUMENTATION/ ISRAEL (WASHINGTON UNIV.): HEAVY NUCLEI EXPERIMENT
INVESTIGATOR   KOCH (SACLAY) AND PETERS (DANISH SPACE RESEARCH
                 INSTITUTE): COSMIC RAY ISOTOPE EXPERIMENT
                 JACOBSON (JPL): GAMMA RAY LINE SPECTROMETER

SPACECRAFT: 3000 kg
                 HEAO-1   SCANNING, SIX MONTH LIFETIME (EXTENDED MISSION
                          UNDER CONSIDERATION
                 HEAO-B   POINTED, 12 MONTH LIFETIME
                 HEAO-C   SCANNING, SIX MONTH LIFETIME

LAUNCH: ATLAS CENTAUR
                 HEAO-1   LAUNCHED AUGUST 12, 1977
                 HEAO-B   1978
                 HEAO-C   1979

TABLE 6

HEAO-1 EXPERIMENT CHARACTERISTICS

Experiment	Major Objectives	No. of Detectors	FOV XY x FOV YZ (FWHM)	Total Area (cm²)	Energy Range
A-1 Friedman (7 large area collimated proportional counters)	Survey entire sky for X-ray sources down to ~10⁻⁴ Crab. Measure spectra, locations, and temporal variations.	4 2 1	1°x4° 1°x1/2° 8°x 2°	8800 4400 2200	0.15 keV to 20 keV 0.15 keV to 20 keV 0.15 keV to 20 keV
A-2 Boldt/Garmire (6 collimated proportional counters)	Measure spectrum and isotropy of diffuse X-ray background. Observe spectral and temporal characteristics of discrete sources.	2 1 3	1½°,3°,6°x3° 1½°,3°x 3° 1½°,3°,6°x 3°	2000 1000 3000	0.2 keV to 3 keV 1.5 keV to 20 keV 2 keV to 60 keV
A-3 Gursky/Bradt (2 high-resolution modulation collimators, star trackers)	Locate stronger X-ray sources to ~5 arc s. Measure structure of extended sources on 0.5 to 16 arc min scales.	1 1	4° x 4° 0.5 arc min modulation collimator 4° x 4° 2 arc min modulation collimator	450 450	1.5 keV to 15 keV 1.5 keV to 15 keV
A-4 Peterson/Lewin (7 scintillation detectors in an active collimator)	Extend spectra of stronger point sources to ~1 MeV. Measure spectrum and isotropy of diffuse X-ray and gamma ray background.	2 4 1	1° x 20° 20° circular 40° circular	220 170 120	10 keV to 200 keV 100 keV to 5 MeV 200 keV to 10 MeV

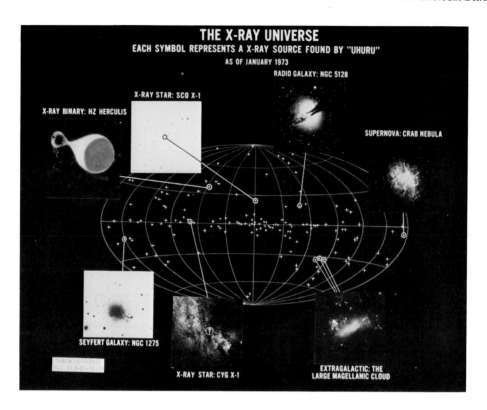

Fig. 7. An early Gamma Ray map of the Galaxy. These observations were
obtained from an experiment on OSO-III.

TABLE 7

WHAT PRESENT MISSIONS WILL TELL US

COSMIC RAYS

    UNMODULATED COSMIC RAY SPECTRUM (?) - VOYAGER
    FIRST MEASUREMENT OF ISOTOPES THROUGH IRON - ISEE-C & HEAO-C
    MEASUREMENT OF COSMIC RAY COMPOSITION TO URANIUM - HEAO-C

X-RAY ASTRONOMY

    SOURCE SURVEY SHOULD LOCATE AT LEAST 1000 SOURCES - HEAO-1
    IMAGING OF GALACTIC X-RAY SOURCES TO FEW ARC SECONDS - HEAO-B

GAMMA RAY ASTRONOMY

    ATTEMPT TO MEASURE LINE SPECTRA FROM CELESTIAL SOURCES - HEAO-C
    APPROXIMATE LOCATION OF GAMMA RAY BURSTS - ISEE/HELIOS/PIONEER VENUS

Some of the areas and specific problems which are likely to be receiving particular scientific attention during the next decade are summarized in Table 4. The matrix structure of Table 4 is intended to give some idea of the relationship between the subdisciplines of X-ray astronomy, γ-ray astronomy and cosmic ray physics and the possible contribution of each subdiscipline to the problems considered.

The next three dedicated High Energy Astrophysics missions will be the three High Energy Astrophysical Observatory (HEAO) missions. HEAO-1 (formerly known as HEAO-A) was launched on August 12, 1977. HEAO-B and C are scheduled to be launched in 1978 and 1979, respectively. These missions are summarized in Table 5 which contains a listing of the specific experiments, experimenters and institutions associated with each mission. More detailed information on the HEAO-1 experiments is given in Table 6. Figures 8, 9, and 10 show the layout of the experiments in the spacecraft for the three missions. It should be noted that HEAO-B, which will be a pointed telescope having an array of focal plane instruments, will have an extensive Guest Investigator program with participants being selected from a competitive evaluation of proposals.

Some of the identifiable pieces of scientific information which are expected to emerge from the currently approved high energy missions are outlined in Table 7. In addition to the information expected from the HEAO missions, Table 7 also illustrates the important point that not all high energy experiments are part of dedicated missions. The first measurements of the unmodulated cosmic ray spectrum may be made with instruments aboard the Voyager spacecraft which were launched in August and September 1977. Several other planetary or interplanetary missions either have contained (Helios) or will contain γ-ray burst detectors which, in conjunction with detectors on earth-orbital satellites, should yield important new information on the location of cosmic γ-ray bursts.

There are currently no approved High Energy Astrophysics dedicated missions beyond HEAO-C. However, numerous studies (eg., A Program for High Energy Astrophysics (1977-1988) - Report of the Ad Hoc Planning Group of the High Energy Astrophysics Management Operations Working Group) have identified a number of areas of investigation which can be addressed by possible new missions. An outline of some of the information which may be obtained from such new missions is given in Table 8.

Substantive plans for a follow-on program for the mid-1980's are now being formulated. As a first step, an Announcement of Opportunity has been issued to solicit proposals for experiments to be flown aboard a Gamma Ray Observatory (GRO). This mission (with a projected launch date in 1983) is currently envisioned as the first of an anticipated five mission program of Astrophysical Observatories emphasizing gamma ray astronomy, X-ray astronomy and cosmic ray astrophysics. The subsequent missions are planned as a cosmic ray observatory, an X-ray astronomy observatory (which would carry mostly non-telescope instruments supplied by individual investigators), a large-area, moderate resolution X-ray observatory and a pointed, grazing-incidence X-ray telescope approximately 1.2 meters in diameter. The order of subsequent missions has not yet been determined; additional Announcements of Opportunity for these missions will be released at appropriate times. It must be emphasized that neither GRO nor the subsequent missions have been authorized or approved.

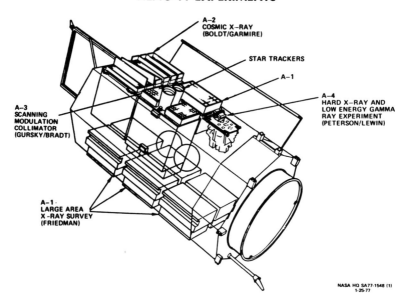

Fig. 8. The HEAO-1 Scientific Instrument Package

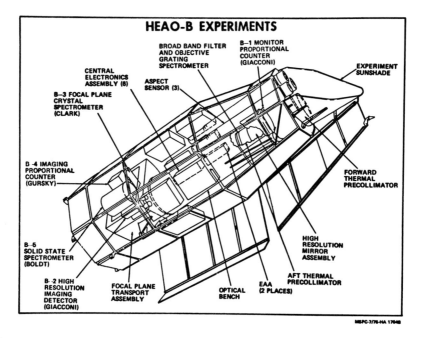

Fig. 9. The HEAO-B Scientific Instrument Package

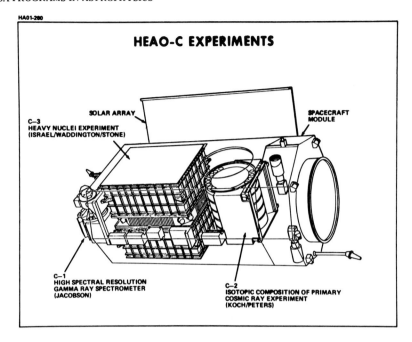

Fig. 10. The HEAO-C Scientific Instrument Package

STS MISSIONS

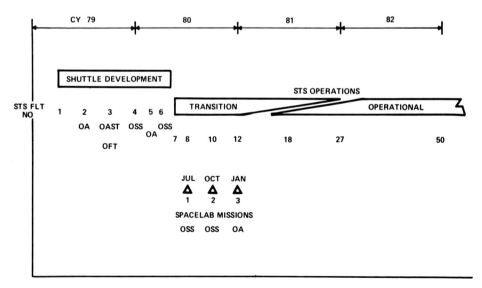

Fig. 11. A cutaway view of the IUE scientific instrument

TABLE 8

WHAT NEW MISSIONS WILL TELL US

_____

COSMIC RAYS
- COSMIC RAY COMPOSITION BEYOND URANIUM
- COSMIC RAY ISOTOPES BEYOND IRON OVER A WIDE RANGE OF ENERGIES
- ENERGY SPECTRA AT HIGH ENOUGH RESOLUTION TO SEPARATE COMPONENTS
- PRESENCE OF ANTIMATTER OR OTHER "STRANGE" PARTICLES IN COSMIC RAYS
- COSMIC RAY GRADIENT IN 3 DIMENSIONS

X-RAY ASTRONOMY
- IMAGING OF EXTRAGALACTIC SOURCES - COSMOLOGY
- DETERMINATION OF NATURE AND ORIGIN OF THE VARIETY OF TRANSIENT
  X-RAY PHENOMENA
- SURVEY AND IMAGING OF EXTREME ULTRAVIOLET SKY

GAMMA RAY ASTRONOMY
- SEARCH FOR SOURCES AT LEVEL OF SENSITIVITY 25 TIMES GREATER THAN
  SAS-2
- MAPPING OF GALACTIC CENTER AT HIGH RESOLUTION
- HIGH RESOLUTION LINE SPECTRA FROM EXPLOSIVE SOURCES
- DETERMINATION OF LOCATION AND ORIGIN OF GAMMA RAY BURSTS

_____

TABLE 9

THE IUE SCIENTIFIC INSTRUMENT

_____

TELESCOPE
    TYPE: CASSEGRAIN
    APERTURE: 45 CM
    FOCAL RATIO: F/15
    IMAGE SIZE: 1 ARC SEC
    ACQUISITION FIELD: 10 ARC MIN DIAMETER

SPECTROGRAPHS
    TYPE: ECHELLE
    ENTRANCE APERTURES:
    (i) A CIRCULAR HOLE, 3 ARC SEC IN DIAMETER
    (ii) AN ELONGATED HOLE, 10 x 20 ARC SEC
    DETECTORS: SEC VIDICON COUPLED WITH A UV-TO VISIBLE CONVERTER

	SPECTROGRAPH 1	SPECTROGRAPH 2
HIGH DISPERSION:		
WAVELENGTH RANGE	1150 – 1950Å	1800 – 3200Å
RESOLVING POWER	10^4	1.5×10^4
LOW DISPERSION:		
WAVELENGTH RANGE	1120 – 2160Å	1750 – 3250Å
RESOLUTION ELEMENT	6Å	7Å

_____

The issuance of Announcements of Opportunity is for planning purposes only, in order to assure timely selection of investigations and to aid in detailed design of the missions.

One important area not covered by any of these missions is Extreme Ultraviolet (EUV) astronomy. During the past few years, preliminary definition studies have been made of a number of possible Explorer-class satellites and a scanning EUV sky survey mission is still under active consideration for flight on an Explorer. This mission would be capable of performing the first unbiased large-scale systematic search for sources emitting radiation primarily in the 100 - 900 Å region of the spectrum, thereby, opening up one of the last remaining unexplored spectral domains. The scientific importance of this region of the spectrum is discussed elsewhere in this volume in the paper by Paresce.

V.   Astronomy

As is true in the case of the other disciplines, future missions in Space Astronomy will both extend the results of previous missions and open up entirely new areas of investigation.

Starting with the approved program, the International Ultraviolet Explorer (a NASA-ESA-United Kingdom cooperative program) which is currently scheduled for launch in January 1978 will continue the work in ultraviolet spectroscopy started with the Orbiting Astronomical Observatories. The characteristics of the IUE scientific instrument are summarized in Table 9. A cutaway of the instrument is shown in Figure 11. Emphasis will be on high resolution spectroscopy in the mid-ultraviolet range. Observations from Copernicus in this wavelength range have been limited to a fairly small number of bright objects. In the IUE low dispersion mode, it will be possible to obtain observations of a few of the brightest quasars. This satellite will be operated entirely as a facility for Guest Investigators. Scientific areas of emphasis in the currently accepted Guest Investigator program include studies of the chemical composition of hot stars, stellar chromospheres and coronae, the abundances of atoms and molecules in the interstellar medium, mass loss from hot stars, and mass transfer in binary systems.

Another largely unexplored region of the spectrum will be opened up by the Infrared Astronomical Satellite (IRAS) whose primary objective will be to perform the first unbiased all sky survey in the 8-120 micron region. A summary description of the instrument and the mission is given in Table 10. The project has been approved for implementation with the launch planned for February 1981. This mission is a cooperative mission with the Netherlands and the United Kingdom. The broad division of responsibility is for the United States to provide the telescope system (which includes the telescope, dewar and the focal plane detectors for the survey), the launch vehicle and launch support services, and the data reduction and analysis facility for processing the survey data. The Netherlands will provide the spacecraft, a low resolution spectrometer, and, in conjunction with the United Kingdom, a ground station and an operations center for command, control, operational planning and quick look data. It is expected that at the completion of the survey more than one million infrared sources will have been detected providing the first comprehensive view of the cool component of the Universe.

Fig. 12. An artist's concept of the Space Telescope

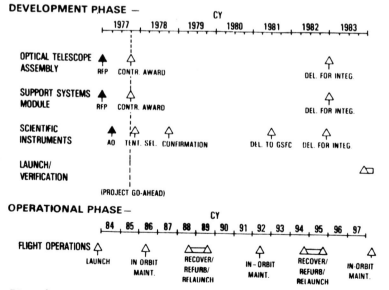

Fig. 13. Planning schedule for the development of the Space Telescope

PERKIN-ELMER OTA BASELINE DESIGN

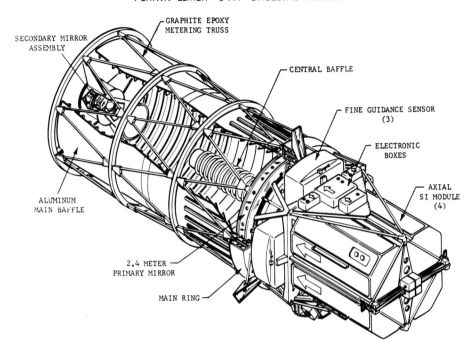

Fig. 14. A schematic drawing of the Optical Telescope Assembly (OTA) for
the Space Telescope

One additional astronomy Explorer is now being studied. It is a
Cosmic Background Explorer (COBE). The objective of this mission will be
to measure both the spectrum and the isotropy of the cosmic background
radiation. A decision as to whether to proceed with this mission will be
made at the completion of the current definition study.

A major step forward in optical space astronomy will take place
in the 1980's with the development of the Space Telescope (Figure 12).
A planning schedule for this mission is shown in Figure 13. Launch is
expected to be in late 1983. This instrument, which will form the heart
of the ultraviolet/optical astronomy program, has been the subject of
intensive study during the past several years. The minimum performance
specifications for the Space Telescope which emerged from these studies
and which are expected to be realized by the actual instrument are
listed in Table 11.

Following approval of the program for a Fiscal Year 1978 new
start, contractors have been selected for the development of the Optical
Telescope Assembly (The Perkin-Elmer Corporation) and the Support Systems
Module (Lockheed Missiles and Space Company, Incorporated). Cutaway
drawings of telescope assembly and spacecraft are shown in Figures 14
and 15.

SSM DESIGN FEATURES

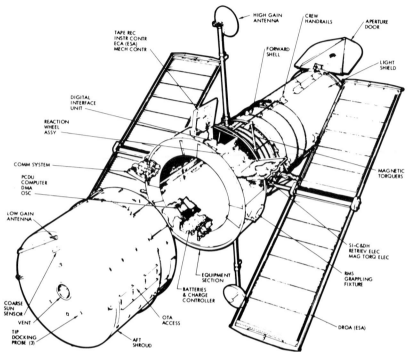

Fig. 15. A schematic drawing of the Support Systems Module (SSM) for the Space Telescope

The Space Telescope will be launched and checked out in orbit by the Space Shuttle, will be periodically visited by the Shuttle for maintenance and repair and replacement of focal plane instruments, and will be brought back to the ground using the Shuttle for major re-furbishment (Figure 16). Schedules shown for maintenance and refurbish-ment are tentative and the actual schedules will depend upon the per-formance of the focal plane instruments and other key system components.

Unlike previous astronomical satellites, the Space Telescope will be a true observatory accommodating a variety of instruments which can be changed and updated as both scientific priorities and instrument or detector capabilities evolve.

As a result of a competitive evaluation of proposals submitted in response to an Announcement of Opportunity, 4 instruments (a wide field camera, high and low resolution spectrographs and a high speed photomer) have been tentatively selected by NASA for inclusion in the first Space Telescope instrument complement. These instruments will be subjected to

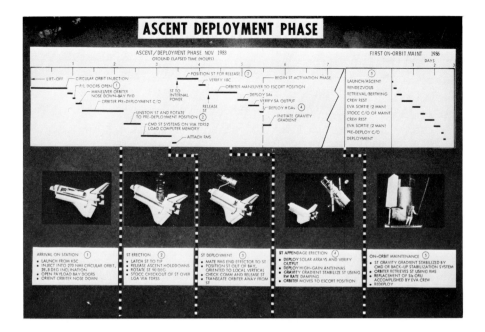

Fig. 16. One possible concept for deployment of the Space Telescope from the Space Shuttle

a one year preliminary design study and the instrument selection will be confirmed (or modified as necessary to meet program constraints) at the completion of that study. In addition to the NASA instruments, a faint object camera and an associated imaging photon counting system will be supplied by the European Space Agency. The total instrument and detector complement are summarized in Table 12.

The extended wavelength coverage, fine angular resolution, and faint source detection capabilities of the Space Telescope and its associated instrumentation will make it the most powerful astronomical instrument which has ever been built. A comparison between the capabilities of the Space Telescope and several ground-based optical telescopes is shown in Figure 17. It will be used to attack a wide variety of frontier problems in Astrophysics particularly in the areas of extragalactic astronomy and observational cosmology. A list of representative possible research programs is given in Table 13. A more detailed discussion of many of these programs is contained in the symposium of the Large Space Telescope – A New Tool for Science (Proceedings of the AIAA 12th Aerospace Sciences Meeting, Washington, D.C., January 1974).

VI.   Spacelab and the Space Shuttle

As was noted in Section I, the major component of the space program during the 1980's will be the NASA Space Shuttle and Spacelab which is being developed by the European Space Agency. Detailed des-

TABLE 10

INFRARED ASTRONOMICAL SATELLITE (IRAS) INSTRUMENT & MISSION DESCRIPTION

TELESCOPE: 0.6m F/10 CASSEGRAIN SYSTEM CRYOGENICALLY COOLED (SUPERFLUID
            HELIUM)

FOCAL PLANE INSTRUMENTATION:
  (1) ARRAY OF DETECTORS TO COVER 8-15μ 15-30μ 30-50μ AND 50-120μ
       BANDS FOR PRIMARY SURVEY
  (2) DUTCH INSTRUMENT PACKAGE: LOW RESOLUTION SPECTROMETER (6-23
         MICRONS), CHOPPING BROADBAND PHOTOMETER (60 AND 100 MICRONS),
         SHORT WAVELENGTH CHANNEL (5-8 MICRONS)

POSITIONAL ACCURACY: BETTER THAN ±0.5 ARC MIN (2σ) FOR THE 10 AND 20μ
                     CHANNELS AND ± 1 ARC MIN FOR THE 50 AND 100μ
                     CHANNELS

OBSERVATIONAL REDUNDANCY ON TIME SCALES OF SECONDS, HOURS, WEEKS AND
MONTHS

SKY COVERAGE: SURVEY GOAL IS TO COVER AT LEAST 80% OF THE SKY IN A
                4-MONTH PERIOD FOLLOWING THE START OF THE SURVEY WITH A
                MINIMUM OF 50% IN THE INITIAL 30 DAY PERIOD FOLLOWING
                CHECKOUT IN SUCH A MANNER THAT EACH AREA IS SEEN ON AT
                LEAST 2 CONSECUTIVE ORBITS

SENSITIVITY: LIMITED IN PRINCIPLE BY ZODIACAL DUST EMISSION (NOISE
                EQUIVALENT FLUX DENSITIES OF 0.006, 0.01, 0.01 AND 0.02
                Jy AT 10, 20, 50 AND 100μ)

SURVEY OUTPUT:
  (1) PRINTED CATALOGUE OF 10,000-30,000 "MOST INTERESTING" SOURCES
  (2) COMPLETE CATALOGUE OF ALL POINT-LIKE SOURCES FOUND IN SURVEY
       (10^6 - 10^7 SOURCES) ON MAG. TAPE

---

criptions of the Shuttle and Spacelab are given elsewhere in this
volume. It is possible to identify two separate and distinct ways in
which the Shuttle will be used for Astrophysics. These are:
  (1) As a launch and recovery vehicle
  (2) As an experiment platform
The first category has already been discussed in this paper in connection with the Space Telescope. With respect to the second type of
usage, some specific examples of capabilities of the Shuttle/Spacelab
system which should be exploitable for the conduct of research in Astrophysics are listed in Table 14.

     Many of these capabilities (eg., use of film, calibration of in-
strumentation both before and after flight, optimization of either
single specialized instruments or groups of complementary instruments

TABLE 11

MINIMUM PERFORMANCE SPECIFICATIONS
SPACE TELESCOPE

---

* VERSATILE, LONG LIFETIME OBSERVATORY, SHUTTLE COMPATIBLE

* CAPABLE OF ACCOMMODATING A VARIETY OF INSTRUMENTS

* APERTURE — AT LEAST 2 METERS (6.56 FEET) IN DIAMETER

* OPTICAL QUALITY — EQUIVALENT TO A 2 METER (6.56 FOOT) DIFFRACTION LIMITED SYSTEM

* OPTICAL SYSTEM — MUST WORK EFFICIENTLY FROM ABOUT 1150 Å TO 1 MM

* SCIENTIFIC INSTRUMENT COMPLEMENT — MINIMUM OF FOUR/LAUNCH

* CAPABLE OF USING SCIENTIFIC INSTRUMENT ENTRANCE APERTURE COMPARABLE IN SIZE TO IMAGE

* CAPABLE OF MEASURING 27TH MAGNITUDE POINT SOURCES WITH S/N = 10 IN 4 HOURS

* CAPABLE OF MEASURING EXTENDED SOURCES OF SURFACE BRIGHTNESS $23.0^m/SEC^2$ WITH S/N = 10 IN 15 HOURS

TABLE 12

SPACE TELESCOPE
TENTATIVE FIRST PAYLOAD

INSTRUMENT	DETECTOR
WIDE FIELD CAMERA	CHARGE COUPLED DEVICE (CCD) 1600 x 1600 PIXEL FORMAT
FAINT OBJECT SPECTROGRAPH	DIGICON (1 x 512)
HIGH RESOLUTION SPECTROGRAPH	DIGICON (1 x 512)
HIGH SPEED PHOTOMETER	IMAGE DISSECTOR TUBE
FAINT OBJECT CAMERA (ESA)	IMAGING PHOTON COUNTING SYSTEM (ESA)
ASTROMETER	PART OF FINE GUIDANCE SYSTEM

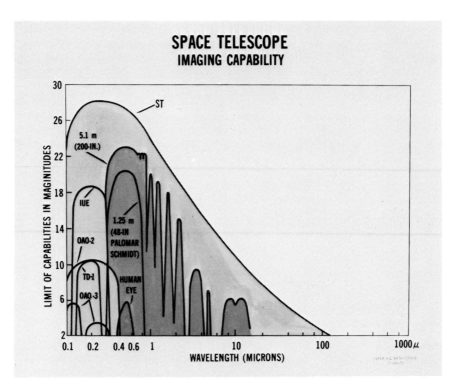

Fig. 17. A comparison of the projected performance of the Space Telescope with the performance of several satellite telescopes and
ground-based optical telescopes

for specific experiments or observations) should be relatively easily
realized and early payloads have, in fact, been selected to exploit
some of these capabilities. It is also clear, however, that realization
of all the potential advantages of the Shuttle will require the acquisition of considerable experience in working with a new and complex
system.

A tentative schedule for the first few years of Shuttle operations
is shown in Figure 18. It should be noted that although the first 6
flights will be devoted largely to tests of the Orbiter itself, there
will be some limited opportunities for experiments as well. Payload
management responsibilities for the Orbital Flight Tests have been assigned to the NASA Office of Applications (flight 2), to the Office of
Aeronautics and Space Technology (flights 3 and 6) and to the Office of
Space Science (flight 4). Flight 5 may be used for a Skylab revisit.

The transition to the routine operational phase of the Shuttle
will begin in July 1980 with the launch of the first Spacelab mission
which will be the first flight of the Manned Module. The payload for
this joint NASA/ESA mission has been tentatively selected with the

TABLE 13

REPRESENTATIVE SPACE TELESCOPE RESEARCH PROGRAMS

· EXTENSION OF THE DISTANCE TO WHICH WELL UNDERSTOOD MEASURES OF
    DISTANCES OF GALAXIES CAN BE APPLIED BY A FACTOR OF THREE TO FIVE

· MEASUREMENT OF DEVIATIONS FROM A LINEAR DISTANCE - REDSHIFT RELATION
    TO UNDERSTAND GEOMETRY OF THE UNIVERSE

· DETECTION AND STUDY OF GALAXIES FORMED EARLY IN THE EVOLUTION OF THE
    UNIVERSE

· MEASUREMENT OF SPECTRA OF HIGH REDSHIFT QUASARS TO DETERMINE PRIMEVAL
    HELIUM ABUNDANCE AND THUS PROVIDE INFORMATION ON PHYSICAL CONDITIONS
    EARLY IN THE UNIVERSE

· MEASUREMENT OF SIZES AND SPECTRA OF NUCLEI OF ACTIVE GALAXIES TO
    DETERMINE THEIR PHYSICAL CHARACTERISTICS

· POSSIBLE DETECTION OF PLANETS AROUND NEARBY STARS

· MONITORING OF ATMOSPHERIC MOTIONS OF VENUS, MARS, JUPITER AND SATURN
    TO STUDY PLANETARY METEOROLOGY

· EXTENSION OF SEARCH FOR OPTICAL COUNTERPARTS OF RADIO AND X-RAY
    SOURCES TO SIGNIFICANTLY FAINTER OBJECTS

· DISCOVERY OF NEW TYPES OF SOURCES AND PROCESSES

---

major emphasis being in the scientific and technical areas of space
plasma physics, physics of the upper atmosphere, space medicine, and
space processing and materials science although a few small astrophysics
experiments will be carried as well.

Spacelab 2 (which will be the first flight of the pallet-only
version of Spacelab) is currently scheduled for an April 1981 launch.
The announced emphasis of this mission was in astrophysics and solar
physics and the payload which has been tentatively selected (Table 15)
does, in fact, reflect this emphasis. Final flight assignment will be in
the spring of 1978 at the completion of the payload definition study
now underway. An artist's conception of the Spacelab 2 mission is shown
in Figure 19. In addition to making valuable scientific observations,
the Spacelab 2 payload is expected to test Shuttle/Spacelab performance
in a number of ways (eg., the solar instruments will provide a stringent
test of the ESA - provided Instrument Pointing System.

These first missions are pioneering ones in which important
operational precedents will undoubtedly be set and as such they are
important steps in learning to work with the Shuttle. While the announced primary objective of these flights is an engineering verification

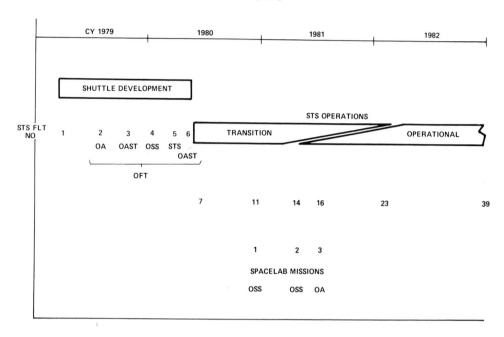

Fig. 18. A tentative flight schedule for the first 4 years of operation
of the Space Shuttle

of the Spacelab itself, clearly an important part of this verification
must be a demonstration that Spacelab and the Shuttle can be used as
tools for carrying out meaningful scientific research.

Looking beyond Spacelabs 1 and 2, it is currently anticipated
that a phased approach will be taken in the development of astrophysics
experiments. Initial emphasis will be placed on the development of
rocket and balloon class payloads by individual Principal Investigators.
Investigations of this type will be selected competitively. Later on in
the program, at least in Solar Physics and Astronomy, emphasis will
probably be placed on the development of larger facility class instruments which may be used by a broad segment of the scientific community.
A number of facility class instruments have been studied by competitively selected Facility Definition Teams. In Solar Physics, feasibility
studies have been made of a meter-class visible and ultraviolet telescope, a hard X-ray imaging facility and an EUV/XUV/Soft X-ray facility.
In astronomy, feasibility studies have been completed of a general
purpose ultraviolet/optical telescope (STARLAB) and of a meter-class
cryogenically cooled infrared telescope (SIRTF). More detailed studies
of at least some of these facilities will be undertaken during the next
few years.

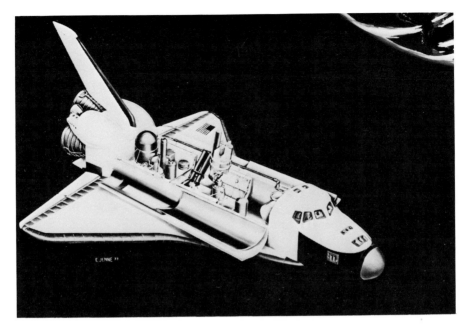

Fig. 19. An artist's concept of the Spacelab 2 mission

TABLE 14

SHUTTLE CAPABILITIES TO BE EXPLOITED FOR EFFECTIVE USE OF SPACELAB
FOR ASTRONOMY/ASTROPHYSICS

- CAPABILITY FOR CARRYING LARGE MASSES

- CAPABILITY OF OPTIMIZING SCIENTIFIC INSTRUMENTS OR COMBINATIONS OF
  INSTRUMENTS FOR PARTICULAR OBJECTIVES

- CAPABILITY FOR RESPONDING RAPIDLY AND EFFICIENTLY TO NEW EXPERIMENTAL
  OR THEORETICAL IDEAS OR PARTICULAR OBSERVATIONAL OPPORTUNITIES

- UTILITY FOR EXPERIMENTAL PROGRAMS HAVING SPECIFIC, LIMITED OBJECTIVES

- CAPABILITY FOR EMPLOYING INSTRUMENTATION USING LIMITED LIFE EX-
  PENDABLES OR REQUIRING PHYSICAL RETURN OF DATA AND EQUIPMENT
  (EG. USE OF FILM, CRYOGENS, DEGRADEABLE OPTICAL COATINGS: POST
  FLIGHT CALIBRATION OF DETECTORS)

- CAPABILITY FOR EVOLUTIONARY DEVELOPMENT OF INSTRUMENTATION AND
  REFLIGHT OF UPGRADED INSTRUMENTATION

- AVAILABILITY OF LONG OBSERVATION AND INTEGRATION TIMES COMPARED WITH
  ROCKETS OR BALLOONS

- PARTICIPATION OF MAN WHERE APPROPRIATE

TABLE 15

TENTATIVE SPACELAB 2 PAYLOAD

DISCIPLINE	TITLE	PI/INSTITUTION
LIFE SCIENCES	VITAMIN D METABOLITES AND BONE DEMINERALIZATION	H.K.SCHNOES/U.WISCONSIN
	INTERACTION OF OXYGEN AND GRAVITY INFLUENCED LIGNIFICATION	J.R.COWLES/U.HOUSTON
SPACE PLASMAS	EJECTABLE PLASMA DIAGNOSTICS PACKAGE	S.D.SHAWHAN/U.IOWA
	PLASMA HOLES FOR IONOSPHERIC AND RADIO ASTRONOMY STUDIES	M.MENDILLO/BOSTON U. A.V. DA ROSA/STANFORD U.
IR ASTRONOMY	SMALL HELIUM-COOLED IR TELESCOPE	G.G. FAZIO/SMITHSONIAN ASTROPHYSICAL OBSERVATORY
HIGH ENERGY ASTRO-PHYSICS	ELEMENTAL COMPOSITION AND ENERGY SPECTRA OF COSMIC RAY NUCLEI	P.MEYER, D.MULLER/U. CHICAGO
	HARD X-RAY IMAGING OF CLUSTERS OF GALAXIES AND OTHER EXTENDED X-RAY SOURCES	A.P. WILLMORE/U.BIRMINGHAM-ENGLAND
SOLAR PHYSICS	SOLAR MAGNETIC AND VELOCITY FIELD MEASUREMENT SYSTEM	A.M.TITLE/LOCKHEED PALO ALTO
	SOLAR CORONAL HELIUM ABUNDANCE SPACELAB EXPERIMENT	A.H.GABRIEL/APPLETON LAB.ENG J.L.CULHANE/U.COLLEGE LONDON
	SOLAR UV HIGH RESOLUTION TELESCOPE & SPECTROGRAPH	G.E.BRUECKNER/NAVAL RES.LAB.
UPPER ATMO-SPHERE	SOLAR UV SPECTRAL IRRADIANCE MONITOR	G.E.BRUECKNER/NAVAL RES.LAB.
SPACE TECH-NOLOGY	IN-ORBIT CALIBRATION OF MESA LOW G ACCELEROMETER	M.A.MELDRUM/BELL AEROSPACE TEXTRON
	DYNAMICAL AND THERMAL PROPERTIES OF SUPERFLUID HELIUM IN ZERO GRAVITY	P. MASON/JET PROPULSION LABORATORY

TABLE 16

POSSIBLE FUTURE MISSIONS

- SOLAR POLAR MISSION (ANNOUNCEMENT OF OPPORTUNITY ISSUED - POSSIBLE
  JOINT NASA/ESA MISSION)

- SMM REFLIGHTS (SOLAR SYNOPTIC STUDIES)

- SOLAR PROBE MISSION

- HEAO FOLLOW-ON PROGRAM
    - 1.2 METER X-RAY FACILITY (STUDY TEAM ESTABLISHED)
    - LARGE AREA MEDIUM ANGULAR RESOLUTION X-RAY OBSERVATORY
    - COSMIC RAY OBSERVATORY
    - γ-RAY OBSERVATORY (ANNOUNCEMENT OF OPPORTUNITY ISSUED)

- GRAVITY PROBE B (ORBITING GYROSCOPE EXPERIMENT)

- VLBI SATELLITE

- SPACE TELESCOPE REFURBISHMENT

- GRAVITY WAVE DETECTOR

- INFRARED INTERFEROMETER

- ADVANCED DEDICATED RELATIVITY MISSION

There are at least two reasons for following this phased approach.
The first practical consideration is that current budget constraints
will probably preclude the development of large facility class instruments during the early years of Shuttle operations. Second, as indicated
earlier, it will be necessary to gain experience operating experiments
on early Spacelab flights before making major hardware commitments to
large facilities. There are a number of potentially major problems such
as simplification of mission operations and reduction of the costs and
time required for payload integration which must be solved before it
will be possible to take full advantage of the capabilities of the
Shuttle outlined earlier. The phased approach will also result in a
gradual build-up from mixed discipline payloads to discipline dedicated
flights as an inventory of experiments and facilities is accumulated.
Current plans are for the first discipline dedicated missions to take
place in late 1982 or early 1983.

VII. Summary and Future Prospects

The program which has been described in this paper offers the
prospect of major, exciting advances in a number of different areas

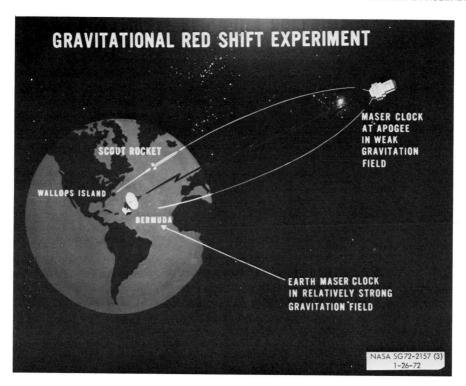

Fig. 20. The gravitational redshift experiment (Gravity Probe A)

during the next 5-7 years. The first sky surveys should be made in one
or more of the remaining unexplored regions of the spectrum. There is
the hope of obtaining a definitive answer to the question of the origin
of solar flares. The first search will be made for celestial γ-ray lines.
The first high spatial resolution X-ray images of a significant number
of X-ray sources will be obtained. Detailed optical observations will be
made of sources which are too faint ever to be observed from the ground.

　　A number of these missions are cooperative ones and it is likely
that in the future international cooperative programs will continue to
play a major role with the Space Shuttle/Spacelab program being the
prime example of such a collaboration.
　　Because of the long lead time required for the planning, design
and development of increasingly complex space experiments, work has
already begun on missions for the period of the mid-1980's. A list of
some missions which have been discussed as possibilities for the longer
term program is given in Table 16. This list is not intended to be either
exhaustive or restrictive. The amount of work already done on the missions on this list varies considerably. At one extreme, low level studies
of the Solar Polar (formerly known as the Out-of-Ecliptic) Mission were
begun in 1968 and the mission is now in a relatively advanced stage of

definition. Proposals for investigations have been solicited by both ESA and NASA for the mission (which is a potential cooperative one) and are being evaluated. NASA participation in the mission has not been authorized or approved and, as in the case of GRO, the issuance of the NASA Announcement of Opportunity has been for planning purposes only. At the other extreme, several mission titles merely represent the first glimmerings of ideas and little, if any, substantive definition work has been done on these missions.

One area which has not been heavily emphasized in the past and which may be receiving increasing attention in the future is the use of space for experiments on relativity and the nature of gravitation. A first step in this direction was taken in June 1976 with the suborbital flight of a hydrogen maser clock on a Scout rocket (Figure 20). The purpose of this flight was to obtain a measurement of the gravitational redshift. The hydrogen maser clock data have been analyzed during the past year and the original experiment goal of an accuracy of 200 parts per million was reached with no apparent deviation from the predictions of general relativity. A more detailed analysis is now in progress. A second experiment which has been under consideration for some time and for which substantial work has been done in seeing whether hardware can be developed to make measurements of the necessary precision is the orbiting gyroscope experiment. Other tests of the theory of gravitation using space techniques appear possible and work is now underway to develop a comprehensive plan for a possible program on relativity and gravitation during the next decade.

I would like to thank Mr. Mark Gotay for his invaluable assistance in helping to compile the material upon which this review is based. The aid of L. Dondey, G. Esenwein, R. Halpern, F. Martin, A. Opp, S. Sofia, W. Taylor and A. Timothy is also gratefully acknowledged. This paper is based on an invited lecture given at the meeting on "Physics and Astrophysics from Spacelab" in Trieste, Italy in September 1976. The manuscript was originally submitted in November 1976. At the request of the editors, a modest revision and updating was made in November 1977.

GAMMA RAY ASTROPHYSICS

Carl Fichtel, Goddard Space Flight Center
David Arnett, University of Chicago
Jonathan Grindlay, Center for Astrophysics, Cambridge, MA
Jacob Trombka, Goddard Space Flight Center

I. INTRODUCTION

The significance of γ-ray astrophysics lies in its ability to reveal the explosive, high energy, and nuclear phenomena of the universe. Thus, being the indicators of change and evolution, γ-rays provide special insight into the "how" and "why" of the physical processes which govern our universe. Although these features had long been recognized, the low intensity of γ-rays both in absolute terms and relative to the cosmic rays caused the first detection of celestial γ-rays to be delayed until the 1960's when low energy diffuse γ-ray emission was seen by Metzger et al. (1964) and clear evidence for high energy galactic plane γ-ray emission was obtained with the aid of the OSO-3 satellite (Kraushaar et al., 1972). Now, particularly with the results of the SAS-2 (e.g. Fichtel et al., 1975; Fichtel et al., 1977) and, more recently COS-B (e.g. Bennett et al., 1977) satellites γ-ray astrophysics has proceeded from the discovery phase to the exploratory phase, and the long awaited promise of γ-ray astrophysics is beginning to be realized.

P. L. Bernacca and R. Ruffini (eds.), Astrophysics from Spacelab, 145-241.

The results of SAS-2 and COS-B have shown the rich character of
the galactic plane diffuse emission with its potential for the study of
spiral arms, the cosmic ray matter distribution, the galactic center,
clouds, the origin of cosmic rays, dynamics of the cosmic-ray gas,
and other features of the galaxy. Moreover, with the observations of
discrete sources, some associated with supernovae and pulsars and others
such as (195,+5) apparently not associated with radiation at other
wavelengths, a new class of emission, from known objects and very
likely from unknown objects, postulated as early as the late 1950's,
is now appearing. The diffuse background is now seen clearly to
have a component which, at least on a crude scale, is isotropic. The
determination of the detailed nature and origin of this radiation with
its remarkable energy spectrum is one of the greater current astrophysical challenges, especially in view of the possible cosmological significance. The medium energy γ-ray range, for which there has not yet been
an opportunity for real exploration, but which has shown its potential
in balloon flights despite the atmospheric background, should provide
detailed information on cosmic ray electrons and add greatly to our
knowledge of sources. In the low energy region, there has been the
detection of short bursts of γ-rays whose origin remains a mystery and
the observation of solar γ-ray lines which reveal the nature of the flare
accelerated particles through their interactions with the ambient material
For the future, the measurement of γ-ray lines from supernovae offers
the possibility of answering the very important questions associated
with nuclear synthesis. In general, of all the electromagnetic spectrum
γ-ray astronomy measures most directly the presence and dynamic effects

of the high energy cosmic ray particles, whose high pressures play a major role in galactic dynamics, element synthesis, antimatter-matter annihilation, and particle acceleration.

In addition, since γ-rays have a very low interaction cross section, they have a very high penetrating power and can reach the earth from essentially any part of the galaxy or universe. Even a high-energy γ-ray passing through the diameter of the central plane of the galactic disk has only about a 1% chance of interacting for a typical path. A high-energy γ-ray can also travel to the earth from any part in the universe with less than a 1% probability of interacting on the average, although naturally there is a redshift. Therefore, these photons retain the detailed imprint of directional and temporal features imposed at their birth, even if they were born deep in regions opaque to visible light and x-rays, or at times far back in the evolutionary history of the universe.

As suggested in the opening paragraph, γ-ray astronomy offers the most direct means of studying the strong interactions associated with nuclei, thereby supplementing our knowledge of the universe which is revealed at lower frequencies through electromagnetic processes. The most dramatic and violent forces in nature are associated with the nuclear interactions; thus, with γ-ray astrophysics there should be the best chance of learning about the forces of change in astrophysics ranging from those in compact objects, such as neutron stars and black holes, through supernovae, cosmic ray pressure effects in the galaxy, galactic structures, other galaxies (and especially the extraordinary ones like radio galaxies, Seyfert galaxies, and possible QSO's), and

perhaps cosmology through either primordial black hole emission or the
study of the diffuse γ-ray emission. These subjects will be explored in
the next four sections with the aim of revealing what is known and its
astrophysical significance, as well as suggesting what future measurements
should be able to reveal. In the final section, the role of the shuttle
will be explored with particular emphasis on the types of detectors which
might be flown on the SPACELAB and on free flyers launched from the shuttl
The general survey of the literature for this paper was carried through
1977; however, an occasional later reference was added in the final draft.

II. DIFFUSE GALACTIC RADIATION

This section begins with a brief summary of the current experimental
results. Then, as a background for an understanding of the significance
of the γ-ray observations and their interpretation in terms of current
galactic models, the possible production mechanisms for galactic
γ-rays are reviewed. Finally, there is a discussion of galactic models
in light of the γ-ray results and what future γ-ray observations should
be capable of contributing to the picture of the galaxy.

(A) General Experimental Picture

As noted in the introduction, the first certain detection of high
energy   γ-rays was the observation of the intense radiation from
the galactic plane with a counter telescope flown on OSO-3. The
improved angular accuracy and sensitivity of the SAS-2 satellite showed
that the galactic plane γ-radiation was strongly correlated with galactic
structural features, especially when the known γ-ray sources were
subtracted from the total radiation. The distribution of high energy

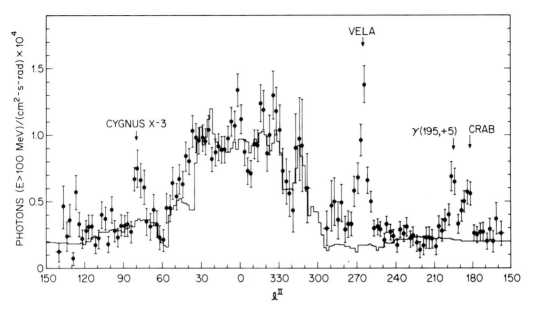

Fig. 1 – Distribution of gamma rays with energies above 100 MeV along
the galactic plane. The SAS-2 data are summed from –10° to
+10° in galactic latitude. Error bars shown are statistical
only. The absolute normalization is uncertain by an additional
10%. Locations of discrete sources outside the galactic center
region are shown by arrows (Fichtel, Kniffen and Thompson,
1977). The solid line histogram represents a theoretical
interpretation of the data to be discussed in the text.

(E > 100 MeV) γ-ray intensity along the galactic plane is shown in Figure 1
(Kniffen, Fichtel and Thompson, 1977)   in bins which are 2.5° wide in
latitude. Notice that the emission from the region $310° < \ell < 50°$ is
particularly intense relative to the remainder of the galactic plane.
This contrast is much greater than that seen in other indicators of
galactic structure such as the atomic hydrogen column density deduced

from the 21 cm measurements, for example. The early results of COS-B confirm these general features (Bennett et al., 1977).

When examined more closely, the longitudinal and latitudinal distributions appear generally correlated with galactic structural features, with maxima occurring at galactic longitudes of about 215°, 330-335°, 340-345°, 0°, and 25-35° (Fichtel et al. 1975 and Kniffen, Fichtel and Thompson 1977) in general agreement, for example, with the galactic center itself and location of tangents to galactic arm segments (Simonson 1976). These results and their theoretical inter-pretation have led to the hope that γ-ray astronomy, particularly with the high penetrating power of γ-rays, may ultimately lead to a much improved picture of the Galaxy.

Even at high latitudes ($10° < |b| < 90°$) a major portion of the γ-radiation observed is now believed to be galactic on the basis of its correlation with the matter and synchrotron radiation (Fichtel et al. 1977). On a coarse scale, the high latitude γ-ray intensity is reasonably well represented by the expression

$$I_\gamma = C_1 + C_2/\sin|b|. \tag{1}$$

When examined more closely, the correlation between the γ-ray emission and both the galactic matter distribution and the radio continuum radia-tion, thought to have its origin in synchrotron radiation, has been found to be quite good, Figures 2 and 3 show the γ-ray intensity as a function of the atomic hydrogen column density deduced from 21 cm data and the 150 MH_z continuum measurements (Fichtel, Simpson, Thompson, 1977). It would be even more meaningful to compare the γ-ray emission to the total matter distribution rather than just the atomic hydrogen

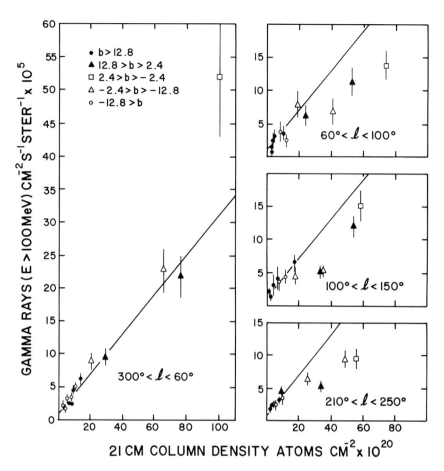

Fig. 2 - Distribution of γ-ray $(E_\gamma \geq 100 \text{ MeV})$ intensity as a function
of atomic hydrogen column density deduced from 21 cm radio
data for the indicated longitude intervals. The solid line
is the one derived from the high-latitude $(|b| > 12.8°)$ data
alone. (Fichtel, Simpson and Thompson, 1977.)

column density, but unfortunately little is known about the local

molecular hydrogen distribution. It should be noted that for direct

comparison to the γ-ray data, the other measurements must be "defocused"

to take into account the lesser angular resolution of the γ-ray obser-

vations. In these figures, points nearer the origin correspond to

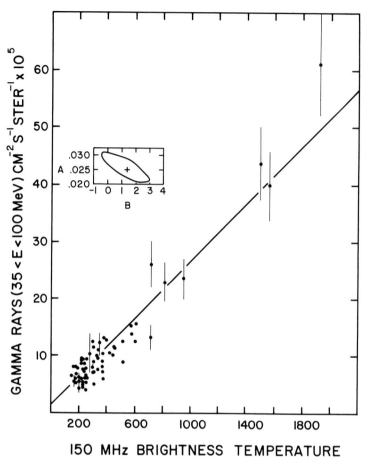

Fig. 3 - Gamma-ray intensity (35 MeV < E < 100 MeV) as a function of
150 MHz brightness temperature. Only typical error bars
are shown because of the high data density of the figure.
The estimated uncertainty in the 150 MHz brightness temperatures is approximately 15%. The straight line is the best
fit to all the data points. Inset: acceptable (1σ above
the minimum χ^2) values of the fitted parameters in the
equation $I_\gamma = A\ I_{150} + B$. (Fichtel, Simpson and Thompson,
1977.)

the highest latitude γ-ray observations. The straight line shown

on each sub-figure is the least squares fit to the combined observations

from all the sky intervals with $|b| \geq 12.8°$. In view of the thickness

of the matter disk, the latitude of 12.8° corresponds to a distance of

about 500 pc to the point of one scale height from the plane in the

case of atomic hydrogen. The fact that the straight line in Figure 2

does not pass through (0,0) is one indication of the apparently isotropic

component to be discussed in section III.

In the general anticenter region, the low $|b| < 10°$ latitude points

fall below the straight lines. These low latitudes correspond to a sub-

stantial contribution from regions lying at distances from the galactic

center greater than the radius from the galactic center to the Sun. In

contrast in the galactic center region, the γ-ray intensity within a

few degrees of b=0 lies above an extrapolation of the same straight line

(see Fichtel, Simpson and Thompson, 1977, for details). These results

are consistent with the relatively high contrast seen between the center

and anticenter noted earlier.

Because of limited energy resolution the energy spectrum cannot

be determined in fine detail. However, if a power law spectrum is

assumed for the galactic component, a differential spectral index of

1.7 $\pm$0.3 and intensities above 35 and 100 MeV of $(6.9 \pm 1.7) \times 10^{-26} N_{HI}$

and $(3.3 \pm 0.8) \times 10^{-26} N_{HI}$ photons $cm^{-2} s^{-1} ster^{-1}$ respectively are obtained.

These correspond to intensities of $(8.3 \pm 2.1) \times 10^{-5}$ and $(4.0 \pm 1.0) \times 10^{-5}$

photons $cm^{-2} s^{-1} ster^{-1}$ for a typical region with a galactic latitude

of about 15°. If a combination of a nuclear interaction ("π°") and

power law spectrum (of the Compton or bremsstrahlung type spectral

index) is assumed, very similar results are obtained primarily
because a relatively large power law component must be assumed. The
high latitude galactic energy spectrum is consistent with that deduced
for the galactic plane emission (Fichtel et al. 1975; Bennett et al. 1977),
supporting the concept of a common source mechanism.

(B) Gamma Ray Origin Mechanisms

Before discussing the general picture of the galaxy and the role
γ-rays play in revealing its structure and dynamics, it is important to
review the basic production mechanisms. The possible contributors to
γ-ray emission are cosmic-ray nucleon interactions with the interstellar
gas leading at high energies principally to γ-rays through π° decay and
at low energies to lines from the decay of excited nuclei, bremsstrahlung
γ-rays from energetic cosmic ray electrons, Compton emission of cosmic-
ray electrons interacting with the starlight, infrared, and blackbody
radiation, the synchrotron radiation of high-energy electrons interacting
with the interstellar magnetic fields, the decay of excited nuclei
injected into the interstellar medium, and point sources. The effect
of the latter is difficult to determine at present; estimates on the
basis of observed sources are difficult to make. The absence of sources
at high latitudes and the smooth connection and consistency with local
 high latitude) γ-ray emission may suggest it is small but, without
much more sensitive γ-ray measurements with better angular resolution,
the contribution of point sources must remain an open question.

Source functions may be calculated for most of the other processes
in the solar vicinity where the relevant parameters are reasonably well
known. The energy region above about 10 MeV will be considered first,

with this discussion being followed by a treatment of the low energy

γ-ray lines. To show the contrast between the high energy (E>100 MeV) region

and the medium energies, intensities are shown in Table 1 for both the

energy range above 100 MeV and the 10 to 30 MeV range. This table is

largely taken from the work of Fichtel et al. (1976), with the neutral-

pion-decay source function being that of Stecker (1973) as updated by

Fichtel, Simpson, and Thompson (1977). For each of the last four

production mechanisms, two numbers are given with the first being the

result predicted for the electron spectrum given, by Daugherty, Hartman,

and Schmidt (1975) using traditional demodulation assumptions, cross

sections given by Koch and Motz (1959) and the secondary electron

spectrum calculated by Ramaty (1974), and the second being that implied

by the γ-ray data if the electron component is primarily responsible

for the low energy γ-ray emission through bremsstrahlung or Compton

emission (Fichtel, Simpson and Thompson, 1977). It is important to

note that the γ-rays being considered here are produced predominantly

by electrons in the energy range 10 to a few hundred MeV, where there

are no direct measurements of the demodulated electron spectrum for the

case of the bremsstrahlung radiation, and by electrons above about 50 GeV,

where the spectra is relatively poorly known, for the case of the Compton

radiators. The Compton scattering term includes the recent work on

the infrared radiation interactions of (Piccinotti and Bignami, 1976).

The contribution of helium and heavier atoms is included, and they increase

the bremsstrahlung radiation by a factor of 1.55 because of the $Z(Z + 1)$

dependence on charge. For this table a local value of 1.15 atoms/cm^3

equivalent (i.e. atomic and molecular hydrogen) is assumed based on

the recent work of Savage et al. (1977).

Table 1

Source Functions in the Solar Vicinity

Source Mechanism	Value of Source Function (cm^{-3} s^{-1})	
	10–30 MeV	> 100 MeV
Cosmic-ray nucleon matter interactions	$(0.7-1.1) \times 10^{-26}$	$(14-22) \times 10^{-26}$
Electron bremsstrahlung	$(13-39) \times 10^{-26}$	$(4-12) \times 10^{-26}$
Compton scattering (starlight and infrared)	$(1.2-3.6) \times 10^{-26}$	$(0.4-1.2) \times 10^{-26}$
Compton scattering (3 K)	$(1-3) \times 10^{-26}$	$(0.2-0.6) \times 10^{-26}$
Synchrotron radiation	$(0.7-2) \times 10^{-30}$	$(0.3-0.4) \times 10^{-30}$

Table I indicates that cosmic ray nucleon and electron interactions

with matter when combined dominate in the galactic plane; however,

if there is a thick cosmic ray disk as many currently believe, or a

very high electromagnetic radiation density in the central galaxy,

Compton radiation becomes quite important.

Along a line of sight in the direction (ℓ,b) for the range of

parameters being considered here, the gamma-ray intensity produced by

interactions between cosmic rays and matter can be shown (e.g. Fichtel

et al., 1976) to be given by the expression:

$$I(E_\gamma, \ell, b) = \frac{1}{4\pi} \int dr \ [S_{\gamma_n}(E_\gamma, r=0) g_n(r, \ell.b) f_m(r, \ell, b)$$
$$+ S_{\gamma_{e_p}}(E_\gamma, r=0) g_e(r, \ell, b) f_m(r, \ell, b) \qquad (2)$$
$$+ S_{\gamma_{e_s}}(E_\gamma, r=0) g_n(r, \ell, b) f_m^2(r, \ell, b)]$$

where f_m is ratio or the total interstellar gas density at a distance

r from the sun in the direction (ℓ,b) to that at r=0, S_{γ_n} represents the

gamma rays produced per second in interactions of nucleonic cosmic rays

(with the intensity and spectral distribution in the solar vicinity with the interstellar gas, and $S_{\gamma e_p}$ and $S_{\gamma e_s}$ are similar functions for primary and secondary cosmic ray electrons, respectively. g_n and g_e express the spatial variation with galactic position of the ratio of the primary cosmic-ray nucleon and electron components respectively to their interstellar value in the solar civinity. Implicit in this approach is the assumption that the spectral shape of each component is unchanged throughout the galaxy. The assumption is reasonable, as long as the source spectral shape in the solar vicinity is typical of that throughout the galaxy and energy losses remain within certain limits. The latter condition is true for the energy region and intensity levels relevant to the consideration here.

It is seen readily from equation (2) that, if the cosmic ray primary electron component is proportional to the cosmic ray nucleon component, as is commonly assumed, and if the secondary electron component is small compared to the primary electron component locally as the positron data suggest , the γ-ray intensity is simply proportional to the integral over the product of the ratios of the cosmic ray intensity and matter density to their local values.

Hence equation (2) may then be rewritten in the approximate form

$$I(E_\gamma,\ell,b) = \frac{1}{4\pi} \int dr\ s_{\gamma_{cr}}(E,r=0)\ g_{CR}(r,\ell,b)\ f_m(r,\ell,B) \qquad (3)$$

where the subscript cr refers to the combination of the cosmic ray electrons and protons, and $S_{\gamma_{cr}}$ is then the sum of the relevant source functions.

With regard to the matter, as has already been noted, a reasonably accurate estimate of the atomic hydrogen column density in any direction can be obtained from the 21 cm measurements; however, the current status of the knowledge of the other major component, namely the molecular

hydrogen, is much less clear. A detailed understanding of the molecular

hydrogen density or even the column density does not exist. From the

Copernicus data on the local interstellar gas densities, Savage et al.

(1977) estimate the fraction of the gas in molecular form to be at least

0.25 and possibly as high as 0.5. The large-scale survey of Gordon

and Burton (1976) indicates that the fraction of gas in molecular

form may be about 0.5 for an average distance of 10 kpc from the

galactic center, lower beyond this radius, and much higher in the

central galactic region.

In addition to the uncertainty in S_γ due to that in the molecular

hydrogen density, there is the cr uncertainty in the electron brems-

strahlung contribution due to the lack of knowledge of the electron

cosmic ray spectrum in the appropriate energy range (10 to several

hundred MeV).

Whereas table 1 might suggest that the Compton radiation is

relatively small, as mentioned before it is in fact likely to be

fairly important because of the large scale height thought to exist

for the cosmic ray electrons relative to that for matter.

Following the work of Fichtel, Simpson and Thompson (1977)

and assuming a gaussian electron distribution perpendicular to the

plane with a scale height of $\sqrt{2}$ x 0.75 kpc, the Compton radiation

resulting from cosmic ray electrons interacting with the blackbody

radiation is given by the following equations:

$$I_{C_{bb}} (E > 100 \text{ MeV}) = 0.07 \times 10^{-5} [\sin|b|]^{-1} \text{ photons cm}^{-2}\text{ster}^{-1}\text{s}^{-1} \quad (4)$$

$$I_{C_{bb}} (35\text{MeV} \leqslant E < 100\text{MeV}) = 0.12 \times 10^{-5} [\sin|b|]^{-1} \text{photons cm}^{-2}\text{ster}^{-1}\text{s}^{-1}. \quad (5)$$

The combined contribution from the infrared radiation and starlight

is much more difficult to estimate not only because the average photon

density in the plane is not very well known, but also its variation

with height above the plane is poorly known.   It appears that the

effective scale height is smaller than that estimated for the electrons,

so that even though the combined source function in the plane is

slightly larger, the net contribution is smaller.   It is estimated

to be:

$$I_{C(s,ir)}(E>100\text{MeV}) = 0.03\text{x}10^{-5}[\sin|b|]^{-1}\text{photons cm}^{-2}\text{ster}^{-1}\text{s}^{-1} \quad (6)$$

$$I_{C(s,ir)}(35\text{MeV}<E<100\text{MeV}) = 0.04\text{x}10^{-5}[\sin|b|]^{-1}\text{photons cm}^{-2}\text{ster}^{-1}\text{s}^{-1}. \quad (7)$$

Hence, the combined estimate for Compton radiation is:

$$I_{C}(E>100 \text{ MeV}) = 0.10\text{x}10^{-5}[\sin|b|]^{-1} \text{ photons cm}^{-2}\text{ster}^{-1}\text{s}^{-1}. \quad (8)$$

$$I_{C}(35\text{MeV}<E<100\text{MeV}) = 0.16\text{x}10^{-5}[\sin|b|]^{-1} \text{ photons cm}^{-2}\text{ster}^{-1}\text{s}^{-1}. \quad (9)$$

The uncertainty in these two numbers is quite large, for the reasons

already given.

Both the Compton and the bremsstrahlung spectra are power laws with

similar spectra, and both radiations would enhance the 35 to 100 MeV

intensity relative to that above 100 MeV compared to the "π° spectrum".

The spectrum observed by Fichtel et al. (1975), Bennett et al. (1977), and

Fichtel, Simpson, and Thompson (1977) are consistent with a large electron

contribution.

In order to calculate the γ-radiation from the Galaxy, mass and

photon distributions are required.   Since information on both these

subjects is limited (and indeed should ultimately be improved by

combining γ-ray astronomical data with radio, optical, and x-ray data),

specific models based on the data that are available must be assumed
before predictions may be compared to available data.

Turning now to the low energy γ-ray region, there is the interesting
possibility of being able to detect γ-ray lines. These lines may be
formed in two distinctly different ways: (1) the interaction of
low-energy cosmic rays with both interstellar gas and dust grains
leading to excited nuclei which subsequently emit γ rays through
electromagnetic de-excitation of an excited state and (2) the decay of
recently formed unstable nuclei into excited states of daughter
nuclei which in turn emit γ rays.

With regard to the first case, since the excited states are
populated directly by the inelastic collision of interstellar matter
and energetic particles, the strength of the line radiation contains
information about the composition and nuclear excitation conditions
in the radiating gas. Thus, the γ-ray lines can be used as diagnostics
for energetic cosmic gases in a way analogous to the use of optical
lines for more ordinary nebulae. One expects at least some of the
dominant lines formed in this way to be de-excitation of the lower-
lying levels of the most abundant nuclei. Table 2 lists the first
few of these nuclei with their abundance and the energy of the first
excited state; a Population 1 composition is assumed. The most abundant
nuclei tend to be (1) light (in Table I only Fe^{56} has A > 20) and (2)
fairly tightly bound. This means the excitation energy tends to be
high (cf 0^{16}). Thus one expects the main nuclear lines from direct
excitation to lie in the energy range 0.1 MeV<E< 7 MeV, where the lower

Table 2

Important Excited States of Abundant Nuclei

Nucleus	E_i^* (MeV)	Abundance (mass faction)
H	------	7.4×10^{-1}
He^4	------	2.4×10^{-1}
O^{16}	6.14	1.1×10^{-2}
C^{12}	4.43	4.5×10^{-3}
Fe^{56}	0.847	1.3×10^{-3}
Ne^{20}	1.63	1.2×10^{-3}
N^{14}	2.31	9.7×10^{-4}

range is affected by the contributions of higher excited states and

less abundant nuclei. The lines from interstellar grains should be

separable from those from the gas by their much narrower line widths

allowing the two constituents to be studied separately.

The second, and in some respects more interesting, mechanism for

forming excited states is by the decay of parent nucleus

into an excited state of the daughter. The strength of such lines is

essentially independent of the local physical conditions of the matter,

but is determined by the abundance and rate of decay of the parent

nucleus. Such decays often occur to fairly energetic states in the

daughter, so that several   γ-rays of characteristic energies and

relative strengths are emitted. This gives an excellent signature

for the existence of the unstable parent nucleus. Further, the very

fact of its instability and the values of the half-life and signal

strength, provide direct information on its rate of production. If the

half-life of the unstable parent nucleus is long compared to the typical
time interval between production (nucleosynthesis) events in the volume
of space observed, the abundance tends to build up to a steady-state
value, and observed lines provide information about the average in an
ensemble of events (e.g. supernovae and novae). On the other hand, if
the half-life is short (in this sense), the lines tell us about the
closest most recent event. Because nucleosynthesis of unstable nuclei
requires high temperatures and the nuclei must be ejected from the hot,
dense environment for the γ-rays to be seen, likely sources are violent
cosmic explosions: flares, novae, supernovae, and galactic nuclei for
example. Thus, γ-ray lines provide a means for study of cosmic explosions and nucleosynthesis events through their residue in interstellar
space, as well as through direct observation of the explosion itself.

(C)   Galactic Models and Gamma Ray Emission

        In considering the question of galactic γ-ray emission and the role
that γ-rays will ultimately play in understanding the nature of our
Galaxy, several fundamental questions arise at once. Are the cosmic
rays galactic or universal? Is there a major spiral pattern in our
Galaxy? What is the relationship between cosmic rays and clouds?
Gamma-ray astronomy is already providing some information which can
bear on these questions and in the future, through the study of γ-ray
lines, it may provide information on still further questions, such as
those discussed at the end of the last section, IIc. To understand
the significance of the results, it is worthwhile considering briefly
first some theoretical questions regarding the Galaxy and cosmic rays,
and these will now be reviewed.

Whereas the structure of close external galaxies is defined relatively clearly by the distributions of young stars, HII regions, 21 cm
radiation, and in some cases continuum radio emission, in our Galaxy
the necessary observations are complicated by the fact that the solar
system is immersed in the Galaxy far from its center. However, the
distribution of continuum radiation (Landecker and Wielebinski, 1970;
Price, 1974), γ-radiation (Bignami et al. 1975), HII regions (Georgelin
and Georgelin 1976), supernova remnants (Clark and Caswell, 1976), pulsars
(Seiradakis, 1976), infrared emission (Hayakawa et al., 1976), and 21 cm
neutral hydrogen emission (Burton, 1976) are all consistent with the
existence of spiral structure in the Galaxy. In particular, Simonson
(1976) has used the 21 cm measurements, the density wave theory, and an
arm-to-interarm density ratio of 3:1 to construct a model of the overall
spiral pattern of our Galaxy. In the above-mentioned observations which
cover both sides of the galactic center, the spiral arm features in
the 270° to 360° longitude quadrant appear more pronounced than those
in the 0° to 90° quadrant. At present, the CO observations which are
related to molecular hydrogen densities exist only for 0° to 180°
(Scoville and Solomon, 1975; Burton et al., 1975). Although these
measurements do not show conclusive evidence of spiral structure, it
is not clear whether this effect is the result of the relatively
limited observations. As more CO observations are made, this component
should become a very useful probe of spiral structure.

The high energy (> 100 MeV) γ-ray intensity measurements (Fichtel
et al. 1975) do show evidence for spiral structure in our own Galaxy.
There are enhancements along the directions of spiral arm tangents as

noted at the beginning of this section, although improved statistical information is certainly desired. In addition, theoretical analyses (e.g. Bignami and Fichtel, 1974; Paul et al., 1974; Schlickeiser and Thielheim, 1974 ; Bignami et al., 1975; Paul et al., 1975; Stecker et al., 1975; Stecker, 1976; Puget et al., 1976; Paul et al., 1976) have shown that the relative intensity distribution along the plane is about what would be expected if there were cosmic ray matter coupling as originally proposed by Bignami and Fichtel (1974) based on concepts developed earlier by Parker (1966 and 1969). Figure I shows the theoretical prediction of Kniffen, Fichtel and Thompson (1977) for the γ-radiation (E> 100 MeV) compared to the experiment data, under the assumption that the comsic ray density is coupled to the matter density on the scale of arms. The agreement is seen to be reasonably good.

The coupling of the cosmic rays with the matter on the scale of arms is a concept which deserves some further discussion. Consider the assumption that the cosmic rays and magnetic fields are primarily galactic and not universal; then, the fields and cosmic rays can only be constrained to the galactic disk by the gravitational attraction of the matter through which the magnetic fields penetrate (Bierman and Davis, 1960; Parker, 1966, 1969). (If the cosmic rays are universal, their density would normally be expected to be uniform, although there have been some arguments presented that density waves would cause them to be concentrated into arms in the galaxy also). The local energy density of the cosmic rays is about the same as the estimated energy density of the magnetic field and that of the kinetic motion of matter. Together the total expansive pressure of these three effects is estimated to be approximately equal to the maximum that the gravitational attraction can hold in

equilibrium. Assuming the solar system is not in an unusual position

in the Galaxy, these features suggest that the cosmic ray density throughout

the Galaxy may generally be as large as could be contained under near-

equilibrium conditions. Further theoretical support is given to this

concept by the calculated slow diffusion rate of cosmic rays in the

magnetic fields of the Galaxy, the small cosmic ray anisotropy, and the

likely high production rate of cosmic rays. The above considerations

then lead to the postulate that the energy density of the cosmic rays

is larger where the matter is larger on the scale of galactic arms.

Whereas the cosmic rays are likely to be correlated with the spiral

arms in the plane, their scale height perpendicular to the plane is very

probably much larger. The nonthermal continuum radiation in the Galaxy,

which is generally attributed to the synchrotron radiation from cosmic

ray electrons interacting with the galactic magnetic fields (e.g. Ginzburg

and Syrovatskii, 1964, 1965) has a scale height which is large compared

to that of either the galactic atomic or molecular hydrogen. Baldwin

(1967, 1976) estimates the equivalent disk thickness to be about 750 pc,

and some analyses have suggested it is even larger. Significant non-

thermal emission is even seen as high as 2 kpc above the plane. If it

is assumed that the electron density and magnetic field density both

have the same distribution on the average and that it is gaussian, then

since the synchrotron radiation is proportional to the product of the

two, the scale height of each individually is $\sqrt{2}$ x 0.75 kpc, the number

used earlier. As seen before, these considerations lead to the prediction

of a significant Compton component.

It should be reemphasized that the question of a significant point source contribution to the low energy γ-ray intensity is also still open. This is a possibility which is difficult to verify or deny with the limited angular resolution of current experiments. Those sources which have been seen and can be associated with compact objects on the basis of short periodic emission would probably not alone make a major contribution (e.g. Ogelman et al. 1976, and Strong, Wolfendale and Worrall, 1976).

If the Galaxy is examined on a smaller scale than that of the galactic arms, the problem is a very dynamic one because the pressures of the interstellar gas, magnetic fields, and cosmic rays create an inherently unstable situation (Parker, 1966). If this concept is correct, the net result is that the interstellar gas is suspended in the magnetic field in discrete clouds with separations of the order of 10 to 10^2 pc. It is, in fact, observed that the interstellar gas exists mainly in widely separated discrete clouds, many of which have masses which are too small to maintain the cloud in equilibrium by self-gravitation alone. The cosmic ray gas would be expected to bulge significantly between the clouds expanding to relatively large dimensions before finally escaping the galaxy. The radio continuum measurements support this concept as noted earlier, and recent cosmic ray results also suggest that the cosmic rays spend a significant fraction of their lifetime outside the central plane region, where the matter density is very much smaller than that in the galactic field disk (Jokipii, 1976).

The γ-ray energy spectrum supports this concept in that there is substantially more γ-ray emission in the 35 MeV to 70 MeV range relative to that above 100 MeV (Fichtel et al. 1976; Bennett et al. 1977; Fichtel

Simpson and Thompson, 1977) that can be easily explained except by assuming a large scale height perpendicular to the plane, which allows both a significant amount of blackbody Compton γ-radiation and a larger low energy electron component leading to a larger γ-ray bremsstrahlung prediction. Without the larger scale height, the high energy loss in the plane combined with the long lifetime makes it more difficult to maintain a major low energy electron component.

Two predictions of these considerations are (1) that the interstellar gas exists mainly in widely separated discrete clouds, many of which have masses which are too small to maintain the cloud in equilibrium by self-gravitation alone, as is apparently the case, and (2) that the cosmic ray density is uniform on a cloud to cloud scale. The concept of uniform cosmic ray density on a local or cloud-to-cloud level is supported by the analysis of Freier, Gilman, and Waddington (1977) who show that the level of the γ-ray emission can only be explained if the cosmic rays are uniform locally since the predicted γ-ray emission would be much too large for the cosmic rays to be proportional to the matter on the scale of clouds.

It would be clearly very desirable to have more refined γ-ray measurements than currently exist to study in detail the relationship between cosmic rays and clouds, as well as to study further the structure and evolution of the Galaxy on a broad scale.

III. DIFFUSE GAMMA RADIATION FROM HIGH GALACTIC LATITUDES

(A) Introduction

The first evidence for the detection of a general diffuse cosmic γ-radiation came from the work of Metzger et al., 1964 and Clark, Garmire, and Kraushaar, 1968. The radiation was apparently diffuse and did not

show the very large anisotropy evident for the stronger emission from the

galactic plane discovered by the OSO-3 experiment. The degree of isotropy

of this emission suggested that it could be partially extragalactic and

possibly of cosmological significance. The cosmological importance of

the diffuse γ-radiation--or more specifically the lack of a very

intense diffuse radiation--had already been established with the

upper limits for such radiation obtained by Kraushaar and Clark (1962)

with the Explorer 11 satellite. The upper limit for the diffuse

γ-radiation set with that experiment severely constrained the density of

anti-matter relative to matter in the universe except in well separated

regions.

Since these early experimental results, detailed spectra and in-

tensities have become available so that it is now possible to re-examine

many of the models proposed for the diffuse γ-radiation in a new light.

The experimental results will first be reviewed in a historical manner

and then the various models will be discussed. While this presentation

cannot be complete, the major steps in our understanding can at least

be identified and future directions pointed out.

(B) Experimental Results

At low energies the first suggestion that a diffuse radiation field

extended from the x-ray to at least the low energy γ-ray range ($\sim$1 MeV)

was reported by Arnold et al., (1962). Their detector on the Ranger 3

and Ranger 5 moon probes had recorded an $E^{-2.4}$ power law spectrum up

to $\sim$1 MeV and an apparent flattening between $\sim$1 and 2.6 MeV, the maximum

energy recorded. This result was also the first indication that structure (i.e. changes in slope) might be present, and much of the subsequent work, both experimental and theoretical, has been concerned with the possible spectral break or bump, at ~1 MeV. Upper limits were obtained by Peterson (1966) from OSO-1 for any diffuse flux in the range ~0.5-4.5 MeV, though these did not contradict the Ranger result.

At energies >50 MeV, the first results were those of Kraushaar and Clark (1962), whose upper limits from Explorer 11 provided a crucial experimental refutation of the steady state cosmology which would predict a large annihilation flux (Morrison 1958). This upper limit was also able to establish that the product of the intergalactic electron and photon (optical) densities must be < 10% of the galactic values since otherwise the inverse Compton flux predicted by Felten and Morrison (1963) would have been detected. Several other upper limits were reported from early balloon experiments, but the first unambiguous evidence for a diffuse high energy flux came from the OSO-3 satellite experiment mentioned above. Preliminary results were reported by Clark, Garmire and Kraushaar (1968) and in final form by Kraushaar et al., (1972). No spectral information was available from this result, however, and only a single point at $\geq$ 100 MeV energies could be added to the diffuse background spectrum. This point was roughly consistent with an extrapolation of the diffuse x-ray spectrum as given by Schwartz and Gursky (1973).

As more sensitive γ-ray experiments were developed, further reports of positive results appeared. With the detector on the ERS-18 satellite (Vette et al., 1970), renewed evidence for an enhancement in the diffuse spectrum at $\geq$ 1 MeV was presented. This possible flux excess above the x-ray spectrum extrapolation was apparently confirmed by the results

of numerous other experiments in the difficult $\sim$ 1-10 MeV range. These

will not be summarized here, though prominent among them were the Apollo

15 results (Trombka et al., 1973) and data from the balloon-borne double

Compton telescope (Schronfelder et al., 1975).   That both the deep

space (e.g. Apollo 15) data and results from balloon experiments (e.g.

Fukuda et al., 1975) gave such similar excesses above the extrapolated

power law spectrum seemed to confirm the reality of the spectral feature

and prompted several models to be developed as discussed below. However,

after careful re-analysis of the Apollo 15 and 17 data, Trombka et al.,

1977a) have now concluded that most of their reported excess was due to

an incomplete subtraction of the detector background induced by cosmic

ray activation in the NaI crystal. Daniel and Lavakare (1975) had also

suggested this possibility, and criticized many of the balloon results in

this low energy range because of the uncertainties in the subtraction of

the atmospheric background; their own results, as well as others (e.g.

Share et al., 1974), showed no significant excess.

At higher energies ($\geqslant$ 35 MeV), the OSO-3 point (Kraushaar et al.,

1972) has been augmented by the positive results from SAS-2(Fichtel,

Kniffen, and Hartman, 1973 and Fichtel et al., 1977). At these energies.

the experimental systematic uncertainties are quite small, much less than

at low energies, and the balloon results (e.g. upper limits by Hopper

et al., 1973) have generally been consistent. The result of the final

analysis of all the SAS-2 data at high galactic latitudes ($10° < b^{II} \leq 90°$)

by Fichtel et al., (1977) clearly reveals the relative contributions of

the galactic disk and the isotropic, possibly extragalactic, component of the diffuse emission. Referring back to equation (1), whereas the second

term, $C_2/\mathrm{Sin}|b|$, dominates at high energies ($\gtrsim$ 100 MeV), in the 35 MeV

$\lesssim$ E $\lesssim$ 100 MeV interval the contribution from the much steeper isotropic

Component C_1, possibly a true extragalactic component, is comparable or

larger at high latitudes. We shall discuss the implications of this

spectrum in the next section.

The most recent total spectrum and summary of the latest results

for the diffuse cosmic γ emission is shown in Figure 4. Although

the composite spectrum appears to have a discontinuity at $\sim$ 3 MeV, it

might be kept in mind that the extrapolation of the diffuse x-ray

spectrum, $dI/dE \simeq 2.3 \times 10^2 (E/1 \text{ keV})^{-2.4}$ photons/cm^2 sec keV (which

fits the x-ray data reasonably well above $\sim$40 keV (Schwartz and Gursky

1973) is within $\sim$ 1 or 2 σ of all the plotted points. This is true

even for the total high energy intensity measured by SAS-2, although

the curve is not a good fit to the steeper spectrum of the isotropic

component. The strongest test for a two-component diffuse radiation,

other than different spatial distributions, is whether the hump at $\sim$1 MeV

exists, since a steepening spectrum at high energies is not an unexpected

result for a naturally occurring energetic charged particle induced

γ-ray spectrum.

Although the diffuse spectral measurements are now reasonably

self-consistent, the degree of spatial isotropy is still poorly known.

The x-ray spectrum through $\sim$ 100 keV is known (cf. Schwartz and Gursky

1973) to be isotropic to within $\sim$ 5%. At low γ-ray energies

($\sim$ 1 MeV), Trombka et al., (1977) estimate that the anisotropic component

from galactic sources does not exceed 20% of the total flux. At high

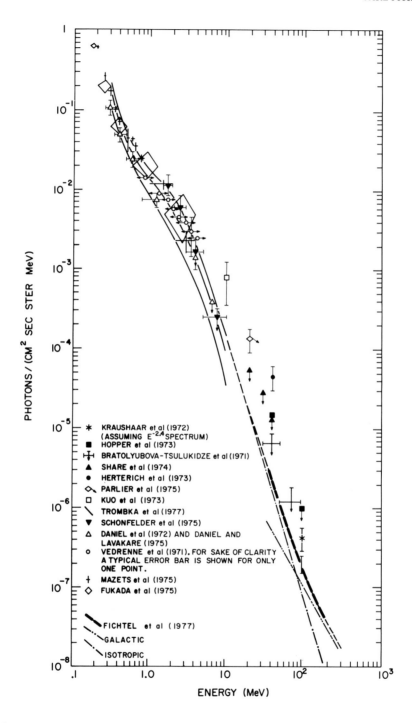

Fig. 4 – Energy spectrum of the diffuse, apparently isotopic γ-radiation.

energies (35-100 MeV), the center-to-anticenter ratio for $20°<|b|<40°$

was measured to be 1.10 $\pm$0.19 and the perpendicular to the galactic

plane intensity to that in the $20°<|b|<40°$ region to be 0.87 $\pm$0.09;

each consistent with isotropy to within errors (Fichtel, Simpson, and

Thompson 1977). Clearly more precise measures of the isotropy

are desired, but no evidence for a major anisotropy exists.

(C) Models for the Diffuse Gamma Ray Background

When the measurements of the diffuse hard x-ray background and the

OSO-3 γ-ray ($\sim$ 100 MeV) flux (Clark et al., 1968) suggested a single

power law description, non-thermal models to unify the entire spectrum

were proposed. Brecher and Morrison (1969) developed an inverse Compton

model whereby high energy electrons leaking from both radio and normal

galaxies scatter from the 2.7° K microwave radiation, giving rise to

the power law x-ray and γ-ray spectra. This model has been criticized

by some both on energetic grounds, since the required electron number

could arise from normal galaxies only if trapping times are low, and

from the point of view that rapid leakage of the required low energy

($\sim$10 MeV) electrons through galactic magnetic fields would seem unrealistic.

Non-thermal bremsstrahlung of low energy electrons (< 10 MeV) interacting

with low density intergalactic gas has also been proposed (Silk and

McCray, 1969). This would give a smooth power law in the low energy

γ-ray range as well as the flattening observed at lower ($\sim$ 40 keV)

X-ray energies, but requires a large ad hoc intensity of subrelativistic

intergalactic cosmic rays produced in a burst at z $\sim$ 10. These and other

models for the low energy spectrum are reviewed by Silk (1970).

With the early evidence (Arnold et al. 1962, Vette et al. 1970) for the flattening of the diffuse spectrum at ~ 1 MeV, and especially the early Apollo 15 results (Trombka et al. 1973), the theoretical emphasis shifted to account for this special feature. Bui-Van and Hurley (1974) extended the Brecher-Morrison inverse Compton model to include second-order Compton scatterings which add a spectral flattening at $\gtrsim 1$ MeV. However, since this model requires the same high loss rate of low energy electrons from galaxies as the Brecher-Morrison picture, the same difficulties apply, including the expected smoothing (for any inverse Compton model) of the spectral break at X-ray energies if this is due to the break in radio galaxy electron spectra (Cowsik and Kobetich 1972).

Alternate attempts to explain the apparent feature at ~ 1 MeV include red-shifted π° decay cosmological models. The first of these models, in which the γ-ray diffuse background is dominated by a cosmological source process different from any responsible for the X-ray spectrum, was proposed by Stecker (1969). It attributes the ~ 1 MeV feature to π° decay from intergalactic cosmic rays interacting with intergalactic gas integrated back to $z_{max} \approx 100$. The major problems with this picture are 1) the predicted bulge in the spectrum exceeds the most recent observations (cf. Figure 4) 2) z_{max} is an arbitrary parameter and 3) the required cosmological cosmic ray density is very high and a new class of cosmic ray sources is required. Most significantly, however, the recent SAS-2 spectral parameters for the isotropic high latitude flux (Fichtel et al., 1977) at ~ 35 to 100 MeV are rather steep (α of $dI/dE = AE^{-\gamma}$ being $2.85^{+0.50}_{-0.35}$) for this model which would instead predict a steepening at ~ 7 GeV (Fazio and Stecker 1970). Despite these difficulties, Montmerle (1977) has found that the cosmological cosmic

ray hypothesis can also account for the observed ^{7}Li abundance although

^{7}Li/^{6}Li ratio is subject to uncertainties in nuclear cross sections.

A more promising model, and one which is still consistent with the

data, was proposed by Stecker et al., (1971). This is the annihilation

hypothesis in which for a baryon-symmetric cosmology (cf. Omnes 1969 and

references therein) with equal amounts of matter and anti-matter, the

flattening above 1 MeV is the red-shifted, non-absorbed annihilation

spectrum resulting from matter-antimatter interactions at the edges

of superclusters of galaxies. The spectral feature is thus not sharp,

but is rather due to baryon-antibaryon annihilation over all z. The high

energy cutoff at $\sim$ 1 GeV is due to the nuclear rest mass, whereas at low

energies the spectral feature is limited to $\geqslant$ 1 MeV by the opacity of

the universe. That is, at z > 10 the universe is opaque to sources of

photons much below 1 MeV (which would in turn arise from annihilation

at z > 10^3, because of the Compton scattering and pair production from

intergalactic electrons and protons, respectively (Arons and McCray 1969).

This opacity combines with the pair production collisions of γ rays

with the 2.7°K background radiation, which leads to a "cut-off" of the

observed spectra at

$$E_c \simeq [2.6 \times 10^2/(1 + z_{max})]^{2.06} \text{ GeV} \qquad (10)$$

for sources at z_{max} (Fazio and Stecker 1970), giving a "window" from

$\sim$ 1 MeV to 10^4 MeV for observing cosmological γ rays as shown in

Figure 5 (Stecker and Puget 1974). Although as we have pointed out above,

the reality of the spectral flattening is now less certain, the steep

isotropic spectrum that is distinguished at $\sim$ 35 MeV from the galactic

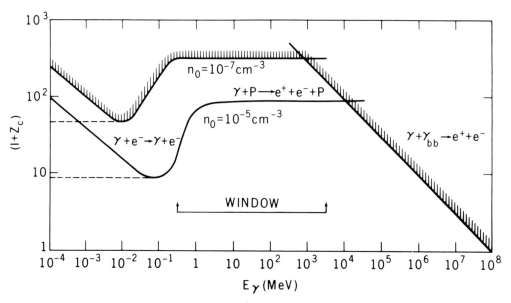

Fig. 5 - The redshift at which the universe becomes opaque to photons

given as a function of observed gamma-ray energy (Stecker

and Puget, 1974). Gamma-rays originating at all redshifts

below the curve can reach us unattenuated with the energy indi-

cated. The two curves on the left side of the figure are

for pair production. The right-hand curve results from atten-

uation of gamma-rays by interactions with the microwave black-

body radiation.

component (Fichtel et al., 1977) currently provides strong support for
the baryon-symmetric models. Assuming this model is correct more detaile
spectra could further constrain n_o. The differences in γ- ray opacity
between open and flat universe models and thus the observed spectrum are
negligible, however (Stecker 1975b). It must be kept in mind also, in
regard to the baryon symmetric big bang theory, that there is substantial
doubt as to whether the required separation into regions of matter and

antimatter which ultimately must become superclusters of galaxies can

occur (e.g. Steigman, 1974).

The data are still not sufficiently precise to exclude other

models, however. One model which fits the spectrum through at least

$\sim$ 30 MeV (but not the steepening at higher energies) is that of Strong

and Wdowczyk (1975) who analyzed the electromagnetic cascades and inverse

Compton spectrum resulting from pair production electrons produced in

cosmic ray-blackbody photomeson interactions at $z \sim 15$. Since the

comparison with the observed spectrum is very dependent on the uncertain

cosmological starlight spectrum, it is possible that better agreement with

the data could be obtained for different assumed evolutionary corrections.

A more easily tested class of models would have the diffuse γ-

ray background arise as the integrated effect of point sources. Strong

et al., (1976) have shown that, whereas only a small percentage of the

isotropic diffuse radiation could be attributed to normal galaxies,

Seyferts and other active galaxies may provide a large protion of the

γ-ray diffuse radiation, although the energy spectrum may be a problem.

Recent x-ray observations (e.g. Tananbaum et al., 1977) indicate that

$\sim$ 20% of the Type I Seyferts are now detectable as 2-6 keV x-ray sources,

but that they would account for only $\sim$ 20% of the low energy x-ray back-

ground. However, since the x-ray spectra of these objects are usually

very hard, it is possible that they could make a significant contribution

to the hard x-ray and low energy γ-ray background--at least through

several MeV (and the possible spectral bump) where the brightest individual

sources such as Cen A are in fact detected (Hall et al., 1976). Further

discussion of this possibility is given in Section V.

Rocchia, Ducros and Goffet (1976) have also proposed a point source super-position model, but they have concentrated on explaining the observed spectral break at ~ 25-40 keV and the possible flattening at ~ 1 MeV. They invoke Compton cooling of the usually hard source spectra by an assumed population of low energy electrons in optically thick regions around the compact sources. This effectively re-distributes photons (from ~ 1 MeV) towards low energies, and, for the observed break at ~ 30 keV, cosmological sources at z_{max} ~ 3 must be assumed. Thus direct detection of these sources is unlikely although the predicted Compton cooling spectra could be looked for in local sources.

Finally, the primordial black holes postulated by Page and Hawking (1976) can evaporate (via a particle tunneling process) at an accelerating rate that ends in an explosive γ-ray burst. The time for this to occur is inversely proportional to the black hole mass such that ~ 10^{15} gram primordial black holes formed at the Big Bang would just now be exploding. The explosive burst yields γ-rays whose average energy and duration may typically be ~ 100 MeV and ~ 10^{-7} sec though these are highly model dependent. Nevertheless, it is quite certain these events would be significantly harder and faster than the γ-ray bursts observed (Klebesadel, Strong and Olsen, 1973) at ~ 200 keV. Such events redshifted over all z could, however, possibly provide the diffuse background, and the spectral shape (below ~ 120 MeV) would then constrain the primordial black hole mass spectrum and particularly the evaporation process. The most significant test of this model and upper limit for the local density of primordial black holes has been the optical light flash and atmospheric Cerenkov experiment of Porter and Weekes (1977). This limited the density of objects providing bursts at $\gtrsim$ 200 MeV and durations $< 10^{-7}$ sec to be $\leq 10^9 \text{pc}^{-3}$ and (for the

assumed burst parameters) would seem to eliminate primordial black holes

as being a significant source of the diffuse radiation. Given the

uncertain mass spectrum and burst characteristics to be expected for

these objects, however, they cannot yet be totally ruled out.

(D)   Conclusions and Future Tests

The high latitude diffuse γ-ray background can be regarded as firmly

established and most probably extragalactic. However, the detailed spectral

shape (cf. Figure 4) from ~ 0.1 - 1000 MeV remains poorly known, as

does the degree of isotropy which is also of critical importance in

selecting possible models. Evidence for the reality of the spectral

flattening at ~ 1 MeV is now (Trombka et al., 1977) much less compelling

than previously, and it is entirely possible that a single source spectrum

applies from the x-ray through the γ-ray energy range. Inverse Compton

models would then be favored since small changes in spectral slope can

be accommodated. However, if the hump near 1 MeV is confirmed and

measured more precisely by subsequent experiments, then the case is much

more compelling that the x-ray and γ-ray backgrounds are dominated

by different processes.

Two classes of models would then appear worthy of further study.

Whereas the low energy x-ray background is not dominated by the assumed

contribution of active galaxies and Seyferts, their high rate of

detectability and hard spectra may make them a significant contribution

at ~ 1 MeV. The apparent flattening at ~ 1 MeV and steepening above 10 MeV

may then reflect the spectra of the individual sources and also include

the effects of nuclear line emission at   1-5 MeV as possibly observed from

Cen A (Hall et al., 1976).   The second and equally promising general model

class for a different origin of the γ-ray background is the baryon symmetr

cosmology interpretation of Stecker, Morgan and Bredekamp (1971).

Here the high penetrability and the possibly unique cosmological infor-

mation content of the cosmic γ-rays in the $\sim$ 1 MeV to 10 GeV range

have several implications.   The baryon symmetric cosmology developed

by Stecker and Puget (1974) suggests that: a) clumping on the scale of

super clusters of galaxies is obtained b) annihilation turbulence can

trigger galaxy formation c) the galaxy formation epoch can be estimated

($z \simeq 60$) and d) the observed mean densities and angular momenta of

galaxies can be derived consistent with the annihilation rate required

for the background flux.   Clearly more sensitive observations with the

Space Shuttle generation of γ-ray instruments are urgently needed

to distinguish between these models.

IV. LOCALIZED SOURCES

(A) Gamma-Ray Observations in our Solar System

New types of information are now being provided by observations in

the γ-ray energy region of the Sun, Moon, and Planets. Of particular

interest in the field of solar and planetary physics is the detection of

γ-ray lines during electromagnetic deexcitation of nuclei.

In the last two decades, significant observational data have been

obtained from both space flight programs and meteorite studies allowing

certain constraints to be imposed on the theoretical models for the

origin and evolution of the solar system. Further, various theoretical

approaches can now be evaluated in terms of these observational tests,

and more rigorous models can be developed. Both chronology and present

day dynamics of the Sun and solar systems can be critically examined
with γ-ray astronomical observations. First, consider the study of the
Sun itself which, in addition to the natural interest it has for us, is
the star that the astronomer can study in greatest detail.

Gamma-ray astronomy, as applied to the observation of the Sun, has
the specific objective of considering high energy processes that take
place in the outer region of the Sun's atmosphere and the relation of
these phenomena to the basic problems of solar activity. A measurement
of the spectra of discrete γ-ray line emission will reveal the detailed
dynamics and time structure of solar flares and energetic particle
acceleration and release. Such measurements should also yield qualitative, and in some cases quantitative information on composition of specific ambient or transient nuclides in the outer regions of the Sun's
atmosphere. That such observations are indeed possible was confirmed
when γ-ray lines associated with solar spectra were observed during two
events in 1972 from the OSO-7 satellite (Chupp et al. 1973 and Chupp,
Forest and Suri, 1975).

The following appear to be justifiable conclusions to be derived
from an analysis of these γ-ray solar flare observations: γ-ray lines
are produced by the interaction of solar flare accelerated particles
with ambient nuclei in the solar atmosphere; the number and energy
spectrum of the accelerated nuclei in the flare region can be determined;
comparison can be made between electron and ion acceleration in flares,
thereby providing information on the charged particle acceleration process.
Finally, from the determination of an upper limit for the He/H ratio of
a few times 10^{-5}, an upper limit on the photospheric ^{3}He abundance can

be obtained, which has direct implications for theories of the solar interior and cosmology.

Regarding solar γ-ray observations to be considered for the future, and especially the Shuttle program, high energy resolution observations seem most promising. Three problems of interest are: formation of the 0.511 MeV line in solar flares, determination of energetic particle spectra by observation of γ-ray emission due to various excitation levels of a given nuclear species, and the use of Doppler shifting of discrete γ-ray line emission due to anisotropies of charged particles in solar flares.

The γ-ray line produced at 0.51 MeV in solar flares is believed to be the result of either the free annihilation of positrons with electrons or the formation and decay of positronium. The positrons are produced as a result of the decay of π^+ mesons and the ecay of radioactive nuclei produced in nuclear reactions of flare-accelerated particles with constituents of the solar atmosphere (Lingenfelter and Ramaty 1967; Ramaty and Lingenfelter 1973; Ramaty, Koslovsky, and Lingenfelter, 1975). The formation of the 0.51 MeV line depends on the source of the positrons, on the propagation of the positrons in the solar atmosphere, on the density and temperature of the ambient medium in which the positrons decelerate,and on annihilation since the positrons may annihilate freely or form a ground state of positronium. Observation of the 0.51 MeV line and the γ-ray continuum down to lower energies will yield information concerning these properties. The spectral shape is affected in the following ways: broadening of the 0.51 MeV line, relative intensity of emission of the 0.51 MeV line with respect to other γ-ray lines, the magnitude of the continuum below 0.51 MeV coming from the triplet

positronium decay, and the temporal variation of the intensity of the

0.51 MeV. Detailed analysis of the variation of the spectral shape as

a function of these parameters can be found in papers by Ramaty,

Kozlovsky and Lingenfelter (1975) and Crannell, Joyce and Ramaty (1976).

Gamma-ray line emission is evidence that a given nuclear species

with a corresponding nuclear energy level has been excited by particles

with kinetic energies above the excitation threshold. A number of

studies have been carried out to determine whether or not such excita-

tion could be used to determine the energetic particles spectra

(Lingenfelter and Ramaty 1967; Dolan and Fazio 1965). Recent calcula-

tions have been made of the intensity of the 15.11 MeV γ-ray line

relative to that of the 4.44 MeV γ-ray line for the excitation and the

subsequent de-excitation of the corresponding states of ^{12}C (Crannell,

Ramaty, and Crannell 1977). The results indicate that the relative inten-

sities   in the two lines are particularly sensitive to the spectral

distribution of energetic particles which excite the corresponding

nuclear energy levels. Further,the background at the 15.11 MeV level

should be rather low. In terms of detector requirements high detector

efficiency and large detection areas will be required; however, high

energy resolution is not as important.

Finally, anisotropies of charged particles accelerated in solar

flares can be studied by observing Doppler shifts in selected γ-ray

lines. Examples of nuclei which may be of interest in this case are

^{12}C and ^{16}O, which emit 4.4 and 6.1 MeV γ-rays respectively. These

discrete line emissions are excited by protons with energies greater

than 10 MeV. Such high energy interactions will impart kinetic
energy to the nucleus. Because the γ-rays are emitted in a time
interval after the interaction which is short compared with the slowing
down time of the nucleus, any directional anisotropy in the primary
exciting particles would cause a Doppler shift in the energy of the
lines. Detailed calculation of this effect has been carried out by
Ramaty and Crannell (1976) for the 6.1 MeV of ^{16}O. It was found that
if the accelerated particles were isotropic the line will be centered
at 6.129 MeV, and its full width at half-maximum would be about 100 keV.
However, for particle anisotropies that might be produced during a solar
flare, the line is shifted to lower energies by about 30 to 40 keV.
Observation of such broadening and energy shifts are possible utilizing
high resolution solid state detectors. Because of the low intensity of
the expected discrete line flux, arrays of detectors will be needed and
extremely careful calibration of the detectors in the array will
be required.

Gamma-ray astronomy observations also find important application in
studying the development of the planets out of the primitive solar nebula.
For those planets where the small atmosphere and trapped radiation
environment do not interfere significantly with the γ-ray emission,
orbital measurements can be conducted. The terrestrial planets are
examples of such systems. Asteroids and comets can also be studied.
Elemental surface composition can be inferred from observations of
the discrete line emission. This emission can be mainly attributed to
natural radioactivity (Th, U, and K) and the primary and secondary
cosmic-ray induced activity producing identifiable emissions from H, O,
Si, Al, Mg, Fe, and Ti.

In acquiring an understanding of the geology of a planetary body,
a knowledge of the total chemical composition and of the variation of
the surface composition of the body are of fundamental importance. The
overall composition will be related to the mechanism of accretion and
accumulation leading to planetary formation. The distribution of
elements has also been affected by geological processes which have been
operative during planetary evolution.

A number of missions have been flown to the Moon and Mars from
which such information has been obtained (e.g. in the American space
program Apollo 15 and Apollo 16, and in the Russian program a number
of lunar missions, Mars 4 and Mars 5). The data from the Apollo 15 and
16 x-ray and γ-**ray** spectrometer have been most throughly analyzed
(Adler et al. 1972; Metzger et al. 1973; and Trombka et al. 1977a).

Figure 6 shows a smoothed calculated spectrum characteristic of
those obtained with the γ-ray spectrometer on the two Apollo missions.
The tremendous wealth of information on the major elemental surface
composition can be from such a spectrum. Here the poorer resolution (NaI(Tl) spectrometer was used. With higher resolution solid state
detectors even more information should be obtained.

The Apollo 15 and 16 data indicate that the Moon has a global
aliminum or plagioclase rock crust which is the result of the first major
geochemical event on the Moon's geological evolution after its formation.
By outlining variations of the distribution of uranium, thorium, and potassium, the γ-ray information suggested that large basin-forming events
were capable of creating the geochemical provinces by the ejection from

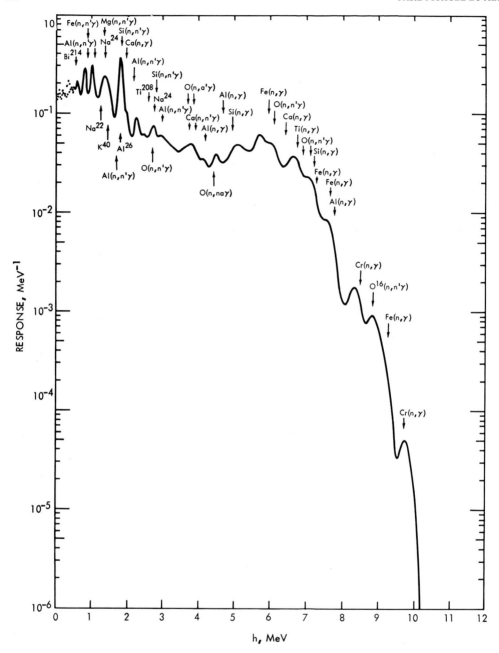

Fig. 6 - Calculated spectrum characteristic of those obtained with the
γ-ray spectrometer on Apollo 15 and 16 (e.g., Trombka et al.,
1977b).

depths of ten or more kilometers. The depletion of volatile materials relative to refractory materials was found to hold true on a global basis, since the K/Th ratio determined from the Moon was significantly lower over most of the Moon than that on Earth. The measurement of radioactive elements Th, U, and K obtained from Mars 4 and 5 indicate that the soil overflow on the surface of Mars is basaltic in nature.

This type of information from solar, lunar, and Martian γ-ray emission spectrum can be obtained for other planetary surfaces. The results can also be used in understanding the type of information that may be obtained when discrete γ-ray line spectra are measured which may characterize either galactic or extragalactic sources. Both radioactive decay and inelastic collision produce nuclei in excited states. Therefore, for example, radioactive γ-ray sources in space may be associated with events of nucleosynthesis. The average intensity of these γ-ray fluxes will be small. Thus, large arrays of cryogenically cooled solid state detector systems properly collimated will be required for performing such measurements. These are again a most important type of experiment to be carried aboard Spacelab missions.

Finally, the Jovian radiation environment is an excellent excitation source for producing characteristic x-rays from the surfaces of the Galilean satellites. Characteristic emission due to excitation of Mg, Al, Si and Fe should be produced. Although rather low in intensity such x-ray fluxes may be observed near the vicinity of the Earth. Thus it may be possible to observe such emissions from Spacelab orbits with large array x-ray telescopes and spectrometers.

B. Exploding Stars

  1. Supernovae

    a. <u>Continuum Emission</u>. After the explosion high energy electro-
magnetic radiation may be produced by an uncovered pulsar or by the
interaction of the ejected debris with the interstellar medium. In
addition there may be high energy radiation associated with the explosion
itself, especially with high temperatures.

    It appears that most of the radiation from type II supernovae is
thermal radiation. To produce copious γ-radiation of this sort
requires a high temperature. The observational data is well represented
by low temperatures (T $\leqslant$ 20,000 °K). Thus, kT $\leqslant$ 10 eV, which certainly
is not favorable for γ-ray emission. A detailed discussion for SN 1969c
is given by Falk and Arnett (1976). Most of the luminosity seems to be
due to the diffusive release of imprisoned radiation by an expanding
plasma. There is a "first burst" due to the arrival of the supernova
shock at the stellar surface. It is not clear just how high the temperatures get in this brief stage (Δt $\leqslant$ 1 day). The calculations give
$T_{max} \simeq$ 40,000 °K; it is unlikely that this is off by orders of magnitude. Consequently it appears that type II supernovae do not release
very much of their energy as γ-ray continuum radiation.

    There is a simple reason for this result. A massive star of, say,
10 $M_\odot^*$ develops a large radius, r $\simeq$ 5 x 10^{13} cm, after helium burning.
The observed luminosity at peak, which looks like a blackbody, has
L $\simeq 10^{43}$ ergs^{-1} so if we allow for some expansion,

---

* $M_\odot \equiv$ solar mass.

$$\sigma T_e^4 = L/4\pi r^2 \leqslant 10^{43}/\pi \times 10^{28} \tag{11}$$

or T $\leqslant$ 50,000 °K. Doppler shifts in absorption lines of up to about 10^9 cm/sec are observed. The radius doubles in a time

$$t \simeq r/v \simeq 5 \times 10^{13}/10^9 \simeq 5 \times 10^4 \text{ sec} \simeq 1/2 \text{ day}, \tag{12}$$

and in two days $r \simeq 2 \times 10^{14}$ cm so $T_e \lesssim 25{,}000$ °K and it continues to cool. When the photospheric temperature drops below about 6,000 °K the opacity decreases due to recombination and one sees in to deeper, hotter regions. Thus T_e decreases below 6,000 °K only slowly.

When will the temperature throughout the envelope drop low enough to cause transparency? First the initial (post shock) temperature must be estimated. For 10 $M_\odot$ and an average velocity of 6,000 km/sec gives a kinetic energy of 3.6×10^{51} ergs, a value in accord with that estimated as necessary to explain the nature of galactic supernova remnants. This energy can fill a sphere of radius $r = 5 \times 10^{13}$ cm with blackbody radiation at a temperature $T \simeq 1.0 \times 10^6$ °K.

Is the expansion approximately adiabatic? It seems possible; observed supernova luminosities, integrated over the outburst give energies of the order of 10^{50} ergs. The diffusion time is

$$t \simeq 3r^2/\lambda c \simeq 3r^2 \, \rho\varkappa/c \tag{13}$$

Most of the stellar mass is in an extended, almost constant density envelope; so

$$\rho \simeq 3M/4\pi r^3 \tag{14}$$

and the diffusion time is

$$t \simeq 2 \times 10^8 \text{ sec} \left[\frac{10^{14} \text{ cm}}{r} \right] \left[\frac{\varkappa}{0.4 \text{cm}^2/\text{g}} \right] \quad \frac{M}{M_\odot} \quad . \tag{15}$$

This equals $t \sim r/v$ only for $r \simeq 10^{16}$ cm so the dominant effect is expansion.

For quasi-adiabatic expansion, $\rho \sim T^3$ so we have

$$(T/10^6 \text{ °K}) \simeq 5 \times 10^{13} \text{cm}/(t*10^9 \text{cm/sec}), \tag{16}$$

so for $T \simeq 6000$ °K,

$$t \simeq 80 \text{ days.} \tag{17}$$

After this time a significant fraction of the envelope becomes transparent. (This is sufficiently long to hide many γ-lines; see Table 1 below). More complex calculations support these simple arguments.

It is not clear how high the energy densities associated with the "burst" in type I supernovae become. They may be similar to SN II. The Morrison-Sartori (1969) model requires fairly hard (UV) radiation in large amounts (energies $> 10^{52}$ ergs). While Lasher (1975) has explained the shape of the visual light peak of SN I, his models do not yet explain the hard UV pulse. It is not at all obvious that a hard UV pulse, even if it exists, means that significant γ-radiation will result. Such radiation is difficult to rule out entirely for all (as yet unspecified) model. Recently Falk (1978) has examined the burst in a realistic and well-zoned red giant and finds only about 10^{50} ergs of hard photons.

Colgate (1975) assumes that the supernova shock has a high temperature precursor. If instead the matter is accelerated by radiation pressure (Falk and Arnett 1976) then the reason for expecting a strong

γ-ray burst disappears. Until the physics of the "peak" in SN I is well understood and agreed upon it seems wise to regard the theoretical situation as unclear.

Because of many underlying similarities it is beginning to appear that both SN I and SN II may be the result of explosions in extended stars. More condensed objects may explode and give rise to higher effective temperatures and γ-radiation. Such events would not correspond to those events astronomers term supernovae.

b. <u>Line Emission</u>. The last stable nucleus with Z = N is ^{40}Ca. In stars the thermonuclear synthesis of heavier nuclei tends to form Z = N nuclei which are unstable toward electron capture or positron emission. Radiative decay from excited states of the daughter nucleus gives rise to γ-ray lines. This process is particularly important because a lot of mass--the iron group--is formed this way. Similarly nuclear processing in proton-rich or neutron-rich environments can also produce unstable nuclei whose decay may give γ-lines.

i. <u>The 'typical zone" approach</u>. To be useful for the experimenter a theory of such processes should be able to predict γ-line emission. To date most predictions have relied on detailed analysis of thermonuclear processing in "typical zones". One specifies a set of initial conditions (temperature, density, and composition) and an expansion time scale, then solves the coupled nonlinear equations (reaction network) for the evolution of the abundances. For a full review of the γ-line problem from this viewpoint see Clayton (1973).

The astrophysical aspects of a single zone approximation are clearly oversimplified. Stars are not homogeneous; they have structure.

Table 3.   Gamma Line Prospects for a "Typical" Supernova

Nucleus	$X_\odot$	Progenitor	$\tau_{1/2}$	Atoms/SN
^{56}Fe	1.3×10^{-3}	^{56}Co	77d	3.0×10^{54}
^{56}Co	1.3×10^{-3}	^{56}Ni	6.1d	3.0×10^{54}
^{48}Ti	2.3×10^{-6}	$(^{48}$Cr$)^{48}$V	16d	6.2×10^{51}
^{44}Ca	1.9×10^{-6}	$(^{44}$Ti$)^{44}$Sc	48 yr	5.6×10^{51} (?)
^{60}Ni	2.0×10^{-5}	$\left\{ \begin{matrix} ^{60}\text{Fe} \\ ^{60}\text{Co} \end{matrix} \right.$	$\left. \begin{matrix} 3\times10^{5}\ \text{yr} \\ 5.26\ \text{yr} \end{matrix} \right\}$	4.4×10^{52} (?)
^{22}Ne	1.2×10^{-4}	^{22}Na	2.6 yr	(?)
^{238}U (example)	1.3×10^{-10}	(r-process)	4.5×10^{9} yr	1.3×10^{47}
^{26}Mg	8.6×10^{-5}	^{26}Al	7.3×10^{5} yr	4.5×10^{50}

Further, stars of different mass behave in very different ways. The
net result of all this complexity may be different in some important
details from a set of typical zones which reproduce some of the
dominant features. Further, the half-lives of the promising radioactive nuclei are usually sufficiently short that the experimenter
should look for the closest, most recent object; in this case individual
variations could be of prime importance.

    ii.   The Stellar Model Approach. Some preliminary work which
attempts to go beyond the one zone approach has just been completed.
The evolution of the cores of stars of mass $10 \leqslant M_\odot \leqslant 95$ have been
evolved to dynamical instability. They all develop a nickel-iron core
which exceeds the Chandrasekhar mass and contracts toward the neutron

star state (or beyond). The remaining matter is loosely bound in a surrounding mantle. It is assumed that this mantle is explosively ejected from the star and that this process corresponds to at least some observed supernovae (see Arnett 1975). The circumstantial evidence for this point of view is fairly strong.

Recently investigation of the neutrino transport during the hydrodynamic collapse of the stellar core has shown that neutrinos are strongly trapped due to the large neutral current opacity (Arnett 1977), suggesting that the explosion mechanism is a reflected shock. This idea has been followed up by Van Riper (1978) who has shown that such explosions can occur, depending upon the equation of state used. Hydrodynamic calculations of these shocks as they traverse a fairly realistic stellar model ($M_\alpha = 8 M_\odot$) have been done, and the nucleosynthesis in the ejected matter determined. These preliminary results allow a reassessment of the γ-ray line problem. The results are in Table 3.

The ^{56}Fe-^{56}Co-^{56}Ni group. These nuclei are produced by the $\mathcal{C}$-process (complete oxygen-silicon burning) in high abundance. The amount ejected from the star depends on the shock energy (more energy, more iron), but is about $0.1 M_\odot$ or so for "typical" values ($E \sim 10^{51}$ ergs). For lower energies much iron can be left in the condensed remnant (neutron star or black hole). Observation of these γ-lines may be difficult; in a "typical" supernova this matter may not become transparent for a few months (see above).

The ^{48}Ti-^{48}V-^{48}Cr group. These nuclei are also formed by the e-process; the comments above for A=56 apply here as well.

The ^{44}Ca-^{44}Sc-^{44}Ti group. Although an adequate abundance of ^{44}Ca

gas formed as ^{44}Ti in constant temperature quasi-equilibrium calculations, better analysis (include freeze out) suggests they are destroyed before ejection. While it seems plausible that the observed ^{44}Ca is formed as ^{44}Ti, the correct conditions have not yet been recognized.

$\underline{^{26}\text{Mg-}^{26}\text{Al}}$. Formed in both hydrostatic and explosive carbon burning (Arnett, 1969), this is one of the best prospects (Arnett 1976; Ramaty and Lingenfelter 1977). An excess of ^{26}Mg in the Ca-Al rich inclusions in the carbonaceous meteorite Allende has been interpreted as strong evidence for ^{26}Al in primitive solar system matter (Lee, Papanstassiou and Wasserburg 1976). The upper limit for ^{26}Al line radiation from the Crab (Leventhal, MacCallum and Watts, 1977) is

$$\frac{M\ (^{26}\text{Al})}{M_\odot} \leq 0.013\ (d_{\text{Crab}}/2\ \text{Kpc})^2, \qquad (18)$$

where d_{Crab} is the distance to the nebula.

Carbon burning produces $^{26}\text{Al}/^{27}\text{Al} \sim 10$, so $M(^{27}\text{Al}) \leq 13 M_\odot$ but both the predicted and observed abundances of the products of carbon burning in the Crab are orders of magnitude below this. In young supernova remnants, the large velocities smear the line; however, in older remnants better signal to noise might be obtained with higher energy resolution (see Ramaty and Lingenfelter 1977).

The $\underline{^{60}\text{Ni-}^{60}\text{Co-}^{60}\text{Fe group}}$. The mechanism for production of the observed ^{60}Ni is an open question. If explosive carbon burning had a large neutron flux, ^{60}Fe could be made, giving a γ-line prospect. However actual stellar hydrodynamic calculations suggest that (1) the neutron flux during carbon burning is too small, and (2) most ^{60}Ni is made as ^{60}Zn. The decays take less than a half-hour, so that the γ-rays are not likely to escape the star.

The ^{22}Na-^{22}Ne pair.   The helium zones in the massive star models

thus far calculated do not get hot enough for $^{14}N(\alpha,\gamma)$ $^{18}F(\alpha,\gamma)$ ^{22}Na

to occur significantly during the explosion.   Perhaps novae

are a better site.

2. Novae.   The theoretical understanding of novae is very compli-

cated, and it is not yet clear which of several models represents the

best picture of a novae.   Hence, a theoretical treatment of novae will

not be given here.   It is worth noting, however, that the results of

Leventhal, MacCallum and Watts (1977) is already constraining theoretical

models of novae Clayton and Hoyle 1974; Sparks, Starrfield and Tulan

1976).   For example, rapid CNO burning produced Na^{22} which gives rise

to a γ-ray line at 1275 keV, and the present experimental upper limit

is already a factor of three below the original prediction (see

Leventhal, McCallum, and Watts 1977).

C. Pulsars

At this relatively early stage in the development of γ-ray

astronomy, it is quite unexpected that five, and possibly six, γ-ray

pulsars (E > 30 MeV) with radio counter-parts would already be

identified (Ogelman et al. 1976 and D'Aimco et al. 1978).

The discovery of radio pulsars by Hewish et al. (1968) initiated an

extensive search for these objects, which has led to the detection of

well over a hundred radio pulsars whose periods range from about

0.03 to several seconds.   However, in spite of extensive searches

in the x-ray energe range, only one of the known radio pulsars,

PSR 0531+21, has been detected with certainty at these frequencies.

Hence, the discovery of γ-radiation from pulsars is particularly remarkable.

For several, now well-established,reasons including their very fast and extremely constant periods, pulsars are generally accepted as being neutron stars, as originally suggested by Gold (1968 and 1969). There is not, however, general agreement on the specific mechanism responsible for the radio and γ-ray emission although curvature radiation in the

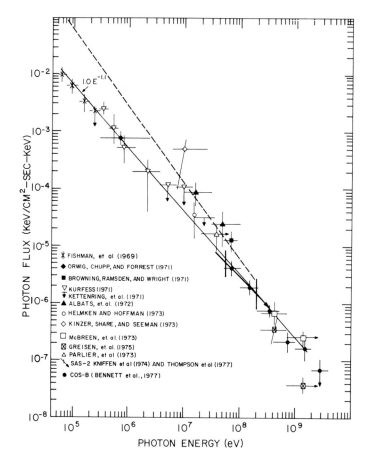

Fig. 7 - Energy spectrum of the pulsed γ-ray emission from PSR 0531+21, compared to the unpulsed flux shown by a dashed line. The dashed line is based on X-ray data and the results of SAS-2. The solid line connected the pulsed X-ray results near 10 keV with the pulsed SAS-2 results at 100 MeV.

strong magnetic fields near the neutron star, synchrotron radiation,

and combinations of the two are often a part of the current theoretical

hypotheses.

The first discrete source of high-energy γ-radiation that was

observed is PSR 0531+21 in the Crab nebula, the fastest (that is,

shortest-period) pulsar known.   In this case the observation of pulsed

emission at the same period as, and in phase with, the radio pulsar

clearly established the identification of the source.   The observations

of this source are summarized in Figure 7 except for those at very high

energies ($> 10^{11}$ eV) reported by Fazio et al. (1972) and Helmken, Grindlay,

and Weeks (1975).   The most striking feature seen in this figure is the

transition from a predominantly unpulsed flux at low energies to a pre-

dominantly pulsed flux at higher energies.   This transition may be

understood, qualitatively at least, by a model in which the pulsed

emission comes from synchrotron radiation near the pulsar while the

unpulsed component comes from regions of lesser magnetic field strength,

probably by a combination of scattering of electrons and interaction of

high-energy protons and the ambient gas.

Helmken, Grindlay and Weeks (1975) and Grindlay, Helmken and

Weeks (1976) reported a flux of γ-rays above 10^{11} eV detected with the

10-m reflector at Mt. Hopkins which yielded evidence for a time-

variable pulsed flux of these high-energy γ-rays from PSR 0531+21.

The intensity of the pulsed radiation has also been reported to

have changed significantly in the energy range 0.8 to 2 GeV by Greisen

et al. (1975), in particular by at least a factor of five between Octo-

ber 1971 and July 1973, in marked contrast to the lack of any significant

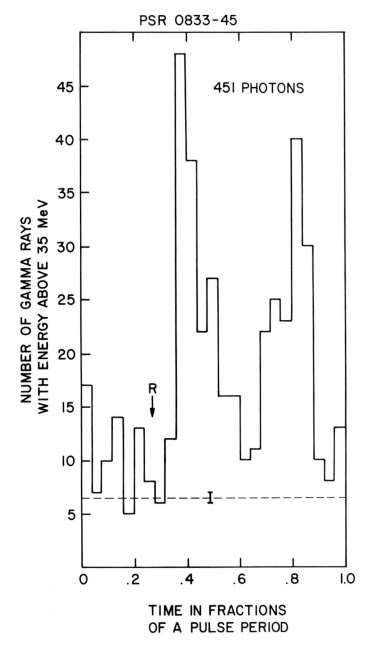

Fig. 8 - Photon distribution as a function of the phase of the radio
         pulser period (Thompson et al., 1977a). The "R" and arrow
         indicate the phase of the radio peak.

(30% or more) variation in the total emission in the 1 to 10 keV range over the same period as observed by UHURU. The possible intensity variation of the high energy γ-ray flux PSR 0531+21 clearly deserves further study.

Turning next to Vela, Thompson et al. (1975 and 1977b) reported that strong excess emission above that of the galactic plane was observed by SAS-2 from the Vela region corresponding to approximately (12 $\pm$1.2) x 10^{-6} photons/cm^2 ster. sec. above 100 MeV during the three periods it viewed this region, February 14-21, 1973; February 21-28, 1973; and April 3-9, 1973, and that a major fraction of the γ-ray emission (> 35 MeV) is pulsed at the radio period from PSR 0833-45. The results of an analysis of the Vela pulsar are shown in Figure 8. The pulsation occurs over the entire energy region examined, from 30 to over 200 MeV. The pulsed emission consists of two peaks, one following the radio peak by about 13 msec, and the other by 48 $\pm$2 msec. Pulsed emission from Vela has also been seen by Albats et al. (1974), and Bennett et al. (1977).

The luminosity of the pulsed emission above 100 MeV from Vela is about 0.1 that of the pulsar PSR 0531 +21 in the Crab nebula, whereas the pulsed emission from Vela at optical wavelengths is less than 2×10^{-4} that from the Crab. The relatively high intensity of the pulsed γ-ray emission and the double peak structure, compared to the single pulse in the radio emission, suggest that the high energy γ-ray pulsar emission may be produced under different conditions from that at lower energies, contrary to the situation existing for the Crab pulsar. The differences between the pulsed radiation from the Crab and Vela are summarized in Figure 9. Although the analysis for pulsation is not yet

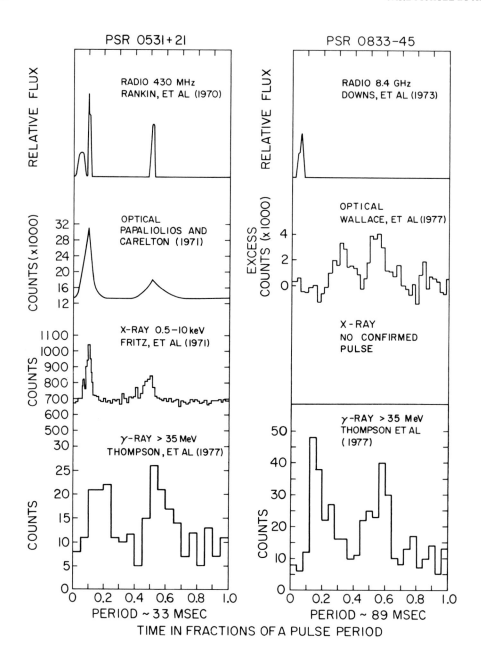

Fig. 9 - Comparison of the pulse structure and phase at radio, optical,
X-ray and γ-ray energies for NP0532 and PSR 0833-45, based on
representative experiments as shown (Fichtel 1977).

complete, the COS-B results have already shown that the γ-ray spectrum

from Vela extends beyond 1 GeV (Bennett et al. 1976 and 1977).

In addition to PSR 0531+21 and PSR 0833-45, Ogelman et al. (1976)

reviewed the data from the SAS-2 high energy (> 35) γ-ray experiment

for pulsed emission from each of 73 radio pulsars which were viewed by

the instrument and which have sufficiently well-defined period and period

derivative information from radio observations to allow for γ-ray

periodicity searches. When the γ-ray arrival times were converted to

pulsar phase using the radio reference timing information, two addi-

tional pulsars, PSR 1747-46 and PSR 1818-04, showed positive effects,

each with a probability less than 10^{-4} of being a random fluctuation

in the data for that pulsar. Subsequently, two other pulsars,PSR 1022-09

and PSR 0740-28, have been seen with the COS-B γ-ray telescope

(D'Amico et al., 1978).

In some pulsar models (e.g. Ostriker and Gunn, 1969; Ruderman and

Sutherland 1975), the current value of $P/(2P)$ is representative of the

pulsar age. For the Crab pulsar, this apparent age is reasonably con-

sistent with the known age of the pulsar. The analysis by Ögelman et

al. (1976) of existing γ-ray data in this light suggests that the

threshold for detection is just being approached by the SAS-2 and COS-B

γ-ray instruments, and that further developments in this field in the

near future could be extremely rewarding. Even a γ-ray telescope with

only a factor of ten greater sensitivity could improve our knowledge of

γ-ray pulsars substantially.

D.   Other Point or Localized Sources

In addition to the sources which are associated with pulsars and supernovae, several other γ-ray sources now seem clearly established or are at least possible candidates.

A very clear source has been reported by Kniffen et al. (1975) at ($\ell \sim 195°$, b $\sim$ +5) based on the SAS-2 data. A search for γ-ray pulsations from pulsars in the region did not show any statistically significant signal. Further, a study has shown that it is very improbable that this increase is associated with the recently-discovered satellite galaxy (Simonson, 1975) in the galactic anticenter region (core position ℓ = 197.3±0.6°, b = 2.1±0.5°) both on the basis of location and an analysis of position error, and on the basis of the intensity of emission (Fichtel, 1977). Both the SAS-2 (Thompson et al., 1977a) and COS-B (Masnov et al., 1977) results indicate a possible 59s modulation in the intensity of this source. Maraschi and Treves (1977) have, therefore, speculated that this source may be either a slowly-rotating accreting neutron star surrounded by a cloud thick to X-rays or a freely-precessing neutron star.

Energetic (E > 35 MeV) γ-rays have also been observed from Cyg X-3 with the SAS-2 γ-ray telescope. They are modulated at the 4.8^h period observed in the X-ray and infrared regions, and within the statistical error are in phase with this emission. The flux above 100 MeV has an average value of $(4.4±1.1) \times 10^{-6}$ photons $cm^{-2} s^{-1}$. If the distance to Cyg X-3 is 10 kpcs, this flux implies a luminosity of more than 10^{37} ergs s^{-1} if the radiation is isotropic and about 10^{36} ergs s^{-1} if the radiation is restricted to a cone of one steradian, as it might be in a pulsar. Reports of γ-ray emission from Cygnus X-3 have also been

made by Galper et al. (1975) and Vladiminsky et al. (1975).

Many localized sources have been   reported recently by Hermsen

et al. (1977) and D'Amico et al.(1978) from COS-B observations. It is

not clear yet whether these are mostly point sources or emission from

localized regions, such as clouds or galactic arm tangents, in some

cases. Future observations of some of these regions should determine

which are actually point sources, and correlated radio measurements

should help to determine if some of them are pulsars. Other possibilities

for γ-ray sources have been reported in the literature by Frye et al.

(1971) and Stephanian et al. (1975).

E.   Low Energy Gamma Ray Bursts

The nature of the γ-ray burst sources discuvered by Klebesadel,

Strong and Olsen (1973) remains an unsolved problem in γ-ray astronomy.

These are certainly point sources, and they are detected primarily at

energies $\gtrsim 0.1$ MeV with total fluxes of $\sim 10^{-5}$ to 10^{-3} erg cm^{-2} and

durations usually of $\lesssim 10^2$ sec. From matching burst profiles, which

show significant structure on time scales ~ 0.1 sec, loci of possible

arrival directions have been established by multiple satellite

detections over long baselines. Whereas the arrival directions

inferred in this way from some of the original burst data have relatively

large uncertainties due to limitations in timing accuracy, recent

determinations for several bursts using the detectors on the Helios

spacecraft and others with ~ 1 AU separation have yielded bands of

position only several arcmin wide, though many degrees long (Cline,

1978). No known X-ray or γ-ray "steady" sources are included in these latest positions, nor are any other known "unusual" classes of astronomical object (e.g. white dwarfs or pulsars). However, for a meaningful identification with any object, much more accurate ($<$few arcmin^2) positions are needed for γ-burst sources. These may be forthcoming with experiments planned for the Gamma Ray Observatory.

The distribution of the number of events as a function of the observed flux for the $\sim$ 80 γ-bursts now recorded show some indication for a spectral index for the flux of $\alpha < 3/2$, or perhaps $\alpha \sim 1$ as would be expected for a galactic distribution of the sources. However, the small number of the few smallest ($\sim 10^{-6}$ erg cm^{-2}) bursts so far detected as well as large uncertainties in the total burst energy flux (given the often limited energy information and spectral differences between events) preclude a firm conclusion that the γ-burst sources are not isotropic (Cline and Schmidt (1977) and White et al., 1977).

V. OTHER GALAXIES

A. Introduction

Extragalactic γ-ray astronomy appears at the threshold of many discoveries when more sensitive detectors become available in the coming Space Shuttle era. This claim can be made because a) our Galaxy is a normal one and known to be a γ-ray emitter; so at least similar normal galaxies of the local group should be detected, and b) the high energy processes evident in the nuclei of active galaxies, together with the existing γ-ray detection of three of them, virtually assures that other members of this class will be detected. Though results to date are limited, the existing ones will be summarized and their implications

discussed. Models that can be tested by future observations will be

mentioned and future directions discussed.

B. Present Observations and Implications

In Section III it was noted that, if the apparent flattening of the

diffuse spectrum at E ~ 1 MeV and the marked steepening for E > 10 MeV

are confirmed, then it is likely that this radiation is not an extrapola-

tion of the x-ray background, but is due to either the summation of point

sources or the annihilation spectrum calculated (Stecker 1977; Stecker

and Puget (1972) for a baryon symmetric cosmology. Clearly observations

of individual galaxies are the crucial test of the first hypothesis.

To date three galaxies beyond our own has been shown with certainty

to be γ-ray sources. NGC 5128, the active galaxy associated with the

powerful radio source Cen A, was the first to be seen. It is an interesting

object for detailed study in that its relative proximity (~ 5 Mpc) makes

it the ideal laboratory for the study of the active galaxy phenomenon.

Its nucleus has been detected from radio through very high energy γ-rays;

the spectrum is summarized by Grindlay (1975). In the γ-ray range, Cen

A was first detected at very high energies ($\gtrsim 3 \times 10^{11}$ eV) by ground-

based observations using a modified version of the atmospheric cerenkov

technique (Grindlay et al., 1975). This high energy flux particularly,

as well as the radio and medium-high energy x-ray spectra, seem best

interpreted by a Compton-synchrotron model of the central source

(Grindlay 1975; Mushotzky 1976). The model of Grindlay requires a two

component compact source: component I has a radius, R, $\leqslant$ light days,

electrons with energies, E_e, between ~ 10 MeV and 3 GeV and a magnetic field B of about 2 gauss, and component II with R ~ 130 light days,

E_e between ~ 50 MeV and 10^4 GeV and B ~.01 gauss. Extensive observations

of x-ray variability on these two time scales summarized by Beall et al.

(1977) and Lawrence, Pye and Elvis (1977) lends support to this

two-component picture. This model specifically predicts that the inverse

Compton hard x-ray spectrum should extend to ~ 10 MeV and then

decrease sharply.

Indeed no high energy (E > 35 MeV) γ-ray flux has yet been detected

through upper limits (Fichtel et al. 1975) from SAS-2 I (> 100 MeV)

$\leqslant 1.5 \times 10^{-6} ph. cm^{-2} sec^{-1}$ are consistent with the model. At low γ-ray

energies, the detection of a flat continuum and possible line emission

through ~5 MeV (Hall et al. 1976) is also consistent with the model and

established Cen A as the first extragalactic source in this energy range.

The Mushotzky model involves the adiabatic expansion of an electron

cloud with the resulting electron synchrotron energy losses leading to

inverse Compton emission. This model agrees with the data reasonably

well except that it does not predict any very high energy γ radiation

Recently γ rays have been detected from both NGC 4151, a Seyfert

galaxy (DiGocco et al. 1977; Graml, Penningsfeld and Schönfelder, 1978),

and from 3C273, a quasar (Swanenburg et al. 1978). The low energy

measurements of the γ radiation from NGC 4151 together with the x-ray

results and the high energy γ-ray upper limits obtained from SAS-2 indi-

cate a marked steepening of the spectrum at a few MeV. There is also

a suggestion of a change in spectral shape for 3C273 between the x-ray

and high energy γ-ray regions.

It seems worth pursuing the possibility that active galaxies are

likely to be luminous γ-ray sources at least through several tens of

MeV. Although clearly more sensitive observations in the energy range 0.5 to > 100 MeV are crucial for establishing the detailed spectral shape of active galaxies, based on existing x-ray and γ-ray spectra, it is very likely that many close active galaxies will ultimately be detected in γ rays. The detection, however, will require a factor of ~3 to 10 increase in sensitivity over previous experiments. Some of these may indeed be detected in the low energy γ-ray range by the detectors on HEAO-C. It will be very interesting to see if the integrated emission of these objects can account for the spectral shape of the diffuse high latitude γ radiation.

C. Models and Future Observations

Since the SAS-2 data (Fichtel et al. 1975) show that our galaxy emits $\sim 5 \times 10^{38}$ erg s^{-1} in photons with $E_\gamma > 100$ MeV, M31 should also be detectable at a level of $\sim 10^{-7}$ photons/cm^2 sec, which should be detectable by high energy γ-ray experiments which have been proposed. The dominant π° and bremsstrahlung contribution to the galactic flux (e.g., Stecker 1977; Kniffen, Fichtel and Thompson 1977) indicate that the medium-to-high energy range should be most productive in the detection of nearby normal galaxies. Such observations would enable detailed comparisons between total γ-ray luminosity and directly observed galactic structure and gas content. This will provide important constraints on galactic structure and evolution models. Strong and Wdowczyk (1975) have also calculated that ~10% of the diffuse background radiation might be due to integrated emission from "normal" galaxies, although the contribution is more likely less than this figure. Direct observations can test these predictions.

Inverse Compton models, such as that described above for Cen A, appear most promising for the radiation process dominating the hard x-ray and γ-ray spectra of compact nuclei of active galaxies. Elvis et al. (1977) point out that self-Compton scattering of the infrared non-thermal (Synchrotron) IR spectra of Seyfert galaxies can account for their x-ray emission, and Mushotzky, Baity and Peterson (1977) arrive at similar conclusions for the hard x-ray (∼100 keV) spectra they have already detected from several active galaxies. These models would predict spectra extending to at least ∼10-20 MeV, and possibly beyond 100 MeV, depending on the IR spectrum cutoff. They would also predict source variability on time scales appropriate to the inferred source size (∼ light days).

The nature of the central object in the nuclei of active galaxies remains one of the most exciting problems in astrophysics. Fabian et al. (1976) and Meszaros and Silk (1977) have provided recent versions of a class of models which continues to satisfy the data. This class is based on accretion onto a massive ($\sim 10^8$ M_0) black hole in the galactic nucleus. Both of these models would predict thermal spectra extending up to at least ∼1 MeV. Lovelace (1976) has developed a dynamo model which can account for the apparently diametrically directed beams of high energy cosmic rays that supply the lobes of double radio sources surrounding the nuclei of active galaxies. Particles are accelerated to high energies in the electric fields generated in an accretion disk surrounding a massive black hole and may be expected to convert an appreciable fraction of their luminosity ($\sim 10^{44}$ erg s^{-1})

into high energy γ-rays by collisions (and subsequent π° decay) with

the ambient medium, as well as curvature radiation.

Regardless of the detailed model, it seems certain that cosmic

rays are being accelerated somehow by the central objects (be they

massive black holes, spinars, etc.) in active galaxies and quasars and

that gravitational energy is ultimately involved. The future detection

of the spectra of the γ-rays, which are almost certainly being

produced there, will clarify the nature and perhaps detailed physics

of the central sources.

VI. THE SHUTTLE, SPACELAB AND THE FUTURE

A. Introduction

Many of the instruments needed to make the next step in γ-ray

astronomy are necessarily quite large, and it is fortunate that the

Shuttle with its large capability is now coming into being. The

Spacelab is attractive to γ-ray astronomers from two points of view:

first, it provides the opportunity to fly relatively large instruments,

hopefully fairly frequently, to make specific studies of selected objects

or regions of the galactic plane in great detail and to study newly-

discovered phenomena with specialized instruments. The pointing

requirements are modest and can be satisfied by very simple platforms, or

in some cases by the Shuttle itself. Secondly, while also obtaining data,

Spacelab missions permit the thorough study in a space environment of

the response of limited versions of the very large complex instruments

to be flown on free flyers. The instruments to be flown on free flyers

are ones related to experiments whose objectives are sufficiently broad
as to need several months to more than a year to complete their observing
program, or ones which need to be away from the large mass of the Shuttle

In the next paragraphs some of the instruments used in γ-ray
astronomy will be described. To place the current status in context,
some historical background will be given in some cases. Following this
description of the γ-ray instrument, there will be a short discussion
of the promise of the Shuttle era.

B.   Gamma-Ray Detectors

The detection methods for astrophysical γ-rays are considerably
different than those characteristic of the more classical systems of
optical astronomy. Methods based on the wave properties of the incident
radiation are not possible. The γ-radiation is detected through its
interaction with matter wherein the γ-ray photon is removed individually
from the incident beam in a single event. The designs of the γ-ray
telescopes necessarily depend strongly on the character of these interactions. In general, γ-ray instruments, as a result, look more like
charged particle detectors than optical telescopes.

There are basically three processes by which γ-rays may interact
with matter. At photon energies below 0.5 MeV, the most probable interaction is the photo-electric effect. The photon is completely absorbed
by, and all its energy is transferred to an atomic electron. This
electron is ejected from the atom with a kinetic energy equal to the
difference between the photon energy and the binding energy of the
electron in the atom. In the energy region from 1 to 5 MeV, the main
form of photon interaction is the Compton interaction. An incident

photon interacts with an electron, but its energy is not completely

absorbed. Part of the energy goes to an electron as kinetic energy and

the rest is scattered as a "degraded" or lower energy photon. Energy

and momentum are conserved in the process.

A third interaction process which is possible for γ-rays above

1.02 MeV is that of pair production. Here the photon is absorbed in

the electric field of the nucleus, producing both a positive and

negative electron. The photon energy goes primarily into the rest mass

of the two electrons (0.51 MeV each) and their kinetic energies with a

very small part going to the kinetic energy of the nucleus. This process

cannot occur, of course, for photons with energies less than 1.02 MeV,

but increases in probability with photon energy.

Both the photoelectric effect and pair production increase relative

to the Compton effect with increasing atomic number of detector material.

A detector containing high Z elements, therefore favors the complete

absorption of γ-rays, since the photoelectric effect leaves easily

absorbed radiation and pair production leaves two electrons. The Compton

process, on the other hand, since it involves only partial absorption,

more readily allows for some of the γ-ray energy to escape.

The relative magnitude of the various interaction processes described

above for a given γ-ray energy region strongly affects the choice of

detector materials and design. In the following, a number of detector

designs will be discussed.

1. The Low ($\sim$0.3 to 10 MeV) Energy Region

In this section, possible designs of detector systems for per-

forming observations of the differential energy distribution, of the

anisotropy, of the temporal variations, and of the spatial source
distribution are considered. It is rather difficult, if not impossible,
to design a single detector system to perform all of the types of measure
ments just enumerated. Before the total system can be discussed, an
understanding of the central γ-ray detectors is required.

a. Detectors. Two basic types of detectors are used: scintillation and solid state. Gamma rays and x rays passing through matter
transfer part or all of their energy to electrons mainly by the
processes mentioned in the introduction. These secondary electrons
dissipate their energy in turn by ionization or excitation of molecules.
In the case of the scintillation detector, a phosphor is used in which
the de-ionization and de-excitation of the molecules results in the
emission of a fluorescent radiation. The number of photons produced (the
intensity) is proportional to this energy loss. Since the direction
of these photons are random, the scintillator is generally surrounded
by a reflector which maximizes the number of photons collected on a
photosensitive cathode. This cathode is part of a photomultiplier.
The γ-ray interaction is finally observed as a voltage pulse at the
anode of the photomultiplier tube and the amplitude of the voltage
output is approximately proportional to the energy loss in the scintillation detector.

The process in solid-state material is initially the same, but
collection of the analogue voltage output is different. Gamma radiation
ionizes the detector material thus producing electrons with kinetic
energy which is transferred to electrons in the valence bands of the
solid state materials. An applied electric field sweeps the charges

excited from the valence bands through the detector material so that an
electrical pulse is obtained corresponding to the detection of a single
γ-ray photon. The number of secondary electrons excited from the valence
band to the conduction band will be proportional to the kinetic energy
imparted to primary electrons produced by the γ-ray in its initial interaction with the crystal. Thus the electrical output is detected as a
voltage pulse whose magnitude is proportional to the amount of energy
lost as ionization by the γ-ray. The solid state detector, usually Ge,
used for γ-ray spectroscopy must be operated at cryogenic temperature.

The major difference in use of the two types of detectors is
reflected in their energy resolution capability. The energy resolution
in Ge solid state material is about one or two orders of magnitude better
than that for NaI(Tl) scintillation detectors ($\sim$ 2 keV resolution at
about .661 MeV for Ge and $\sim$ 50 keV for NaI(Tl). Resolution is energy
dependent and thus the improvement does depend on energy.

Once the voltage pulse is produced at the output of the detector
system this pulse is amplified, and, by the use of an analogue to digital
converters, energy loss (or pulse height) distribution are obtained.
The shape of the detector response to a monoenergetic γ-ray is not a
line, but can be made up of lines and continuum depending on the initial
energy of the γ-ray and detector size and material. Details of the
spectral shape can be found in a paper by Neiler and Bell (1968).

The choice of detector materials will strongly depend on the type
of measurement required. The higher density scintillation detector
will have significantly lower partial absorptions and thus lower continuum. Ge detectors and solid state detectors are usually smaller and

less dense and will produce responses with more partial absorption
effects. Thus, when one is interested in measuring the shapes of continuous distributions as if the case for measurement of the diffuse
spectrum, one would prefer using scintillation detectors. There is
the least distortion of the continuum by this method. On the other
hand, if discrete lines spectra are to be determined with high accuracy,
the Ge detectors are far superior.

b) Background. The optimum design, location, and data analysis
of spectral information for γ-ray astronomy experiments are strongly
influenced by the levels of background radiation produced by interaction of the spacecraft and detectors materials with the space particle
environment and the natural radioactivity present.

The most important sources of background are the following:

(1) Atmospheric secondary neutrons and γ-rays.

(2) Spacecraft background which can be considered due to three
sources: natural background, cosmic ray primary and secondary radiation,
and trapped particle primary and secondary radiation.

(3) Induced activity in the detector due to the ambient proton and
neutron flux in the spacecraft environment.

(4) Charge particle interactions in the detector

Details of these components are considered by Trombka et al. (1977a).
The fourth effect can be almost completely eliminated by active shielding
methods (Harrington et al. 1974).

The background particularly plagues the low energy region (γ-ray
energies up to $^\sim$8 MeV). Measurements made early in the Space Shuttle
and Spacelab flight programs can help define the electromagnetic and

particle fluxes. Both passive and active shields can be designed and
tested during spaceflight prior to starting detailed astrophysical measurements. Information obtained from these earlier missions, both manned
and unmanned, is also important in itself.

c)   "Telescopes"---Measurement of angular distribution and
        source direction.

There are basically four techniques used for determining
the angular flux distribution: collimation utilizing active and passive
shielding, Compton telescopes, shaped crystals, and the gammascope.

In the collimator type of system, the detector crystal
is placed inside a large shield. The shield absorbs some of the γ-radiation
coming from any direction except through the collimator. The shields
can also be used in the so-called active manner; that is dense scintillators are used as shields, but operate in an anticoincidence manner.
Thus, if a γ-ray is detected in the shield it is rejected so that only
γ-radiation coming through unshielded sections (i.e. through the
collimator holes is detected. Examples of such systems are described
by Matteson (1974) and Chupp (1976).

The Compton telescope utilized two detectors separated by a fixed
distance which determines the angular resolution. Only γ-rays which
produce a Compton scatter in the first detector and then an interaction
in coincidence in the second detector are recorded. The amount of
energy lost in each detector is also recorded. In this way the direction and some energy information concerning the incident γ-radiation
can be determined. Examples of such telescopes are given by Chupp (1976).

A detector can be shaped so that the counting efficiency strongly

depends on the angular relationship between the incident γ-ray flux and
the principal axis of the detector. As the detector is rotated, the
count rate will be modulated depending on the anisotropy of the flux or
the presence of a source in some given direction. The modulation of
the count rate as a function of crystal rotation is then used to measure
flux isotropy   or source position. This system is described in the
work of Trombka et al. (1974).

Finally, the gammascope has been used mainly in medical application,
but should find application on Spacelab flights. In this type of imaging
device a multi-hole collimator (either active or passive) is placed
in front of a large diameter thin scintillation detector. The crystal
is made thin so as to minimize the number of interactions a given γ-ray
will undergo in the crystal. In fact the ideal situation is that only
a single interaction will occur. The maximum light intensity for a
given interaction will be observed at the site of this interaction.
A series of photomultiplier tubes is used to determine the distribution of light intensity for each interaction. By centroid methods
the position of maximum intensity can be determined. This in turn
indicates the position in the crystal where the γ-ray interacted and
because of the collimator should allow for γ-ray imaging. Discussion
of the gammascope used for radioisotope scanning in medicine is given
by Herstel (1972).

Data interpretation and background problems for the systems
described above are extremely complex. The use of any of the systems
described above or combinations of them during manned Spacelab missions
shows great promise since human intervention can greatly simplify the

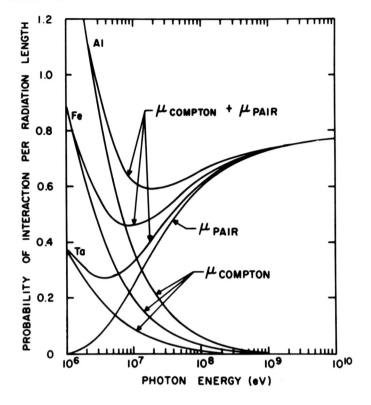

Fig. 10 - Conversion probability per radiation length for the Compton
and pair production processes for three different materials.

system design and allow for careful tuning and calibration of the
spectrometers and telescopes. Also, the system are often large
requiring the weight lifting capability of the shuttle.

2. The Medium (7 to 50 MeV) and High (> 50 MeV) Energy Regions .

In the detection of γ-rays, the transition from the low energy to
medium energy range at about 5 to 10 MeV is a very important one. As
illustrated in Figure 10, with an appropriately designed detector using
high z material the predominate interaction changes from the Compton

process to e^+, e^- pair production. There are many advantages related to the pair production interaction for the design of a detector system. First, all of the energy is converted from electromagnetic form to charged particle form at one time so there is no need to be concerned about an additional attempt to estimate the secondary photon's properties as is necessary in the Compton interaction regime. Since both secondary products are charged particles their properties can be measured with charged particle detectors and hence the energy and direction of the primary γ-raycan be estimated in a straightforward manner. The electron and positron have a relatively strong memory of the parent γ ray's direction, particularly in the high energy ($>$ 50 MeV) region. Finally, the pair production interaction is unique, and hence a very positive γ-ray identification exists so that with relatively sophisticated detector systems the background can be made to be effectively non-existent.

Whereas the γ-ray instruments to be used on Spacelab will almost certainly be of the more sophisticated pictorial variety, it would be inappropriate not to mention that these evolved only after a long development period in the 1960's. Prior to their development several scintillator-Cerenkov detector telescopes were developed, and one flown on OSO-3 by Kraushaar et al. (1972),   led to the first certain detection of celestial γ-rays. The instrument consisted of a scintillator converter sandwich, a lucite Cerenkov counter, and a combination NaI and metal sheet energy discriminator surrounded by a plastic scintillator anticoincidence system. The results of this experiment and the pictorial spark chamber telescopes of several types being flown

on balloons in the 1960's indicated that the intensities of the γ-rays
were sufficiently small both in absolute terms and relative to the
cosmic rays that larger detectors which could unambiguously identify
the γ-ray were required. Further, they had to be flown outside the
earth's atmosphere because the secondary flux from even the small
amount of residual atmosphere above the highest altitude research
balloons obscured all, but the few strongest sources.

The next major step in high energy γ-ray astronomy occurred with
the γ-ray telescope flown on SAS-2 (Derdeyn et al. 1972 and Fichtel et
al. 1975), from which the first detailed picture of the galactic plane
emerged revealing its structure and several previously unknown point
sources, as well as the two component diffuse radiation. The SAS-2
telescope used a four-element scintillator, directional Cerenkov trigger
system, but its most important feature was a 32 level magnetic core
spark chamber system which provided an electronically digitized picture
of each γ-ray event in three dimensions. The spark chambers were inter-
leaved with thin (0.03 radiation lengths), high z metal plates to provide
conversion material for γ-rays, but minimal electron scattering before
information on the initial electrons' direction was obtained. This
picture then provided both the unambiguous identification of the γ-ray
and information on its energy and arrival direction. Figure 11 provides
a schematic diagram of the SAS-2 detector which is shown because it is
the model for the most likely Spacelab instruments in the medium and
high energy range. More recently a very similar instrument has been
flown on COS-B (Bennett et al. 1976). The major differences are that
the COS-B instrument has an energy measuring scintillator below the

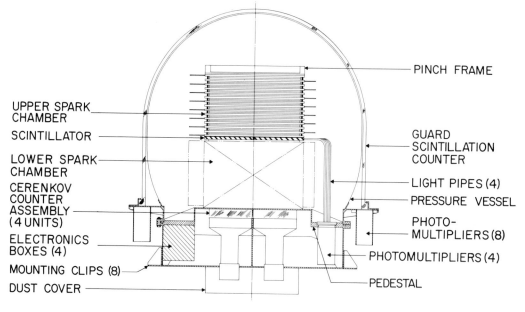

UPPER SPARK CHAMBER

SCINTILLATOR

LOWER SPARK CHAMBER

CERENKOV COUNTER ASSEMBLY (4 UNITS)

ELECTRONICS BOXES (4)

MOUNTING CLIPS (8)

DUST COVER

PINCH FRAME

GUARD SCINTILLATION COUNTER

LIGHT PIPES (4)

PRESSURE VESSEL

PHOTO-MULTIPLIERS (8)

PHOTOMULTIPLIERS (4)

PEDESTAL

SAS-B GAMMA RAY EXPERIMENT

Fig. 11 - SAS-2 γ-ray telescope.

Cerenkov counter, but does not have spark chambers between the scintillator and the Cerenkov elements.

The γ-ray instruments appropriate for Spacelab will most likely
have several improvements over the earlier instruments including better
energy and angular resolution, and probably two or more different
instruments together will be able to cover a much broader range in
energy than has been possible thus far. However, by far the most
important step to be taken is an increase in sensitivity. The major
limitation which prevented SAS-2 and COS-B from making even a greater
contribution than they have was the limited number of photons collected.
Major increases in the collecting area are essential for the next steps

in high energy γ-ray astronomy. Medium energy γ-ray astronomy, which
has not yet really had an opportunity, should make a very important
step with Spacelab. In the medium energy range, the intensities are
high enough that pictorial detectors in the 0.5 to 1.0 m^2 class can
make a major contribution both in studies of point sources and the
electromagnetic diffuse radiation.

For a point source study, the solid directional Cerenkov system
has sometimes been replaced with a gas Cerenkov system on balloon
instruments. It has the advantages of simplicity for large sizes and
a natural low energy threshold. For unambiguous γ-ray identification
a pictorial element would still be required.

Driven both by the desire to have a very good direction measurement (The solid directional Cerenkov counts have a relatively poor
"backward" rejection efficiency.) and to be able to include easily a
a good energy measurement device, several groups have developed time-of-
flight systems, with a space quality time-of-flight system having already
been developed by Ross (1977). The basic two component scintillator
system replaces the scintillator-Cerenkov system and can be used in a
multi-element array to obtain the desired solid angle area. The time-
of-flight system, as the name implies, measures the direction by
determining the time difference between the two scintillators. There
are several different approaches which have relative merits and disadvantages depending on the specific application. One system, for
example, is the Charpak time-of-flight one, which has the advantage of
somewhat improved directionality for small distances, but the disadvantage of more power and lower long-term reliability.

Another important aspect of future experiments is a good energy measurement. The large weight capability of Spacelab will allow the addition of large crystals to measure accurately the energy of the γ-ray and hence the energy spectra of point sources and the diffuse radiation in various directions, both of which are very important in our further understanding of the production mechanisms.

It is important to remember that the low intensities of γ-rays mean that only certain experiments are appropriate for Spacelab application. However, not only are these of substantial interest for the scientific results they will produce, but also from the standpoint of instrument development for future free-flying satellites.

3.   The Very High Energy ($> 10^{11}$ eV) Region

Whereas at least, for the moment, this region of energy seems to be limited to being studied from detector systems on the ground, because of the interrelationship of this energy range with the others, it seems appropriate to discuss the detection techniques briefly.

At these energies, $\gtrsim 10^{11}$ eV cosmic γ-rays may be detected optically by the atmospheric Cerenkov radiation produced in the accompanying extensive air shower, EAS, (Jelley 1958). The disk of optical Cerenkov photons ($\sim$ 150 m radius, $\sim$ 2 m thickness) produced in such a shower is sufficiently dense ($\sim$ 5 photons m^{-2}) that it may be detected as a $\sim$ 10 nsec light flash at the focus of a large optical light collector such as the 10 m reflector at Mt. Hopkins Observatory. Since the optical Cerenkov photons are largely emitted within $\sim$ 2° of the primary direction, single (or coincidence arrays) of optical collectors have been pointed directly at suspected sources in either drift scan or tracking modes in the early searches for cosmic γ-ray sources at $\geqslant 10^{11}$-10^{12}eV

(e.g. Chudakov et al. 1965; Weekes et al. 1972).

Though positive results on the Crab pulsar (Porter, Delaney and

Weekes, 1974; Erickson, Fickle and Lamb, 1976) and the periodic (4.8 hr.)

X-ray source Cyg X-3 (Stephanian et al., 1975) have been reported,

it was clear that higher sensitivity and some method of cosmic ray EAS

rejection was needed.

O'Mongain et al. (1968) developed one such method which relied on the

expected (Rieke 1969) very fast (< 3 n sec) Cerenkov pulse component

for γ-rays EAS (from electrons above the shower maximum) but

not for cosmic ray events. Positive effects on the pulsar PSR0950+08

were reported.

However, the major study of cosmic ray vs. γ-ray EAS and their

Cerenkov light distributions was conducted by Grindlay (1971) and

resulted in the so-called "double beam" technique. This enabled partial

identification of cosmic ray EAS by separate detection of the muon

(and nuclear) cores vs. electron maximum in each event. EAS with a muon

core component could be rejected from the analysis for a γ-ray flux from

a given source. This technique resulted in detection of a variable phased

flux from the Crab pulsar at $\geqslant$ 800 GeV (Grindlay, Helmken and Weekes,

1976) and the first detection of an extragalactic source (Cen A)

at very high energies (Grindlay et al. 1975). A possible variable

flux from the Vela pulsar was also detected in the latter experiment.

More recent evidence for time variable emission at > 500 GeV from

the Crab pulsar and PSR0950+08 have been reported by Gupta et al. (1977)

using a multi-coincidence system. Thus it is clear that pulsars and active galaxies are detectable at very high energies, but that more

sensitive techniques are needed for detection of all but the most

important sources.

Detection of angular and spatial structure in the shower light

pool by using a multiple detector array version of the double beam

technique is perhaps the best approach (Grindlay 1976). Weekes and

Turver (1977) have proposed a system incorporating multiple detectors

on ≥ 2 separated large reflectors that would enable the Cerenkov light

structure to be measured (and cosmic ray EAS partially rejected at

high energies), as well as allow the energy threshold to be reduced

below 100 GeV. Since average proton-initiated EAS produce almost no

detectable Cerenkov photons at these lowest energies, the cosmic ray

background would be reduced to nearly the ($\sim$ 1%) cosmic ray electron

spectrum.

Thus the entire γ-ray spectrum from $\sim$ 30 to 10^4 GeV could be explored by

ground-based experiments with $\geq$ 10 times the sensitivity of previous

experiments and would complement proposed satellite experiment surveys

through $\sim$ 10 GeV.

C. Summary

It is hoped that the impact that γ-ray astronomical observations

will have on our understanding of the dynamic processes involved in the

evolution of our solar system, galaxies, and the universe has been made

clear. Information can be obtained on the most energetic interactions

originating in our galaxy and beyond. These types of observations pro-

vide the most direct data relating to the study of the largest transfers

of energy occurring in astrophysical processes. Furthermore, studies carried out in γ-ray astrophysics should yield the most direct measurement of the presence and dynamic effects of the energetic charged cosmic ray particles and of element synthesis. Gamma rays may survive billions of years and still reveal their source. Thus, this portion of the electromagnetic spectrum presents a window for obtaining information on the early history of our universe.

Because of the complexity of the instrumentation required and the low intensity of the fluxes, progress in the field has been rather slow. Furthermore, almost all observations except for those made at the highest energies must be obtained above the earth's atmosphere and away from the high intensity trapped radiation belts. The pioneering works carried out during the OSO-3, SAS-2, Apollo, and COS-B missions have been outlined above. These results indicate that indeed many of the goals for obtaining astrophysical results in high energy γ-ray astronomy are attainable.

Spaceflight opportunities during the Spacelab missions will provide a most logical extension of the research in this field. Gamma astrophysical measurements should enter the domain of detailed investigation during this time period. Advanced design spectrometers and telescopes will be available for use on the Spacelab. Such spaceflight opportunities will allow for short turn-around times wherein results obtained on earlier flights can be incorporated into later missions. In some instances, Spacelab will be a proving ground for large instruments to be flown on satellites. Further with Spacelab it will be possible to build large area detectors which will be capable of greatly improving our knowledge since at present one of the major hurdles to overcome in γ-ray astronomy is the low intensities.

This hurdle, however, can be overcome fully only with a large Shuttle-launched satellite. The goals of this Gamma Ray Observatory might reasonably be expected to be: (i) A study of the dynamic, evolutionary forces in compact objects such as neutron stars and black holes, as well as γ-ray emitting objects whose nature is yet to be understood; (ii) A search for evidence of nucleosynthesis--the fundamental building process in nature--particularly in the environment of supernovae; (iii) The study of intense γ-ray bursts of many types, whose origins remain a mystery; (iv) The exploration of our Galaxy in the γ-ray range particularly with regard to regions difficult to observe at other wavelengths, the origin and dynamic pressure effects of the cosmic rays, and structural features particularly related to high energy particles; (v) The study of the nature of other galaxies in the high energy realm and especially the extraordinary ones such as radio galaxies, Seyfert galaxies, and possible QSO's; (vi) The study of cosmological effects through the detailed examination of the diffuse radiation and the search for primordial black hole emission.

These scientific goals require a set of individual experiments that include the capability to carry out: (i) A survey of γ-ray sources and diffuse emission with sensitivities of 10^{-7} to 10^{-8} photons/cm^2sec and energy resolution around 15 percent at energies above $\sim$ 30 MeV; (ii) A survey of γ-ray sources and diffuse emission with sensitivities around 10^{-5} photons/cm^2sec and energy resolution around 25 percent at energies between 0.1 and 40 MeV; (iii) Detection and identification of nuclear γ-lines in the 0.1 to 10 MeV energy region with an energy resolution of < 0.4 percent and sensitivities of the order of 10^{-5}

photons/cm^2sec. The initial subjects of observation will be the inter-

stellar medium and supernova shells; (iv) Observations of γ-ray bursts,

studies of their spectral and temporal behavior, and locations of the

sources to 0.1°. The technology to build the appropriate instruments

exists, and the development has already progressed to the point where

construction of flight units can occur.

Following this step, there should be more detailed studies of the

actual   objects observed with instruments designed for specific objectives.

The necessary instruments for this subsequent γ-ray satellite should

evolve to a significant degree from the Spacelab missions.

Acknowledgement.

We wish to thank Drs. Robert Hartman and David Thompson for their

careful reading of the text and helpful comments.

REFERENCES

Adler, I., Trombka, J., Gerard, J., Lowman, P., Schmadebeck, R.,

    Blodgett, H., Eller, E., Yin, L., Lamothe, R., Osswald, G.,

    Gorenstein, P., Bjorkholm, P., Gursky, H., and Harris, B. 1972,

    Science, 177, 256.

Albats, P., Frye, G. M., Jr., Zych, A. D., Mace, O. B., Hopper, V. D.,

    and Thomas, J. A. 1972, Nature, 240, 221.

Albats, P., Frye, G.M., Jr., Thomson, G. B., Hopper, V. D., Mace, O. B.,

    Thomas, J. A., and Staib, J. A. 1974, Nature, 251, 400.

Arnett, W. D. 1969, Ap. J., 157, 1369.

Arnett, W. D. 1975 , Ap. J., 195, 727.

Arnett, W. D. 1976, 8th Texas Symposium, in press.

Arnett, W. D., 1977, Ap. J. 218, in press.

Arnold, J. R., Metzger, A. E., Anderson, E. C., and Van Dilla, M. A.
    1962, J. Geophys. Res., 67, 4876.

Arons, J., and McCray, R. 1969, Ap. J., 158, L91.

Baldwin, J. E. 1967, I.A.U. Symposium No. 31, Radio Astronomy and the
    Galactic System, 337.

Baldwin, J. E. 1976, in The Structure and Content of the Galaxy and
    Galactic Gamma Rays, NASA CP-002, p. 189.

Beall, J., Rose, W., Graf, W., Price, K., Dent, W., Hobbs, R., Conklin,
    E., Ulich, B., Dennis, B., Crannel, C., Dolan, J., Frost, K., and
    Orwig, L. 1977, Ap. J., in press.

Bennett, K., Bignami, G. F., Boella, G., Buccheri, P., Burger, J. J.,
    Cuccia, A., Hermsen, W., Higdon, J., Kanbach, G., Koch, L., Lichti,
    G. G., Masnou, J., Mayer-Hasselwander, H. A., Paul, J. A., Scarsi,
    L., Shukla, P. G., Swanenburg, B. N., Taylor, B. G., and Wills,
    R. D. 1976, in The Structure and Content of the Galaxy and Galactic
    Gamma Rays, NASA CP-002, p. 45.

Bennett, K., Bignami, G. F., Buccheri, R., Hermsen, W., Kanbach, G.,
    Lebrun, F., Mayer-Hasselwander, H. A., Paul, J. A., Piccinotti, G.,
    Scarsi, L., Soroka, F., Swanenburg, B. N., and Wills, R. D. 1977,
    in Proceedings of the 12th ESLAB Symposium, Frascati, European
    Space Agency, 83.

Bennett, K., Bignami, G.F., Boella, G., Buccheri, R., Hermsen, W.,
    Kanbach, G., Lichti, G.G., Masnou, J.L., Mayer-Hasselwander, H.A.,
    Paul, J.A., Scarsi, L., Swanenburg, B.N., Taylor, B.G., and Wills,
    R.D.   1977, Astron. Astrophys. 61, 279.

Bierman, L., and Davis, L. 1960, Zs. F. Ap., 51, 19.

Bignami, G. F., and Fichtel, C. E. 1974, Ap. J. (Letters), 189, L65.

Bignami, G. F., Fichtel, C. E., Kniffen, D. A., and Thompson, D. J. 1975, Ap. J., 199, 54.

Bratolyubova-Tsulukidze, L. I., Grigorov, N. L., Kalinkin, L. F., Melioransky, A. S., Pryakhin, Ye. A., Savenho, I. A., Yufarkin, V. Ya. 1971, Geomagnetism and Aeronomy (Soviet), 11, 585.

Brecher, K., and Morrison, P. 1969, Phys. Rev. (Letters), 23, 802.

Browning, R., Ramsden, D., and Wright, P. J. 1971, Nature, 232, 99.

Bui-Van, A., and Hurley, K. 1974, Ap. J. (Letters), 188, L51.

Burton, W. B. 1976, in The Structure and Content of the Galaxy and Galactic Gamma Rays, NASA CP-002.

Burton, W. B., Gordon, M. A., Bania, T. M., and Lockman, F. J. 1975, Ap. J., 202, 30.

Chudakov, A., Dadykin, V., Zatsepin, V., and Nesterova, N. 1965, Cosmic Rays, Proc. P. N. Lebedev Inst., No. 26, 99.

Chupp, E. L., Forrest, D. J., Suri, A. W., Tsai, C., and Dumphy, P. O. 1973, Nature, 241, 333.

Chupp, E. L., Forrest, D. J., and Suri, A. W. 1975, Proc. IAU COSPAR Sym. 68, Solar Gamma-, X-, and EUV Radiations, ed. by S. Kanes, p. 341.

Chupp, E. L. 1976, Gamma Ray Astronomy, D. Reidel Publishing Co., 260.

Clarm, D. H. and Caswell, J. L. 1976, Mon. Nat. R. Soc., 174, 267.

Clark, G. W., Garmire, G. P., and Kraushaar, W. L. 1968, Ap. J. (Letters), 153, L203.

Clayton, D. D. 1973, Explosive Nucleosynthesis, ed. Schramm, D. N. and Arnett, W. D. (Univ. Texas Press, Austin), 264.

Clayton, D. D., and Hoyle, F. 1974, Ap. J. (Letters), 181, L101.

Cline, T.L.   1978, private communication.

Cline, T.L., and Schmidt, W.K.H.   1977, Nature 266, 749.

Colgate, S. A.   1975, Ann. N. Y. Acad. Sci., 262, 34.

Cowsik, R., and Kobetich, E. J. 1972, Ap. J., 177, 585.

Crannell, C. J., Ramaty, R., and Crannell, H. 1977, Proc. of 12th ESLAB
    Symposium, ESA SP-124, 213.

Crannell, C. J., Joyce, G., and Ramaty, R. 1976, Ap. J., 210, 582.

D'Amico, N. and the COS-B Collaboration, 1978, 6 th European Cosmic Ray
    Symp., Kiel.

Daniel, R. R., Joseph, G., and Lavakare, P. J. 1972, Ap. and Space Sci.,
    18, 462.

Daniel, R. R., and Lavakare, P.J. 1975, 14th Intl. Cosmic Ray Conf., 1, 23

Daugherty, J. K., Hartman, R.C., and Schmidt, P.J. 1975, Ap. J., 198, 493.

Derdeyn, S. M., Ehrmann, C. H., Fichtel, C. E., Kniffen, D. A., and
    Ross, R. W. 1972, Nucl. Instr. and Methods, 98, 557.

DiCocco, G. Boella, G., Perotti, F., Stiglitz, R., Villa, G., Baker, R.E.,
    Butler, R.C., Dean, A.J., Martin, S.J., and Ramsden, D. 1977,
    Nature 270, 319.

Dolan, J. F., and Fazio, G. G.   1965, Rev. of Geophys., 3, 319.

Downs, G. S., Reichley, P. E., and Morris, G. A. 1973, Ap. J., 181,
    L143.

Elvis, M., Macacaro, T., Wilson, A., Ward, M., Penston, M., Fosbury, R.,
    Perola, G. 1977, MNRAS, in press.

Erickson, R., Fickle, R., and Lamb, R. 1976, Ap. J., 210, 539.

Fabian, A., Maccagni, D., Rees, M., and Stoeger, W. 1976, Nature, 260,
    683.

Falk, S.W.   1978, to be published.

Falk, S. W., and Arnett, W. D. 1976, Ap. J., 210, 733.

Fazio, G. G., Helmken, H. F., O'Mongain, E. P., and Weeks, T. C. 1972,

   Ap. J. (Letters), 175, L-117.

Fazio, G., and Stecker, F. 1970, Nature, 226, 135.

Felten, J. E., and Morrison, P. 1963, Phys. Rev. Letters, 10, 453.

Fichtel, C. E. 1977, Space Science Reviews, 20, 191.

Fichtel, C. E., Hartman, R. C., Kniffen, D. A., Thompson, D. J.,

   Bignami, G. F., Ögelman, H. B., Özel, M. E., and Tümer, T. 1975,

   Ap. J., 198, 163.

Fichtel, C.E., Kniffen, D.A., and Hartman, R.C.   1973, Ap. J. (Letters)

   186, L99.

Fichtel, C. E., Kniffen, D. A., Thompson, D. J., Bignami, G. F., and

   Cheung, C. Y. 1976, Ap. J., 208, 211.

Fichtel, C. E., Kniffen, D. A., and Thompson, D. J. 1977, Proceedings

   of 12th ESLAB Symposium, Frascati, ESA SP 124, 95.

Fichtel, C. E., Hartman, R. C., Kniffen, D. A., Thompson, D. J.

   Ögelman, H. B., Özel, M. E., and Tümer, T. 1977, Ap. J. (Letters),

   217, L9.

Fichtel, C. E. Simpson, G. A., and Thompson, D. J. 1978, Ap. J. 222

   833.

Fishman, G. J., Harnden, F. R., Jr., Johnson, W. N. III, and Haymes,

   R. C. 1969, Ap. J.(Letters), 158, L61.

Freier, P., Gilman, C., and Waddington, C. J. 1977, Ap. J., 213, 588.

Fritz, G., Meekins, J. F., Chubb, T. A, Freidman, H., and Henry,

   R. C. 1971, Ap. J., 164, L55.

Frye, G. M., Jr., Albats, P. A., Zych, A. D., Staib, J. A., Hopper,
  V. D., Rawlinson, W. R., and Thomas, J. A. 1971, Nature, 231, 372.

Fukada, Y., Hayakawa, S., Kashara, I., Makino, F., and Tanaka, Y. 1975,
  Nature, 254, 398.

Galper, A. M., Kirillov-Ugryumov, V. G., Kurochklin, A. V., Leikov,
  N. G., Luhkov, B. I., and Yurkin, Yu. T. 1975, Proc. 14th International Cosmic Ray Conference, Munich, Vol. 1, p. 95.

Georgelin, Y.M., and Georgelin, Y. P. 1976, Astron. and Astrophys,
  49, 57.

Ginzburg, V. L., and Syrovatskii, S. I. 1963, The Origin of Cosmic
  Rays, Oxford, Pergamon Press.

Ginzburg, V. L., and Syrovatskii, S. I. 1965, Ann. Rev. Astron. 3, 297.

Gold, T. 1968, Nature, 218, 731.

Gold, T. 1969, Nature, 221, 25.

Gordon, M. A., and Burton, W. B. 1976, Ap. J., 208, 346.

Gram l , F., Penningsfeld, F. P., and Schönfelder, V. 1978, Proc. of
  Symp. on Gamma Ray Spectroscopy in Astrophysics, NASA–Goddard Space
  Flight Cntr., Aug. 1978, NASA TM 79619, p. 207.

Greisen, K., Ball, S. E., Jr., Campbell, M., Gilman, D., and Strickman,
  M. 1975, Ap. J., 197, 471.

Grindlay, J. 1971, Smithsonian Astro. Obs. Spec. Rept. No. 334.

Grindlay, J. 1975, Ap. J., 199, 49.

Grindlay, J. 1976, in The Structure and Content of the Galaxy and
  Galactic Gamma Rays, NASA CP-002, p. 81.

Grindlay, J., Helmken, H., Brown, R. H., Davis, J., and Allen, L. 1975,
  Ap. J. (Letters) 197, L9.

Grindlay, J., Helmken, H., and Weekes, T. 1976, Ap. J., <u>209</u>, 592.

Gupta, S., Ramana Murthy, P., Sreckantan, B., and Tonwar, S. 1977,
    Preprint.

Hall, A., Meegan, C., Walraven, G., Djnth, F., and Haymes, R. 1976,
    Ap. J., <u>210</u>, 631.

Harrington, T. M., Marshall, J. H., Arnold, J. R., Peterson, L. E.,
    Trombka, J. I., and Metzger, A. E. 1974, Nucl. Instr. Methods, <u>118</u>,
    401.

Hayakawa, S., Ito, K., Matsumoto, T., Ono, T., and Uyama, K. 1976,
    Nature, <u>261</u>, 29.

Helmken, H. and Hoffman, J. 1973, Proc. 13th Internat. Cosmic Ray Conf.,
    <u>1</u>, 31.

Helmken, H. F., Grindlay, J. E., and Weeks, T. C. 1975, 14th Internat.
    Cosmic Ray Conference, <u>1</u>, 123.

Hermsen, W., Swanenburg, B. N., Bignami, G. F., Boella, G., Buccheri, R.,
    Scarsi, L., Kanbach, G., Mayer-Hasselwander, H. A., Masnou, J. L.,
    Paul, J. A., Bennett, K., Higdon, J. C., Lichti, C. G., Taylor, B.G.,
    and Wills, R. D. 1977, Nature, <u>269</u>, 494.

Herstel, W. 1972, Adv. Electronics & Electron Physics, <u>33B</u>, p. 1041.

Herterich, W., Pinkau, K., Rothermel, H. and Sommer, M. 1973, 13th
    Internat. Cosmic Ray Conf., <u>1</u>, 21.

Hewish, A., Bell, S. J., Pilkington, J. D., Scott, P. F., and Collins,
    R. A. 1968, Nature, <u>217</u>, 709.

Hopper, V. D., Mace, O. B., Thomas, J. A., Albats, P., Frye, G. M., Jr.,
    Thomson, G. B., and Staib, J. A. 1973, Ap. J. Letters, <u>186</u>, L55.

Jelly, J. 1958, <u>Cerenkov Radiation and Its Applications</u>, New York,
    Pergamon Press.

Jokippi, J. R. 1976, Ap. J. 208, 900.

Kettenring, G., Mayer-Hasselwander, H. A., Pfefferman, E., Pinkau, K.,
    Rothermel, H., and Sommer, M. 1971, Proc. 12th Internat. Conf. on
    Cosmic Rays, 1, 57.

Kinzer, R. L., Share, G. H., and Seeman, N. 1973, Ap. J., 180, 547.

Klebesadel, R. W., Strong, I. B., and Olsen, R. A. 1973, Ap. J. (Letters),
    182, L85.

Kniffen, D. A., Fichtel, C. E., and Thompson, D. J. 1977, Ap. J. 215,
    765.

Kniffen, D. A., Hartman, R. C., Thompson, D. J., Bignami, G. F.,
    Fichtel, C. E., Ögelman, H., and Tümer, T. 1974, 251, 397.

Koch, M. W., and Motz, J. W. 1959, Rev. Mod. Phys., 31, 920.

Kraushaar, W. L., and Clark, G. W., 1962, Phys. Rev. (Letters), 8, 106.

Kraushaar, W. L., Clark, G. W., Garmire, G. P., Borken, R., Higbie, R.,
    Leong, V., and Thorsos, T. 1972, Ap. J. 177, 341.

Kuo, Fu-Shong, Frye, G. M., Jr., and Zych, A. D. 1973, Ap. J. (Letters),
    186, L51.

Kurfess, J. D. 1971, Ap. J. (Letters), 168, L39.

Landecker, T. L., and Wielebinski, R. 1970, Australian J. Phys. Suppl.,
    16, 1.

Lasher, G. 1975, Ap. J., 201, 174.

Lawrence, A., Pye, J., and Elvis, M. 1977, MNRAS, in press.

Lee, T., Papanastassiou, D. A., and Wasserburg, G. J. 1976, Ap. J.
    (Letters), 211, 107.

Leventhal, M., MacCallum, C., and Watts, A. 1977, Ap. J., 216, 491.

Lingenfelter, R. E., and Ramaty, R. 1967, High Energy Nuclear Reactions
    in Astrophysics, ed. B.S.P. Shen (New York; Benjamin), 99.

Lovelace, R. 1976, Nature, 262, 649.

Matteson, J. L. 1974, Proceedings Conference on Transient Cosmic Gamma-
    and X-Ray Sources, I. B. Strong, Ed. LA-5505-C, Los Alamos Scien-
    tific Laboratory, p. 237.

Masnou, J.L. Bennett, K., Bignami, G.F., Buccheri, R., Caraveo, P.,
    D'Amico, N., Hermsen, W., Kanbach, G., Lichti, G.G., Mayer-Hasselwander,
    H.A., Paul, J.P., Swanenburg, B.N., and Wills, R.D. 1977, Proceedings
    of the 12th ESLAB Symposium, Frascati 24-27 May, 1977, ESA SP124.

Maraschi, L., and Treves, A. 1977, Astron. and Astrophys. 61, L11.

Mazets, E.P., Golenetskii, S.V., Il'inskii, V.N., Gur'yan, Yu. A.,
    and Kharitonova, T.V. 1975, Ap. and Sp. Sci. 33, 347.

Mazets, E.P., Golenetskii, S.V., and Il'inskii, V.N. 1977, Astrophys.
    Letters 18, 155.

McBreen, B., Ball, S.E., Jr., Campbell, M., Greisen, K., and Koch, D.
    1973, Ap. J. 184, 571-580.

Meszaros, P., and Silk, J. 1977, Astron. and Astrophys., in press.

Metzger, A.E., Anderson, E.C., Van Dilla, M.A., and Arnold, J.R.
    1964, Nature 204, 766.

Metzger, A. E., Trombka, J. I., Peterson, C. E., Reedy, R. C., and
    Arnold, J. R. 1973, Science, 1979, 800.

Montemerle, T. 1977, Ap. J., 216, 620.

Morrison, P. 1958, Nuovo Cimento, 7, 858.

Morrison, P., and Sartori, L. 1969, Ap. J., 158, 541.

Mushotzky, R. 1976, Ph.D. Thesis, U. of Calif., San Diego.

Mushotzky, R., Baity, W., and Peterson, L. 1977, Ap. J., 212, 22.

Neiler, J. H., and Bell, P. R. 1968, Alpha-, Beta-, and Gamma-Ray
    Spectroscopy, Vol. 2, ed. Kai Siegbaum, North Holland Pub. Co.,
    p. 245.

Ögelman, H., Fichtel, C. E., Kniffen, D. A., and Thompson, D. J. 1976,
    Ap. J., 201, 584.

Omnes, R. 1969, Phys. Rev. Letters, 23, 38.

O'Mongain, E., Porter, N., White, J., Fegan, D., Jennings, D., and
    Lawless, B. 1968, Nature, 219, 1348.

Orwig, L. E., Chupp, E. L., and Forrest, D. J. 1971, Nature Phys. Sci.,
    231, 171.

Östriker, J. P., and Gunn, J. E. 1969, Ap. J. 157, 1395.

Page, D. N. and Hawking, S. W. 1976, Ap. J., 206, 1.

Papaliolios, C. and Carelton, N. P. 1970, The Crab Nebula, ed. R. D.
    Davies and E. G. Smith), D. Reidel, Dordrecht, 142.

Parker, E. N. 1966, Ap. J., 145, 811.

Parker, E. N. 1969, Space. Sci. Rev., 9, 654.

Parlier, B., Agrinier, B., Forichan, M., Leroy, J. P., Boella, G.,
    Maraschi,L.,Buccheri, R., Robba, N. R., and Scarsi, L. 1973,
    Nature Phys. Sci., 242, 117.

Parlier, B., Forichon, M., Montmerle, T., Agrinier, B., Boella, G.,
    Scarsi, L., Niel, M., and Palmeira, R. 1975, 14th Internat. Cosmic
    Ray Conf., 1, 14.

Paul, J., Cassé, M., and Cesarsky, C. J. 1974, Proceedings of 9th ESLAB
    Symposium, ESRO SP-106, 246.

Paul, J., Cassé, M., and Cesarsky, C. J. 1975, Proc. 14th Internat. Cosmic Ray Conf., 1, 59.

Paul, J., Cassé, M., and Cesarsky, C. J. 1976, Ap. J., 207, 62.

Peterson, L. E. 1966, in Space Res., Proc. Internat. Space Science Symp., 6, 53, MacMillan, N. Y.

Piccinotti, G., and Bignami, G. F. 1976, Astron. and Astrophys., 52, 69.

Porter, N., Delaney, T., and Weekes, T. 1974, Proc. ESLAB Symp. on Gamma Ray Astronomy, Frascati, 295.

Porter, N., and Weekes, T. 1977, Preprint.

Price, R.M. 1974, Astr. and Astrophys. 33, 33.

Puget, J. L., Ryter, C., Serra, G., and Bignami, G. 1976, Astron. and Astrophys., 50, 247.

Ramaty, R. 1974, High Energy Particles and Quanta in Astrophysics, ed. by F. B. McDonald and C. E. Fichtel, MIT Press, Cambridge, MA, Chapter III.

Ramaty, R. and Crannell, C. J. 1976, Ap. J., 203, 766.

Ramaty, R., Kozlosky, B., and Lingenfelter, R. E. 1975, Space Science Reviews, 18, 341.

Ramaty, R., and Lingenfelter, R. E. 1973, High Energy Phenomena on the Sun, ed. R. Ramaty and R. G. Stone, NASA Sp-342 (Washington, NASA) p. 301.

Ramaty, R. and Lingenfelter, R. 1977, Preprint.

Rankin, J. M. Comella, J. M., Craft, H. D., Jr., Richards, D. W., Campbell, D. B., and Counselman, C. C. III, 1970, Ap. J., 162, 707.

Rieke, G. 1969, Smithsonian Astro. Obs. Sp. Report No. 301.

Rocchia, R., Ducros, R., and Goffet, B. 1976, Ap. J. 209, 350.

Ross, R. 1977, to be published.

Ruderman, M. A., and Sutherland, P. G. 1975, Ap. J., 196, 51.

Savage, B. D., Bohlia, R. C., Drake, J. F., and Budich, W. 1977, "I.
A Survey of Interstellar Molecular Hydrogen", Wisconsin Astrophysics
Preprint Number 46.

Schlickeiser, R., and Thielheim, K. O. 1974, Astron. and Astrophys.,
34, 169.

Schönfelder, V., Lichti, G., Daugherty, J., and Moyano, C. 1975, 14th
Internat. Cosmic Ray Conf., 1, 8.

Schramm, D. N. 1975, Ann. N. Y. Acad. Sci., 262, 65.

Schwartz, D. A., and Gursky, H. 1973, in Gamma Ray Astrophysics
(eds. F. W. Stecker and J. I. Trombka) NASA SP-339, U. S. Govt.
Printing Office.

Scoville, N. Z., and Solomon, P. M. 1975, Ap. J. (Letters), 199, L105.

Seiradakis, J. 1976, in The Structure and Content of the Galaxy and
Galactic Gamma Rays, NASA CP-002, p. 265.

Share, G. H., Kinzer, R. L., and Seeman, N. 1974, Ap. J., 187, 511.

Silk, J. 1970, Space Sci. Rev., 11, 671.

Silk, J., and McCray, R. 1969, Astrophys. Letters, 3, 59.

Simonson, S. C., III 1975, Ap. J. Letters, 201, L103.

Simonson, S. C., III 1976, Astron. & Astrophys., 46, 261.

Sparks, W. M., Starrfield, S., and Trurian, J. W. 1976, Ap. J., 208,
819 and references therein.

Stecker, F. W. 1959, Ap. J. 157, 507.

Stecker, F. W. 1973, Ap. J., 185, 499.

Stecker, F. W. 1975a, in <u>Origin of Cosmic Rays</u> (J. Osborne and A. Wolfendale, eds.) Reidel, 267.

Stecker, F. W. 1975b, Phys. Rev. Letters, <u>35</u>, 188.

Stecker, F. W. 1976, in <u>The Structure and Content of the Galaxy and Galactic Gamma Rays</u>, NASA CP-002, 315.

Stecker, F.W. 1977, Ap. J. <u>212</u>, 60.

Stecker, F. W., Morgan, D. L., and Bredekamp, J. 1971, Phys Rev. Letters, <u>27</u>, 1469.

Stecker, F. W., and Puget, J. 1972, Ap. J., <u>178</u>, 57.

Stecker, F. W., Solomon, M. M., Scoville, N. Z., and Ryter, C. E. 1975, Ap. J., <u>201</u>, 90.

Stecker, F.W., and Puget, J. 1974, <u>Proc. Ninth ESLAB Symp.</u> (Frascati), ESRO SP-106, 147.

Steigman, G. 1974, International Astronomical Union Symposium No. 63, Reidel Pub. Co., 374.

Stepanian, A. A., Vladimirsky, B. M., Neshpor, Yu.I., and Fomin, V. P. 1975, Astrophys. Space Sci., <u>38</u>, 267.

Strong, A. W., and Wdowczyk, J. 1975, Proc. 14th Internat. Cosmic Ray Conf., <u>1</u>, 20.

Strong, A. W., Wolfendale, A. W., and Worrall, D. M. 1976, J. Phys. A: Math. Gen., <u>9</u>, 1553.

Swanenburg, B. N., Bennett, K., Bignami, G. F., Caraveo, P., Hermsen, W., Kanbach, G., Masnou, J. L., Mayer-Hasselwander, H. A., Paul, J. A., Sacca, B., Scarsi, L., and Wills , R. D. 1978, Nature, <u>275</u>, 298.

Tananbaum, H., Peters, G., Forman, W., Giacconi, R., Jones, C., and Avni, Y. 1977, submitted to Ap. J. (Letters).

Thompson, D. J., Fichtel, C. E., Kniffen, D. A., and Ögelman, H. B.
    1975, Ap. J. (Letters), 200, L79.

Thompson, D.J., Fichtel, C.E. Hartman, R.C., Kniffen, D.A., and Lamb,
    R.C. 1977a, Ap. J., 213, 252.

Thompson, D.J., Fichtel, C.E., Kniffen, D.A., and Ogelman, H. B.
    1977b, Ap. J. (Letters) 214, L17.

Trombka, J. I., Arnold, J. R., Adler, I., Metzger, A. E., and Reedy,
    R. C. 1977b, Proceedings of the Soviet-American Conference on
    the Cosmo-Chemistry of the Moon and Planets, Lunar Science Institute
    (in press for publication Fall 1977).

Trombka, J. I., Dyer, C. S., Evans, L. G., Bielefeld, M. J., Seltzer,
    S. M., and Metzger, A. E., 1977a, Ap. J., 212, 3, Part I, 925.

Trombka, J. I., Metzger, A. E., Arnold, J. R., Matteson, J. L., Reedy,
    R. C., and Peterson, L. E. 1973, Ap. J., 181, 737.

Trombka, J. I., Vette, J. I., Stecker, F. W., Eller, E. L., and Wildes,
    1974, 117, 99.

Van Riper, K. 1978, Ap. J.

Vedrenne, G., Albernhe, F., Martin, I., and Talon, R. 1971, Astr. and
    Ap., 15, 50.

Vette, J. I., Gruber, D., Matteson, J. L., and Peterson, L. E. 1970,
    Ap. J. (Letters), 160, L161.

Vladimirsky, B. M., Neshpor, Yu.I., Stepanian, A. A., and Fomin, V. P.
    1975, Proc. 14th Internat. Cosmic Ray Conference, 1, 118.

Wallace, P. T., Peterson, B. A., Durdih, P. G., Danziger, I. J.,
    Manchester, R. N., Lyne, A. G., Goss, W. M., Smith, F. G., Disney,
    M. J., Hartley, K. F., Jones, D.H.P., and Wellgate, G. W. 1977,
    Nature, 266, 692.

Weeks, T., Fazio, G., Helmken, H., O'Mongain, E., and Rieke, G.   1972,

    Ap. J., <u>174</u>, 165.

Weekes, T., and Trever, K.   1977, preprint.

White, A.S., Ryan, J.M., Wilson, R.B., and Zych, A.D.   1977, U. of

    Calif., Riverside, preprint UCR/lGPP-77/32.

ASTROPHYSICS IN THE EXTREME ULTRAVIOLET

Francesco Paresce
Space Sciences Laboratory
University of California
Berkeley, CA 94720
USA

ABSTRACT

The current status of extreme ultraviolet (EUV) astrophysics is
reviewed. Seven compact emitters of EUV radiation have been identified
in the < 5% of the sky surveyed randomly so far and the existence of a
diffuse EUV background firmly established. Our discussion will focus on
the quality of astronomical information that can be gleaned from a combined x-ray, EUV, far UV and optical study of these objects and on the
impact EUV observations of the Sun and a large number of classes of
compact objects will have on our understanding of the structure and evolution of evolved stars.   The diffuse EUV background may be of even greater
interest as it probably reveals the distribution of the hot coronal gas
responsible for the interstellar OVI absorption and the source of heating
and ionization of the interstellar medium. Finally, future mission possibilities are analyzed as to their potential for increased understanding
of the EUV universe.

I.   Introduction.

The extreme ultraviolet spans almost exactly one decade of the
electromagnetic spectrum from 100 to 1000 Å or 120 to 12 eV approximately.
As is well known, photons of this energy interact very strongly with
matter since the great majority of known elements have outer electron
binding energies in this range. As a result: 1) the earth's atmosphere
is very opaque at EUV wavelengths due to N_2, O_2, and O photoabsorption
(the e^{-1} depth for vertically penetrating EUV radiation in the atmosphere
is roughly 150 km), 2) EUV photons have great difficulty reflecting from
or being transmitted through conventional optical surfaces or materials
and 3) even minute amounts of hydrogen or helium spread over astronomical
distances can be enough to destroy most of the EUV photons making their
way through the interstellar or circumstellar material.

The present underdevelopment of EUV astrophysics is mainly the
result of these last two points. As Aller correctly noted in 1959, the
mean free path of a EUV photon in the interstellar medium (ISM) cannot

243

P. L. Bernacca and R. Ruffini (eds.), Astrophysics from Spacelab, 243-289.
Copyright © 1980 by D. Reidel Publishing Company.

be much larger than 10^{18} cm resulting from the ratio of approximately
1 atom cm^{-3}, the average hydrogen density in the galaxy as inferred from
21 cm observations, and 10^{-18} cm^2, the typical photoionization cross
section of matter. Since no objects excepting the planets are known to
exist within this small distance from the sun, it was easy to conclude
at the time that EUV astrophysics did not have a very bright future.

In recent years, however, it has become increasingly clear that 1
atom cm^{-3} is not a representative value for the local n_H in the vicinity
of the sun although it probably is a good estimate for an average n_H over
the few kiloparsecs mean free path of 21 cm photons. Hydrogen is not
uniformly distributed in space (Heiles and Jenkins, 1976, for example)
but highly clumped leaving a large region of very low density. That the
sun happens to reside in such a region is obvious from the results displayed in Table I taken from the work of Cash et al., 1979 and representing the H column densities derived from observations of interstellar
absorption at Lα for stars within 100 pc of the sun. Typical densities
range from a low of $\simeq$ 0.01 cm^{-3} to a high of 0.1 cm^{-3}, in all cases very
much less than the average galactic density mentioned above. As one
goes further out the H density may increase slightly perhaps to an average of 0.1 or 0.2 atoms cm^{-3}. Because of its nonuniformity, however,
one cannot exclude the possibility that the interstellar medium is
crisscrossed by a number of tunnels of low density extending for many
hundreds of parsecs in certain directions.

With these numbers one might expect a remarkable improvement in the
EUV mean free path. That this is the case is demonstrated in Figure 1
wherein we have plotted the distance at which attenuation of EUV radiation reaches 90% as a function of wavelength (from Cruddace et al. 1974)
for possible   values of n_H of 0.03, 0.1 and 0.2 cm^{-3}. We are probably
safe in using 0.03 atoms. cm^{-3} out to $\simeq$ 100 pc typical density around the
sun so that we note from this graph that even at the ionization edge of
hydrogen at 912 A, EUV photons have a fair chance of traveling 7 or 8
parsecs and in the directions in which this density is maintained they
could travel as far as 1 kpc if they have an energy of $\sim$ 100 eV. This
means that at the poles we could almost see out of the galaxy in the
EUV, an undreamed of possibility only a few years ago.

The arguments we have just presented effectively remove obstacles
resulting from ISM opacity to potential astronomical observation in the
EUV provided the instrumental difficulties mentioned in point 2 can be
surmounted. At first, a number of attempts to find EUV sources were
made from sounding rockets using crude, insensitive instrumentation
covering small areas of the sky (Henry et al. 1975a, b; Henry et al.
1976a, Riegler and Garmire 1975). In these surveys no sources were found
and even though the upper limits were used to place very interesting
constraints on such diverse phenomena as the mass accretion rate onto the
white dwarf in U Geminorum systems and the space density of O subdwarfs,
EUV astronomy remained a subject of rather limited interest.

TABLE 1

Star	Distance (pc)	$n_{HI}(cm^{-3})$	Reference for Column Density of HI
Sun	–	.11	Freeman et al. (1979)
α Cen	1.34	.06 – .30	McClintock et al. (1978)
ε Eri	3.3	.06 – .20	McClintock et al. (1978)
ε Ind	3.4	∿ .1	McClintock et al. (1978)
α CMi	3.5	.09 – .13	Anderson et al. (1978)
β Gem	10.8	.02 – .15	McClintock et al. (1975)
α Boo	11.1	.02 – .15	McClintock et al. (1975)
α Aur	14	.04 – .05	McClintock et al. (1978)
α Tau	21	.02 – .15	McClintock et al. (1975)
α Leo	22	.02 / .01	Rogerson et al. (1973) / Kondo et al. (1978)
α Eri	28	.07	Rogerson et al. (1973)
α Gru	29	.09 – .18 / .18	York (1976) / Kondo et al. (1978)
HR 1099	33	.003 – .007	Anderson & Weiler (1978)
η UMa	42	0.005	York (1976)
G191 – B2B	47	> .03	Cash et al. (1979)
σ Sgr	57	< 0.17	Bohlin et al. (1978)
HZ 43	62	< 0.013	Auer and Shipman (1977)
α Pav	63	< 0.1	Bohlin (1975)
β Cen	81	0.13	Bohlin et al. (1978)
β Lib	83	0.06 – 0.13	York (1976)
ζ Cen	83	< 0.39	Bohlin et al. (1978)
α Vir	87	0.037	Bohlin et al. (1978)
Feige 24	90	.02 – .05	Cash et al. (1979)
λ Sco	100	< 0.078	Bohlin et al. (1978)

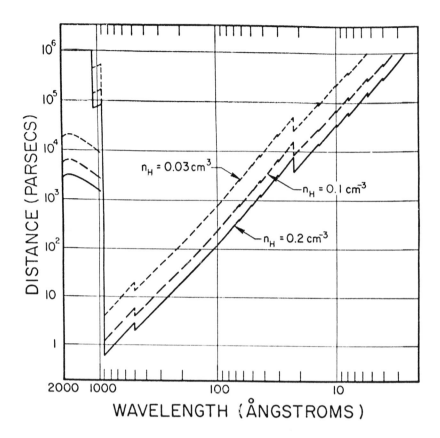

Figure 1.   Distance at which the attenuation of the incident
             radiation reaches 90% as a function of wavelength.
             An ionized interstellar medium of normal composi-
             tion is assumed.

     Only with the development of more sophisticated and sensitive tech-
niques did the search bear fruit. In fact important results were obtained on the very first attempt with an instrument specifically designed
for EUV astronomy research from an orbiting platform: the EUV telescope
on the Apollo spacecraft launched on July 15, 1975. This experiment
discovered the first five compact EUV sources and established the EUV
background spectrum in the 50–300 Å range. Results from this mission
have been summarized by Paresce, (1977) and Bowyer et al. (1977) for the
compact sources, Bowyer, (1979) for the white dwarfs and Malina (1979).
Subsequently a number of experiments have been carried out resulting in
more detailed analyses of the ASTP sources and the discovery of, possibly, two more sources. The total area of the sky surveyed so far in a
rather haphazard way is less than ∿ 5% of the whole sky. Table II lists
all the EUV astronomy experiments attempted as of December, 1980 and
Table III lists the detected or claimed sources.

TABLE 2

Compendium of EUV Astronomy Experiments to December, 1979

Author	Group	Observation Date	Bandpass (Å)	Result/Spacecraft
A. Compact Sources				
Riegler and Garmire (1975)	Caltech	22 Oct., 1971	75–800	Stellar upper limits/rocket
Henry et al. (1975a)	UCB	9 June, 1972	135–475	Search of North Galactic Pole/rocket
Henry et al. (1975b)	UCB	10 Feb., 1973	145–515	Limits on RS And and U Gem/rocket
Henry et al. (1976b)	UCB/UCL	4 Feb., 1974	44–165	Possible detection of VW Hyi/rocket
Troy et al. (1975)	NRL	8 Oct., 1971	912–1050	8 Stellar fluxes/rocket
Paresce (1977) Malina (1979) and references therein	UCB	20–22 July, 1975	55–780	Discovery of 5 sources/Apollo–Soyuz
Cash et al. (1978b, 1979a)	UCB	19 Nov., 1976	250–700	Upper limits on 4 stars/rocket
Malina et al. (1979a)	UCB	10 Apr., 1978	170–500	First spectrum of HZ 43/rocket
Sagdeev et al. (1979)	SRI/USSR	July, 1978	600	Possible detection of Feige 4/Prognoz 6
Brune et al. (1979)	JHU	17 Feb., 1977	912–1050	5 Stellar Spectra/rocket
Shemansky et al. (1979)	Kitt Peak	Nov., 1977	912–1050	Detection of Cygnus Loop/Voyager
B. Diffuse Background				
Paresce and Bowyer (1976)	UCB	10 Feb., 1973	775–1050	Upper limits/rocket
Cash et al. (1976)	UCB	25 Nov., 1974	44–1050	Detection/rocket
Stern and Bowyer (1979) Paresce and Stern (1979)	UCB	20–22 July, 1975	55–780	Detection and upper limits/Apollo–Soyuz
Sandel et al. (1979)	Kitt Peak	Nov., 1977	600–1050	Detection/Voyager

TABLE 3

Catalog of Detected or Claimed Stellar EUV Sources

Name	RA	Dec	Distance	Flux ergs cm^{-2} s^{-1}			Type	Reference
				500–780 Å	170–620 Å	50–150 Å		
HZ 43	13^h14	+29°22'	65±15 pc	-	4×10^{-9}	1.1×10^{-9}	DA White Dwarf	Lampton et al. 1976
Feige 24	2^h32	+ 2°12'	80+50 -25	-	3×10^{-9}	-	DA White Dwarf	Margon et al. 1976c
Proxima Centauri	14^h26	-62°28'	1.3	::	-	7×10^{-10}	Flare Star	Haisch et al. 1977
SS Cygni	21^h40	+43°21'	30–50	-	-	9×10^{-11}	Dwarf Nova	Margon et al. 1978
HD 192273?	20^h14	-69°45'	-	7×10^{-10}	-	-	B Star	Cash et al. 1978b
VW Hyi?	4^h40	-71°	-	-	-	1.3×10^{-9}	U Gem	Henry et al. 1976b
Feige 4?	0^h17	+13°36'	70 pc	6×10^{-11}	-	-	DB White Dwarf	Sagdeev et al. 1979

In the following sections we shall describe briefly the present status of the results of the observations of these sources and the diffuse background.

II.   Compact Sources

HZ 43.

The hot DA white dwarf HZ 43 (Humason and Zwicky 1947; Greenstein and Sargent 1974) was discovered to be a strong EUV source on July 22, 1975 by the EUV telescope on ASTP (Lampton et al. 1976). The ASTP spectrum of this source in the EUV is shown in Figure 2 where as a function of wavelength we have plotted the source intensity in milliflux units for the three positive detections in the Al + C(170-620 Å), Par (55-150Å), and Par + Be (113-150 A) channels and the upper limit in the Sn (500-780 Å) channel. The light horizontal lines represent the flux levels determined by a recent rocket observation of the soft x-ray source that is coincident with HZ 43 (Margon et al. 1976b). It is clear from this figure that we are dealing with a true EUV source and not the tail of an optical or x-ray source. The emission peaks at about 300 Å at a level of more than 30 mfu and contains a total energy flux in the 170-620 Å band of 4×10^{-9} ergs cm^{-2} sec^{-1}. This spectrum is consistent with an object radiating approximately 97% of its total luminosity in the ASTP bandpass.

Optical observations of HZ 43 were carried out using the Robinson-Wampler image tube scanner at the focus of the Lick 120" telescope with 8 Å resolution (Margon et al. 1976a). Both the white dwarf HZ 43 and the composite system were observed between 4000 and 8000 Å yielding by subtraction the spectrum of the red dwarf companion HZ 43B. This permits the classification of HZ 43A as a typical DAwk with no sign of helium and HZ 43B as a dM 3.5e. The trigonometric parallax derived from 22 plates of the Allegheny Observatory yields a distance of 63 ± 15 pc for the system. With the numerous monochromatic visible magnitudes of HZ 43A obtained with the image scanner, one can construct a diagram of the type shown in Figure 3. A monochromatic magnitude of a black body at any wavelength yields a locus in a solid angle-black body temperature plane such as the one drawn in Figure 3. In this figure five such curves for five different wavelengths are drawn using for solid angle the dimensionless ratio stellar radius to stellar distance. The five wavelengths were carefully selected from regions free of absorption features and of contamination from the red dwarf. Although they are not quite as coincident as one would expect from a purely theoretical point of view, they all lie in a narrow strip almost parallel to the temperature axis. It is clear how difficult it would be to determine its black body temperature from these curves alone. This reflects the basic fact that only about 0.1% of the total luminosity of the star is being emitted in the visible band.

We now add the EUV observations we have just discussed on this graph in the form of a locus obtained from the intensities of five black

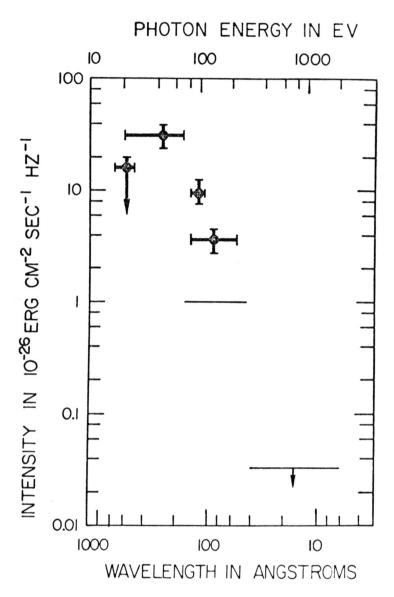

Figure 2.   Spectral intensities derived from the ASTP EUV
             telescope data on HZ 43.

body models representing the extremes of the allowable values of the
adjustable parameters. The EUV observations define a much narrower range
of allowed temperatures. We can do even better if we combine the optical
and EUV data since the two must be consistent, which means that the star
must lie in the small region of intersection of the various curves. In
this way the data just described is used to determine the black body

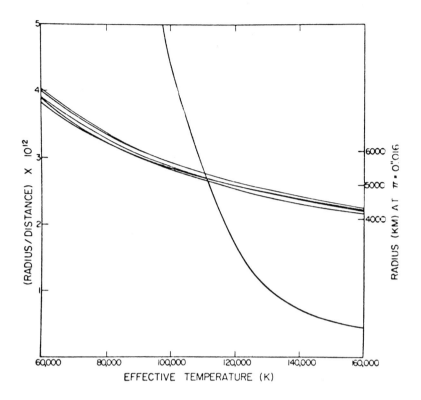

Figure 3.   Solid angle - effective temperature loci for the
             ASTP EUV data on HZ 43 (near vertical curve) and
             for the optical monochromatic flux measurements
             of HZ 43 (near horizontal curves). The inferred
             stellar radii at the optimum parallax are also
             indicated.

temperature of HZ 43A to be 110,000 ± 10,000 °K and, if we use the distance determined by the trigonometric parallax, we can also say that its
radius is 5000 ± 1000 km and therefore that its luminosity is $\sim$7 L_0.
This is not a typical underluminous white dwarf!

     The hydrogen column density to this object is one of the free
parameters in the black body emission model that can now be determined
unambiguously. This method yields a hydrogen column density N_H of
4 x 10^{18} cm^{-2} corresponding to an average number density n_H of 0.02 atoms
per cm^3, or a striking and direct confirmation of our earlier expectations summarized in §I.

Although a black body thermal emission model is a simple and convenient way to parameterize the observed emission rate it cannot strictly
apply in our case since ionization edges at 912, 504 and 228 A, will
distort the emission profile. To do things correctly one needs to construct a set of white dwarf model atmospheres to compare to the existing
data. This has been done by Durisen et al. (1976) using high gravity
solar abundance atmospheres and finding compatibility with objects having
a slightly higher effective temperature of about 125,000 °K. Solar
abundances are not appropriate, however, since HZ 43 is a DA white dwarf
and thus might be deficient in helium and metals. With this in mind,
Auer and Shipman (1976) have constructed metal-poor atmospheric models
and determined effective temperatures for HZ 43 A ranging from 55,000 to
70,000 °K or slightly less than the black body temperatures. Other
determinations of this parameter (Wesselius and Koester, 1978; Greenstein
and Oke, 1979) using far UV data from ANS and IUE yield temperatures
falling in this range. As Malina (1979) has pointed out, the major uncertainties arise about both the reliability of the models and the exact
value of the helium and metal abundance. The data most sensitive to T_e
and abundance are the continuum slope, intensity and possible discontinu-
ities in the EUV spectrum of this source.

To resolve these issues, HZ 43 was extensively reobserved between
170 and 500 Å with 15 Å resolution by Malina et al. (1979a) using a
second generation imaging telescope coupled to a EUV spectrometer
(described in Malina et al. 1979b and in § III) and by Bleeker et al.
(1979) in the 44-150 Å range with proportional counters. The spectrum
observed by Malina et al. (1979b) and discussed in detail in Malina
(1979) is marked by a prominent absorption edge at 225 Å ± 15 Å due to
He^+ with a ratio $F_\lambda(230 \text{ Å})/F_\lambda(220 \text{ Å}) = 0.5 \pm 0.13$. No other emission or
absorption features are observable at the level of 10 Å equivalent width
or more. The observed discontinuity requires a column density of singly
ionized helium of $2.6-5.3 \times 10^{17}$ ions cm^{-2} corresponding to a photospheric
density between 10^{-5} and 3.10^{-5} in units of n(He)/n(H) for a 60,000 °K
atmosphere. Thus this measurement represents the first determination of
the helium abundance in a DA white dwarf. The result necessarily entails
that diffusion processes in pure hydrogen atmospheres are not as efficient
in separating elements than heretofore believed and that competing
processes could be important. Another scenario more compatible with the
observations might envisage a rich hydrogen envelope over a helium rich
atmosphere.

The results of these latest observations are plotted together with
measurements in the optical and UV in Figure 4 while in Figure 5 we show
some recent attempts at predicting the observed spectrum. As is clear
from this latter figure, current theoretical white dwarf model
atmospheres adequately reproduce the optical and UV data but fail, often
by large margins, to account for the EUV and soft X-ray spectral shape
and intensity. Possible sources of this discrepancy have been discussed
extensively by Malina (1979) and include model limitations, sensitivity
to surface gravity variations, pressure broadening effects, abundance
gradient effects and uncertainties regarding metal abundances.

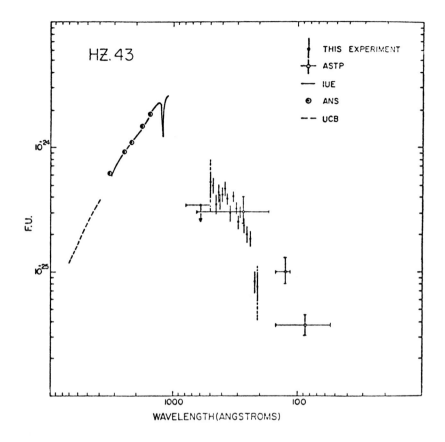

Figure 4.   Optical and EUV spectrum of HZ 43. Data is from UCB
             (Margon et al. 1976); ANS (Wesselius and Koester
             1979); EUV (Greenstein and Oke 1979); and ASTP
             (Lampton et al. 1976). The data from Malina et al.
             1979 are shown with ± 1σ error bars (solid error bar)
             and a typical ± 20% estimated systematic uncertainty
             (dotted extension to error bars).

A possible caveat is also represented by the uncertain abundance of He^+
in the ISM. Perhaps not all the observed discontinuity at 228 Å is
due to photospheric He^+.

     From a physical point of view the computed stellar radius of 5,000
km for HZ 43 should be considered quite normal, placing it well into a
fully degenerate configuration. It also implies a mass between about
0.8 and 1.05 solar masses (Margon et al. 1976a) depending on interior
composition, putting HZ 43 on the classical white dwarf cooling at a
constant radius curve for its mass and radius (Salpeter 1971 & Figure 9).

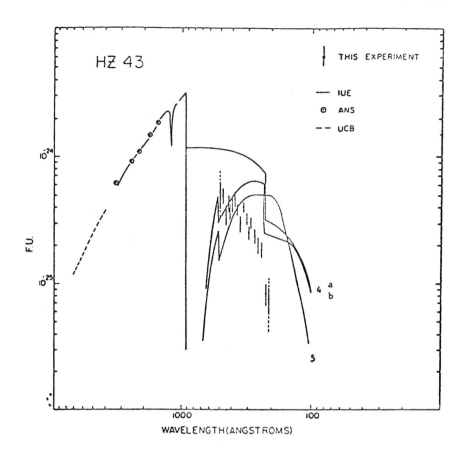

Figure 5.   Trace helium models from Heise and Huizenga (1979)
             and Auer and Shipman (1977). Data same as for
             Figure 4. Curve 4a (upper curve): T_{eff} = 60,000°
             K, He/H = 3 x 10^{-5}, no ISM. Curve 4b (lower curve):
             T_{eff} = 60,000° K, He/H = 3 x 10^{-5}, neutral ISM N_H =
             7.7 x 10^{18} cm^{-2}. Curve 5: Suggested Heise and
             Huizenga (1979) model with neutral ISM N_H = 10^{18} cm^{-2},
             ionized ISM N_H = 5 x 10^{18} cm^{-2}.

Feige 24.

     Another very blue white dwarf in the direction of Feige 24 in the
constellation Cetus was discovered by the EUV telescope on ASTP (Margon
et al. 1976c). As with the HZ 43 observation there is a clear signal in
the Al + C(170-620 Å) channel well above background (statistically significant at the 45σ level) when looking straight at the intended target
Feige 24 but, in stark contrast to HZ 43, there is no detectable signal
in any other channel. Its absence in the Sn (500-780 Å) channel is
understandable from interstellar absorption considerations but its absence

in the harder channels must mean that its spectrum falls off much more steeply than HZ 43's. This effect can be seen in Figure 6 where the inferred fluxes and upper limits (boxes) from this star are plotted as a function of wavelength. Thus, although it turns out it has about the same flux as HZ 43 around 300 Å, it is only 10% or less as bright at the shorter wavelengths.

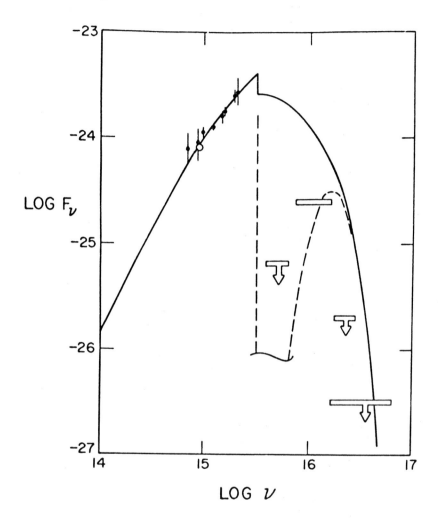

Figure 6.   Spectral intensities (boxes) derived from the ASTP EUV telescope data on Feige 24. Optical and UV data are from Oke (1974, open circle) and Holm (1976, filled circles). The solid line corresponds to a pure hydrogen stellar model having T_e = 60,000° K, log g = 8 without interstellar attenuation. The broken line shows the effect of this last parameter.

With this data we can constrain parameters of simple emission models
with the result (Margon et al. 1976c) that for a thermal emission model,
T must be less than $\sim 90,000$ °K and n_H larger than $\sim 2 \times 10^{18}$ cm^{-2} at the
90% confidence level. Accurate optical spectrophotometry of Feige 24
confirms that the white dwarf is a DA white dwarf with a very similar
spectrum to HZ 43A. On the surface this means that we must compare our
data directly to a set of pure hydrogen, high gravity model atmospheres
of the type described earlier in connection with HZ 43. This task was
carried out by Margon et al. (1976c) and the results shown as a solid
line (no interstellar absorption) or a broken line (with interstellar
attenuation) in Figure 6.

These lines represent the best fit to all the $\lambda > 170$ Å points
available including the optical and UV data Oke (1974) and Holm (1976)
for $T_e = 60,000$ °K, log g = 8 and $n_H = 4 \times 10^{18}$ cm^{-2}, and corresponds
to a stellar radius of 17,000 km at a distance of 100pc, implying a mass
of ~ 1 M_0. Unfortunately it is also clear even from a superficial view
of Figure 6 that the just cited model is grossly inconsistent with the
$\lambda > 170$ A upper limits. No way out of this discrepancy seems possible as
long as we use the pure hydrogen models implied by the DA classification
or to the optical identification of the EUV source. Models cooler than
50,000 °K have insufficient flux to explain the 170 to 620 Å observations yet models hotter than 30,000 °K have more 100 Å flux than is
observed. As in the case of HZ 43, by adding a minute amount of ionized
helium that absorbs strongly below about 170 Å we can depress the < 170 Å
part of the expected spectrum leaving the rest of the star's spectrum
unchanged. Introducing a helium abundance of $\sim 2.10^{-4}$ with respect to
hydrogen at 60,000 °K into the atmosphere allows a fit to the <170 A data
without violating the lack of HeII lines in the optical spectrum. Further
model atmosphere work is clearly called for, however, to precisely determine the appropriate parameters. We have emphasized this point here to
illustrate how sensitive to composition EUV observations of these objects
actually are.

Cash et al. (1979) using a second generation imaging EUV telescope
subsequently observed Feige 24 in the 250-350 Å band without detecting
it, indicating most of the flux from this star lies between 170 and 280
Å to be consistent with the Apollo observations. Bowyer (1979) has discussed in detail the extraordinary optical appearance of Feige 24. The
work of Leibert and Margon (1977) and Thorstensen et al. (1978) suggests
that Feige 24 is a white dwarf-red dwarf binary system with the observed
optical emission lines originating from the reprocessing of EUV radiation
from the white dwarf by the atmosphere of the red dwarf. Feige 24 lies
somewhat to the right of HZ 43 on the HR diagram displayed in Figure 9 .

Sirius B. and Feige 4.

The detection of soft x-rays from the Sirius system by Mewe et al. (1975)
has sparked a long history of theoretical and experimental attempts at
understanding the origin of this emission. This situation has been
recently reviewed by Lampton and Mewe (1979) and Bowyer (1979). The EUV

telescope on ASTP failed to detect it but the results were marginally
consistent with a Sirius B photosphere at T > 32,000 °K (Shipman 1976;
Shipman et al. 1977). An observation of Sirius by Cash et al. (1978)
with an imaging EUV telescope on a sounding rocket, however, yielded
much more stringent limits on the EUV emission that effectively rule out
any photospheric participation in the soft X-ray emission mechanism.
Lampton and Mewe (1979) have shown that all the observations below 912 Å
are compatible with a standard corona on Sirius B. Their calculations
together with Shipman's and the observations are summarized in Figure 7.
Since both Sirius A and B are now known to emit X-rays, with the majority
coming from Sirius A, (Bowyer 1979), we may be forced to go back to the
original suggestion of Mullan (1976) that Sirius A may have a corona
similar to the Sun. It probably is an understatement to claim that the
physics of this source is not well understood presently, especially
considering the possible source variability implicit in Bunner's
(1977) and Lampton et al.'s (1979) soft X-ray results.

Another white dwarf reported recently to be emitting in the EUV is
Feige 4 (Sagdeev, Kurt and Bertaux 1979). According to Bowyer (1979)
this star is between 30 and 70 pc distant. This means that it must be
bright indeed to be observable at ∿ 600 Å as claimed by the authors,
since ISM opacity is quite high at these wavelengths (Cruddace et al.
1974). A detailed analysis needs to be performed to ensure proper
rejection of far UV radiation from γ Peg, a very bright (m_v=2.9) nearby
star before definitive conclusions can be made at this point.

Proxima Centauri

The dwarf M5e UV Ceti-type flare star Proxima Centauri was observed
twice with the EUV telescope on Apollo-Soyuz (Haisch et al. 1977). It
was not detected in any bandpass during the first observation but was
strongly detected in the 50-150 Å during the second. Unfortunately, a
snow storm in New Zealand prevented simultaneous optical observations, so
that it is not known whether the star was flaring optically. However,
the variation in the EUV flux argues that it was a flare. Assuming an
optically thin flare at a temperature between 10^6 °K and 10^8 °K, an
emission measure of 4.6 x 10^{52} cm^{-3} and a total thermal X-ray emission
of ∿ 10^{30} ergs s^{-1} can be derived.

Since Mullan (1976) has computed a theoretical scaling theory for
stellar flares assuming thermal radiation from a hot plasma as the
radiation source it would be instructive to compare our results to this
theory. The main obstacle to this endeavor is represented, not surprisingly, by the absence of simultaneous optical data. The fundamental
parameter to be compared is the ratio L_x/L_{opt}, the X-ray luminosity to
the optical luminosity, for the flare. One flare on UV Ceti is unfortunately the only known flare observed simultaneously in the X-ray
and the optical (Heise et al. 1975) and for which an unambiguous determination of L_x/L_{opt} can be made. For Prox Cen we must rely on its past
optical flare history to derive L_{opt}. Using this analysis it is
found that while observations of UV Ceti and YZ CMi seem to be reasonably

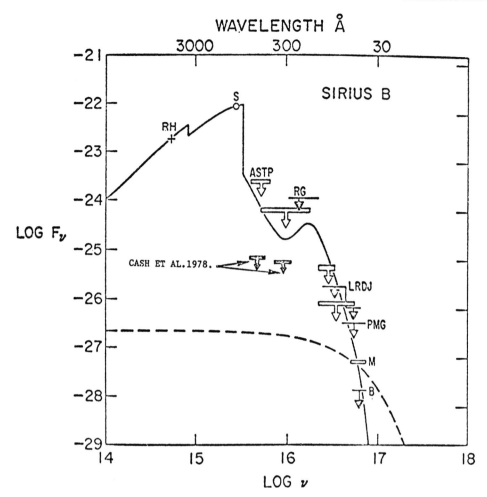

Figure 7.   Experimental data on the spectrum of Sirius B
             plotted with the model atmosphere of Shipman, 1976
             (solid line) and the pure bremsstrahlung corona of
             Lampton and Mewe, 1979 (dashed line). The 3-sigma
             EUV upper limits of Cash et al. (1978) rule out a
             photospheric origin for the soft X-ray flux.

compatible with Mullan's theory, the ASTP observation of Prox Cen
is definitely not. Either the observed EUV flare was an extraordinarily
bright event or the L_x/L_{opt} ratio for stellar flares is a highly variable
and as yet unpredictable parameter.

SS Cygni

    The fourth positive detection of an EUV source from Apollo-Soyuz
is that of SS Cygni, the well-studied irregular cataclysmic variable that

undergoes as much as three magnitudes variability quasi-periodically. Clear detections were made on the three different occasions it was in the instrument field of view both in the 55-150 Å and the 113-150 Å channels (Margon et al. 1978). Fortunately, in this case excellent simultaneous optical data on SS Cygni are available from the work of Lyutyj (1976), who showed that at the time of the ASTP observations the object was at its maximum optical light at approximately 9th visual magnitude (at minimum it reaches about 12th magnitude). On the fist two of the three observations, SS Cyg has a flux of $9 \cdot 10^{11}$ ergs cm^{-2} s^{-1} incident at Earth while on the third the EUV flux dropped by a factor of ~ 40 while L_{opt} remained roughly constant. As the intensity of the soft x-ray and EUV flux at visible light maximum is now known to vary by more than two orders of magnitude, the connection between the visible and EUV outbursts is obviously much more complex than heretofore imagined. Margon et al. (1978) show that a number of emission regions are needed to explain the observations. Specifically one needs at least a $T \sim 15,000$ °K blackbody to explain the visible spectrum and a hot ($\sim 2.5 \times 10^5$ °K), free - free emitter ($n_e \sim 4 \times 10^{14}$ cm^{-3}) to account for the x-ray, EUV and infrared emissions. The latter is such that it would not significantly perturb the observed visible spectrum.

The dwarf novae AE Aqr, Z Cha and VW Hyi

All three objects were observed by ASTP and VW Hyi possibly detected by Henry et al. (1976 b) at a level of 5-10 x 10^{-10} ergs cm^{-2} s^{-1} in the 44-165 Å band. The ASTP observations yield only upper limits (Margon et al., 1978) and simultaneous optical observations indicated that each source was in a relatively quiescent state during all four measurements. Thus the basic question of whether all dwarf novae (including SS Cyg) have detectable quiescent soft x-ray /EUV emission must await further observations.

Pulsars

The first EUV observations of radio pulsars, (Greenstein et al, 1977) although they were not detected with the EUV telescope on the ASTP mission, provide yet another example of the usefulness of this young branch of astronomy. There has never been an unambiguous detection of thermal radiation from a radio pulsar. Such an observation would be of great interest because it would provide a direct comparison of data with the theory of neutron star structure. It now seems fairly certain that the typical pulsar is not a strong thermal x-ray emitter; upper limits on x-ray emission from a variety of pulsars have typically constrained log T to less than 6.7. If the effective temperature of a typical pulsar is in the range $10^5 < T < 10^6$ °K, the bulk of the thermal radiation will appear in the EUV and for this reason included in the ASTP observing program were the three radio pulsars we mentioned, known to be nearby because of their small dispersion measures. In no case is there any evidence in the data for a signal in excess of background in

any of the five ASTP channels for any of the three pulsars. These count
rates can be used to derive upper limits to the incident flux at Earth
from the pulsars, the most sensitive ones being for PSR 192+10, of order
of 10% of the flux observed from HZ 43. Unfortunately the conversion of
these limits into intrinsic luminosities and pulsar temperatures depends
intimately on distance, interstellar hydrogen density n_H and radius R.
Since none of these quantities are well known, the upper limits one can
obtain must be parameterized by n_H and R. The maximum value of effective
temperature permitted by the data in the most favorable cases (i.e. n_H
$\sim$ 0.01 cm^{-3}, R = 100 km) corresponds to $T_e \gtrsim 1.6 \times 10^5$ °K a far more
stringent limit than can be set with x-ray data alone under those circumstances. In the worst cases ($n_H \sim 1$ cm^{-3}, R = 100 km) the limits are
much weaker than those already set by the x-ray data. For two pulsars
(PSR 1133+16 and 1929+10) the upper limits approach but do not violate
the expected temperature from cooling or frictional heating theory.
Clearly more detailed observational (especially more sensitive EUV
observations) and theoretical work is necessary on this fascinating
subject.

The source in Pavo

Finally there is evidence in the ASTP data for a new source in the
constellation Pavo (Cash et al., 1978b). The potential EUV source,
observed only in the 500 - 780 Å channel, has an intensity therein of
35 $\pm$ 9 mJy. Positional overlap suggests a possible identification with
the ultraviolet star HD 192273 and recent observations of this star
(Shipman and Wegner, 1979) suggesting it is a component of a low mass
binary adds some strength to this hypothesis. Until more accurate and
complete EUV data is available on this and other putative sources (VW Hyi
and Feige 4 for example) not much else can be said about them except that
they most probably represent the tip of the iceberg of EUV sources.

III. Diffuse background

Although the 44 - 60 Å component of the cosmic diffuse background
radiation was discovered almost a decade ago (Bowyer et al., 1968) it is
still not well understood. This radiation is spatially anisotropic,
in contrast to the diffuse x-ray background at shorter wavelengths. Also,
the flux is detected in directions where the mean free path due to
interstellar photoelectric opacity is known to be small. These facts
indicate that this radiation is distinct from the hard x-ray background
and has a relatively local galactic origin. Various investigators have
searched for correlations of the intensity of this radiation with numerous
other phenomena (e.g. neutral interstellar hydrogen column density, far-
UV absorption lines of hot interstellar gas, low-frequency radio features
etc.) with ambiguous results. Most of the difficulty is the lack of a
complete spectrum from 44 to 1,000 Å and of an all sky survey made with
one instrument over a relatively short period of time.

Cash, Malina and Stern (1976), using a rocket-borne proportional
counter/beryllium filter combination, reported the first unambiguous

measurement of the EUV background intensity between 100 - 150 Å, confirming the speculation of Yentis et al. (1972). This experiment was
limited by the large field of view of the detectors and by the short
exposure times available.

The ASTP EUV observations in the 50 - 620 Å range reported by
Stern and Bowyer (1979) and Paresce and Stern (1979) represent a
tremendous increase in the data available to examine this problem. A
combination of data acquired during operations at or near the preselected
EUV targets, during a number of great circle scans and maneuvers en-
compasses a remarkably representative sample of all galactic longitudes
and latitudes. The observations were made primarily with the 2.4° -
diameter FOV, which was the best spatial resolution ever used for such
a survey, and were all acquired by the same instrument over a period of
time that was short compared to many solar and terrestrial phenomena.
The data also extend to longer wavelengths than previously used for this
work to take advantage of the sharply increasing fluxes due to the steep
spectrum. Finally, data were acquired at a wide variety of different
geocentric, geomagnetic, and solar coordinates, enabling a careful
search to be made for noncelestial flux contributions that have been a
problem in many previous brief experiments.

Stern and Bowyer (1979) have discussed the spatial variation of the
observed background. To within the counting statistics of the observations the intensity seems reasonably uniform as a function of longitude
and latitude. Unlike some previous claims for the 44 - 60 Å background
(Gorenstein and Tucker, 1972), no simple dependence with galactic latitude is found. Also, these results are in sharp contrast to the 44 -
60 Å surveys (Davidsen et al., 1972) in which high-to-low latitude
intensity contrasts of a factor of 5 are not uncommon. Thus, these data
are interpreted as evidence that spatial isotropy may be gradually res-
tored as observations are shifted to longer and longer wavelengths.

A natural (although tentative) interpretation of this phenomenon
is as follows. As the wavelength of observation increases, the mean
free path due to interstellar photoelectric absorption decreases. If
the radiation is in fact due to large-scale galactic features such
as thermal emission from hot gas clouds or radio loops, the line of
sight at these longer wavelengths will reach only the very nearest such
features, which will in turn have the largest solid angles as viewed
from Earth. As fewer emitting features are superposed, a large-scale
intensity isotropy would be expected to emerge. Further analysis of the
EUV background observations will be needed to verify the plausibility
of this concept.

The spectrum of the EUV background between 50 and 600 Å has been
reported and analyzed in detail by Paresce and Stern (1979). In Figure
8 we show the average spectrum as determined from an examination of data
from the Apollo-Soyuz diffuse background survey and of earlier observations obtained with rocket-borne proportional counters. This spectrum
is analyzed by means of a hot interstellar plasma emission model wherein

the plasma is either isothermal or characterized by a differential emission measure with a power-law temperature distribution. In the isothermal case, physically realistic fits are obtained only for a narrow range of model parameters near $T \simeq 10^{5.5}$ K, and emission measure, EM, $\simeq 10^{-2}$ cm^{-6} pc. The allowed range of parameters is consistent with those for the OVI absorbing gas as reported by Jenkins (1978). Assuming a common origin for both the emission and absorption, a filling factor f for the hot gas $\simeq 0.1$ at an interstellar medium pressure p/k $\lesssim 10^4$ cm^{-3} K is derived. The power law distribution models are consistent with the observed EUV/soft x-ray diffuse fluxes provided the spectral index $\alpha \lesssim 1.5$ for a plasma in the range 10^5 K $\lesssim T \lesssim 10^7$ K. The upper limit to the allowed values of α is increased if a high temperature cut-off to the distribution is introduced. For example, for $T_{max} = 10^{5.8}$ K, $\alpha \lesssim 0.5$, which is more consistent with recent OVI absorption results.

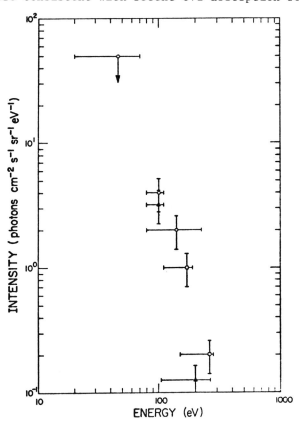

Figure 8.   Observed EUV/soft x-ray background fluxes in photons cm^{-2} s^{-1} sr^{-1} eV^{-1} as a function of photon energy in eV. Apollo-Soyuz results are shown as open circles, Berkeley sounding rocket results as filled triangles and the Wisconsin observations are indicated by the open squares.

IV.   Solar EUV Astrophysics.

     During the past eight years a number of advances have been made in
the observations and understanding of the EUV emission from the Sun.
They are related to a variety of old and newly discovered phenomena such
as coronal holes, bright points, active regions and flares. The results
of these studies have been reported extensively in the literature (see
e.g. Kane and Donnelly 1971; Noyes et al. 1975; various articles in
Kane 1975; Bohlin 1976). It is clearly beyond the scope of the present
review to describe them in any detail. Here we present briefly only
the highlights of perhaps the most dramatic of the various solar phenomena,
viz. the highly transient EUV emission from solar flares.

     Flares are localized brightenings on the Sun occupying an area
$\lesssim$ 0.1% of the solar disk. EUV flares represent a relative increase of
$\sim$ 0.1% to $\sim$ 100% above the quiet time background EUV radiation flux from
the whole solar disk. A highly sensitive detector is therefore required
to make measurements of the majority of solar flares. Two approaches
have been used to achieve this: (1) Indirect measurements of broad
band solar flux through its effect on the Earth's ionosphere (Sudden
Frequency Deviation). This procedure has been described in detail by
Donnelly (1973). It is useful primarily for the transient flare radiation
and gives an integral flux over the 10-1030 Å range. Its main advantages
are high sensitivity, high time resolution ($\lesssim$ 1 sec) and nearly continuous (day time) monitoring of the Sun with ground based instrumentation.
Its disadvantages are the lack of spectral and spatial resolution. (2)
Direct measurements with instruments on-board spacecraft: line as well
as broad-band spectra have been measured in this way (see e.g. Noyes
et al. 1975). The principal advantage of this method is good spectral
resolution. Until recently the principal disadvantage had been the
relatively poor time resolution ($\gtrsim$ 5 sec) especially in experiments with
high spatial resolution. This inadequacy is expected to be removed in
the future experiments such as those to be flown on NASA's Solar
Maximum Mission and Spacelab.

     The role of the EUV measurements in the understanding of the overall
solar flare process, particularly the impulsive phase, has been reviewed
by Kane (1974). EUV emission in the 10-1030 Å range represents a major
part of the total energy radiated during the impulsive phase of a solar
flare. The variations in this transient radiation have a time constant
$\sim$ 1 sec. The total duration varies from 10 to a few hundred seconds.
The observed magnitude of the peak radiation flux varies from one flare
to another and is in the range 0.01 - 3 ergs cm^{-2} sec^{-1} at 1 A.U. A
preliminary analysis indicates that the occurrence frequency distribution
of flares with 10-1030 Å flux > F ergs cm^{-2} sec^{-1} is consistent with
$N(>F) \sim F^{-\alpha}$ where $\alpha \sim 1$. The impulsive rise in EUV emission occurs
essentially simultaneously at all levels in the solar atmosphere between
the transition region and the corona (Noyes et al. 1975), indicating a
rapid energy transport from the location of the primary energy release
to the entire EUV source. A model consistent with these EUV observations
and other simultaneous flare emissions such as hard x-ray, microwave and

optical radiation, assumes that electrons up to a few hundred keV energy
are accelerated in the lower corona during the impulsive phase of solar
flares. A large fraction of these electrons propagate downwards towards
the photosphere and, as they lose energy along their paths, they produce
additional ionization and heating of the different layers in the solar
atmosphere. The subsequent free-free, free-bound and bound-bound emissions contribute to the observed EUV radiation.

    EUV observations have contributed a great deal to the development
of the above picture of the impulsive phase of a flare. Inpulsive solar
flare emissions, particularly the hard x-ray bursts, have an appearance
similar to the cosmic γ-ray bursts, first reported by Klebesadel et al.
(1973). Detectors designed primarily to measure hard x-rays from the
Sun have measured the cosmic hard x-ray emission simultaneously with some
of the cosmic γ-ray bursts (see e.g. Kane and Anderson 1976). It has
been suggested that the cosmic γ-ray bursts are produced in "super flares"
occurring on magnetic white dwarfs (Stecker and Frost 1973). EUV
emission simultaneous with cosmic γ-ray bursts has not yet been observed.
If and when such cosmic EUV bursts are observed they will certainly throw
new light on otherwise little understood phenomena such as cosmic γ-ray
bursts.

V.   Future Observations and Their Impact on Astronomy.

    In retrospect the ASTP mission proved far better than anticipated,
but it did, in the beginning, present a number of shortcomings that could
have in principle distorted for many years our picture of the EUV universe.
For one thing, as we noted earlier, this was a pointed mission so that
outright guesses as to possible EUV emitters had to be made on the basis
of questionable theories concerning EUV emission mechanisms of known optical
or x-ray sources. It should be obvious from our discussion that what was
really needed was a simple and sensitive all-sky photometric survey to
locate any possible EUV sources, whether associated with an optical object or not. For another, going this way implied looking at a very large
number of possible candidates to try to maximize the chance of seeing
that class of EUV emitter. Only a handful of objects were observed with
any kind of accuracy, maybe at most two or three per class. Thirdly, the
mediocre pointing capabilities of the Apollo spacecraft which was certainly not built with arc second astronomical observations in mind severely
reduced the overall sensitivity of the telescope especially since
measurements in the EUV are background limited in the near Earth environment. For this reason, and others having to do with the not ideal viewing
conditions under which many objects were observed, a priori chances of
being able to pick out even a strong EUV source from the background were
reduced substantially. Consequently the mere fact that five out of
approximately 30 sources observed were actually detected can be ascribed
either to incredible luck or more probably to the surprisingly large space
density of EUV sources. If this last conjecture is true it might be
instructive to consider the possible classes of objects one should see
with even a modestly powerful instrument of the type that exists today,
provided it is used in the way we shall outline in § VI.

White dwarfs.

We have already mentioned the promise of this area of research. Considering the optical observational bias towards high velocity objects, even the most pessimistic assessment of the number of white dwarfs available for study in the EUV is surprisingly high. But what effect would this have on our knowledge of the universe? For this, let us consider the idealized Hertzprung-Russell diagram shown in Figure 9 and adapted from Salpeter's (1971) Figures 1 and 3. A rough boundary of the region of the H-R diagram that is of interest to EUV astronomy can be simply obtained by plotting on this figure the two vertical lines corresponding to the temperatures for which black body radiation reaches its maximum at the approximate wavelength limits of the EUV range ($\approx$ 37,000 °K for 1000 Å and $\approx$ 370,000 °K for 100 Å). Thus it is inside the region bounded by these two solid vertical lines shown in Figure 9 that most objects emitting predominantly in the EUV ought to be found. Apart from being physically large, this area of the H-R diagram is particularly significant in that it is very sparsely populated observationally speaking and that it is traversed by normal stars in the later stages of their evolution. The solid lines on this figure represent schematically the observed main sequence, red giant and horizontal branches for globular clusters together with an approximate delineation of the outer boundaries of the observed planetary nebulae nuclei and white dwarfs. The solid curves correspond to a few theoretical evolutionary tracks for planetary nebulae nuclei from Salpeter (1971) for stars consisting mainly of carbon and oxygen and having masses of 1.02 and 0.75 $M_\odot$ The effect of adding even slight amounts of helium in an envelope around the star is quite evident. The fact that the theoretical tracks do not quite match the positions of the observed nuclei should not embarrass us excessively, since even slight variations in stellar composition can bring them into line with the optical sources (such is the sensitivity of these objects to this parameter) and we are not even sure that the nuclei studied in the optical region are really the only ones observable.

A remarkable absence of objects in the region between the planetary nebulae nuclei and the white dwarfs is immediately apparent. This makes it very difficult to associate the two classes of objects as most classical theories of the late stages of stellar evolution would like. In earlier treatises of the subject (cf. O'Dell 1968) this region was actually populated by a number of hot white dwarfs, but a recent sizeable downward revision of temperatures has pushed these back into the area in the lower left hand corner as shown. Now we are in the position after ASTP and the dawn of EUV astronomy to bring at least two of them back into the gap; HZ 43 and Feige 24. These stars' approximate positions are indicated by the crosses and HZ 43's cooling curve is shown as a dotted line leading towards the classical white dwarf region and perhaps coming from the typical Harmon -Seaton sequence for planetary nebulae nuclei. This should not be taken to suggest that this is the actual evolutionary sequence since other possibilities are available, especially evolution from the OB subdwarf region (Weidemann, 1968). Nevertheless

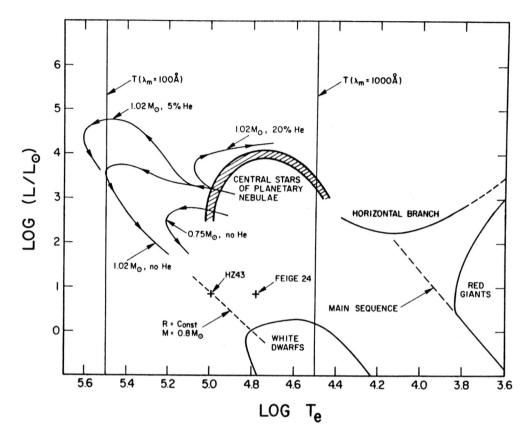

Figure 9.   Luminosity – effective temperature diagram adapted
             from Salpeter (1971). The approximate observed
             locations of the main sequence, red giant branch
             and the horizontal branch for globular clusters
             are shown together with the mean L – Te sequence of
             the known central stars of galactic planetary nebulae
             and the rough boundary of the known white dwarfs.
             The approximate positions of HZ 43 and Feige 24 are
             indicated with the crosses and the dashed line
             running through HZ 43 corresponds to a typical
             constant radius cooling curve for its mass and
             radius. Theoretical evolutionary tracks for
             stars consisting mainly of carbon and oxygen
             for various masses and helium abundance from
             Salpeter (1971) are also shown. The vertical
             lines define the region of maximum probability of
             detecting thermal EUV sources for temperatures
             yielding maximum flux at 100 and 1000 Å.

it is abundantly clear how crucial for an understanding of this stage
of a star's lifetime it is to probe this particular area of the H-R
diagram with accurate observations.

Three related major unsolved problems are the exact chemical composition and structure of a hot white dwarf's atmosphere the calibration
of the brightness-temperature scale for T > 50,000 °K, and the exact
cooling rate of white dwarf configurations (neutrino cooling might be
important in this stage). EUV observations, as we have given abundant
preview when describing the HZ43 and Feige 24 observations, are peculiarly
suited to answering these questions. We have seen from Feige 24 for
example, what small amounts of helium will do and from HZ 43 how much
more precise a temperature one obtains. Now one can go back to the
optical objects and perhaps reclassify many of them on this basis.
Finally, after a reasonably complete all-sky survey, one can start to
compute the average space density of hot white dwarfs in the manner
described by Henry et al. 1976a, Wesemael (1978) and Koester (1978) and
to thereby deduce the evolutionary time scales and consequently the unknown cooling rates of typical full degenerate white dwarf configurations.
This effect is illustrated in Figure 10 taken from Koester (1978) showing
the expected number of white dwarfs within 100 pc. of the sun with and
without neutrino cooling. The expected number reaches 100 at T $\sim$ 40,000
°K.

Since these objects may have a considerable effect on the surrounding
interstellar gas and might contribute to the diffuse EUV background, this
study would also have substantial impact on our understanding of the
structure of the interstellar medium and of the diffuse background. If
there are enough hot objects of this type around, we might have an
alternate explanation of the observed diffuse emission at 120 Å
(Stern and Bowyer, 1980).

Stellar coronae

To be observable in the EUV, stars need not necessarily lie in the
vast region of the H-R diagram bounded by the vertical lines in Figure 9.
Cooler stars with hot coronas such as the Sun will do nicely, provided
they are not too far away and the coronal emission is strong enough.

Observation of stellar coronae in the EUV promises to be one of the
most exciting fields of study during the Spacelab era. It is surprising,
but the basic source and mechanisms for the heating of even the solar
chromosphere and corona are at present the subject of such intense
debate that they can only be considered unknown (e.g. Cram 1977). A
basic problem is that detailed studies of a single star must be supported
by observations of a wide selection of stars for the identification of
the critical parameters, and the great difficulty of observing stellar
coronae has thwarted such studies. The recent development of UV space
astronomy has allowed the study of stellar chromospheric features. With
the development of EUV astronomy, there finally exists a method for direct observation of coronae and we can look forward to break-throughs
in our understanding of this phenomenon.

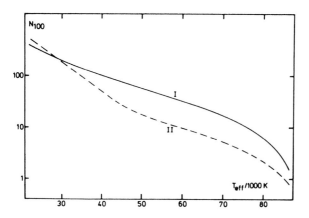

Figure 10.   Number of white dwarfs hotter than T_{eff} within
             100 pc.

    Despite the obvious difficulties, a few theorists have attempted to
predict intensities of stellar coronae. The first were Landini and
Fossi (1973), who used the detailed stellar coronal model predictions
of deLoore (1970) to predict the radio and x-ray intensities of main
sequence and subgiant stars as a function of spectral type. Subsequent
x-ray observations from rockets (Cruddace et al. 1975; Margon et al.
1974) and the HEAO -1 satellite have now detected the existence of
stellar coronae with temperatures between 5×10^5 K and 10^7 K. The
discovery that Capella has an isothermal corona of temperature 10^7 K
(Cash et al. 1978c) had an immediate impact on theories of stellar
coronae. All the models, including Landini and Fossi (1973) and Hearn
(1975), relied on gas gravitationally bound to the star, but Capella
for example cannot gravitationally bind any gas at temperatures greater
than 1.5×10^6 K. It seems likely that the coronal emission of Capella
comes from magnetically confined gas, in which case we may expect that
the corona of Capella can be a testing ground for newer theories of
the solar corona.

    A significant topic of debate with regard to the solar corona, is
the question of how mass motions, which presumably power the corona, can
dissipate their energy. The classical mechanism (Schwarzschild 1948)
was that the mass motions sharpen into supersonic shock waves as they
climb into the upper and less dense regions of the atmosphere. Direct
observations of mass motions in the sun, however, showed that the mass
always moves at subsonic velocities. (Boland et al. 1975). This
caused some theoretical re-evaluation of the shock processes, and in
particular, McWhirter et al. (1975) showed that the growth of shock
waves would be inhibited by dissipation due to thermal conductivity
and viscosity. Furthermore, it has now been shown that non-LTE effects
can make a great difference in the evolution of the perturbations
(Kneer and Nakagawa 1976). Even worse, it is well known that magnetic

fields are a dominant factor in solar gas motions (Parker 1977;
Withbroe and Noyes 1977), but none of the above theories have attempted
to include these effects. Recently, Rosner et al (1978) have developed
an entirely new tack in that the coronal temperatures can be explained
through anomalous dissipation of plasma currents in the solar magnetic
field. Rosner et al. (1978) and Craig et al (1978) have independently
proposed a scheme for energy balance in the solar corona based on the
geometry of magnetic flux tubes. Since the magnetic fields are the
source of confinement, the predictions of the new theory (the Loop
theory) are markedly different from those that have gone before. The
Loop theory has also been successful in explaining the isothermal
character of the solar corona. More recently the simple Loop Theory has
been attacked because it fails to explain the temperature of single
loops. The models proposed to model individual Loops in detail predict
different overall properties of stellar coronae which can be tested by
spectroscopic observations of other stars.

A primary observing goal in EUV spectroscopy of nearby stars is to
find the distribution of gas temperatures in the chromospheres and
coronae. These gases should radiate the bulk of their thermal emission
in lines, although significant continuum emission also occurs. In the
case of the Sun, extensive observations in the EUV exist with high
spectral resolution. In Figure 11 we show a composite solar EUV
spectrum of relative line intensities in 5 Å bins obtained by combining
recent observations by Malinowsky and Heroux (1973), Firth et al. (1974)
and Behring et al. (1976). Because of the structure of atomic ionization
levels, there is a gross inverse correlation between the wavelength of
emitted lines and the temperature of the emitting region. At the wavelengths longer than 400 Å the strongest lines are characteristic of
temperatures of several hundred thousand degrees. Through these observations, we should learn much about the relationship between the chromosphere and corona, and the role played by the transition region. We
should be able to test the magnetic loop theory for stellar coronae
through the observed temperature distribution. As can be seen in
Figure 11, moderate resolution is sufficient to clearly distinguish
line structure. Other types of objects may possess a hot corona, especially active dwarfs that show strong He 10830 Å emission that is
probably coronally associated in some way (ε Eridani, 61 Cyg, 70 Oph
for example) and red giants and subgiants as suggested by Hills (1973),
provided they are losing mass sufficiently rapidly. Moderately hot
white dwarfs also could possess hot coronas and their observation in
the EUV might produce some intriguing results (Muchmore and Bohm, 1978).

A wilder speculation in this context concerns the possibility of
observing coronas of subdwarfs in the EUV. As Schwarzschild (1958) has
pointed out, a measurement of the helium abundance in a subdwarf could
constitute an essential clue for cosmological problems in that it would
tell us something about the initial abundance of this element in the
galaxy. Since no nuclear transmutations have altered the helium
abundance in subdwarfs from their birth at the beginning of the galaxy,
the measurement would be quite direct. Unfortunately their spectral

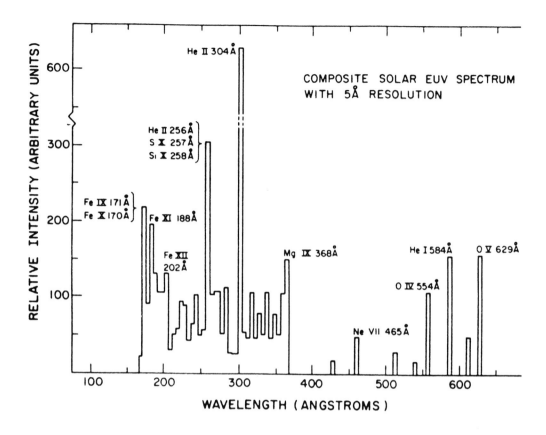

Figure 11. Composite solar EUV spectrum.

types are too late to permit classical optical determination of the
helium content of the photosphere, but if these objects possess hot
coronas, neutral and ionized helium might be observable (as in the
case of the Sun) through their strong resonance emissions at 584 and 304
Å.

Another important class of nearby stars which show strong evidence
for coronal and chromospheric activity is the group of binary systems
known as RS CVn variables.

A typical RS CVn system (Popper and Ulrich 1977) consists of a
hotter component (late F or G) on or just above the main sequence and
a cooler, evolved component (generally IV) in a close binary orbit (1-14
day periods). The cooler component exhibits chromospheric activity by
the presence of strong and variable emission lines CA II H and K, and
occasionally of Hα. These systems are also charaterized by a small
but pronounced dip in the light curve which migrates toward decreasing
binary phase and varies in amplitude on 10-30 year timescales. Other
peculiarities include random changes in the binary period, UV excesses,
quiescent radio emission and tremendous radio flaring. Recent HEAO-1

observations (Walter et al. 1978) have shown that these systems are also strong quiescent x-ray sources.

The canonical model (Hall 1972) has 40% of one hemisphere of the cooler component covered with starspots. An equatorward drift of the spots (analogous to that of the sun), coupled with differential rotation in the star, provides for the observed wave migration. The period changes are due to mass loss through coronal holes in the hemisphere free of spots. The vastness of the spotted area implies great surface activity, which leads to an active chromosphere and the observed flaring behavior. The large amount of energy deposition, and the ready availability of magnetic flux in the spotted regions, leads to the large, bright corona implied by the x-ray observations. In summation, these are similar to the active sun, but many orders of magnitude more active.

Observations with the HEAO-1 spacecraft have shown that ten of the RS CVn systems, and possibly an eleventh, are sources of soft x-ray sources.

The soft x-ray emission (.28 keV) arises in a plasma of temperature 10^7 K and volume emission measure $\sim 10^{54}$ cm^{-3}. Typical soft x-ray luminosities are $\sim 10^{31}$ erg s^{-1}. These parameters are in accord with the predictions of the Loop model of stellar coronae (Rosner et al. 1978), and tend to rule out steady state coronal models.

EUV observations of these active systems will provide the most direct measurements of the properties of gas in the lower corona/upper chromosphere. The 10^7 K gas which produces the x-radiation must be supported by cooler gas underneath. In analogy to the active sun, RS CVn systems would be expected to produce copious amounts of EUV radiation. Due to their high space density, many of these systems are close to the sun and hence will be observable with little interstellar alteration.

Dwarf novae

This is the U Geminorum-type very short binary consisting of a red dwarf spilling matter on to a close hot white dwarf companion. These exhibit quasi-periodic outbursts of several magnitudes and sometimes at maximum, rapid optical flickering. Dwarf novae must be similar to the binary x-ray emitters Her X-1 and Cen X-3, except that instead of a neutron star we are dealing in this case with a white dwarf. This similarity can be pushed a little further to calculate the typical dwarf nova spectral intensity, by realizing that since the masses of the two systems are probably very similar (≈ 1 M$_\circ$) the ratio of the energies must equal the ratio of the radii. Adopting R(White Dwarf) $\simeq 10^9$cm, R(neutron star) $\simeq 10^6$cm, and a characteristic spectrum of a binary x-ray source as kT $\simeq$ 12-15 keV, we obtain kT(White dwarf) $\simeq$ 12-15 eV with characteristic temperatures in the $10^4 - 10^5$ °K range. Consequently we should expect them to be very bright in the EUV, and the observation of SS Cygni on ASTP bears out this expectation. But the exciting point here is that although there is only one known x-ray binary in a 2 kpc radius, within

300 pcs of the Sun there are at least 60 known U Geminorum type variables
with characteristic flickering outbursts. Even if we can observe a small
fraction of these in their main emitting range, the physics of the out-
burst mechanism and of the low level oscillations both in these and the
x-ray systems they closely resemble will have received a very
significant impulse.

Flare stars.

Flare stars are late red dwarfs (dMe classification), some, but
not all, in the binary systems showing characteristic rapid increases
in optical flux of a few magnitudes lasting for about half an hour one
to ten times a day. There are theoretical reasons to believe that they
are intrinsically powerful sources of soft x-ray and EUV radiation, but
the most compelling arguments for this are those deduced from a combined
radio/optical study of a giant flare from a typical member of this class,
UV Ceti (Kahn 1974). These calculations indicate the presence of a
2×10^6 °K plasma and a total energy dissipation of about 10^{30} ergs
sec $^{-1}$ during the flare. If such conditions are typical of flare stars,
they should be easily detected in the EUV. Again, this conjecture has
turned into reality with the detection of Prox Cen on the ASTP mission.
But many others should be easily observable (UV Ceti, AD Leo, Wolf 424
and EV Lac, just to name a few within 5 pcs of the Sun), up to 500 A
for the closest ones. Recently two such objects have been observed in
the soft x-ray region (Heise et al. 1975), so that simultaneous EUV
and x-ray observations could be an exciting reality in the next few years.

Planetary nebulae and ultraviolet stars.

Being right in the allowed region for EUV detection in the H-R diagram of Figure 9, they are prime candidates for observation in this
region of the spectrum. Those that we know of are obviously only the
ones that are observable in the optical; but as we mentioned earlier,
very little of the total luminosity of these objects ends up in that
region to start with. More than 90% is radiated in the EUV and the
only problem here is the attenuation of the surrounding nebula, if it
exists and if it is uniform enough. Recent calculations (Robbins and
Bernat 1974) indicate that the continuum opacity of these nebulae might
be quite low leaving plenty of room for EUV photons from the nucleus
to escape. Quite a few known highly luminous planetary nebulae exist
within 500 pc of the Sun: NGC 7293, 6853, 650, 246, 3587, and VV 127,
with the last one emitting 3.3×10^{37} ergs sec $^{-1}$ in the EUV, easily
measurable with even modest exposures.

Our search need not be restricted to a limited few of the optically
identified planetary nebulae, since as Hills (1972) has pointed out, the
nebula at some point expands to invisibility leaving the nucleus the
same copious EUV emitter as before (if anything, brighter, because of
the absence of absorption in the nebula). But now we cannot identify
it easily as planetary since there is nothing to draw attention to its
peculiarity. Hills has named these objects UV stars and discussed

extensively their properties (Hills 1973, 1974) showing that, if they exist, they should have luminosities on the order of 10^{35} ergs sec^{-1} and temperatures around 10^5 °K. Assuming these are lost planetary nebulae nuclei that will in time appear as white dwarfs (a possible evolutionary sequence that might or might not be distinct from the hot white dwarf phase we discussed earlier) it is a straightforward task to compute their density. About 300 UV stars should be found within about 200 pcs of the Sun if these considerations are correct. Again, neutrino cooling might upset this picture. We should stress here that only a survey experiment could locate them, as they lack, at least a priori, an optical identification.

O-B Subdwarfs.

In an alternative evolutionary sequence, stars bypass the planetary nebulae stage completely and pass through the hot O-B subdwarfs which possess inert helium cores and substantial hydrogen burning shells (Trimble 1973). They should have T in the 10^4-10^5 °K range, and luminosities up to 10^{37} ergs sec $^{-1}$. How they can be distinguished, if at all, from the UV stars is still not clear but they do have one distinguishing feature: a few of them have already been identified optically (Greenstein and Sargent 1974) as typical sdOs. These include HZ 22 and 44 with luminosities greater than 10^{35} ergs sec $^{-1}$, and Abell 31, Feige 34, and BD+28°4211 probably having temperatures greater than 50,000 °K. But there may be many more optically unidentified stars of this type.

Neutron stars.

Another class of objects that might in principle heavily contribute to the population in the EUV "region" of the H-R diagram shown in Figure 9 are dead or cooling pulsars. It is in fact strange that pulsars seem to disappear after their period lengthens past a 4 second threshold. Ostriker, Rees and Silk (1970) have proposed that these "dead" stars actually far outnumber the known radio objects, and may be common members of the solar neighborhood, with space densities approaching 3×10^{-2} pc^{-3}. Not all, perhaps, will be efficient EUV emitters, since to be such the neutron stars must accrete interstellar matter at a rapid rate. This rate is a very rapidly decreasing function of increasing relative velocity of the star with respect to the interstellar gas, so that only those stars moving very slowly with respect to the gas can hope to be heated to temperatures of order 10^5 °K by this effect. But even if only 10% of dead pulsars have the necessary low velocities this would still entail, if the initial estimates are correct, 10^4 objects of this type within 100 pcs of the Sun.

Furthermore it is not certain at this point what the black body emission spectrum of a normal neutron star looks like. The major un- certainty is the cooling rate, and recent work (Greenstein and McLintock 1974) has indicated that it might be low enough to permit detection of post-supernova neutron stars having temperatures of $\sim 3 \cdot 10^5$ °K and

luminosities of order 10^{33} ergs sec^{-1} ! As we observed earlier in
connection with our upper limits on three nearby pulsars, the X-ray
data may be a poor indicator of these parameters, since it utilizes
only the rapidly decreasing tail of the appropriate blackbody emissivity
curve. Consequently, sensitive EUV observations (or even upper limits)
might constrain cooling or equation of state theories to a greater
degree than crude measurements reported from the ASTP mission
(Greenstein et al., 1977).

Main sequence close binaries.

Another way of producing EUV emission with a system that is not
itself particularly hot is by transferring gas between two close binary
companions. Since close binaries are so common in the solar neighborhood (N > 3 x 10^{-3} pc^{-3}), they should be given more than glancing consideration here. When the transferred gas thermalizes near the surface
of the primary, it may produce considerable EUV emission. Temperatures
of 10^5 or 10^6 °K and luminosities of order 10^{31} erg sec^{-1} are typically
expected. Observations of close binaries frequently yield evidence
for gas clouds, rings or streams indicating that mass transfer at the
required rate may be very common. Recent soft X-ray and far UV observations of Capella (a spectroscopic binary consisting of two G stars)
seem to confirm this possibility (Mewe et al. 1975; Catura et al. 1975;
Dupree 1975). No great difficulty in observing such objects out to
many tens of parsecs should be encountered with even presently available
instrumentation. Observations of this type might go far in elucidating
the complex physical mechanisms that control the dynamics of mass
accretion of stars in the main sequence.

The diffuse EUV background.

The presence of OVI absorption lines in the spectra of hot stars
has been interpreted (but see Weaver et al. 1977) as evidence for
the existence in interstellar space of a pervasive, tenuous coronal
gas of temperature T > 10^5 K. There is at present considerable uncertainty regarding both the spatial extent and temperature distribution
of this gas (Jenkins 1978). It has been suggested (Cox and Smith
1974) that the soft X-ray background is due to the same medium,
although the temperature regimes to which each method are most sensitive
are fundamentally different (T > 10^6 K for the X-ray regime, T < 10^6 K
for the OVI regime). This hypothesis has been questioned (Jenkins 1978;
Tanaka and Bleeker 1978) and there is growing evidence that the two
techniques are sampling different gases or emitting regions.

The extreme ultraviolet (EUV) could represent the crucial link
between the two temperature regimes. As Paresce and Stern (1979) have
pointed out, the observation of the OVI absorbing interstellar material
in emission would determine the filling factor of the gas and rule out
a stellar-related emission mechanism.

The measurement of the level and shape of the diffuse soft EUV

background spectrum is of crucial importance. In addition a sensitive,
moderate spatial and wavelength resolution survey of the sky would be
of fundamental importance for understanding the extent and distribution
of the emitting plasma and the absorbing clouds out to $\sim$ 1 kpc of the
sun. EUV photons have sufficient energy to ionize, dissociate and heat
the interstellar medium. Although it is likely that a large fraction
of the ISM heating and ionization is due to shock wave interactions
with the hot coronal gas and the conduction interfaces of cooler clouds
(McCray and Snow, 1979; McKee and Ostriker, 1977), the effect of UV
photons on the smaller scale in the transition region between the cold
cloud cores and the hotter conduction interfaces may be quite important
(Meszaros, 1974, Grewing, 1975). For example, the proposed heating
mechanisms of diffuse interstellar clouds require UV photons to
photoeject electrons from grains (Watson, 1972;de Jong, 1977) or to
excite H_2 to vibrationally bound levels (Stecher and Williams, 1973).
Absorption of UV radiation by H_2 controls the abundance of this and
other molecules in cool clouds (Hollenbach et al. 1971 Jura, 1974).

   Most of the EUV radiation in the solar neighborhood probably originates in O and B stars (Jura 1974; Grewing 1975; Torres-Peimbert et al.
1974) within 1 kpc of the Sun. Just below the HI ionization threshold
at 912 Å, the mean free path of UV photons is small (see Figure 1) and
the radiation field is characteristic of a smaller region around the Sun
out to $\simeq$ 100 pc or so. The field in this casewill be dominated by the
combined effects of the $\approx$10 early B stars to be found at these distances
from the Sun (Allen 1973) and hot white dwarfs, subdwarfs, novae and
other possible EUV sources discussed in this article. Enough radiation
below 912 Å will seriously affect the He/H ratio in the ISM and observations of the interstellar/interplanetary wind (Grewing 1975; Blum
and Fahr 1976).

   Because of the interest generated by these considerations, a number
of theoretical or empirical determinations of the expected energy density
in the EUV have been recently attempted. For heating, ionization and
dissociation of ISM atoms and molecules the important parameter is clearly
the mean intensity of the EUV radiation field or the equivalent energy
density U_λ. A compilation of the results of the calculations of U_λ in
ergs $cm^{-3} Å^{-1}$ has been given by Paresce and Jakobsen (1980).

   The computations are hampered by many of the uncertainties encountered in the context of establishing the total luminosity of the galaxy
since they depend in one way or the other on an integration of all the
possible galactic sources of EUV radiation. Specifically, the uncertainties refelct our incomplete knowledge of the flux and space density of
hot stars and of the distribution and scattering characteristics of dust
grains. The calculations are, in principle, very sensitive to the exact
location in space but as long as the chosen spot is well removed from
the vicinity of very hot stars and associations, this effect, in practice,
is not very important (Habing 1968).

   A completely different way of approaching the subject was proposed

by Grewing (1975). Assuming the fractional ionization of C, N, Mg, Al,
Si and S in the ISM as determined by the UV telescope on Copernicus,
is due to photoionization, Grewing computes the required relative EUV
radiation field and converts it to an absolute spectrum by requiring
consistency with Witt and Johnson's (1973) computed field at 1900 Å.
The energy density obtained in this self-consistent way thus fulfills
several important requirements. The estimate is completely independent
of assumptions as to the source of the emission, the ionization of H to
the correct level corresponding to the intercloud medium in the vicinity
of the Sun and the maintenance of the observed ionization of the heavier
elements, are both accomplished simultaneously with a spectrum which
resembles that of a B1.5 star diluted by a factor of $\approx 10^{15}$, and by an
ISM having a local electron density n_e of 0.2 ± 0.1 cm^{-3}. The absolute
values of the flux depend linearly on n_e and a vertical scaling to match
observations rather than calculations would be more appropriate.

Unfortunately no observations of EUV interstellar radiation density
exist at present. The measurement requires coverage of a fair portion
of the observable sky in order to estimate the flux from all directions.
Measurements of the field intensity of a small number of view directions,
as most of the observations performed up to now tend to be (Paresce and
Bowyer 1976; Sandel et al. 1979), set no interesting constraints on
U_λ unless one knows a priori the spatial variation of the incident
radiation at that wavelength.

VI.   Future Experimental Possibilities

A systematic study of the EUV sky will require substantially better
sensitivities than what was available on the ASTP mission. The EUV
telescope on this mission does not represent the ultimate in instrument
sensitivity and resolution and we can foresee a number of improvements.
We should point out that we wish not only to increase the efficiency of
EUV instrumentation but also minimize the noise in the regions where it
represents the limiting factor to instrumental sensitivity. On this
basis we can identify two areas where considerable improvement can be
expected: observing conditions and detector design.

Let us first consider how we could improve our viewing conditions.
As the constraints are set principally by the geocoronal and interplanetary background we must find particular configurations of the telescope
position and pointing that minimize its contribution. In Figure 12 we
show the expected intensity of a typical night EUV sky at about 500 km
altitude excluding astronomical compact sources but including the diffuse
background contributions discussed in §III and the terrestrially or
interplanetary associated emission lines one must contend with. There
are two categories of the latter that must be considered separately
because of their morphological differences. The first and dominant
contribution in certain regions of the EUV spectrum are the expected or
observed ionic emission lines of He$^+$ at 304 and 256 Å, of O$^+$ at 834,
719 and 485 Å, of N$^+$ at 776 and 746 Å, O^{++} at 834, 703, 508, 600 and
304 Å and finally N^{++} at 991, 772, 764 and 686 Å. All these lines are

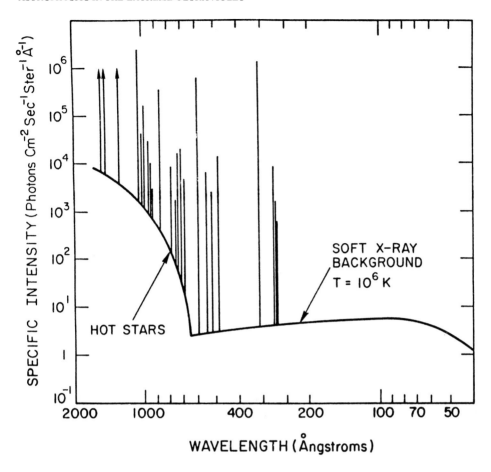

Figure 12.   Typical expected specific intensity of the back-
             ground night sky radiation field with contributions
             from all sources excepting possible compact astronomical
             objects.

reasonantly scattered into the night hemisphere by the sunlit ionized
components of the Earth's plasmasphere.

   Any broadband astronomical observation in this critical region is
bound to be severely affected by these emissions. Not only is this
background quite high (it can reach 10^6 photons cm^{-2} sec^{-1} strdn^{-1})
but also extremely variable depending on spacecraft position in the
orbit and viewing direction.   There is one way to reduce it considerably,
however, and that is to observe as far away as possible from the
terminators and with the telescope field of view completely contained
in the shadow cone of the Earth.   In this way plasmaspheric ions in the
viewing path are not sunlit and the resulting emission is only due to
a small residual multiscattered and interplanetary component that is at

least two orders of magnitude less intense than the maximum background
previously quoted. Obviously the whole sky cannot be covered at once
in this configuration since the dark circular EUV patch is approximately
60° in diameter around the antisolar point. If it can be contrived to
observe in the patch every day of the year, however, a substantial area
of the sky around the ecliptic can be explored at maximum instrument
sensitivity in those regions where ionic emission lines dominate. For
coverage of the rest of the sky towards the ecliptic poles with broad
band low resolution instruments having comparable sensitivities we
would have to resort to observations from a polar orbiting platform.

The background contribution from emission lines of the neutral
elements H (1026, 973, 950, 938, 931 and 926 Å) is different because
1) the geocoronal emission source is evenly distributed around the
Earth out to very large distances 2) the optical depth is very high
and 3) interplanetary contributions are relatively large. Little if
anything can be done about minimizing the hydrogen contamination above
900 Å solely by selecting the appropriate viewing conditions. These
emissions are strong (10^5 to 10^6 photons cm^{-2} sec^{-1} strdn^{-1}) and
uniformly distributed in space. A very shallow minimum is encountered
at local midnight. The interplanetary H contribution peaks in summer
and is minimum in the winter but this variation is only about a factor
of two (Thomas and Krassa 1971). Although there is considerable
uncertainty about the magnitude of the terrestrial helium emission
(Freeman et al. 1976) it is most probable that interplanetary emissions
dominate (Paresce et al. 1974b). These can be quite intense (nearly
10^6 photons cm^{-2} sec^{-1} strdn^{-1} at 584 Å) and highly sensitive to
season (Paresce and Bowyer 1973) with a factor of ten variation from
winter peak to summer minimum common. Consequently if one wishes to
observe in the 500 to 600 Å range with broad band coverage encompassing
these lines for continuum observations or because one wants to detect
helium coronal lines, the best observing condition is at local midnight
viewing in the antisolar direction in summer.

Another but more insidious source of extraneous counts are particles
and particle generated photons. Excepting in the South Atlantic Anomaly
and the auroral regions, the direct effect of particles both charged and
neutral can be minimized by appropriate magnetic and electrostatic
shielding. A subtler problem might be encountered at 910 Å where an
oxygen recombination line peaking in the tropical arcs near the equator
could seriously contaminate measurements in this very delicate region
around the H ionization edge if observations are carried out at or
around ± 20° geomagnetic latitude at any longitude. Sensitive
observations should be avoided in or near the auroral regions as they
represent almost certainly a strong source of EUV photons due to charged
particle impact on the upper atmospheric constituents. Daytime observations are usually ruled out owing to the intensity and richness of
the expected and observed dayglow spectrum (Christensen 1976). Atmospheric absorption should also be minimized by selecting an orbit or
an apogee that is high enough to reduce absorption to tolerable levels
even when viewing towards the horizon but low enough to avoid the radi-

ation belts. This optimum altitude is in the range 400 to 500 km,
approximately. It is interesting to note than many of the constraints
on optimal EUV observational configurations listed here were in one way
or another violated by a large majority of the ASTP observations of
compact sources (the worse case being the observations of Wolf 424 in
daytime!). This leaves one with the distinct feeling that many of the
sources which were observed on the ASTP flight and yielded only upper
limits might have been detected under more appropriate conditions.
Summarizing we can conjecture that an improvement of at least a factor
of ten in sensitivity can be obtained in certain EUV bands for any instrument provided the viewing conditions are carefully selected.

   Now let us see what room for improvement exists at the instrument
level itself. The ability of a photon detector to detect and measure
the brightness of any object can be expressed in terms of the statistical
uncertainty in the intensity estimators of the object calculated from
the image output data (Lampton et al. 1975, Paresce et al., 1978).
The detectability of any source can be evaluated by constructing
the estimator I* of its brightness I and calculate its variance or mean
square deviation var I*. The signal is recovered by subtracting from
the on source count an estimate of its background derived from the off
source count as in the ASTP observations yielding:

$$I^* = \frac{1}{AQT} \, [n_s - \frac{\Omega_s}{\Omega_B} \, n_B]$$

where A = detector collecting area
      Q = system quantum efficiency averaged over a given wavelength band
      T = source accumulation time
      n_s = expected count rate on source
      n_B = expected count rate on background
      Ω_s = solid angle occupied by source
      Ω_B = solid angle available for background determination

   Under not very restrictive conditions, usually well met by most
detector systems, the variances from the two terms in this equation
add, yielding:

$$var \; I^* = \frac{I}{AQT} + (1 + \frac{\Omega_s}{\Omega_B}) \, \Omega_s \frac{B+Rf^2/AQ}{AQT} + (1 + \frac{\Omega_s}{\Omega_B}) \frac{\Omega_s}{w} \frac{E^2}{(AQT)^2}$$

where B = background brightness in the given wavelength band
      R = inherent detector background count rate
      f = focal length of the optical system
      w = elementary solid angle corresponding to one image element
      E = the root mean square readout noise of the detector per image
          element

   We shall not discuss all the implications of this last fundamental
relation which must of course be minimized for optimal detector performance but restrict ourselves to a brief consideration of its qualitative
behaviour. Every term reflects the obvious and well expected fact that

accurate detection and measurement of any object demands large AQT products;   long observations with large effective areas.   Beyond this dependence, however, what is of crucial importance to us is the appearance of the image format options that control the variables w, Ω_S and Ω_B. The fact that they appear in this last equation offers the possibility of maximizing the ultimate sensitivity of photometric detection and analysis systems through appropriate detector design, format choice and image processing.   The first term in the last equation we shall consider negligible in comparison to the others since it dominates only for bright objects and we wish to maximize detectability of faint objects. The second term shows that the consequences of the external and internal backgrounds B and R can be minimized by decoupling Ω_S from Ω_B i.e., minimizing Ω_S and maximizing Ω_B.   Nonimaging collectors of the ASTP type in which $\Omega_S = \Omega_B$ then must be substantially inferior in terms of sensitivity to imaging systems whose final images can be tailored to precisely the source extent with a background field as large as possible.   We should expect on this basis alone considerable improvement in EUV sensitivity over the light bucket techniques by selecting imaging systems, since collecting areas are usually limited by more mundane and independent constraints such as available rocket or satellite space or fabrication difficulties.

Independently of the image format, we can obtain more performance out of any detector system if we use as the primary sensing element in the imaging system a pulse counting image detector for which $E^2 = 0$ and the last term in that var I* equation vanishes.   This is because pulse counting systems utilize digital memories for which the readout error is negligible in comparison with analog image memories such as those used in the MOS, CCD or vidicon integrating sensors.   A final variable not implicityly appearing in our discussion and not yet mentioned very often is wavelength resolution.   Although it is true that the best way to discover sources is by a relatively broad band high sensitivity detector system, in particular areas of the EUV spectrum unwanted background can be further and in some cases dramatically reduced by using adequate spectral resolution to avoid the EUV airglow emission lines.   In many cases even moderate resolutions of order 20 - 30 Å are sufficient for this purpose.   They can only be obtained however, with some sacrifice in effective collecting areas but this can be more than made up by the increased signal to noise ratio.

With these general remarks we have made a case for imaging optics coupled to pulse counting image detectors as the most promising technique for the highest possible sensitivity study of EUV astronomical emissions. We shall now present some simple instrumental configurations that we feel could advance considerably our knowledge of EUV astronomy in the Spacelab era.   The first experiment we shall describe is a broad band imaging EUV sensor which could be used to conduct a survey of the entire sky throughout the EUV range.   The sensitivity of this experiment is limited only by the information capacity or data rate which can be conveniently handled by a typical ground data receiving network.

A possible experiment configuration that fulfills the aims set
forth in the preceding paragraph and satisfies the many experimental
and mission constraints discussed in this section would make use of a
spinning satellite whose spin axis is controlled to point to the Sun.
Consequently a telescope or telescopes mounted with their axes perpendicular to the spin axis would sweep the entire sky systematically during
the first six months of operation. Three separate telescopes each of
which could have ∿100 cm^2 effective collecting area and a 625 resolution
element sensor could be easily accommodated on a typical size satellite
platform. A multi-telescope array of this kind each with its own broad
band filter has several advantages over a larger single telescope: it
eliminates the need for a moving filter wheel to provide broad band
energy discrimination, has higher sensitivity, and it has higher redundancy against detector and filter failure. This type of experimental
configuration also leaves ample room for a fourth telescope of ∿350 cm^2
effective area pointed in the direction antiparallel to the spin axis.
The antisolar direction is precisely the one identified earlier as the
one that minimizes external background and thus maximizes survey sensitivity in a small area of the sky. Another reason for increased sensitivity
is the threefold increase in exposure time for the telescope owing to the
small area of sky covered. This area can be increased somewhat by
slightly offsetting the view direction from the axis and causing it to
perform circles in the sky around the ecliptic. This telescope might be
equipped with either a filter wheel or a filter array at the focal plane.

The most promising optical configuration for this particular application consists of a Wolter-Schwarzschild Type I telescope (a slightly modified paraboloid-hyperboloid grazing incidence mirror combination). These
telescopes can be fabricated out of a single surface and do not have the
severe alignment problems of the Kirkpatrick-Baez (1948) telescopes. An
important design criterion amply satisfied by this type of optics is that
good images be achieved over a wide field i.e., across approximately the
central 5 degrees with a reasonable effective collecting area. The rms
blur circle radius increases to about 10 arcminutes at 2- 2.5° off axis.
This is probably the best performance one can expect from any of the presently available optical configurations.

The next important item is filters and the material and thickness
are both calculated to achieve maximum sensitivity for the circumstance
under consideration. This is done by using an algorithm to evaluate the
minimum detectable flux for any given EUV experiment for a wide range
of possible filter materials. From a careful quantitative study using
this maximization procedure we have selected the filters most suitable
for this purpose and the passbands they define are shown in Figure 13.

The effective area plotted in this last figure is computed assuming
the filters in question are coupled to a microchannel plate detector-
resistive anode combination pulse counting image sensor (Ranicon). It is
described in detail by Lampton and Paresce (1974). This type of sensor
has maximum and stable quantum efficiency in the EUV (Mack et al. 1976),
very low internal background, photon counting capability, large diameter

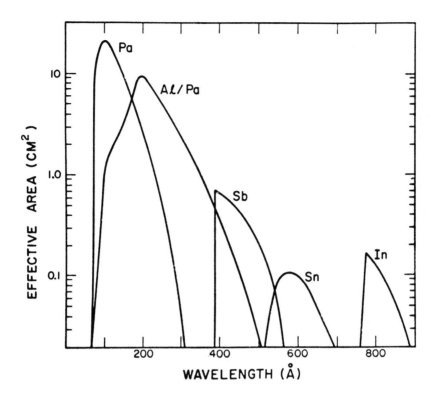

Figure 13. Effective area (effective collecting area x filter
            transmission x detector quantum efficiency) of the
            pointed experiment with optimized filters in place
            as a function of wavelength. Detector is the Ranicon.

(up to 75 mm) working field sizes, modest power requirements and excellent resolution (up to 10^6 picture elements). With this sensor the
sensitivity of any instrument is maximized independently of viewing
conditions or image formatting as it permits the last term in the var
I* relation to be vanishingly small.

    One qualitative advantage of this detecting technique for any ap-
plication but especially for manned pointed missions such as might be
conceived for Spacelab is the ability of an observer to preview and
interact with the image as it is being accumulated. An on-line live
TV display would be an immense benefit to the observer both in setting
up on an object and in making the most efficient use of valuable observing time. Image quality and signal to noise ratio would be visible in
real time and each exposure could be discontinued when sufficient information had been gathered. The separation of the detection and storage
functions is an added advantage permitting, for example, electronic
digital guiding, anticoincidence against energetic charged particles
and adaptability to a variety of image formats.

We can best summarize all the arguments presented in this section up to now by illustrating in Figure 14 the possible sensitivity of this kind of survey in terms of a minimum detectable flux. We wish to stress here that the attainment of these sensitivities is not pushing the outer limits of present feasibility but is actually comfortably within present day state of the art proven and reliable techniques. The amazing improvement of many orders of magnitude over the ASTP telescope is due to a wise and studied choice of optimum optical, sensor, filter and mission operations configurations. In this figure we also show the sensitivities of experiments in neighboring regions of the spectrum and for comparison purposes the observed and implied intensity of the first EUV source, HZ 43. In terms of energy flux incident at Earth one can conceivably obtain EUV observations of most of the celestial sphere at a sensitivity two orders of magnitude better than the National Geographic/Palomar Observatory Sky Survey in the optical band. With this type of instrument configuration all classes of sources discussed in § III are detectable to distances exceeding approximately 100 parsecs; many hundreds of sources would be potentially in range of such an observatory. Although it would not be optimized for diffuse background observations it would nevertheless represent a considerable improvement in every wavelength band it covers.

In the softer EUV range between approximately 1000 and 600 Å it would be very useful to complement the survey type broad band observations we have just mentioned with sensitive spectral measurements in selected regions of the sky. This could be accomplished with a simple normal incidence concave grating spectrometer of the type developed by Kumar et al. (1974). These are simple, extremely sensitive instruments that should give meaningful results even with moderate exposure times such as can be obtained on typical Spacelab flights (several thousand seconds) and resolutions of order 10 Å roughly.

This type of instrument in principle is capable of either detecting the soft EUV diffuse background or at the very least placing highly significant and severe constrainst on its intensity and spectrum. Thus detecting and studying both the soft and hard EUV background from Spacelab should be, with these instruments, relatively straightforward.

Bigger and more sophisticated payloads of the type described in and flown by Cash et al.(1978) and Malina et al. (1979a) having much larger collecting areas and resolution approaching 1 Å are currently being studied at Berkeley, Utrecht and Mullard Space Sciences Laboratory in the United Kingdom. The higher throughput and resolution would be made possible by the much larger capacity of the NASA space shuttle and the joint NASA-ESA Spacelab on which this instrument could be placed in the not too distant future as this symposium has made abundantly clear. The Space Research Laboratory at Utrecht has been actively investigating and testing the properties of EUV transmission gratings that would be used to yield the high resolutions required. This research has yielded very encouraging results. A similar amount of effort has recently been devoted to the improvement of the Ranicon detector both for what regards

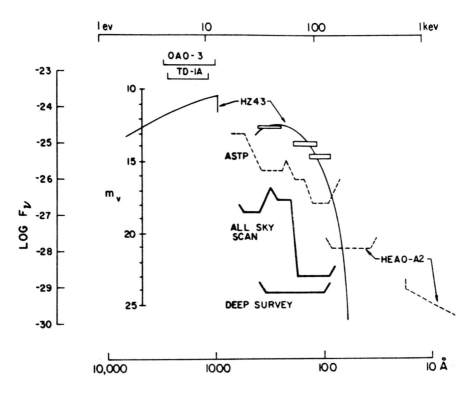

Figure 14.   Minimum detectable flux in ergs cm^{-2} sec^{-1} Hz^{-1}
              for a EUV sky survey as a function of wavelength.
              The heavy lines represent the sensitivity of the
              scanning (all sky survey) and the pointed (deep
              survey) telescopes described in the text. The
              wavelength bands are defined by filters of Sb,
              Al-C and Parylene in the all sky survey and a
              filter of Al-Parylene for the pointed instrument.
              The sensitivities of the ASTP, OAO-3, TD-1A and
              HEAO-2 are also shown for comparison. The ob-
              served and inferred spectrum of HZ 43 is also
              drawn in to illustrate the capabilities of the
              survey.

the microchannel plate itself as the resistive anode and the associated
readout electronics. As for the former recent advances have yielded
MCPs sensitive to their full geometric area, having lower noise, better
pulse height distributions and quantum efficiencies (see Lampton, 1976
for a review of this situation). As for the latter, a very recent
modification of the basic anode structure has permitted an exceptional
improvement in system resolution (Wijnaents van Resandt et al., 1976;
Lampton and Malina, 1976) to the point that spatial resolution in a
Ranicon is now solely limited by the MCP center-to-center spacing and
bore diameter.

To conclude, we think it is safe to say that the future of EUV astrophysics dawns bright and promising. If we do not take the full advantage of this fact it will not be due to technical or natural limitations but rather to the ever shrinking size of our national space budgets and aspirations.

Acknowledgments.

This survey could not have been written without the fundamental contributions of the entire University of California, Berkeley, EUV-ASTP team consisting of, besides the authors, Drs. Stuart Bowyer, Michael Lampton, Bruce Margon and Robert Stern. Dr. Sharad Kane contributed the section on solar EUV astrophysics. It is also a pleasure to thank Dr. Roger Malina and Dr. Webster Cash for their valuable contributions especially in the last section of this survey. We are also grateful to the Accademia Nazionale dei Lincei, Rome, Italy, and to NASA contract NAS9-13799 for partial support.

References

Allen, C.W., 1973, in Astrophys Quantities, London, Athlone Press
Aller, L.H., 1959, P.A.S.P. 71, pp. 324
Anderson, R.C. and Weiler, E.J., 1978, Ap. J., 224 pp. 143
Anderson, R.C., Henry, R.C., Moos, H.W., and Linsky, J.L., 1978,
        Ap. J., 226, pp. 883
Auer, L., and Shipman, H.L., 1977, Ap. J., 211 pp. L103
Behring, W.E., Cohen, L., Feldman, V., and Doschek, G.A., 1976,
        Ap. J., 203 pp. 521
Bleeker, J., et al., 1978, Astr. Ap., 69 pp. 145
Bohlin, H.D., 1976 in Physics of Solar Planetary Environments, Vol. I,
        (ed. D.J. Williams) A.G.U. Publ., pp 114
Bohlin, R.C., 1975, Ap. J., 200, pp. 402
Bohlin, R.C. Savage, B.D., and Drake, J.F., 1978, Ap. J. 224, pp. 132
Boland, B.C. et al., 1975, MNRAS, 171, pp. 697
Bowyer, S., Field, G. and Mack, J. 1968, Nature, 217, pp. 32
Bowyer, S. Margon, B. Lampton, M., Paresce, F. and Stern, R., 1977, in
        ASTP Summary Science Report, NASA SP - 412, 1, pp. 49
Bowyer, S., 1979, in Proceedings of IAU Colloquium No. 53, in press
Brune, W.H. Mount, G.H., and Feldman, P.D., 1979, preprint
Bunner, A., private communication
Cash, W., Malina, R., and Stern, R., 1976, Ap. J. (Letters), 204 pp. L7
Cash, W., Bowyer, S., and Lampton, M., 1978a, Ap. J., 221 pp. L87
Cash, W., Bowyer, S., Freeman, J., Lampton, M. and Paresce, F., 1978b,
        Ap. J. 219, pp. 585
Cash, W., Bowyer, S., Charles, P., Lampton, M. Garmire, G., and Riegler,
        G., 1978c, Ap. J., 223, pp. L21
Cash, W., Bowyer, S. , and Lampton, M. 1979a, Astr. Ap., in press
Catura, R.C., Acton, L.W., and Johnson, H.M., 1975, Ap. J. (letters), 196,
pp. L47

Christensen, A.B., 1976, Geophys. Res. Letters, 3, pp. 221

Cox, D.P. and Smith, B.W., 1979, Ap. J., 189, pp. L105

Craig, L., McClymont, A., and Underwood, J., 1978, Astr. Apl, 70, pp. 1

Cram, L. E., 1977, Astr. Apl, 59, pp. 151

Cruddace, R., Paresce, F., Bowyer, S., and Lampton, M., 1974, Ap. J.,
    187, pp. 497

Cruddace, R. Bowyer, S., Malina, R., Gargon, B., and Lampton, M., 1975,
    Ap. J., 202, pp. 69

Davidsen, A., Shulman, S., Fritz, G., Meekins, J., Henry, R.C., and
    Friedman, H., 1972, Ap. J, 177, pp. 629

de Jong, T., 1977, Astr. Ap., 55, pp. 137

de Loore, C., 1970, Ap. Space Sci. 6, pp. 60

Donnelly, R.F., 1973, in High Energy Phenomena on the Sun, (ed. R. Ramaty
    and R. G. Stone) NASA SP-342, pp. 242

Dupree, A.K., 1975, Ap. J. (Letters), 200, pp. L27

Durisen, R.H., Savedoff, M.P., and Van Horn, H.M., 1976, Ap. J. (Letters),
    206, pp. 449

Firth, J., et al., 1974, MNRAS, 166, pp. 543

Freeman, J., Paresce, F., Bowyer, S., and Lampton, M., 1976, Ap. J.,
    208, pp. 747

Freeman, J., Paresce, F., and Bowyer, S., 1979, Ap. J., 23, pp. L37

Gorenstein, P. and Tucker, W., 1972, Ap. J., 176, pp 333

Greenstein, G., and McLintock, J.E., 1974, Science, 185, pp. 487

Greenstein, G., Margon, B., Bowyer, S., Lampton, M., Paresce, F., Stern,
    R., and Gordon, K., 1977, Astr. Ap., 54, pp. 623

Greenstein, J.L., and Sargent, A., I., 1974, Ap. J. Suppl., 28 pp. 157

Greenstein, J., and Oke, J., 1979, Ap. J., 229, pp. L141

Grewing, M., 1975, Astr. and Ap., 38, pp. 391

Habing, H.J., 1968, Bull. Astr. Inst. Netherlands, 19, pp. 421

Haisch, B., Linsky, J., Lampton, M., Paresce, F., Margon, B., and Stern,
    R., 1977, Ap. J. (Letters), 213 pp. L119

Hall, D.S., 1972, PASP, 84, pp. 323

Hearn, A.G., 1975, Astr. Ap., 40, pp. 355

Heiles, C., and Jenkins, E.B., 1976, Astr. Ap., 46, pp. 333

Heise, J., Brinkman, A.C., Schrijver, J., Mewe, R., Gronenschild, E.,
    den Boggende, A., and Grindlay, J., 1975, Ap. J. (Letters), 202,
    pp. L73

Heise, J., and Huizenga, H., 1979, Astr. Ap., in press

Henry, P., Cruddace, R., Paresce, F., Bowyer, S., and Lampton, M., 1975a,
    Ap. J., 195, pp. 107

Henry, P. Cruddace, R., Lampton, M., Paresce, F., and Bowyer, S., 1975b,
    Ap. J. (Letters), 197, pp. L117

Henry, P., Bowyer, S., Lampton, M., Paresce, F., and Cruddace, R., 1976a,
    Ap. J., 205, pp. 426

Henry, P., Bowyer, S., Rapley, C.G., and Culhane, J.L., 1976b, Ap. J.
    (Letters), 209, pp. L29

Hills, J.G., 1972, Astr. and Ap., 17, pp. 155

Hills, J.G., 1973, Astr. and Ap., 26, pp. 197

Hills, J.G., 1974, Ap. J., 190, pp. 109

Hollenbach, D., Werner, M., and Salpeter, E., 1971, Ap. J. 163, pp. 165

Holm, A.V., 1976, Ap. J. (Letters), 210, pp. L87

Humason, M.L., and Zwicky, F., 1947, Ap. J., 105, pp. 85
Jenkins, E.B., 1978, Ap. J., 220, pp. 107
Jura, M., 1974, Ap. J., 191, pp. 375
Kahn, F.D., 1974, Nature, 250, pp. 125
Kane, S.R., 1974, in Coronal Disturbances, (ed. G. Newkirk, Jr.) IAU
      Symp. 57, D. Reidel Publ. Co., Dordrecht, The Netherlands, pp. 105
Kane, S.R., 1975, Solar Gamma-, X-, and EUV Radiation, IAU Symp. 68,
      D. Reidel Publ. Co., Dordrecht, The Netherlands
Kane, S.R., and Anderson, K.A., 1976, Ap. J., 210, pp. 875
Kane, S.R., and Donnelly, R.F., 1971, Ap. J., 164, pp. 151
Kirkpatrick, P., and Baez, A.V., 1948, J. Opt. Soc. Am., 38, pp. 766
Klebesadel, R.W., Strong, I.B., and Olson, R.A., 1973, Ap. J. (Letters),
      182, pp. L85
Kneer, F., and Nakagawa, V., 1976, Astr. Ap., 47, pp. 65
Koester, D., 1978, Astr. Ap., 65, pp. 449
Kondo, Y., Talent, D.L., Barker, E.S., DuJour, R.J., and Modisette, J. L.,
      1978, Ap. J., 220, pp. L97
Kumar, S., Paresce, F., Bowyer, S., and Lampton, M., 1974, Appl Opt.,
      13, pp. 575
Lampton, M., and Paresce, F., 1974, Rev. Sci. Instrum., 45, pp. 1098
Lampton, M., Bowyer, S., and Paresce, F., 1975, UCBSSL No. 579 pp. 75
Lampton, M., 1976, Proceedings of the IAU Colloguium No. 40, Paris-
      Meudon Observatory Publications.
Lampton, M., and Malina, R. 1976, Rev. Sci. Instrum, 47, pp 1360
Lampton, M., Margon, B., Paresce, F., Stern, R., and Bowyer, S. 1976,
      Ap. J. (Letters), 203, pp. L71
Lampton, M., and Mewe, R., 1979, Astr. Ap., 78, pp. 104
Lampton, M., Tuohy, I., Garmire, G., and Charles, P., 1979, in (COSPAR)
      X-ray astronomy, ed. Baity and Peterson, Pergamon, Oxford, pp. 125
Landini, M., and Monsignori Fossi, B.C., 1973, Astr. and Ap., 25, pp. 9
Leibert, J., and Margon, B., 1977, Ap. J., 216, pp. 18
Lyutyj, V.M., 1976, Astr. J. Letters, Acad, Nau,, USSR, 2, pp. 116
Mack, J., Paresce, F., and Bowyer, S., 1976, Appl. Opt, 15, pp. 861
Malina, R.F., 1979, Ph.D. Thesis, University of California, Berkeley
Malina, R.F., Bowyer, S. and Paresce, F., 1979a, in (COSPAR) X-ray
      Astronomy, ed. Barty and Peterson, Pergauion, Oxford, pp. 287
Malina, R.F. Bowyer, S., Finley, D., and Cash, W., 1979b, in Proceedings
      of the SPIE, pp. 184
Malinovsky, M. and Heroux, L., 1973, Ap. J., 181, pp. 1009
Margon, B., Mason, K., and Sanford, P., 1974, Ap. J., 194, pp. L75
Margon, B., Liebert, J., Gatewood, G., Lampton, M., Spinrad, H., and
      Bowyer, S., 1976a, Ap. J., 209, pp. 525
Margon, B., Malina, R., Bowyer, S., Curddace, R., and Lampton, M., 1976b,
      Ap. J. (Letters), 203, pp. L25
Margon, B., Lampton, M., Bowyer, S., Stern, R., and Paresce, F., 1976c,
      Ap. J. (Letters), 210, L79
Margon, B., Szkody, P., Bowyer, S., Lampton, M., and Paresce, F., 1978,
      Ap. J., 224, pp. 167
McRay, R., and Snow, T., 1979, Ann. Rev. Astr. Ap., 17, pp. 213
McKee, C.F. and Ostriker, J.P., 1977, Ap. J., 218, pp. 148
McLintock, W., Linsky, J.L., Henry, R.C., Moos, H. W., and Gerola, H.,
      1975, Ap. J., 202, pp. 165

McLintock, W., Henry, R.C., Linsky, J.L. and Moos, H.W., 1978, Ap. J.
    225, pp. 465
McWhirter, R., Thoneman, P. and Wilson, R., 1975, Astr. Ap., 40, pp. 63
Meszaros, P., 1974, Ap. J., 191, pp. 79
Mewe, R., Heise, J., Gronenschild, E., Brinkman, A.C., Schrijver, J.,
    and den Boggende, A., 1975, Nature, 256, pp. 711
_____, 1975b, Ap. J. (Letters), 202, pp. L67
Muchmore, D., and Bohm, K., 1978, Astr. Ap., 69, pp. 113
Mullan, D.J., 1976, Ap. J., 209, pp. 171
Noyes, R.W., Foukal, P.W., Huber, M.C.E., Reeves, E.M., Schmahl, E.J.,
    Timothy, J.G., Vernazza, J.E., and Withbroe, G.L., 1975, in
    Solar Gamma-, X-, and EUV Radiation, (ed. S.R. Kane) IAU Symp. 68
    D. Reidel Publ. Co., Dordrecht, The Netherlands, P. 3.
O'Dell, C.R., 1968, Planetary Nebulae, (ed. C.R. O'Dell and D.E.
    Osterbrock) Dordrecht, The Netherlands, D. Reidel Publ. Co.
Oke, J.B., 1974, Ap. J. Suppl. 27, pp. 21
Ostriker, J.P., Rees, M.J., and Silk, J., 1970, Astrophy. Letter, 6,
    pp. 179
Paresce, F., and Bowyer, S., 1973, Astr. and Ap., 27, pp. 399
Paresce, F., Bowyer, S., and Kumar, S., 1974b, Ap. J. (Letters), 188
    pp. L71
_____, 1976, Ap. J. 207, pp. 432
Paresce, F., 1977, Earth and Extraterr, Sci., 3, pp. 55
Paresce, F., Bowyer, S., Cash, W., Lampton, M. and Malina, R., 1978,
    (COSPAR) New Instrumentation for Space Astronomy, ed. van der Hucht
    and Vaiana, Pergauion, Oxford, 1, pp. 77
Paresce, F., and Stern, R., 1979, Ap. J., in press
Paresce, F., and Jakobsen, P., 1980, UCB preprint
Parker, E., 1977, Ann. Rev. Astr. Ap., 15, pp. 45
Popper, D.M. and Ulrick, R.K., 1977, Ap. J., 212, pp. L131
Kiegler, G.R., and Garmire, G., 1975, Astr. and Ap., 45, pp 213
Robbins, R.R., and Bernat, A.P., 1974, Ap. J., 188, pp. 309
Rogerson, J.B., York, D.G., Drake, J.F., Jenkins, E.B., Morton, D.C.,
    and Spitzer, L., 1973, Ap. J., 181, pp. L110
Rosner, R., Tucker, W., and Vaiana, G., 1978, Ap. J., 220, pp. 643
Sagdeev, R., Kurt, V., and Bertaux, J., 1979, IAU Circular 3261
Salpeter, E.E., 1971, Ann. Rev. Astr. Ap. 9, pp. 127
Sandel, B.R., Shemansky, D.D., and Broadfoot, A.L., 1979, Ap. J.,
    227, pp. 808
Schwarzschild, M., 1948, Ap. J., 107, pp. 1
Schwarzschild, M., 1958, Structure and Evolution of the Stars, Dover,
    New York
Shemansky, D.E., Sandel, B.R., and Broadfoot, A.L., 1979, Ap. J., 231
    pp. 35
Shipman, H.L., 1976, Ap. J. (Letters), 206, pp. L67
Shipman, H., Margon, B., Bowyer, S., Lampton, M., Paresce, F., and
    Stern, R., 1977, Ap. J. 213, pp. L25
Shipman, H., and Wegner, G., 1979, Nature, 281, pp. 126
Stecher, T.P., and Williams, D.A., 1973, M.M.R.A.S., 161, pp. 305
Stecker, F.W., and Frost, K.J., 1973, Nature Phys. Sci., 245, pp. 70
Stern, R., and Bowyer, S., 1979, Ap. J., 230, pp. 755

Stern, R., and Bowyer, S., 1980, Ap. J., in press.

Tanaka, Y. and Bleeker, J.A., 1977, Space Sci. Rev. 20, pp. 815

Thomas, G., and Krassa, R.F., 1971, Astr. and Ap., 11, pp. 218

Thorstensen, J., Charles, P., Margon, B., and Bowyer, S., 1978,
    Ap. J. 223, pp. 260

Torres, Peimbert, S., Lazcano-Aranjo, A., and Peimbert, M., 1974,
    Ap. J., 191, pp. 401

Trimble, V., 1973, Astr. and Ap., 23, pp. 281

Troy, B.E. Johnson, C.Y., Young, J.M. and Holmes, J.C., 1975,
    Ap. J., 195, pp. 643

Walter, F., Charles, P. and Bowyer, S., 1978, 83, pp. 1539

Watson, W.D., 1972, Ap. J., 176, pp. 103

Weaver, R., McCray, R., Castor, J., Shapiro, P. and Moore, R.,
    1977, Ap. J., 218, pp. 377

Weidemann, V., 1968, Ann. Rev. Astron. Astrophys., 6, pp. 351

Wesemael, F., 1978, Astr. Ap., 65, pp. 301

Wesselius, P., and Koester, D., 1978, Astr. Ap., 70, pp. 745

Wijnaendts van Resandt, R.W., den Harink, H.C., and Los, J., 1976
    J. Phys. E., 9, pp. 503

Withbroe, G., and Noyes, R., 1977, Ann. Rev. Astr. Ap., 15, pp. 363

Witt, A.N. and Johnson, M., 1973, Ap. J. 181, pp. 363

Wolff, S.C., Pilachowski, C.A., and Wolstencroft, R.D., 1974, Ap. J.
    (Letters), 194, pp. L83

Yentis, D.J., Novick, R., and Vanden Bout, P., 1972, Ap. J., 177,
    pp. 365

York, D. G., 1976, Ap. J., 204, pp. 750

ULTRAVIOLET SPECTROSCOPY OF THE OUTER LAYERS OF STARS

Theodore P. Snow, Jr., Laboratory for Atmospheric and Space
Physics, University of Colorado at Boulder and
Jeffrey L. Linsky,[*] Joint Institute for Laboratory
Astrophysics, National Bureau of Standards and University
of Colorado, Boulder, Colorado   8039

I.   INTRODUCTION

In this article the discussion is limited to extended atmospheres
and circumstellar envelopes, for which fundamental advantages are
offered by ultraviolet observations. Thus, stellar winds, properties of
Be and shell stars, and stellar chromospheres and coronae are included,
while observations of stellar photospheres are not.

The presence of extended atmospheres in a variety of stars has been
inferred from visual-wavelength spectroscopy over several decades.
Among the hot stars, emission lines due to hydrogen, helium, carbon, and
nitrogen among others have been understood to form in non-LTE conditions
in the outer layers of stars. Among the late-type stars, the existence
of chromospheres was indicated by the presence of Ca II emission, and
the existence of coronae was suspected by analogy with the sun.

Because most atoms and ions in a low-density environment tend to be
in their ground state, and because the transitions out of the ground
level for most of them occur in ultraviolet wavelengths, extended atmospheres can be directly observed primarily in the ultraviolet. This
fact has provided the impetus for the development of a number of space-
borne telescopes over the past decade, and has in turn led to a pheno-
menal increase in our understanding of the outer layers of stars.
Rocket observations by Morton (1967), Carruthers (1968), and Stecher
(1968) were the first to reveal the existence of high-velocity winds
from OB stars and the importance of these winds in transporting material
away from stars and returning it to the interstellar medium. Ultraviolet balloon and rocket experiments, and especially Copernicus and the
International Ultraviolet Explorer, have led, through observations of
resonance-line emission, to a broadened understanding of the distribution and character of chromospheres and coronae in cool stars.

[*]Staff Member, Quantum Physics Division, National Bureau of Standards.

P. L. Bernacca and R. Ruffini (eds.), Astrophysics from Spacelab, 291-316.

This article will briefly review the recent results and will
outline important observational work still needed, both for the early-
type stars (§ II) and for cool stars (§ III). Several useful observations possible with instruments which are compatible with the Spacelab
format will be described.

II.   THE HOT STARS

A. Historical Perspective and Ground-Based Data

    Visual emission lines are observed in a variety of early-type
stars, and in many cases they represent evidence for mass outflow, showing velocity shifts (which may be variable) with respect to the rest
frame of the star. Only in the most luminous stars, however, is the
presence of strong winds evident from the visual-wavelength lines. When
the density in the outflowing material is sufficiently high, a P Cygni
profile may develop; this consists of an emission line roughly at the
rest wavelength, with superposed, blueshifted absorption. Profiles of
this type are named for the prototype early B supergiant star in which
some 137 such features are seen, with absorption velocities ranging
between -11 and -298 km s^{-1} (Struve and Roach 1939). Beals (1951)
developed a generic classification scheme for a variety of profiles, and
reviewed the visual spectroscopic data on a number of early-type supergiants.

    While the early results clearly indicated the presence of
outflowing material in extreme cases, it was not clear except for a few
examples (Wilson 1958) that the material actually escaped, nor was it
suspected that the phenomenon occurred in any but the hottest, most
luminous objects such as the Of stars and a few special cases like Wolf-
Rayet stars and P Cygni itself. In a series of papers Hutchings (1970
and references cited therein) showed, by correlating line formation
energy density with velocity, that as the temperature and density
decrease outward from an Of star, the velocity of expansion increases
(this had been found earlier by Struve 1935). This, and the fact that
infalling material is not usually seen, made it appear that the outflowing material accelerates to velocities sufficient to escape the star
entirely. It remained for UV experiments to confirm this by directly
observing material moving outward at speeds in excess of the escape
velocity.

    In the Be stars, which are dwarfs of spectral classes B0 to B6 or
so which show Balmer lines in emission (and occasionally other lines as
well), the emitting gas is usually at or near rest with respect to the
stellar frame. In many cases sharp circumstellar absorption lines also
appear, and these shell lines may show small velocity shifts. Oe stars
are similar to Be stars but are somewhat hotter, extending to spectral
types as early as O8 (Frost and Conti 1976). All of the emission and
shell absorption line phenomena can be variable, either randomly or
periodically. Be star envelopes are generally thought to form from

ejected stellar material, although there is some controversy over the means by which this ejection occurs. It is likely that rotation plays a role since most (and possibly all) Be stars are rapid rotators. In one view (e.g., Underhill 1966; Marlborough, Snow, and Slettebak 1978), the circumstellar material results from mass loss via stellar winds, while in a competing model this material arises from mass exchange in binary systems (e.g., Kriz and Harmenec 1975). Certainly many Be stars are binaries, but it is not clear that all are. The infrared excesses observed in many of them can be accounted for by free-free emission from ionized circumstellar gas, consistent with the single-star mass-loss hypothesis; or by stellar continuum radiation from late-type companions, consistent with the binary-star model. The books by Hack and Struve (1970) and Slettebak (1976) provide full discussions of these alternative views.

B.   Recent Ultraviolet Results

The first UV rocket spectra showed that a number of resonance lines in OB supergiants have widely-displaced absorption components, with maximum velocity shifts as large as -3000 km s^{-1} (Morton 1967; Carruthers 1968; Stecher 1968). Furthermore, mass-loss rates of order 10^{-6} M$_\odot$ yr^{-1} were derived from crude models of the outflow (e.g., Morton 1967); this is of sufficient magnitude to be important for stellar evolution (e.g. several articles in Conti and de Loore 1979, and references cited therein).

Only the most extreme stellar winds have sufficient density to produce displaced absorption detectable in visual wavelengths or sufficiently high velocities to be easily seen on the early, low-resolution rocket spectra. As a result of this, the widespread distribution of mass loss among early-type stars was not known until the Copernicus satellite was available, and now the IUE satellite, with its great sensitivity, has extended knowledge of mass-loss parameters to subluminous stars such as the central stars of planetary nebulae. In the HR diagram mass loss is seen to occur in nearly all OB stars with M$_{bol}$ < -6.0. Since the Snow and Morton survey was completed, Lamers and Snow (1978) have found that mass loss is evident in a number of fainter stars, and Heap et al. (1978) have found evidence for strong winds in the subluminous stars. Furthermore, Snow and Marlborough (1976) and Marlborough and Snow (1976) discovered that many Be stars with M$_{bol}$ > -6.0 have high-velocity winds. Therefore, the dividing line between stars that undergo mass loss and those that do not is not a clear-cut function of luminosity alone. This has recently been confirmed by Conti and Garmany (1980) who found large differences in mass-loss rates for stars of similar luminosity.

Besides providing the first comprehensive picture of the distribution of stellar winds in the HR diagram, the Copernicus data have provided substantial information on the physical conditions in the winds, such as the densities and temperatures. A high degree of ionization is present in all cases; ions such as O VI are observed in the

winds which cannot be produced in radiative equilibrium with the photosphere. If these are produced collisionally, as in the models of
Rogerson and Lamers (1975), Lamers and Morton (1976), Lamers and Snow
(1978), and Lamers and Rogerson (1978), then temperatures of order 10^5 K
are required. An alternative model (e.g. Cassinelli, Olsen, and Stalio
1978) hypothesizes a very hot $(\sim 10^{6-7}$ K), thin corona which produces the
observed ionization radiatively. These models will be discussed more
thoroughly in the following section.

The Copernicus data have provided information on a further aspect
of the stellar winds: variability with time. Although visual-wavelength
spectroscopy has shown that many hot emission-line stars have variable
line strengths and profiles (e.g., Rosendahl 1973a,b and references
cited therein), and the Be stars have been known for decades to undergo
transient phenomena (see Slettebak 1976 for extensive discussions), evidence has been lacking for variability among the most extreme mass-
losing stars. Recently, however, Conti and Niemela (1976) found a long-
term change in the Hα profile in ζ Pup (O4If), and York et al. (1977)
found from Copernicus data that the O VI absorption in three stars fluc-
tuates significantly in these of hours. More recently Snow (1977) surveyed some 15 stars for variability. Finding in most cases that at
least some details of the UV P Cygni profiles varied over times of
years, he suggested in a few cases that these changes represented real
variations in the mass-loss rates. New work on δ Ori A and ζ Pup (both
among the three reported by York et al. to have hourly O VI variations)
shows that small fluctuations occur on short timescales in a number of
ions, but no evidence for periodicity has been found (Snow and Hayes
1978; Wegner and Snow 1978; Snow, Wegner and Kunasz 1980). These new
data in both cases show weaker O VI fluctuations than those seen by York
et al. in data obtained 1-3 years earlier. Most likely, these short-
term variations represent random outbursts of material from the stellar
surface, since the time required for a fluid element to traverse the
entire observed region is of order 3 hours. This idea is supported
further by the observation that variations are usually stronger in the
visible wavelength features that form at low levels than in the ultraviolet profiles formed at high levels (cf. Leep and Conti 1979; Snow,
Wegner, and Kunasz 1980). This is consistent with the dissipation of
density enhancements as they flow outward.

It is possible that variability in stellar winds increases toward
the cooler supergiants. This is implied by the more commonly-observed
Hα variability in these stars, and is strongly borne out by the recent
high-resolution balloon ultraviolet data of Lamers, Stalio, and Kondo
(1978), who found drastic variations in the stellar winds of β Ori A
(B8Ia) and α Cyg (A2Ia).

It has already been mentioned that UV observations have revealed
the existence of stellar winds in a number of Be stars (Snow and
Marlborough 1976; Marlborough and Snow 1976; Lamers and Snow 1978).
Besides the high-velocity winds inferred from the extended short-
wavelength wings of lines due to such ions as Si IV, the UV spectra also

reveal a large number of shell absorption lines which may be displaced
from rest (Heap 1976; Marlborough 1977; Snow, Peters, and Mathieu 1979).
These UV absorption lines, which form in a lower-density region than the
visible-wavelength shell lines, generally show larger velocities of
expansion, indicating that the flow is accelerating outward. It is
tempting to speculate that Be stars simply represent mass-losing B stars
with sufficiently slow acceleration in their winds that a thick, nearly
stationary region exists to provide the observed emission and shell
lines. It is likely that the mass loss is catalyzed by rotationally-
induced turbulence, in view of the good correlation between rotational
velocity and both Be phenomena and stellar winds (Snow and Marlborough
1976).

Be stars also show wind variability, as shown by Marlborough, Snow,
and Slettebak (1978) and Slettebak and Snow (1978) for γ Cas and by
Marlborough and Snow (1979) and Doazan, Kuhi, and Thomas (1979) for 59
Cyg.

Since both the mass loss from luminous OB stars and the presence of
extended atmospheres in Be stars apparently can be understood in the
context of radiatively-driven stellar winds, the discussion of theoretical interpretations of these phenomena will be confined to this model.
More complete discussions of alternative models for Be stars can be
found in Slettebak (1976), Huang (1973, 1977), and Marlborough, Snow,
and Slettebak (1978).

C. Theoretical Interpretations

There are a number of competing theoretical interpretations of the
high-velocity stellar winds observed in hot, luminous stars. Each in-
vokes acceleration via radiation pressure, shown by earlier workers
(Lucy and Solomon 1970; Castor, Abbott, and Klein 1975) to be capable of
overpowering inward gravitational forces in hot stars. The differences
among the models arise primarily in the assumed source of the initial
wind acceleration and in the source of the observed ionization. All of
the models are faced with explaining a higher degree of ionization in
the winds than that observed in the underlying photospheres. Three
principal types of models may be characterized (Lamers and Snow 1978) as
hot coronal wind models, imperfect flow models, and radiation-pressure
models, which can be either cool or warm wind models. These models have
been summarized in detail by Cassinelli, Castor, and Lamers (1978) and
by Cassinelli (1979a).

In the coronal model proposed by Hearn (1975a,b), it is hypothe-
sized that the wind is driven by gas pressure in analogy with the solar
wind. For the OB-star winds, this requires temperatures in the coronae
of 10^{6-7} K. Serious constraints are placed on these models by the
observed degree of ionization in the winds, which is not as high as
would be expected for $T > 10^6$ K, and by the soft X-ray fluxes from O
supergiants such as ζ Pup (Mewe et al. 1975) and several stars in the
VI Cygni association (Harnden et al. 1979). It may be possible to avoid

both difficulties if the hot region is thin enough so that the observed
ions exist in a cooler zone outside of it. The final acceleration to
large velocities is due to radiation pressure acting on strong absorption lines in the cool outer corona.

The imperfect flow model (Cannon and Thomas 1977) relies on the
assumption that a non-thermal energy source prevents the star from
establishing a hydrostatically stable structure. The wind itself is
also assumed to be unstable, which results in heating by shocks.

In the cool radiation-driven wind models, the temperature in the
envelope is determined by radiative equilibrium. Since ions are invariably observed in the winds which do not exist in the photospheres, it
is necessary in these models to hypothesize an appropriate flux distribution for the ionizing radiation field. One possibility has been suggested by the models of Cassinelli (1979a, and references cited
therein), who points out that a very thin, very hot coronal zone just
above the photosphere could produce the observed ionization radiatively
without conflicting with the observed soft X-ray limits. These models
have received strong support from the Einstein Observatory detections of
soft X-ray fluxes from OB supergiants (Harnden et al. 1979; Cassinelli
1979b) and from Cassinelli and Olson's (1979) demonstration that the
ionization balance observed in the winds is just what would be expected
from Auger ionizations by soft X-rays from a corona.

In the warm radiation-driven wind models, it is assumed that the
observed ionization is collisional, implying kinetic temperatures in the
winds of order 2×10^5 K. This in turn requires non-radiative heating.
In a series of papers by Lamers and co-workers (Rogerson and Lamers
1975; Lamers and Morton 1976; Lamers and Snow 1978; Lamers and Rogerson
1978), these models have been developed in some detail. One attractive
feature is that heating above the photosphere, which produces ions such
as O VI and N V which do not exist in lower levels, provides a natural
way of enhancing the outward acceleration, since photons capable of
being absorbed by the strong UV resonance lines of these ions are not
blocked by photospheric lines (this advantage also applies to the
coronal models).

Another possibility is radiation-induced non-thermal motions,
suggested by Lamers and de Loore (1976) on the basis of observed turbulent velocities in early-type stars.

All three types of stellar wind models yield rather similar
estimates for the rate of mass loss, since this rate depends primarily
on knowledge of the wind density and velocity at some height and the
ionization balance, but is largely independent of the temperature. For
the well-studied O4If supergiant ζ Pup, the estimated mass-loss rates
range from 3.5×10^{-6} $M_\odot$ yr^{-1} to 9×10^{-6} $M_\odot$ yr^{-1}. These values may be
regarded as representing a rough upper limit on rates of mass loss by
radiatively-driven winds (except for the Wolf-Rayet stars); at the other
end of the scale, Lamers and Rogerson (1978) estimate a rate of about

10^{-8} $M_{\odot}$ yr^{-1}, and Snow and Marlborough (1976) crudely estimated
$M \sim 10^{-10} - 10^{-9}$ $M_{\odot}$ yr^{-1} for the Be star 59 Cyg.

Mass-loss rates of order 10^{-6} $M_{\odot}$ yr^{-1} are significant over the
lifetimes of massive stars, and hence, can strongly influence their
evolution. Bohannan and Conti (1976) have discovered binary systems
containing O stars in which the dynamically-derived masses are significantly less than those implied by the observed luminosities. de
Loore, De Greve, and Lamers (1977); Chiosi, Nasi, and Sreenivasan
(1978); Wilson (1977), Dearborn (1979); and Stothers and Chin (1979)
have computed stellar evolution tracks taking mass loss into account,
finding generally better agreement between observed and computed HR
diagrams than in the case when mass is conserved. These considerations
may be affected by the time variability in the mass-loss rates which was
discussed earlier. It is not clear that the changes in the mass loss
rates are large enough to significantly alter the evolution, however,
being only of order 50% in the two cases where quantitative assessments
have been attempted (Conti and Niemela 1976; Snow, Wegner, and Kunasz
1980).

D. Further Observations

There are three general areas in which new UV data on extended
atmospheres of hot stars are needed: (1) observations of fainter objects, extending the coverage of the HR diagram; (2) data on wavelength
regions not yet well covered, including transitions of important new
ions; and (3) better observations of time variability, including both
shorter timescales than have been covered so far, and simultaneous observations of different ions. These three types of new observations are
described more fully in the following.

1. UV observations of faint objects.

Copernicus and the IUE together have provided a fair representation
of the variety of objects with winds. Good coverage of both the main
sequence and the supergaint branch O4 to A1 exists, although with many
gaps and regions represented by only one or two objects. Coverage in
the giant branch is still very spotty. Several Wolf-Rayet stars have
been observed as well, and the coverage of Be stars, central stars of
planetary nebulae, binary X-ray sources, sub-luminous hot dwarfs, peculiar Be stars, and a variety of unusual objects is improving.

In order to fully outline the basic parameters of early-type stars
with extended atmospheres and winds, however, it is important to explore
as many different types of objects as possible. Besides improving our
understanding of the phenomena, this could also be useful in studying
their impact on stellar evolution.

It may be that the Be stars represent a phase in the evolution of
nearly all B dwarfs, but the statistical sample available to current UV
instruments is insufficient to support this hypothesis. Furthermore,

since UV spectroscopy is much more sensitive to the presence of circumstellar absorbing material than are visible-wavelength data, it is quite
possible that a wider-ranging survey of low-excitation UV lines, such as
those of Fe III (see Snow, Peters, and Mathieu 1979), would reveal a
number of stars with extended atmospheres which were previously unrecog-
nized. Indeed Copernicus data have shown the existence of circumstellar
shell absorption lines in at least one B dwarf not known to be a Be or
shell star.

The peculiar Be stars show evidence for the presence of infrared-
emitting circumstellar dust, and also are observed to have forbidden
Fe II emission lines. It would be interesting to analyze the physical
conditions in these shells by measuring the UV absorption lines as well
as by searching for emission lines (which have been seen in the UV so
far in only one Be star, γ Cas [Marlborough, Snow, and Slettebak 1978]).

A study of A, F, and G supergiants would be quite useful in
determining whether the mass-loss rates go through a minimum in the
stellar temperature regime where there is a transition from radiatively-
driven winds to coronal winds as suggested by microturbulence determinations (e.g. de Loore 1970).

Another class of object that is of particular interest is
represented by the central stars of planetary nebulae. It is clear that
the nebulae are formed by ejection of material from the central stars,
but very little is known about the process. While it is usually thought
that this ejection is gradual, it seems probable that objects with effective temperatures of order 10^5 K might also have high-velocity winds,
and this is confirmed by IUE data (Heap et al. 1978).

For the purposes discussed here, a resolving power of about 10^3 is
needed in order to delineate the P Cygni lines due to strong winds, and
$R = 10^4$ would be preferable for study of UV shell absorption lines,
which tend to be narrow.

2. New wavelength regions.

In order to carry out a detailed model calculation of stellar wind
parameters of a given star, it is necessary to have observed line profiles representing as wide a range of ionization potentials and excitation levels as possible.

Some of the early ultraviolet rocket experiments (e.g., Morton
1967) obtained UV spectra over a wide range of wavelengths, and more
recently the Skylab (Henize et al. 1975) and Orion 2 (Gurzadyan 1975)
objective grating spectrographs have obtained numerous continuous spectra extending quite far into the ultraviolet, but none of these achieved
sufficient spectral resolution to allow a detailed analysis of any but
the grossest properties of the stellar winds. A few more recent rocket
instruments (e.g., Heap 1976) and the Balloon-Borne Ultraviolet Spectrograph (BUSS; Kondo et al. 1979; de Jager et al. 1979) have obtained

sufficient resolution, but were limited either to only a very few objects or to a small wavelength region, or both. Copernicus has surveyed
mass-loss and shell-line characteristics of a number of relatively
bright objects, but for the most part adequate data could be obtained
only between 1100 Å and 1250 Å, including the resonance lines of S III
(which rarely shows P Cygni effects), Si III, N V, and the 1175 Å mul-
tiplet of C III (Snow and Morton 1976). Only for relatively bright
objects can Copernicus obtain good data on other important ions such as
Si IV ($\lambda\lambda$ 1393, 1402), O VI ($\lambda\lambda$ 1032, 1037), and S IV ($\lambda\lambda$ 1062, 1073).
Other critical species, such as C IV (1550 Å), He II (1640 Å), N III
(990 Å), and S VI ($\lambda\lambda$ 933, 944), are virtually unobservable with Copernicus due to its low sensitivity (or high noise) in the appropriate
wavelengths. The IUE has now provided coverage of the longer-wavelength
regions inaccessible to Copernicus, but the important region between the
Lyman limit and 1000 Å remains largely unexplained.

It would therefore be quite useful to launch a UV spectrometer
capable of observing these uncharted spectral regions. While it would
be desirable to do so for fainter stars than Copernicus is capable of
scanning, even at the Copernicus magnitude limit such an instrument
would provide important data. A resolving power of 10^3 would be adequate.

3. Observations of time variations.

In order to achieve a full understanding of the dynamics of stellar
winds and expanding envelopes, it is necessary to take into account
variability in the wind or envelope structure. The presence and nature
of such variability can provide important clues regarding the origins of
the winds or shells, and furthermore, may affect any conclusions regarding the long-term behavior of the winds. For example, in estimating
the influence of mass loss on stellar evolution it is necessary to know
not only the present mass-loss rate, but also the time over which it is
relevant.

Because the Copernicus spectrometer is a scanning instrument, it is
impossible to monitor very short timescale fluctuations, nor can simultaneous observations of different wavelengths be carried out. Furthermore, some ions which are evidently particularly sensitive indicators of
fluctuations in the wind, such as Si IV ($\lambda\lambda$ 1393, 1402) and C IV ($\lambda\lambda$
1549, 1551), are accessible to Copernicus in only a very few objects.
The IUE can cover many wavelengths at once, but its photo-metric limitations make it difficult to detect and analyze any but the strongest
variations. At least two major projects to study variability with the
IUE are under way, one concentating on OB supergiants (Snow et al. 1980)
and the other aimed at variability in Be stars (Doazan, Kuhi, and Thomas
1979; Doazan et al. 1980).

In order to substantially improve the data on variations in stellar
winds and extended atmospheres, it would be most desirable to launch an
instrument with moderate resolving power (R $\sim$ 10^3 would suffice), high

photometric accuracy, and a multiplexing capability so that the entire
spectrum could be observed at once.   This means that the detector must
be an area detector such as a television camera or CCD array, or a
linear strip upon which the spectrum can be projected.   Such detectors,
now being developed for space astronomy applications, have been used in
the BUSS payload and are planned for IUE; and will, no doubt, be available for future Spacelab missions.

One of the most important unanswered questions has to do with the
phase relationships between changes occurring at different levels in the
winds.   If the variations occur sequentially with increasing height,
then they must be due to outflowing density perturbations.   If on the
other hand, they occur simultaneously, then they are likely due to
variations in the ionizing X-ray flux.

III.   THE COOL STARS

A.   Historical Perspective and Ground-Based Data

Prior to the availability of ultraviolet spectral data progress in
understanding the outer atmospheres of cool stars was limited by the few
spectroscopic diagnostics present in the visible and near infrared spectrum.   These diagnostics included the Ca II H and K lines, He I $\lambda 10830$
and $\lambda 5076$ (the so-called D_3 line), He II λ 4686, hydrogen Hα and Hε, and
several near-ultraviolet lines of Fe II.   These lines are all chromo-
spheric in origin.   Coronal lines do not appear in the visible except in
unusual cases such as novae (e.g., McLaughlin 1960; Herbig and Hoffleit
1975), symbiotic stars (Sahade 1960), X-ray binaries (Baliunas et al.
1977), and the peculiar star R Aqr (Zirin 1976); it is therefore not
surprising that ground-based studies have concentrated on the chromospheres of cool stars.   Recent reviews of stellar chromsopheres and
relevant spectroscopic diagnostics include those by Linsky (1977),
Praderie (1973), and Doherty (1973).   The review of stellar chromospheres by Linsky (1980) will contain references subsequent to the completion of the present paper and a more detailed discussion of several
topics here discussed.

Ground-based studies of stellar chromospheres have addressed a
number of important questions including the following:

1.   The region of the HR diagram in which chromospheres exist.

The hottest stars with known Ca II emission are α Car (F0Ib; Warner
1966) and γ Vir N (F0V; Warner 1968).   Occasionally emission is also
seen in the A7 III star γ Boo (Le Contel et al. 1970).   Early A-type
stars such as Vega have been suggested to have chromospheres based on
the observed 20μ excess (Morrison and Simon 1973) and possible K line
emission (Linsky et al. 1973), but recent measurements of the K line
central intensity (Freire et al. 1978) argue against chromospheres in
these stars.   K line emission is commonly seen in F, G, and K stars of

all luminosity classes (cf. Bidelman 1954 for a useful bibliography). Emission is also seen in Cepheids (Hollars 1974) and is especially strong in T Tauri stars (e.g., Kuhi 1974) and spectroscopic binaries such as the RS CVn stars (Young and Koniges 1978). In the M supergiants Jennings and Dyke (1972) find an inverse correlation of K line emission with polarization an infrared excess. This suggests (cf. Jennings 1973) that when the grain density is high in circumstellar envelopes, the nonradiative energy input to the outer atmosphere of these stars goes into heating the circumstellar envelope and driving the stellar wind rather than heating a chromosphere.

2. Dependence of chromosphere properties on stellar evolution.

There exist a number of interesting correlations of K line emissions with other observable quantities, which suggest that "chromospheric activity," as measured by the strength of chromospheric emission lines, decreases as stars age. For example, strong K line emission is correlated with high lithium abundance (Herbig 1965), rapid rotation (e.g., Kraft 1967), small velocities perpendicular to the galactic plane (Wilson and Woolley 1970), and distance from the zero age main sequence (Wilson 1963; Wilson and Skumanich 1964), all indicators of stellar youth. Wilson (1966) has reviewed this work. Skumanich (1972) has proposed that the decrease in K line strength, rotational velocity, and lithium abundance with age obey a $t^{-1/2}$ time-dependence scaling law.

Kippenhahn (1973) has discussed a scenario, previously suggested by others, in which chromospheric activity results from enhanced heating of the outer layers of a cool star by dissipation of magnetohydrodynamic waves. The strength of this heating depends on the strength of surface magnetic fields which result from dynamo processes and thus on the stellar rotation rate. Since stellar winds can effectively slow down the rotational velocity of the outer layers of stars cooler than about spectral type F6 (Durney and Latour 1978), it is plausible for chromospheric activity to decrease with age. Additional support for this scenario is given by the well-known correlation of K line emission strength with magnetic fields on the Sun (e.g., Frazier 1971). There are no definitive observations yet of magnetic fields in cool stars, but Boesgaard's (1974) measurements suggest magnetic fields in the active chromosphere stars γ Vir N, ξ Boo A, and 70 Oph.

3. The search for stellar analogues of the solar 22-year cycle.

The Sun has a 22-year magnetic cycle in which the X-ray, and presumably also the extreme ultraviolet flux, are enhanced when the number of sunspots and strength of the field is strongest (White 1977). Surprisingly the strength of the K line has not been monitored over a solar cycle, but Sheeley (1967) suggests that the integrated K line flux may vary by 40 percent over a cycle; and White and Livingston (1978) have begun a careful solar monitoring program.

Since 1967 Olin Wilson has systematically observed main sequence stars to search for variability in the Ca II H and K line emission as evidence for stellar cycles and also stellar rotation by the appearance of plages on the disk. His approach is described by Wilson (1968), and a presentation of his data can be found in Wilson (1978). He has now found several stars of spectral type later than GO V with compeleted cycles, but finds no evidence for periodicity in stars hotter than spectral type GO V or in cooler stars with very strong H and K emissions. Zirin (1976) has also searched for stellar cycles in another diagnostic, the λ10830 line of He I. In his large data set about 74 percent of the G and K stars show variability, but only one star, θ Her, so far shows evidence for a stellar cycle.

4. Line width-luminosity relations.

In their survey of the H and K lines in late-type stars, Wilson and Bappu (1957) found that the K line width W, measured over an interval somewhere between the K_2 emission peak and the K_1 minimum beyond the emission peaks (cf. Wilson 1976), is related to the stellar luminosity according to $W \sim L^{1/6}$. Unfortunately, the great interest in interpreting and extending this width-luminosity relationship (the so-called Wilson-Bappu effect) has so dominated the study of stellar chromospheres that other possibly more interesting empirical results have not been studied to the extent that they merit. The fundamental question in interpreting the Wilson-Bappu effect is whether the width W refers to the Doppler core or to the damping wing. In the former case one is led to the conclusion that chromospheric turbulent velocities increase with stellar luminosity (Goldberg 1957; Fosbury 1973; Reimers 1973; Wilson 1957). An alternative approach is that W refers to a feature in the damping part of the line profile in which case Ayres et al. (1975) showed that $W \sim L^{1/6}$ naturally follows from the increase in mass column density above the temperature minimum needed to retain the same H^- continuum optical depth as the gravity of a star decreases (cf. Lutz et al. 1973). This question remains unresolved. In the meantime other width-luminosity relations have been found for Hα (Kraft et al. 1964; Fosbury 1973; Reimers 1973), the Mg II resonance lines (Kondo et al. 1975a; McClintock, et al. 1975a), and Lα (McClintock et al. 1975a).

B. Recent Ultraviolet and X-ray Results

Experiments flown on Copernicus, ANS, Apollo-Soyuz, and several sounding-rocket and balloon flights have opened up the ultraviolet and X-ray spectrum for the study of cool stars. The significance of this new spectral range lies in the availability of new spectroscopic diagnostics which can effectively sample outer atmospheric regions unobservable in the visible.

Important chromospheric diagnostics in the ultraviolet include the resonance lines of Mg II (h and k at λ2803 and λ2796), H I (λ1216), O I (λλ1302, 4, 6), C II (λλ1334, 1335), and Si II (λλ1808, 1817). Since magnesium is more abundant than calcium, and Mg^+ has a higher ionization potential than Ca^+, the detection of chromospheres in stars hotter than spectral type F0 may prove to be more feasible by searching for emission in the cores of the Mg II h and k lines than in the Ca II H and K lines. Lamers and Snijders (1975) have searched for Mg II emission in A-type stars without success, but Kondo et al. (1975b, 1976) have observed Mg II emission in the early B-type stars γ Ara (B1 Ib) and β Cen (B1 II) and possible emission in four other hot stars. It is not clear whether this emission is indicative of chromospheres or extended envelopes in these stars.

Lα emission has been detected in a number of cool stars, but interstellar absorption removes the core of the line and it is difficult to interpret the remaining flux in the line. Linsky (1977) has summarized the extensive literature on Lα and Mg II observations of cool stars. The high resolution Mg II line profiles are very useful in modelling the 6-8000 K regions of stellar chromospheres, as described below, and also in deriving properties of the circumstellar envelopes of K and M giants and supergiants. The 0 I resonance lines have strong emission in the Arcturus (K2 III) spectrum (Weinstein et al. 1977). Haisch et al. (1977a) have shown that these lines are pumped by Lβ in the same manner as 0 I λ8446 in planetary nebulae (Bowen 1947). It is likely that other ultraviolet emission lines in cool stars may be formed by similar fluorescent processes and thus indirectely measure the flux of otherwise unobservable extreme ultraviolet lines like He II λ304.

In the Sun a geometrically thin transition region is located between the chromosphere and corona, which is heated by thermal conduction from the corona and possibly by other mechanisms (Dupree 1972; McWhirter et al. 1975). Stellar analogues of the solar transition region are difficult to detect by ground-based observations for lack of useful diagnostics of $3-30 \times 10^4$ K plasma in the visible. Using Copernicus, Evans et al. (1975) were able to detect the Si III λ1206 and 0 VI λ1032 resonance lines in α CMi (F5 IV-V). They showed that the star has a transition region which is thicker and less dense than that in the Sun. Vitz et al. (1976) observed ultraviolet lines of Si II-IV, C II-IV, and N V in α Aur (G5 III+G0 III) using a moderate resolution rocket spectrograph. Haisch and Linsky (1976) then showed that these data, as well as the 0 VI λ1032 observation of Dupree (1975), are consistent with a transition region model with a pressure ten times that of the quiet Sun transition region and similar to that of a solar active region (Dupree et al. 1973). Kelch et al. (1978) have criticized this transition region model, however, on the grounds that the pressure exceeds that which they derive for the underlying chromosphere by a factor of 400. Instead, they suggest a lower pressure, greatly extended transition region which differs qualitatively from that of the Sun in that thermal conduction is not an important heating mechanism. The lower pressure of <0.004 dynes cm^{-2} suggested by Kelch et al. (1978) is also consistent with models that Mullan (1976) has computed on theoretical grounds (see below).

   The first direct "detection" of a stellar corona was reported by
Gerola et al. (1974) on the basis of Copernicus observations of the
λ1218 intercombination line of O V in β Gem (KO III).   Subsequent re-
observations failed to confirm this detection and they pointed out in an
Erratum that electronic interference in the Copernicus experiment produced spurious counts while the λ1218 was being observed.   Dupree (1975)
has interpreted her observations of the λ1032 line of O VI in α Aur in
terms of a corona around α Aur A with a temperature $>3 \times 10^5$ K.   This is
consistent with her measurement of an upper limit for the O V λ1218
line, a scaling of the solar coronal temperature, and theoretical arguments (Mullan 1976).   The ultraviolet spectrograph on IUE (International
Ultraviolet Explorer) and UV spectrographs which will presumably fly on
Spacelab will be well suited to search for relatively cool coronae
$(T < 10^6$ K) that may exist in giants and supergiants.   Solar-like hot
coronae $(T > 10^6$ K), which probably exist in dwarf stars (Mullan 1976),
cannot be well studied for lack of suitable diagnostics at $\lambda > 912$ Å.
Such potential diagnostics as [Fe XII] λ1240, 1349 are very weak lines
only observable in the solar spectrum during flares.

   The X-ray spectrum offers an interesting alternative method of
observing stellar coronae.   α Aur has now been observed at 0.25 keV by
ANS (Mewe et al. 1975, 1976) and with a 0.2-1.6 keV broadband channel by
the Catura et al. (1975) rocket experiment.   Upper limits on the X-ray
flux from a number of cool stars have been given by Mewe et al. (1975,
1976), Margon et al. 1974), Cruddace et al. (1975), and Vanderhill
et al. (1975).   Soft X-ray emission has also been detected from the
flare stars YZ CMi and UV Ceti (Heise et al. 1975; Karpen et al. 1977).
Experiments on HEAO-A and HEAO-B should greatly expand the number of
detected stellar coronae.

   For nearby cool stars the extreme ultraviolet spectrum may also be
very useful.   The first such measurement is the Haisch et al. (1977b)
detection of the flare star α Cen C (Proxima Centauri) in the 44-190 Å
bandpass of the EUV telescope on Apollo-Soyuz.   The detected flux was
interpreted as emission from coronal plasma in a flare.

   Finally, it may be feasible to directly measure properties of cool
star winds from high resolution ultraviolet spectra.   An example is
Dupree's (1975) measurement of a blue shift of 20 ± 7 km s^{-1} in the
emission peak of the O VI λ1032 line in α Aur.   This velocity shift
together with a model for the emitting region may lead to an estimate of
the mass loss rate.

C.   Theoretical Interpretations

   In the last few years the increasing availability of observations
of chromospheres, coronae, and winds in cool stars has stimulated theoretical efforts to explain the data.   Two approaches have been pursued:
the computation of semi-empirical models to explain specific line profiles and line fluxes; and purely theoretical models that derive atmospheric properties based on theories of wave generation, propagation,

and dissipation. The confrontation of these two approaches should lead
to very considerable advances in the field.

1. Stellar chromospheres.

   The first detailed model of a stellar chromosphere was computed by
Ayres et al. (1974) for α CMi (F5 IV-V). They derived the chromospheric
structure of the star extending from the temperature minimum at 4750 K
up to 8000 K, where a steep temperature rise occurs due to hydrogen ionization and the loss of the dominant cooling species in the chromosphere
(Thomas and Athay 1961). This model was based on a non-LTE analysis of
the Ca II K and λ8542 line profiles and the flux of the Mg II h and k
lines, in which the transfer equation was solved in the complete redistribution approximation. Milkey and Mihalas (1973) and Shine et al.
(1975) subsequently showed that taking into account coherency effects in
the inner line wings by means of the partial redistribution (PRD) formu-
lation is both more realistic on physical grounds and can account for
limb darkening of the solar H and K lines which complete redistribution
could not explain. Although this approach is now in general use and
permits analysis of both line cores formed in the chromosphere and damping wings formed in the photosphere, the earlier complete redistribution
method is sufficient for analyzing line cores in static atmospheres.

   Chromospheric models have now been constructed for the main
sequence stars: α Cen A (G2 V) and α Cen B (K1 V) by Ayres et al.
(1976), the quiet Sun (Ayres and Linsky 1976), solar plages (Shine and
Linsky 1974), solar flares (Machado and Linsky 1975), and the active
chromosphere stars 70 Oph A (K0 V) and ε Eri (K2 V) by Kelch (1978).
For giant stars models have been constructed for α Boo (K2 III) by Ayres
and Linsky (1975), and for α Aur (G5 III+G0 III), β Gem (K0 III), and
α Tau (K5 III) by Kelch et al. (1978). These models for the upper
photosphere and chromosphere layers are based on PRD analyses of the
Ca II K line profiles and the Mg II h and k line fluxes in most cases.
Several trends have become evident in this work:

   (a) The upper photospheres and temperature minimum regions are
hotter than predicted by the best available radiative equilibrium
models. Theoretical computations of the propagation of acoustic waves
generated in the lower photosphere (e.g., Ulmschneider et al. 1977)
predict that a significant portion of the initial wave energy will be
dissipated in the photosphere itself. On this basis one expects that
the upper photosphere and temperature minimum regions will be hotter
than predicted by radiative equilibrium, but there are no detailed
comparisons of theory and semi-empirical models yet. One potential
problem is that if an atmosphere is inhomogeneous, then analysis of the
Ca II and Mg II resonance lines will lead to a model closer to the
hotter component. Heasley et al. (1978) have argued that the Ayres and
Linsky (1975) model for α Boo is more representative of the hotter
component, and that the CO fundamental vibration-rotation band, which
they analyze, is more representative of cooler component.

(b) Kelch et al. (1978) show that the temperature minimum of the metal-rich star α Cen B (K1 V) is much cooler than either 70 Oph A (K0 V) or ε Eri (K2 V), which have solar abundances. This behavior confirms Johnson's (1973) predictions of CO cooling.

(c) There is a systematic trend of decreasing m_0 (the mass column density at the top of the chromosphere) with increasing gravity. The slope of this relation d log m_0/d log g = -0.5 (Kelch 1978) is consistent with the gravity dependence of hydrogen ionization and the location of m_0 where $T \approx 1.0$ at the head of the Lyman continuum.

(d) The temperature-mass column density distribution $T(m)$ in the chromosphere appears to be different in dwarfs than in giants; in dwarfs $T(m)$ has an initial rise above the temperature minimum and subsequent plateau out to m_0, and in giants the shape is reversed. This qualitative behavior is presumably indicative of the gravity-dependence of wave dissipation properties.

(e) Kelch (1978) has shown that the active chromospheres of 70 Oph A and ε Eri differ from the quiet chromosphere of α Cen B, not in the value of m_0, but in the steepness of the chromospheric rise in temperature immediately above the temperature minimum. If confirmed in subsequent work, this suggests that "chromospheric activity" is correlated with the initial temperature rise, which itself is presumably due to the flux of wave modes in a small range of periods.

It is clear that a great deal of work must be done on additional stars in different regions of the HR diagram before these apparent trends can be confirmed. In particular, active chromosphere stars such as the T Tauri stars (cf. Dumont et al. 1973) and spectroscopic binaries should be modelled, as well as supergiants and dwarf M stars.

An alternative approach to deriving chromospheric properties involves the computation of the mechanical energy flux generated by convective motions deep in the photosphere and the propagation of this energy by various wave modes to outer layers where it is dissipated as heat. Computations of acoustic wave heating, based on the mixing-length theory of convection and the Lighthill theory of turbulent sound generation, include the work of Kuperus (1965), Ulmschneider (1967), Nariai (1969), de Loore (1970), Renzini et al. (1977), and Ulmschneider et al. (1977). Athay (1976) and de Jager (1976) have recently discussed this work and pointed out the large uncertainties in the mechanical flux generation that results from uncertainties in the mixing-length and wave generation theories. Stein and Leibacher (1974) have reviewed these methods in the context of the solar atmosphere. Praderie (1973) has argued that a net mass flux is a necessary condition for nonradiative heating and is thus required for a chromosphere, but to date mass fluxes have not been incorporated into the theoretical models in a self-consistent way.

These theoretical models can be tested, since they predict the total mechanical heating as a function of stellar gravity and effective temperature. Recently, Ulmschneider et al. (1977) have computed the mechanical heating available in a stellar chromosphere after the initial acoustic fluxes computed by Renzini et al. (1977) propagate through the photosphere. The acoustic fluxes obtained by Renzini et al. agree reasonably well with the earlier calculations of de Loore (1970).

Ulmschneider et al. (1977) find that the acoustic energy available at the base of the chromosphere normalized by the total surface luminosity, (i.e., $F_{mech}/\sigma T_{eff}^4$) decreases slowly towards lower T_{eff}, and increases dramatically toward lower gravity (high luminosity). Linsky and Ayres (1978) have compared these predictions with measured Mg II surface fluxes for 31 stars. They argue that the Mg II h and k lines are the largest emitters of line radiation in chromospheres, and that the flux in other important lines should be proportional to the Mg II lines. They find a decrease in $F(Mg\ II)/\sigma T_{eff}^4$ with decreasing T_{eff} but no significant increase in the ratio toward lower stellar gravity, in contradiction with the predicted dependence. Cram and Ulmschneider (1978) have also pointed out that the Ulmschneider et al. (1977) models predict Ca II line widths inconsistent with observations, except for the Sun, and that this discrepancy arises from computed mass column densities at the temperature minima of giant stars which are a factor of 100 too large. These two criticisms, and the inability of the theory to explain chromospheric "activity," suggest that the two basic assumptions of mixing-length convection and the Lighthill theory of turbulent sound generation may be inappropriate for stellar atmospheres.

2. Stellar coronae and winds.

There are as yet no semi-empirical coronal models computed due to the lack of available data. When such data are obtained with IUE, HEAO-A, HEAO-B, and Spacelab we can look forward to a productive dialogue between the theoretical and semi-empirical methods, as is now happening for stellar chromospheres.

In the meantime, theoretical models exist which have yet to be truly tested against observations. these models have been computed by Kuperus (1965), Ulmschneider (1967), de Loore (1970), and Ulmschneider et al. (1977). In addition to the limitations previously mentioned concerning the mixing-length theory and the Lighthill mechanism, further problems may be introduced by terms ignored in the energy balance equation. One usually includes, in some approximation, heating by acoustic waves and cooling by radiation; heat conduction outward into space and downward into the transition region; and the expansion of the stellar wind against gravity. In the Sun there is evidence for heating due to magnetic annihilation (Tucker 1973; Sheeley et al. 1975), and active regions and flares may be heated entirely in this manner. Also there is a close correlation between coronal base pressure, coronal temperature, and the speed of the coronal wind with the divergence of the magnetic field lines (e.g., Altschuler et al. 1972; Sheeley et al. 1976). As

pointed out by Adams and Sturrock (1975), among others, the divergence
of the magnetic field lines in solar coronal holes increases the solar
wind energy loss resulting in lower coronal temperatures and densities.
Conversely, closed magnetic field structures, which characterize solar
active regions, result in decreased energy loss by the wind and higher
coronal densities and pressures. Thus the geometry and strength of magnetic fields in stellar coronae, which are completely ignored in existing stellar corona and wind models, probably play critical roles, and
must be incorporated in order to construct realistic models.

Hearn (1975a) has developed an alternative method for predicting
stellar coronal temperatures and winds given only the gravity and estimates of the coronal base pressures. He also ignores the magnetic field
and the heating terms in the energy balance equation and considers only
the loss terms. The theory is based on the assumption that the most
likely coronal configuration is one in which the total loss for a given
base pressure is a minimum. The physical basis underlying the minimum
flux assumption has been criticized by Vaiana and Rosner (1978), among
others, and Hearn (1977) has extended the theory to include the effects
of magnetic fields on winds and conduction. These points need further
study, but in the meantime the minimum flux assumption leads to a
straightforward method of computing coronal properties which yields
realistic values for the Sun. Using this theory with some modification,
Haisch and Linsky (1976) estimated for α Aur a coronal temperature of
1.2×10^6 K and mass loss of 2×10^{-8} $M_\odot$ yr^{-1}, close to the value derived using the O VI λ1032 line wavelength shift. Mullan (1976) has
computed coronal models for dwarfs and giants using the minimum flux
theory and the assumption that the fraction of the total stellar luminosity used to heat the corona is the same as that for the Sun. He finds
for main sequence stars that the coronal temperature should increase
with effective temperature. In G and K giants he estimates coronal temperatures well under 10^6 K and detectable winds from α Aur and α Boo.
Also he suggests that the transition region in α Boo may be thick as
independently suggested by Weinstein et al. (1977).

The theoretical framework presently used to describe the solar wind
(cf. Hundhausen 1972; Holtzer 1977) can be used to predict properties of
stellar winds. Roberts and Soward (1972), for example, have delineated
the ranges of parameters for which stellar "winds" and stellar "breezes"
can be expected to occur. An important unresolved question is to define
the region of the HR diagram in which radiatively-driven dusty winds are
dominant rather than supersonic solar-type winds.

Finally, a problem should be pointed out in the analysis of
spectral line shifts as diagnostics of flow speed. In the Sun, downward
motions of chromospheric and transition region material are correlated
with bright emission (Frazier 1971; Doschek et al. 1976). Thus the net
velocity shift of a stellar emission line can give the wrong magnitude
and even the wrong sign of the mean flow velocity. Dupree (1975) measured a blue shift of 20 ± 7 km s^{-1} in the O VI λ1032 line of α Aur
which corresponds to a mass loss of $(1.2 \pm 0.4) \times 10^{-8}$ $M_\odot$ yr^{-1} (Haisch and

Linsky 1976) if the blue shift accurately measures the net Doppler motion. With the same caveat, the asymmetric Mg II and Ca II lines of α Boo (Chiu et al. 1977) imply a mass loss of 8×10^{-9} $M_\odot$ yr^{-1}, but this loss may be sporadic.

D. Future Observations

Undoubtedly, a great deal can be learned by observing in the ultraviolet various types of cool stars at both high and low resolution. Much can be said for survey-type experiments when we presently have so very few ultraviolet observations of even the brightest cool stars. However, it is hoped that future observing programs will be dedicated to answering fundamental questions, and to encourage such goal-oriented research, the following general types of observations are proposed:

1. The determination of the physical properties of stellar chromo-
    spheres, coronae, and winds.

For chromospheres we need to know the temperature-mass column density relations for representative active and quiet chromosphere stars throughout the HR diagram. Given such models, it is feasible to seriously address the heating problem. In particular, it is important to know the hottest and coolest stars for which chromospheres exist. Absolute stellar surface fluxes in the resonance lines of Mg II, O I, C I, C II, and Si II, together with high resolution profiles of the Mg II resonance lines are needed to model cool star chromospheres.

A resolving power of 10^3 is adequate for the line flux observations, but a resolution of 3×10^4 is needed to resolve the Mg II line profiles and to identify the circumstellar components. It is also important to know temperatures, densities, flow velocities, and mass-loss rates in coronae for representative active and quiet chromosphere stars throughout the HR diagram. For giants and supergiants the coronal temperature may be considerably less than 10^6 K and, consequently there are many diagnostics available in the ultraviolet. Absolute fluxes in lines of C IV λλ1548, 1550, Si IV λλ1393, 1402, N V λλ1238, 1242, O V λ1218, and O VI λλ1032, 1037 are important for this work. For dwarf stars coronal temperatures in excess of 10^6 K are expected, and only very weak diagnostics such as [Fe XII] λλ1240, 1349 are available for $\lambda > 912$ Å. It is therefore important to search for Mg X λλ610, 625, Si XII λ499, 521, and Fe XV λ285 emission in nearby stars, and to observe XUV and X-ray broad-band and line fluxes, if feasible, in more distant stars. Given high spectral resolution ($R = 3 \times 10^4$), one can try to directly measure wavelength shifts, although these apparent velocities may not be indicative of the true flow velocities (cf. III.D.2). Also, with a resolving power of 3×10^4, coronal line widths can be measured to determine either the local temperature or nonthermal broadening.

2. The determination of the important heating and cooling terms that
    enter the energy balance equation.

    As discussed above, present theoretical methods for computing the
generation and dissipation of mechanical energy are very uncertain and
appear to be in conflict with observations. It may be more productive
to derive radiative losses and temperature structures in the outer
layers (including photospheres) of different kinds of stars semi-
empirically, and to use this information as a guide for future theoretical calculations. This suggests a major effort to obtain absolute
line fluxes and high resolution line profiles, as described above,
necessary to derive semi-empirical models of stellar chromospheres and
coronae. Equally important is the development and application of spectroscopic diagnostics necessary for deriving models from the data.

    For the G-M supergiants, circumstellar envelopes may play important
roles in the energy balance through mass loss and radiative cooling. To
study these envelopes, a resolving power of 1×10^5 may be necessary to
separate individual circumstellar velocity components and to resolve the
widths of Mg II, Fe II, Fe I, and other lines.

3. The nature of stellar activity.

    An important approach is to ask why the Sun has so many chromo-
sphere, transition region, and corona structures, loosely described as
varying "activity," coexisting in the same gravitational field. To
answer this we need to assess the energy balance in these various structures semi-empirically, and then question how is it that the various
magnetic field structures lead to the various energy balance regimes.

    With whatever insight results from this study of solar "activity,"
we can address the more general stellar "activity" questions. In particular, do active chromosphere stars differ from quiet chromosphere stars
only in the temperature gradient at the base of the chromosphere, or are
there other significant differences as well? In the Sun the corona
above active chromosphere regions (plages) is denser and hotter, but the
solar wind in such regions may actually be weaker due to the closed
magnetic field structures. Over what region of the HR diagram, if any,
is there an inverse correlation of stellar activity and reduced mass
loss? Reliable semi-empirical models for the outer layers of active and
quiet stars are needed to answer this question.

4. Long-term variation in solar activity and the total solar radiative
    output (the solar constant).

    One approach is to seek out statistical effects such as the de-
crease of chromospheric emission with age (Skumanich 1972) and properties of stellar cycles. Another approach is to closely study pairs
of stars which are identical except for their outer atmospheres and
estimated ages. From the differences in the temperature structures and
energy balance in such pairs of stars, one may obtain insight into

evolutionary effects. In particular, the α Cen A - Sun pair should be closely studied.

5. Possible modifications of the outer atmospheres of cool stars by comparisons in binary or multiple systems.

From Ca II K line observations, it is evident that spectroscopic binaries, even relatively widely separated ones, are usually active chromosphere stars. It is important to question how the corona and winds of spectroscopic binaries differ from single stars as a function of separation and stellar radii. It is generally thought that when stars form they lose excess angular momentum by forming either a planetary system or a multiple star system. If so, the internal angular momentum distribution and convective envelopes of stars may depend on their past history in complex ways. Also, the existence of nearby stars will produce Roche lobes about stars which could be significantly filled, and/or tidal coupling may alter the generation and transport of mechanical energy in stars (Young and Koniges 1978). This is a particularly intriguing question.

In the decade that has elapsed since the first measurements of stellar ultraviolet spectra, whole new areas of research and of understanding have developed. Where it was previously only suspected that hot stars lose mass and that cool stars, other than the Sun, have chromospheres and coronae, a wealth of data has now established these phenomena as both commonplace and vitally important in understanding the structure of the outermost layers of stars, the evolution of stars, and their interaction with their surroundings.

Despite all of these advances, only the first beginnings have been made toward the ultimate goal of attaining a full physical description of the outer layers of stars, and the next decades should see greater gains than the last.

Helpful discussions with H. Lamers are gratefully acknowledged. This work was supported by grants NGL-06-003-057 and NAS5-23274 from the National Aeronautics and Space Administration to the University of Colorado.

References

Adams, W. M., and Sturrock, P. A. 1975, Ap. J., 202, 259.

Altschuler, M. D., Trotter, D. E., and Orrall, F. Q. 1972, Solar
     Phys., 26, 354.

Athay, R G. 1976, The Solar Chromosphere and Coronae: Quiet Sun
     (Dordrecht: Reidel).

Ayres, T. R., and Linsky, J. L. 1975, Ap. J., 200, 660.

Ayres, T. R., and Linsky, J. L. 1976, Ap. J., 205, 874.

Ayres, T. R., Linsky, J. L., Rodgers, A. W., and Kurucz, R. L. 1976,
     Ap. J., 210, 199.

Ayres, T. R., Linsky, J. L., and Shine, R. A. 1974, Ap. J., 192, 93.

Ayres, T. R., Linsky, J. L., and Shine, R. A. 1975, Ap. J. (Letters),
     195, L121.

Baliunas, S. L., Dupree, A. K., and Lester, J. B. 1977, Bull. A.A.S.,
     9, 298.

Barlow, M. J., and Cohen, M. 1977, Ap. J., 213, 737.

Beals, C. S. 1951, Pub. D. A. O., 9, 1.

Bidelman, W. P. 1954, Ap. J. Suppl. 1, 214.

Boesgaard, A. M. 1974, Ap. J., 188, 567.

Bohannan, B., and Conti, P. S. 1976, Ap. J. 204, 797.

Bowen, I. S. 1947, Pub. Astron. Soc. Pacific, 59, 196.

Cannon, C. J., and Thomas, R. N. 1977, Ap. J., 211, 910.

Carruthers, G. R. 1968, Ap. J., 151, 269.

Cassinelli, J. P. 1979a, Ann. Rev. Astr. Ap., 17, 275.

Cassinelli, J. P. 1979b, private communication.

Cassinelli, J. P., Castor, J. I., and Lamers, H. J. G. L. M. 1978,
     Pub. Astron. Soc. Pacific, 90, 476.

Cassinelli, J. P., and Olson, G. L. 1979, Ap. J., 229, 304.

Cassinelli, J. P., Olson, G. L., and Stalio, R. 1978, Ap. J., 220, 573.

Castor, J. I., Abbott, D. C., and Klein, R. I. 1975, Ap. J., 195, 157.

Catura, R. C., Acton, L. W., and Johnson, H. M. 1975, Ap. J. (Letters),
     196, L47.

Chiosi, C., Nasi, E., and Sreenivasan, S. R. 1978 Astr. Ap., 63, 103.

Chiu, H. Y., Adams, P. J., Linsky, J. L., Basri, G. S., Maran, S. P.,
     and Hobbs, R. W. 1977, Ap. J., 211, 453.

Conti, P. S., and de Loore, C. (eds.) 1979, IAU Symposium 83, Mass Loss
     and Evolution in O-Type Stars (Dordrecht: Reidel).

Conti, P. S., and Garmany, K. 1980, Ap. J., in press.

Conti, P. S., and Niemala, V. 1976, Ap. J. (Letters), 209, L37.

Cram, L. E., and Ulmschneider, P. 1978, Astr. Ap., 62, 289.

Cruddace, R., Bowyer, S., Malina, R., Margon, B., and Lampton, M. 1975,
     Ap. J. (Letters), 202, L9.

Dearborn, D. 1979, IAU Symposium 83, Mass Loss and Evolution in O-Type
     Stars, eds. P. S. Conti and C. de Loore (Dordrecht: Reidel).

de Jager, C. 1976, Mem. Soc. R. Sci. Liege, 9, 369.

de Jager, C., Kondo, Y., Morgan, T., van der Hucht, K. A., and Lamers,
     H. J. G. L. M. 1979, in preparation.

de Loore, C. 1970, Astrophys. Space Sci., 6, 60.

de Loore, C., De Greve, J. P., and Lamers, H. J. G. L. M. 1977, Astr.
     Ap., 61, 251.

Doazan, V., Kuhi, L., and Thomas, R. N. 1979, Ap. J. (Letters), in press.

Doazan, V., Kuhi, L., Marlborough, J. M., Snow, T. P., and Thomas, R. N. 1980, in preparation.

Doherty, L. R. 1973, in Stellar Chromospheres, ed. S. D. Jordan and E. H. Avrett, NASA SP-317, p. 99.

Doscheck, G. A., Feldman, U., and Bohlin, J. D. 1976, Ap. J. (Letters), 205, L177.

Dumont, S., Heidmann, N., Kuhi, L. V., and Thomas, R. N. 1973, Astr. Ap., 29, 199.

Dupree, A. K. 1972, Ap. J., 178, 527.

Dupree, A. K. 1975, Ap. J. (Letters), 200, L27.

Dupree, A. K., Huber, M. C. E., Noyes, R. W., Parkinson, W. H., Reeves, E. M., and Withbroe, G. L. 1973, Ap. J., 182, 321.

Durney, B. R., and Latour, J. 1978, Geophys. Astrophys. Fluid Dynamics, 9, 241.

Evans, R. G., Jordan, C., and Wilson, R. 1975, Mon. Not. R. Astron. Soc., 172, 585.

Fosbury, R. A. E. 1973, Astr. Ap., 27, 141.

Frazier, E. N. 1971, Solar Phys., 21, 42.

Freire, R., Czarny, J., Felenbok, P., and Praderie, F. 1978, Astr. Ap., 68, 89.

Frost, S., and Conti, P. S. 1976, IAU Symposium 70, The Merrill-McLaughlin Memorial Symposium on Be and Shell Stars, ed. A. Slettebak (Boston: Reidel).

Gerola H., Linsky, J. L., Shine, R. A., McClintock, W., Henry, R. C., and Moos, H. W. 1974, Ap. J. (Letters), 193, L107 (Erratum, 1978 Ap. J. (Letters), 218, L32).

Goldberg, L. 1957, Ap. J., 126, 318.

Gurzadyan, G. A. 1975, Space Sci. Rev., 18, 95.

Hack, M., Struve, O. 1970, Stellar Spectroscopy: Peculiar Stars, (Trieste: Obs. Astr. di Trieste).

Haisch, B. M., and Linsky, J. L. 1976, Ap. J. (Letters), 205, L39.

Haisch, B. M., Linsky, J. L., Lampton, M., Paresce, F., Margon, B., and Stein, R. 1977b, Ap. J. (Letters), 213, L119.

Haisch, B. M., Linsky, J. L., Weinstein, A., and Shine, R. A. 1977a, Ap. J., 214, 785.

Harnden, F. R., Branduardi, G., Elvis, M., Gorenstein, P. Grindlay, J., Pye, J. P., Rosner, R., Topka, K., and Vaiana, G. S. 1979, Ap. J. (Letters), in press.

Heap, S. R. 1976, IAU Symposium 70, The Merrill-McLaughlin Memorial Symposium on Be and Shell Stars, ed. A. Slettebak (Boston: Reidel).

Heap, S. R. et al. 1978, Nature, 275, 385.

Hearn, A. G. 1975a, Astr. Ap., 40, 277.

Hearn, A. G. 1975b, Astr. Ap., 40, 355.

Hearn, A. G. 1977, Solar Phys., 51, 159.

Heasley, J. N., Ridgway, S. T., Carbon, D. F., Milkey, R. W., and Hall, D. N. B. 1978, Ap. J. 219, 970.

Heise, J., Brinkman, A. C., Schrijver, J., Mewe, R., Gronenschild, E., den Boggende, A., Grindlay, J. 1975, Ap. J. (Letters), 202, L73.

Henize, K. G., Wray, J. D., Parsons, S. B., Benedict, G. F., Bruhweiler, F. W., Rybski, P. M., and O'Callaghan, F. G., 1975, Ap. J. (Letters), 199, L119.

Herbig, G. H. 1965, Ap. J., 141, 588.

Herbig, G. H., and Hoffleit, D. 1975, Ap. J., 202, L41.

Hollars, D. R. 1974, Ap. J., 194, 137.

Holtzer, T. E. 1977, Solar System Plasma Physics: A Twentieth Anniversary Overview, ed. C F. Kennel, L. J. Lanzerotti, and E. N. Parker (Amsterdam: North Holland).

Huang, S.-S. 1973, Ap. J., 183, 541.

Huang, S.-S. 1977, Ap. J., 212, 123.

Hundhausen, A. J. 1972, Coronal Expansion and Solar Wind (New York: Springer-Verlag).

Hutchings, J. B. 1970, Mon. Not. R. Astron. Soc., 150, 55.

Jennings, M. C. 1973, Ap. J., 185, 197.

Jennings, M. C., and Dyke, H. M. 1972, Ap. J., 177, 427.

Johnson, H. R. 1973, Ap. J., 180, 81.

Karpen, et al. 1977, Ap. J., 216, 479.

Kelch, W. L. 1978, Ap. J., 222, 931.

Kelch, W. L., Linsky, J. L., Basri, G. S., Chiu, H. Y., Chang, S. H. Maran, S. P., and Furenlid, I. 1978, Ap. J., 220, 962.

Kippenhahn, R. 1973, in Stellar Chromospheres, ed. S. D. Jordan and E. H. Avrett, NASA SP-317, p. 265.

Kondo, Y., de Jager, C., Morgan, T., van der Hucht, K. A., and Lamers H. J. G. L. M. 1979, in preparation.

Kondo, Y., Giuli, R. T., Modisette, J. L., and Rydgren, A. E. 1972, Ap. J., 176, 153.

Kondo, Y., Modisette, J. L., and Wolf, G. W. 1975b, Ap. J., 199, 110.

Kondo, Y., Morgan, T. H., and Modisette, J. L. 1975a, Ap. J. (Letters), 196, L125.

Kondo, Y., Morgan, T. H., and Modisette, J. L. 1976, Ap. J., 209, 489.

Kraft, R. N., Preston, G. W., and Wolfe, S. C. 1964, Ap. J., 140, 237.

Kraft, R. P. 1967, Ap. J., 150, 551.

Kriz, S., and Harmenec, P. 1975, Bull. Astr. Inst. Cechoslovakia, 26, 65.

Kuhi, L. V. 1974, Astr. Ap. Suppl., 15, 47.

Kuperus, M. 1965, Recherches Astron. Obs. Utrecht, 17, 1.

Lamers, H. J. G. L. M., and de Loore, C. 1976, in Physique des Mouvements dans les Atmospheres Stellaires, ed. R. Cayrel and M. Steinberg (Paris: C. N. R. S.), p. 453.

Lamers, H. J. G. L. M., and Morton, D. C. 1976, Ap. J. Suppl., 32, 715.

Lamers, H. J. G. L. M., and Rogerson, J. B. 1978, Astr. Ap., 66, 47.

Lamers, H. J. G. L. M., and Snijders, M. A. J. 1975, Astr. Ap., 41, 259.

Lamers, H. J. G. L. M., and Snow, T. P. 1978, Ap. J., 219, 504.

Lamers, H. J. G. L. M., Stalio, R., and Kondo, Y. 1978, Ap. J., 223, 207.

Le Contel, J. M., Praderie, F., Bijaoui, A., Dantel, M., and Sareyan, J. P. 1970, Astr. Ap., 8, 159.

Leep, E. M. and Conti, P. S. 1979, Ap. J., 228, 224.

Linsky, J. L. 1977, in The Solar Output and its Variation, 1975, ed. O. R. White (Boulder: Colorado Assoc. Univ. Press), Ch. 7.

Linsky, J. L. 1980, Ann. Rev. Astr. Ap., 18, in press.

Linsky, J. L., and Ayres, T. R. 1978, Ap. J., 220, 619.

Linsky, J. L., Shine, R. A., Ayres, T. R., and Praderie, F. 1973, Bull. A.A.S., 5, 3.

Lucy, L. B., and Solomon, P. M. 1970, Ap. J., 159, 829.,
Lutz, T. E., Furenlid, I., and Lutz, J. H. 1973, Ap. J., 184, 787.
Machado, M. E., and Linsky, J. L. 1975, Solar Phys., 42, 395.
Margon, B., Mason, K. O., and Sanford, P. W. 1974, Ap. J. (Letters),
    194, L75.
Marlborough, J. M. 1977, Ap. J., 216, 446.
Marlborough, J. M., and Snow, T. P. 1976, IAU Symposium 70, The
    Merrill-McLaughlin Memorial Symposium on Be and Shell Stars, ed. A.
    Slettebak (Boston: Reidel).
Marlborough, J. M. and Snow, T. P. 1979, Ap. J., in press.
Marlborough, J. M., Snow, T. P., and Slettebak, A. 1978, Ap. J., 224,
    157.
McClintock, W., Henry, R. C., Moos, H. W., and Linsky, J. L. 1975a,
    Ap. J, 202, 733.
McLaughlin, D. B. 1960, in Stellar Atmospheres, ed. J. L. Greenstein
    (Chicago: Univ. of Chicago Press), p. 585.
McWhirter, R. W. P., Thonemann, P. C., and Wilson, R. 1975, Astr. Ap.,
    40, 63.
Mewe, R., Heise, J., Gronenschild, E. H. B. M., Brinkman, A. C.,
    Schrijver, J., and de Boggenden, A. J. F. 1975, Ap. J. (Letters),
    202, L67.
Mewe, R., Heise, J., Gronenschild, E. H. B. M., Brinkman, A. C.,
    Schrijver, J., and de Boggende, A. J. F. 1976, Astrophys. Space
    Sci., 42, 217.
Milkey, R. W., and Mihalas, D. 1973, Ap. J., 185, 709.
Morrison, D., and Simon, T. 1973, Ap. J., 186, 193.
Morton, D. C. 1967, Ap. J., 147, 1017.
Mullan, D. J. 1976, Ap. J., 209, 171.
Nariai, H. 1969, Astrophys. Space Sci., 3, 150.
Praderie, F. 1973, in Stellar Chromospheres, ed. S. D. Jordan and E. H.
    Avrett, NASA SP-17, p. 79.
Reimers, D. 1973, Astr. Ap., 24, 79.
Renzini, A., Cacciari, C., Ulmschneider, P., and Schmitz, F. 1977,
    Astr. Ap., 61, 39.
Roberts, P. H., and Soward, A. M. 1972, Proc. Roy. Soc., Ser A, 328,
    185.
Rogerson, J. B., and Lamers, H. J. G. L. M. 1975, Nature, 256, 190.
Rosendahl, J. D. 1973a, Ap. J., 182, 523.
Rosendahl, J. D. 1973b, Ap. J., 186, 909.
Sahade, J. 1960, in Stellar Atmospheres, ed. J. L. Greenstein (Chicago:
    Univ. of Chicago Press), p. 494.
Sheeley, N. R. 1967, Ap. J., 147, 1106.
Sheeley, N. R., Bohlin, J. D., Breuckner, G. E., Purcell, J. D.,
    Scherrer, V. E., and Tousey, R. 1975, Ap. J (Letters), 196, L129.
Sheeley, N. R., Harvey, J. W., and Feldman, W. C. 1976, Solar Phys.,
    49, 271.
Shine, R. A., and Linsky J. L. 1974, Solar Phys., 39, 49.
Shine, R. A., Milkey, R. W., and Mihalas, D. 1975, Ap. J., 199, 724.
Skumanich, A. 1972, Ap. J., 171, 565.
Slettebak, A. (ed.) 1976, IAU Symposium 70, The Merrill-McLaughlin
    Memorial Symposium on Be and Shell Stars (Boston: Reidel).

Slettebak, A., and Snow, T. P. 1978, Ap. J. (Letters), 224, L127.
Snow, T. P. 1977, Ap. J., 217, 760.
Snow, T. P., and Hayes, D. P. 1978, Ap. J., 226, 897.
Snow, T. P., Lamers, H. J. G. L. M., Machetto, F., de Jager, C., and Grady, C. 1980, in preparation.
Snow, T. P., and Marlborough, J. M. 1976, Ap. J. (Letters), 203, L87.
Snow, T. P., and Morton, D. C. 1976, Ap. J. Suppl., 32, 429.
Snow, T. P., Peters, G. J., and Mathieu, R. D. 1979, Ap. J. Suppl., 39, 359.
Snow, T. P., Wegner, G., and Kunasz, P. B. 1980, Ap. J., in press.
Stecher, T. P. 1968, Proc. Symposium on Wolf-Rayet Stars, ed. K. B. Gebbie and R. N. Thomas, NBS Special Publ. No. 307, p. 65.
Stein, R. F., and Leibacher, J. 1974, Ann. Rev. Astr. Ap., 12, 407.
Stothers, R. and Chin, C.-W. 1979, Ap. J., 233, 267.
Struve, O. 1935, Ap. J., 81, 66.
Struve, O., and Roach, F. E. 1939, Ap. J., 90, 727.
Thomas, R. N., and Athay, R. G. 1961, Physics of the Solar Chromosphere (New York: Interscience).
Tucker, W. H. 1973, Ap. J., 186, 285.
Ulmschneider, P. 1967, Zs. Astrophys., 67, 193.
Ulmschneider, P., Schmitz, F., Renzini, A., Cacciari, C., Kalkofen, W., and Kurucz, R. L. 1977, Astr. Ap., 61, 515.
Underhill, A. B. 1966, The Early Type Stars (Dordrecht: Reidel).
Vaiana, G. S. and Rosner, R. 1978, Ann. Rev. Astr. Ap., 16, 393.
Vanderhill, M. J., Borken, R. J., Bunner, A. N., Burstein, P. H., and Kraushaar, W. L. 1975, Ap. J. (Letters), 197, L19.
Vitz, R. C., Weiser, H., Moos, H. W., Weinstein, A., and Warden, E. S. 1976, Ap. J. (Letters), 205, L35.
Warner, B. 1966, Observatory, 86, 82.
Warner, B. 1968, Observatory, 88, 217.
Weinstein, A., Moos, H. W., and Linsky, J. L. 1977, Ap. J., 218, 195.
Wegner, G. A. and Snow, T. P. 1978, Ap. J. (Letters), 226, L25.
White, O. R. 1977, The Solar Output and its Variations, 1975 (Boulder: Colorado Assoc. Univ. Press).
White, O. R., and Livington, W. C. 1978, Ap. J., 226, 679.
Wilson, O. C. 1957, Ap. J., 126, 525.
Wilson, O. C. 1963, Ap. J., 138, 832.
Wilson, O. C. 1966, Science, 151, 1487.
Wilson, O. C. 1968, Ap. J., 153, 221.
Wilson, O. C. 1976, Ap. J., 205, 823.
Wilson, O. C. 1978, Ap. J., 226, 379.
Wilson, O. C., and Bappu, M. K. V. 1957, Ap. J., 125, 661.
Wilson, O. C., and Skumanich, A. 1964, Ap. J., 140, 1401.
Wilson, O. C., and Woolley, R. 1970, Mon. Not. R. Astron. Soc., 148, 463.
Wilson, R. 1958, Pub. Roy. Obs. Edinburgh, 2, 61.
Wilson, W. R. J. 1977, Ph.D. Dissertation, University of Calgary.
York, D. G., Vidal-Madjar, A., Laurent, C., and Bonnet, R. 1977, Ap. J. (Letters), 213, L61.
Young, A., and Koniges, A. 1978, Ap. J., 211, 836.
Zirin, H. 1976, Ap. J., 208, 414.

NEBULAR AND EXTRAGALACTIC ASTRONOMY FROM SPACE
IN THE OPTICAL REGION

Massimo CAPACCIOLI and Sandro D'ODORICO

Asiago Astrophysical Observatory, University of Padova

June 1978

TABLE OF CONTENTS

P. L. Bernacca and R. Ruffini (eds.), Astrophysics from Spacelab, 317-393.
Copyright © 1980 by D. Reidel Publishing Company.

3.   IMPACT OF SPACE OBSERVATIONS ON SELECTED
      ASTROPHYSICAL PROBLEMS

3.1   Supernova remnants

3.2   Stellar and gaseous content of nearby galaxies

3.3   Nuclei of 'normal' galaxies

3.4   Active nuclei

3.5   Outskirts of galaxies

3.6   The extragalactic distance scale

3.7   The Hubble diagram

INTRODUCTION

This review deals with the impact that astronomical observations from outer space will have on the fields of nebular and extragalactic research.   The discussion will be confined to the classical optical region of the electromagnetic spectrum extended to the adiacent UV and IR. In this spectral range the atmosphere of the Earth seriously hampers the capabilities of ground-based telescopes. First, the atmospheric absorption restricts the available spectrum to a narrow window sharply bounded on its ultraviolet side. Second, the low angular resolution caused by atmospheric turbolence, scintillation and differential refraction (seeing) degrades the images of celestial bodies spoiling them of their finest structural details. Third, the sky brightness, magnified by the contribution of the airglow and scattered light, combines with seeing to set in an operative threshold to the faintness of the sources observable from the ground.

Provided that appropriate observational facilities are available, from outside the atmosphere we will gain a much clearer and deeper view of the universe. This is expected to be particularly rewarding for extragalactic astronomy. In fact, most of the basic questions on the nature of the universe are held by extragalactic sources, whose faintness and size challange the capabilities of the largest telescopes on Earth. The work already done from space, which has been reviewed in Section 2, gives just a pale sample of the breakthrough that space facilities to come in the 1980's will produce. We expect that space observations will shed light in fundamental problems such as stellar content and evolution of galaxies, or <u>missing mass</u> and related structure of the outskirts of galaxies. It will be possible to improve the knowledge of the largest and/or more distant sources as clusters of galaxies and QSO's, to push out the mapping of the extragalactic distance scale and ultimately, via a determination of H_o and possibly q_o on which any scientist may agree, to approach a more definite idea of the age, density and shape of the universe we live in. A more detailed discussion of some of these topics is given in Section 3.

In addition to that, with these new powerful tools we expect not only to enlight some of the problems already set up by ground-based telescopes, but also to discover new unexpected phenomena. In the era to come we will witness a jump of our astronomical knowledge comparable to that caused by the introduction in astronomy of the photographic technique. We cannot obviously foresee which phenomena, among those still hidden in the universe, will be discovered; possibly the statistical approach made by Harwit (1975) will give us a hint on which experiments are the most promising in this respect. However, by just looking at the past experience, we immediately realize that improved observational capabilities have always brought something new and unexpected.

1. GROUND-BASED VERSUS SPACE OBSERVATIONS IN THE WIDE OPTICAL BAND

1.1   OBSERVING FROM THE EARTH

Blurring from Earth atmosphere is the main source of degradation of low time resolution astronomical observations from ground-based telescopes. By simply limiting the angular resolution, it prevents detection of both faint stellar sources and faint structural details. The typical luminosity profile of a degraded stellar image is shown in Fig. 1. It results from the composition of E-W photoelectric drift scans of three stars of increasing magnitude (de Vaucouleurs and Capaccioli 1978a). The central cusp (seeing disk) is due to the variable refraction of the atmosphere and is strictly dependent on the atmospheric conditions. Minor contributions come from aberrations of the optical system, telescope guiding and finite aperture convolution effect, i.e. detector resolution. Assuming, as a first approximation, that the central component of the degraded star profile allows a Gaussian representation (see Fig. 1):

$$I(r) = I(0) \exp (-r^2/2\sigma^2) \tag{1}$$

we may take σ as an operative definition of the seeing figure (for a more realistic fit of the central cusp we refer to de Vaucouleurs and

Capaccioli 1978a,b). Direct measurements give typical values of σ larger than 0.″5, even in the best astronomical sites on Earth. This figure sets the limit to the angular resolution achievable with conventional telescopes and techniques. In fact, σ is somewhat indicative of the Nyquist frequency of optimal sampling (Bracewell 1965) and therefore we cannot hope to retrieve very much at resolution levels better than, say, 0.″5. To overcome this limit, several a posteriori mathematical procedures such as classical deconvolution (Burr 1955, Bracewell 1955) or numerical restoration (Brault and White 1971) have so far been developed. A good example

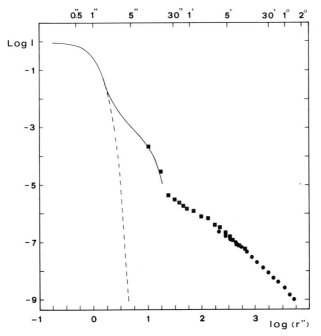

Fig. 1   Luminosity profile across a stellar image seen through the Earth atmosphere. The solid line represents the central cusp encircling most of the star light. The dashed line is the Gaussian interpolation mentioned in the text. Symbols refer to the aureole of the two bright stars α and β Canis Maioris.

of their application to rough ground-based observations has been

given by Arp and Lorre (1976).   Painful and time consuming

procedures such as those used to get the beautifull picture of M87

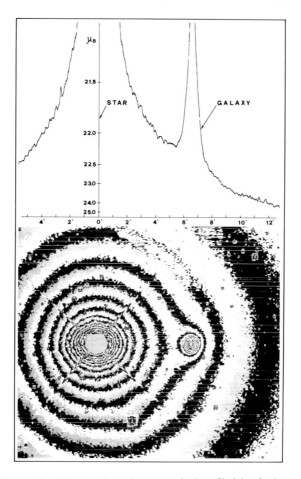

Fig. 2   Joyce-Loebl isophotal map of the field of the galaxy NGC

404. The extended isophotes are due to the scattered light

from the bright nearby star β Andromedae. In the <u>upper</u>

<u>pannel</u> the luminosity profile across the galaxy and the star is

given. It shows that outer regions of NGC 404 fainter than

μ_B = 22.5 mag arc-sec^{-2} are totally embedded in the stellar

aureole.

still require the assumption of several working hypotheses and still give poor results in comparison with current astronomical requirements.

Outside the quasi-Gaussian cusp, the star profile of Fig. 1 presents a faint wing (the aureole), having increasing slope and extending out to very large angular distances. The aureole, produced by atmospheric and instrumental diffraction and scattering (King 1971, Kormendy 1973), removes little energy from the central disk of the stellar image (de Vaucouleurs and Capaccioli 1978b), but it becomes important whenever it contaminates fainter sources. A rather impressive example of that is given in Fig. 2, where the aureole of a very bright star overrides the image of a galaxy. Another example, extremely relevant to extragalactic astronomy, is offered by the possible competition between the light scattered from the bright nucleus of a galaxy and the faint emission of the outer envelope. This consideration brings us to the problems of photometric measurements of both point and extended sources.

Even in the spectral region where atmospheric absorption is not dominant ($3300 \, \text{Å} < \lambda < 7800 \, \text{Å}$), seeing and sky brightness mutilate our chances to measure or even detect faint light sources. In the pure quantum statistics frame, the signal-to noise ratio for a point source is given by:

$$(S/N)_p = (K \, R)^{1/2} \frac{n_p \, t^{1/2}}{(n_p + 2 \pi \alpha^2 \, n_s)^{1/2}} \tag{2}$$

where:

$K = A \times \mathcal{E} \times \Delta\lambda$ is a factor depending on effective collecting area A,
efficiency of the optical system $\mathcal{E}$ and bandwidth $\Delta\lambda$;

R = detective quantum efficiency (Dainty and Shaw 1974);

t = exposure time;

n_p = point source flux in number of photons;

$\pi\alpha^2 n_s$ = number of background photons within the aperture 2α.

The atmospheric turbolence sets $\alpha \simeq \sigma \gtrsim 0\overset{\prime\prime}{.}5$ since, to gather
enough light from the smeared point source, we have to cover the
region interested by the seeing disk. Computed blue limiting
magnitudes (S/N = 10) for a 1 m ground-based telescope as a function
of the exposure time are represented by the dashed curves of Fig. 3
for different seeing conditions. Assumed parameters are: $\Delta\lambda$ = 1000
$\overset{\circ}{A}$, $\mathcal{E}$ = 0.70, R = 0.01. The sky brightness has been set at
μ_s = 22 Bmag arc-sec^2, which is a reasonable estimate even for good
astronomical sites, free from artificial illumination. From the figure
it apparent that the largest telescopes presently available on Earth
can reach a limiting magnitude of 23 or 24 with exposures of the
order of one hour. Beyond this limit the universe stands unknown for
terrestrial observers.

    For extended sources, equation (2) changes into:

$$(S/N)_e = (K\,R)^{1/2}\,(\pi\alpha)^{1/2}\,\frac{n_e\,t^{1/2}}{(n_e + 2n_s)^{1/2}} \qquad (3)$$

Here, clearly, dependence on resolution is reversed; signal-to-noise
ratio increases with aperture 2α. Neglecting possible contamination

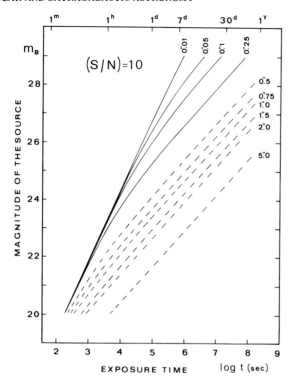

Fig. 3   Limiting magnitudes for detection of point sources with a

signal-to-noise ratio of 10 as a function of the exposure time

and angular resolution α . Assumed parameters are: telescope

aperture 100 cm, $\Delta\lambda$ = 1000 $\mathring{A}$, $\mathcal{E}$ = 0.70, R = 0.01. For the

low resolution ground-based observations (<u>dashed curves</u>) a

sky background of μ_B = 22 mag arc-sec^{-2} has been assumed.

For the high resolution observations (<u>solid curves</u>) a μ_B = 23

mag arc-sec^{-2} typical of a space environment has been used.

from scattered light of nearby bright sources and stretching to the

limit the capabilities of ground-based techniques, it is presently

unconceivable to measure surface brightnesses lower than μ_B = 28-

29 mag arc-sec^{-2}, i.e. about a factor 250 fainter than night sky.

Again, solutions to many fundamental questions lay beyond this
threshold.

   To fully realize the need for improved observational capabilities, it is instructive to compare the limits mentioned in this
paragraph with the data of Fig. 4, which provides a summary of
celestial bodies according to their brightness and size.

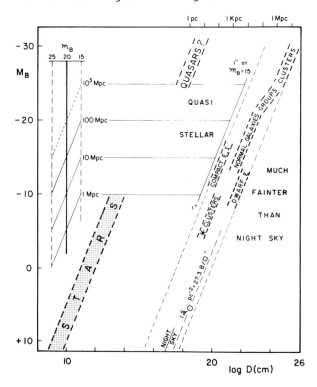

Fig. 4   Distribution of different stellar systems as a function of the
         absolute magnitude M_b and linear diameter D after de
         Vaucouleurs (1973). Normal galaxies and star clusters are
         restricted to a narrow range of surface brightness close to that
         of the night sky on the ground. Optical detection of galaxies
         smaller than 1" or fainter than $1 L_\odot pc^{-2}$ is not possible by
         current techniques.

1.2   OBSERVING FROM ABOVE THE ATMOSPHERE

The most obvious way (even if not unique, as discussed in paragraph 1.4) to bypass the handicap caused by the presence of the Earth atmosphere is to go observing above it. In this paragraph we will outline in a general way the main achievements provided by space astronomical facilities, leaving the scientific discussion of a few selected topics to Section 3.

With proper stabilization and with a focal length matching the detector resolution, we may conceive to reach an overall angular resolution of 0."1 for a medium size telescope flying above the atmosphere of the Earth. The consequent jump with respect to our present knowledge of the structural properties of celestial bodies is schematically represented in Fig. 5. At the distance of M31 (Δ = 820 kpc) we will succeed in detecting details of the size of the Solar System Cometary Reservoir; at the distance of the Virgo cluster[+] (Δ = 20.4 Mpc) nuclei of galaxies will appear resolved at a level of 10 pc.   We will be able to recognize globular clusters surrounding galaxies in the Coma cluster (Δ = 130 Mpc), classify galaxies at least ten times more distant than Coma and study the fuzzy haloes of QSO's. Due to resolution, appropriate space facilities will increase in precision over ground-based astrometry by as much as a factor of 10.

[+] To evaluate extragalactic distances, we assume hereafter H_o = 50 km s^{-1} Mpc^{-1}.

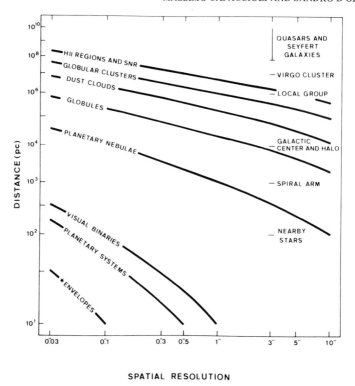

SPATIAL RESOLUTION

Fig. 5   Limiting distances for detection of structural details of galactic and extragalactic sources as a function of the angular resolution.

The improvement in resolution will have a dramatic effect also in those fields where our knowledge rests on spectroscopic techniques. As an example, we sort out one problem from those in dynamical studies of galaxies. Figure 6 shows how structural details of the rotation curve of a galaxy are lost as the angular resolution is degradated. These 'details' are often key features of the rotation curve, which retain the information on the origin of the spiral structure in galaxies. Their loss, due to poor resolution, severely hampers any dynamical approach to the problem.

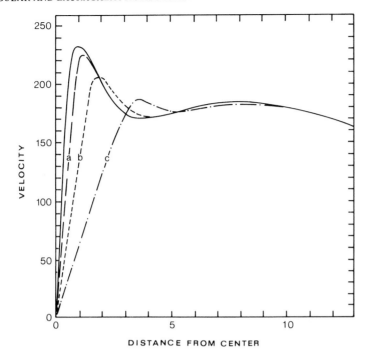

DISTANCE FROM CENTER

Fig. 6   Effect of poor angular resolution on the shape of the inner

rotation curve of a galaxy. The model (solid curve) has been

convolved using a Gaussian point spread function, weighted by

means of an $r^{1/4}$ luminosity profile. The different curves (a, b,

c) have been obtained assuming a 'seeing' of 0.4, 0.8 and 1.6 in

units of the distance from the center of the first peak of the

rotation curve.

The stable point-spread function of a diffraction limited

space telescope is the most desirable property where detection of

faint stellar sources is required. We have already mentioned the

possibility to achieve an overall angular resolution of 0."1, lasting for

a long time. This fact provides by itself a straightforward and

remarkable gain in detection as shown by the solid curves of Fig. 3

(where lowering of background has also been added). Conservatively, we may expect to improve by a factor of at least ten our ability of
looking deep at faint stellar sources. The consequent increase in our
astronomical knowledge can be evaluated syntetically on the basis of
the data of Fig. 4.

Besides resolution, lowering of background flux n_s in extra-
atmospheric observations provides a second improvement to our
ability of detecting faint sources, which is expecially rewarding in
spectral bands contiguous to the optical one. In fact, a significant
contribution to the night sky luminosity on Earth comes from the
airglow. Above it the sky is a factor of two fainter in the blue and the
gain increases markedly toward both UV and IR (Fig. 7). In the case of
a point source, lowering of background luminosity adds to resolution
to diminish the value of the parameter $\pi \alpha^2 n_s$ (formula 2) and,
therefore, to improve the detection threshold of a space telescope
(see Fig. 2). However its main role is paid when extended sources
are concerned. In this case, in fact, the angular resolution is not
relevant and any improvement will solely come by the lowered
background.

Looking at Fig. 7 one might argue that no significant gain is
achieved in the classical photometric bands B and V, i.e. that there is
very little point in observing extended sources from space in the
optical band. This seems not true to us and, to overcome skepticism,
we remind the role paid by Kodak IIIa-J emulsion in recent discoveries. It was enough to gain one magnitude on conventional emulsions
to disclose, for instance, faint haloes around giant elliptical galaxies

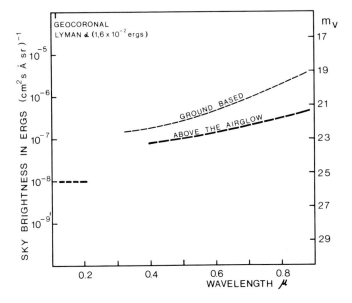

Fig. 7   Trend of the sky brightness as a function of wavelength in

the wide optical band. The magnitude scale in the ordinates

refers to the V band.

(Arp and Bertola 1969, 1971). One more magnitude of gain will allow

to measure their fluxes.

        In concluding this paragraph, we list the main sources of

noise in photometric measurements (at low time resolution) within

the optical band, still retained by telescopes operating from space:

  1)   diffusion and scattering by the optics and residual atmosphere;

  2)   unresolved star and galaxy background;

  3)   zodiacal light and geocoronal emission;

  4)   stray light from the Sun;

  5)   thermal dark current, read-off current and hard radiation

        induced background.

1.3   SPACE TELESCOPES VERSUS NEW GENERATION GROUND-
       BASED TECHNIQUES AND NEXT GENERATION TELESCO-
       PES

We have hitherto commented on the capabilities of optical
telescopes operating from space. At this point it is convenient to
verify the merits of some competitive ground-based techniques. As
far as resolution capabilities are concerned, there is a recently
developed technique, the speckle interferometry, which, according to
the results by Blazit, Romeau, Koechlin and Labeyrie (1977), allows
to reveal morphology of stellar sources at a scale of 0.″015. In other
words, using an appropriate observational technique rather than an a
posteriori restoration of conventional pictures, it is to date possible
to circumvent the traditional limits set by the atmospheric turbolence. However, space high resolution observations will give by far
superior results in comparison with the speckle technique since this
latter is for the present limited to investigations on single stars or
binary systems; it has no chance to disentangle the complex
morphology of crowded extended sources. Moreover, ground-based
speckle interferometry is exceedingly telescope-time consuming; as
shown by Code (1976), it requires an 800^y equivalent time to achieve
a final picture with the same signal-to-noise ratio as a one-orbit
exposure (30^m) with the Space Telescope.

Other possible challangers to the supremacy of optical
telescopes flying above the atmosphere are the next generation
telescopes, i.e. future ground-based gigantic telescopes. Actually,
rapid developments in engeneering and technology have opened the

way for radical new telescope designes which hold promise of much larger apertures. Nowadays, a 25 m diffraction limited telescope is under consideration (Next Generation Telescope, Reports No. 1, 2, 3 and 4, Kitt Peak National Observatory, 1977), which will have a factor of 100 more light gathering power than ST. Therefore, strictly in the optical domain and under reasonable seeing conditions ($\sigma \sim 1"$), its limiting magnitude for a point source will still be lower than that of the finest space facilities presently foreseeble. Moreover, due to the atmospheric blurring, spatial resolution will remain poor unless special techniques, such as speckle interferometry, are used. In the speckle mode a 25 m diffraction limited telescope is estimated to provide the exceptional resolution of 0."003 at $\lambda = 5000$ Å , i.e. more than one order of magnitude better than ST. However, the use of this technique takes back a good deal of the gain in detection given by the large aperture. Moreover, speckle interferometry retains almost unchanged the limits previously mentioned. For extended sources the detection capability of a 25 m telescope operating on Earth will be larger than that of the largest space telescopes only in the B and V bands, where very little gain is capitalized by the absence of the airglow (Fig. 7).

In conclusion, the next generation telescopes seem competitive with the large space telescopes working in the optical band whenever spatial resolution is not needed. The study of crowded objects or extended sources with fine complex morphology is still demanded to space.

1.4   HIGH RESOLUTION AND FIELD OF VIEW IN FUTURE SPACE
TELESCOPES

At this point a comment on the sky coverage achievable with
a space telescope working in the high resolution mode seems in order.
To better set the problem, let us first note that presently available
two-dimensional detectors other than photographic or electronographic emulsions (barely usable in space experiments) have resolution
elements with a typical physical size of the order of a few tens of
μm. For instance, the CCD array developed for the Wide Field
Camera of ST has a pixel size of 15 μm. Therefore, to match the
conventionally assumed resolution of 0".1, a 1 m telescope demands a
focal ratio of at least f/30. At this level the corrected field of view is
usually very small (a few minutes of arc). Moreover, to cover a very
small spot of sky of 1'x1' at a 0".1 angular resolution, a 600x600
equivalent diode frame is requested, which is presently near the limit
of the state-of-the-art of photon counting detector manufacturing.
With that, taking a 30 minutes as the typical exposure time, a one-
year mission would produce as much as about 10000 high quality
pictures but the corresponding sky coverage would be less than that
of a single 48-inch Palomar Schmidt plate (at a standard resolution
level of 2") with a total information of about 10^{11} bits. It is
therefore questionable whether the high spatial resolution achievable
with a space telescope, while per se highly valuable, is unconditio-
nally the only choice for an economical use of the space facilities to
come.

A careful trade-off among field size, angular resolution and limiting magnitude is needed to assess the research effectiveness of the proposed solutions for future space telescopes. However, since the high resolution mode will be fully exploited by the ST, proposals for orbiting survey-type telescopes deserve now special consideration. This is suggested by the long-standing and profitable cooperation shown on Earth between the 200-inch and the Palomar Schmidts. These latters have provided the large, high resolution telescope with a number of candidates (supernovae, peculiar galaxies, QSO's, etc.) deserving detailed investigation. In addition, the exploration of a new spectral band is replete with pleasant surprises (e.g., X-ray and EUV sources). The discovery of unpredicted objects in the UV will come only with an instrument operating in the survey mode, the ST being a blind eye from this point of view.

2. REVIEW OF OBSERVATIONAL RESULTS FROM SPACE

2.1   PERFORMED EXPERIMENTS

In the recent past several astronomical observations have been carried out from outer space. The work done from balloons, sounding rockets, satellites and orbiting laboratories has already given a significant contribution to our understanding of celestial bodies. To date, the major effort has been devoted to fully profit of the newly opened spectral bands, inexplorable with ground-based telescopes.   Thus, the optical band, where a good deal of our astronomical knowlege is presently resting, has been temporarily shelved. Technological limitations such as vehicle stabilization, telescope size and detector sensitivity limited the observations to only fairly bright sources. Under these circumstances past space telescopes could not have been competitive, in the optical band, with the numerous powerful instruments operating from Earth.

Table 1 presents a list of performed space experiments in the wide optical band from satellites and manned space laboratories. Experiments with rockets and high altitude balloons have been included only when they obtained results in the fields treated in this article, that is galactic nebulae and extragalactic objects. The table shows, as it was stressed above, that all but one experiments have been devoted to the exploration of the UV spectral band, unaccessible from the ground. The exception is the Stratoscope balloon flight which was aimed at high angular resolution photography (No. 3 in Table 1). The most effective facilities performed UV wide or intermediate band spectroscopy, and UV medium and high resolution spectroscopy of relatively bright stars, as the TD1 and Copernicus satellites. The extragalactic and extended galactic targets were included only marginally in the observing programs, due to their faintness and/or to the optical layout of the instruments, mainly designed to operate on point-like sources. Some interesting results were also obtained from ad hoc experiments flown on sounding rockets.

The total amount of new data on galactic nebulae and extragalactic sources derived from space observations is still small, but it gives us the flavour of possible future achievements with more dedicated facilities. As remarkable examples, we can quote the UV

TABLE 1 - PERFORMED EXPERIMENTS IN THE WIDE OPTICAL BAND FROM ORBITING SATELLITES, MANNED SPACE VEHICLES, ROCKETS AND HIGH ALTITUDE BALLOONS[+].

No.	Name	Launch	Effective diameter (cm)	Angular resolut.	Field diameter	Spectral resolut.	Main observing modes	Principal investigator
1	OAO-2(WEP)	1968	4x20		on axis	12 Å	wide band photometry, low resolut. scanning imaging	Univ. of Wisconsin-NASA
2	OAO-2(Celescope)	1968	4x30	1.5	2°	WB	high resolution photography	Smithsonian Astr. Obs.-NASA
3	Stratoscope II(8)	1971	91	0".2	1'	WB		Univ. of Princeton-NASA
4	Copernicus	1972	80		on axis	0.05-0.4	spectroscopy	Univ. of Princeton-NASA
5	TD1-S2/68	1972	27.5		on axis	35 Å	low resolut. scanning spectroscopy	U.K. and Belgian team-ESRO
6	TD1-S/59	1972	26		on axis	1.8 Å		Utrecht Obs.-ESRO
7	S-201/Apollo 16	1972	75	2'	20°	WB	large field UV photography	Naval Research Lab.-NASA
8	S019-Skylab	1973	15		4°x5°	31 Å	objective prism	LAS, Marseille-NASA
9	S182-Skylab	1973	3.5	2'	7°x9°	WB	two-color imaging	LAS, Marseille-NASA
10	Orion 2-Soyuz 13	1973	24	5"	5°x5°	8Å, 28Å	objective prism	USSR
11	ANS	1974	22	2.5	on axis	WB	photometry	Univ. of Groningen-NASA
12	Faust	1974	16	2'	7°.5	WB	large field UV photography	LAS, Marseille-CNES, France
13	EUV-ASTP	1975	grazing incid.		on axis	WB	wide band photometry	Space Science Lab., Berkeley NASA-USSR
14	D2B	1975	8	0.5	1°	WB	wide band photometry	CNES, France
15-16	Grating Spectrograph	1974-75	33	17'x24'	on axis	20-35 Å	medium resolution spectroscopy	Goddard Space Flight Center NASA
17	Electronographic cameras	1975	75, 15	3', 30"	20°, 3°	WB	large field UV photography	Naval Research Lab.-NASA
18	FOT	1977	37	2.2	on axis	10-15 Å	low resolut. spectro-photometry	J.Hopkins Univ.-NASA

NOTES: [+] As of December 31, 1977
No.3: balloon flight; No. 12, 15, 16, 17, 18: rocket flights.

spectra of two planetary nebulae, with the measurements of the
intensities of Carbon emission lines and subsequent estimate of the
abundance of this element, and the UV spectrum of 3C 273, of great
relevance to QSO models and to cosmology. In the next two paragraph
we briefly review these results and all other space observations of
interest to the astrophysical problems considered in this article.

2.2   OBSERVATIONS OF GALACTIC NEBULAE

Basic information on the interstellar medium has come from
space observations but the bulk of the data rests on the study of UV
absorption lines from interstellar gas (extensively discussed in
Chapter 12 of this book) and on systematic measurements of
extinction due to the interstellar dust. The region of ionized gas
(diffuse nebulae associated with young stars, planetary nebulae and
supernova remnants) have not been up to date a prime objective of
investigation mainly for the two following reasons. First, the
expected breakthrough in our understanding from the exploration of
the UV emission line spectrum is not spectacular as it was from the
study of the absorption lines of the interstellar gas. Second and most
important, extended emission line objects require instruments with an
ad hoc configuration. A few planetary nebulae have been observed
because they are relatively more compact than other emission
nebulae. Bohlin et al. (1975, 1978) have obtained the UV spectra of
the planetaries NGC 7027 and NGC 7662 with rocket borne telescopes
(experiments 15 and 16 in Table 2). The spectra cover the $\lambda\lambda$ 1400-
2800 $\overset{o}{A}$ region at a resolution of 25 $\overset{o}{A}$. The most interesting result is
the determination of the Carbon abundance, since this element has

strong CIII and CIV emissions in the UV but only faint recombination lines in the visible. Carbon is the major cooling agent in the nebulae and a key element for theories of nucleosynthesis. The observations of the UV continuum are in agreement with the sum of the predicted intensities of the two-photons and Balmer continuum contributions.

The central stars of several planetary nebulae have also been observed. The results are used to check or to extend the classical Zanstra's method to determine the effective temperature of the exciting stars in an optically thick nebula. By improving the knowledge of this parameter, we are able to set more precisely the position of these hot stars in the Hertzprung-Russel diagram. This problem is outside the purpose of this review, but a reference to this work is useful because of its relation to the study of the associated nebulosity. The UV spectra of the central stars of NGC 6543 and IC 2149 are discussed by Boksenberg et al. (1975) and Gurzadian (1975), using data from TD1-S2/68 and the Orion-Soyus experiments respectively.   In a more systematic study, Pottasch et al. (1978) discussed intermediate band observations of about 30 planetary nebulae obtained by ANS.

Very few results are available for HII regions and supernova remnants. The intrumentation flown to obtain the UV spectra of NGC 7662 and NGC 7027 was used by Bohlin and Stecher (1976) to obtain a spectrum of the Orion nebula from 1200 to 2900 $\overset{o}{A}$. The authors found a strong continuum of dust-scattered star-light with only two emission lines, λ 2326 of CII and λ 2470 of OII, well above the continuum. In fact, theoretical predictions indicate that only HII regions at high temperature will show a rich UV emission line spectrum, since most of the lines originate from collisionally excited metastable levels. Carruthers and Page (1976) and Carruthers and Opal (1977a,b) have obtained UV photographs of regions in Gygnus and in Orion with wide field electronographic cameras operated from the surface of the Moon during the Apollo 16 mission and rocket-borne. The cameras were sensitive to the 1050-2000 $\overset{o}{A}$ and 1230-2000 $\overset{o}{A}$ spectral ranges and the images are of good quality. The photographs confirm the spectroscopic results for the Orion region, that is the prevalence of dust scattered light over emission line contribution. The nebular brightness rises in the far UV and this is interpreted as due to an increase in the albedo of the dust in this spectral region. The supernova remnant Cygnus Loop is the most prominent feature in the UV photograph of the Gygnus region. This is not surprising since theoretical calculations predict the presence of strong UV lines in the hot plasma associated to the shock wave originated from a supernova explosion. We will discuss this problem within the scientific rationale for the study of supernova remnants from space in Section 3.1.

2.3   OBSERVATIONS OF EXTRAGALACTIC OBJECTS

Wide band observations of galaxies have been obtained from the satellites OAO-2, TD1-S2/68 and ANS. The results can be found under the following references: Code et al. (1971), Carnochan et al. (1975), van der Kruit and de Bruyn (1976), Wu (1976) and Code and Welch (1977). The galaxies were generally observed in wide bands from 1500 $\overset{o}{A}$ upwards through apertures of large angular size. This introduces a first ambiguity in the interpretation of the data since the contribution of the spiral arm population with respect to the nuclear concentration will be different in galaxies at different distances. So far, data for about twenty galaxies of integrated magnitude 10 or brighter have been published. All morphological types are represented, including a number of Seyfert galaxies. The interpretation of the fluxes in terms of stellar population is not univocal. It appears that late type galaxies have relatively stronger UV fluxes, in agreement with their high content of hot stars. Code and Welch (1977) remark the existence of a large scatter in the UV energy distribution of early type galaxies and discuss the implications on the computation of realistic K corrections to photometry of very distant elliptical galaxies. They also stress that the marked difference in the UV flux relative strength between spirals and ellipticals may have the result that, at large redshifts, a spiral galaxy will be observed as the brightest cluster member.

Integrated spectra are also used to estimate the amount of dust in galaxies (Wu 1977). A marked flux deficiency at λ = 2200 $\overset{o}{A}$ was observed with ANS in NGC 253 and M82, two galaxies with strong absorption lanes. This suggests that the dust has a composition similar to the one found in the Galaxy, where this feature was firstly observed. In a previous study of young associations in the LMC, Borgman et al. (1975) had noted the absence of the 2200 $\overset{o}{A}$ bump and suggested a different composition for the dust in the galaxy. The uniformity in dust properties is still an unsettled question and an important one since there should be a correlation with the differences in chemical composition of the ionized gas found among galaxies. Seyfert galaxies observed with the ANS (Wu 1976) have a behaviour similar to that of normal spirals, with the apparent exception of NGC 4151 for which a steep rise in the UV is reported. At this level of information it seems hazardous to draw any interpretation of this difference.

Two of the most stimulating results in the field of galaxies come from ad hoc experiments. Light et al. (1974) observed the nucleus of M31 with a 0."2 resolution by means of a balloon-borne telescope. The observations indicate that the nucleus is a feature clearly distinct from the bulge of the galaxy, with a size of 0.5 pc. This matter is discussed in details in paragraph 3.3. Quite recently Davidsen et al. (1977) have reported on the first observation of the UV spectrum of the bright QSO 3C 273, made with a rocket-borne telescope. Observations of QSO's are among standard uses of the future Space Telescope and their scientific relevance has been widely discussed in the past. This summary of the results obtained by Davidsen et al. (1977) esemplifies most of the key questions on this subject and some possible answers. The observations of the $\lambda\lambda$ 1200-1700 $\overset{\circ}{A}$ region at 10/15 $\overset{\circ}{A}$ resolution combined with the near UV and visual data give the more extended energy distribution known for a QSO. A ratio: $F(L_\alpha)/F(H_\alpha) = 4$, is observed against a value of 40 predicted by radiative recombination with some enhancement of L_α due to collisional excitation. A spectral index of $\alpha = 0.6$ is found between H_α and L_α, as compared to $\alpha = 1$ between H_α and P_α. Both these results are expected to have a major impact on QSO models, in particular for what concernes the possible presence of internal dust and the nature of the primary energy source. The absence of absorption features shortward of L_α is used to calculate an upper limit of $n_H = 1.5 \times 10^{-12} cm^{-2}$ for the density of intergalactic neutral Hydrogen and to attribute the absorption seen in QSO's at high redshifts to matter at smaller cosmological distances.

2.4   THE INTERNATIONAL ULTRAVIOLET EXPLORER

On January 1978 a new satellite, the International Untraviolet Explorer (IUE) was successfully put into a geosincronous orbit. The satellite is the result of a cooperation between NASA, ESA and the U.K. Science Research Council. It hosts a telescope with an effective diameter of 38 cm and an on-axis <u>echelle</u> spectrograph working from 1150 $\overset{o}{A}$ to 3200 $\overset{o}{A}$ at high (0.1 $\overset{o}{A}$) and medium (6 $\overset{o}{A}$) resolution. The detectors are UV sensitive SEC Vidicon cameras. The targets can be observed through a 3" circular and a 10"x20" elliptical aperture.

The first observations were secured in Spring of 1978 and the telescope has been working since then. Several targets of interest to IM and extragalactic astronomy have been already observed or are in the observing list for the next months. Spectra have been secured for the brightest Seyfert galaxies (NGC 1068, NGC 1275 and NGC 4151), Markarian galaxies (MKN 509 and MKN 297), quasars (3C 273, 3C 282 and 3C 120), and, among galactic objects, for several planetary nebulae and a filament of the Cygnus Loop supernova remnant. IUE is observing at a much better angular and spectral resolution than previous satellites which have looked at extragalactic targets (TD1,

ANS). As stressed in paragraph 2.3 and discussed in 3.4, to understand the nature of the central sources in active galaxies, it is necessary to resolve the central emission from the UV output of the galaxy as a whole. This is simply achieved with the small entrance apertures of IUE spectrograph. The the scientific results will appear starting with 1979. It can be anticipated that they will constitute a major step in our understanding of galactic IM and extragalactic objects and they will represent the basis for the planning of the UV observations of the more powerful Space Telescope.

3. IMPACT OF SPACE OBSERVATIONS
ON SELECTED ASTROPHYSICAL PROBLEMS

3.1   SUPERNOVA REMNANTS

Supernova remnants may be looked at as nature made experiments on the properties of the interstellar medium. As a greedy interloper not admitted to the laboratory console, the observer seeks to obtain both the results of the experiments (density and chemical composition of the IM) and the initial parameters of the probing explosion (total energy, abundance of the ejected material). The latter bears the information on the last stages of the stellar evolution and on element cooking in stellar interior, while the present chemical composition of the IM is related to the metal enrichment of a galactic system as a whole. All these items are fundamental to understand the evolution of galaxies. The study of a SNR is therefore more complex than in the case of an HII region and a planetary nebula. In these latter cases the physical processes that lead to the emission of the spectrum for the gas ionized by UV radiation of hot stars have been known for half a century. With the recent improvements of atomic parameters and with the use of detailed models, the physical conditions (N_e, T_e and ionization structure) and hence the chemical composition of the emitting gas can be determined with reasonable accuracy. On the contrary, for supernova remnants we must built a model where both the initial parameters and the resulting spectrum are unknown and must be determined from the observations.

The basic steps of the evolution of the supernova remnant can be summarized as follows. First, the ejected material sweeps the surrounding medium acting like an expanding piston. In a second stage the kinetic energy of the explosion being transferred to the moving shell of the swept up material, an adiabatic expansion takes place and, finally, as the shock weakens, the temperature drops in the region behind the shock wave and the remnant enters a radiative phase. The radiative cooling gives origin to the characteristic spectrum. Chevalier (1977) has reviewed recent progresses made in this field. Calculations of the emission line spectrum have been made by Cox (1972), Raymond (1976) and Dopita (1977) as a function of the shock velocity, chemical abundances and pre-shock density. These calculations can also be used for the interpretation of the spectra of galactic nuclei where explosive motions are observed. This is the case of the Seyfert nuclei where similar processes (even if more complex than in supernova remnants) may take place (see paragraph 3.4). When the predictions of the models are compared with the spectrophotometric observations of filaments of supernova remnants in the optical region (3500-7000 $\overset{o}{A}$), a general agreement is found but the initial parameters mentioned above cannot be univocally determined.

Observations above the atmosphere offer new and perhaps conclusive opportunities to check the model and hence to define the unknown parameters. In fact, we can use the higher spatial resolution to verify the details of the shock front structure and we can measure the UV lines which are the main cooling agents of the gas at high enough temperatures.   Let us briefly discuss these two points. According to the models of a shock wave quoted above, different ions of the same element are present at different distances from the shock front. By observing an optical filament in the light of emission lines from ions over a large interval of ionization potentials, we will be able to verify the existence of the stratification and its characteristics, which are a function of the expansion velocity and pre-shock density. When the model of Cox (1972) is used for a shock propagating in the medium with $n_o = 1$ cm^{-3}, $B_o = 10^{-7}$ gauss at a velocity of 100 km s^{-1}, we derive a distance of 2.7×10^{-3} pc between the peak of the $[\text{OIII}] \lambda 5007$ $\overset{o}{A}$ and $[\text{OI}] \lambda 6300$ $\overset{o}{A}$ emission lines. For the Cygnus Loop (assumed distance 1 Kpc) this separation corresponds to 0."55, for the Vela nebula to 1" (assumed distance 0.5 Kpc).   In similar computations by Dopita (1977) for a lower expansion velocity, the resulting separation is about twice. Ad hoc computations for other ions and over a larger range of expansion velocities could be easily performed. In conclusion, the resolution of the shock structure is well within the capabilities of medium size telescopes operating above the atmosphere. The program could be carried with the ST since the expected performances both in imaging and spectroscopic mode are superior to these values.

The second aspect is the observation of UV spectra of supernova remnants. Computations of the emission line strengths for a collisionally excited gas at high temperature indicate that various resonance UV lines dominate the cooling. Dopita (1977a,b) predicts relative intensities for several UV lines such as λ 1549 of CIV, λ 1909 of CIII, λ 1335, 2326 of CII. Spectroscopic observations in the UV and subsequent comparison with the predictions will represent a crucial test for the model and will allow an accurate determination of the physical conditions in the shock region. A somewhat more specific goal of the UV observations of the supernova remnants is stressed by Dopita (1977b). From the strong lines emitted in the UV we will be able to measure reliable abundances for Carbon, an element with no prominent emission in the optical region. Determinations of the abundances of the other two elements of the CNO cycle is also possible from UV spectra. The ratio of abundances of primary elements such as C and O to a secondary element like N is related to a very important parameter, the initial mass function of stars, and thus it provides information on stellar evolution (Talbott and Arnett 1973).

3.2   STELLAR AND GASEOUS CONTENT OF NEARBY GALAXIES

A sensible fraction of the observing time of large telescopes is presently devoted to the study of galaxies within 20 Mpc. Now that the original enthusiasm for QSO's and for the multiform fauna of peculiar galaxies has somewhat faded out, astronomers start to realize that many of the most basic questions about the nature of the universe are best investigated through an examination of the content of relatively nearby galaxies. A great deal of systematic programs, such as the search of bright variable stars, the study of globular clusters, the color-magnitude and color-color diagrams of star systems, will have a prominent position in the scheduling of the large telescopes in the coming years. However, there is a number of observational problems where the capabilities of ground-based tele-scopes are stretched to their limits. Any improvement in angular resolution and limiting magnitudes will extend the boundaries of present observations. An instrument such as the Space Telescope which will look at the Magellanic Clouds and at M31 with an angular resolution of 0.''1 and a limiting magnitude of about 29, will realize the astronomers' dream to obtain a complete, deep view of the whole galaxian system. This study could not be carried for the Milky Way due to the position of the Sun near the galactic plane and the presence of the dust. With the ST it will be possible to reach M_V = +4 mag at the distance of M31 and its companions, thus making possible to study the orizontal branchs and the main sequence turn-offs in the HR diagrams of stellar sub-systems of population II. The gain in other fields will be equally remarkable. In galactic globular clusters, for instance, the observer will take advantage of the angular resolution to resolve the central, crowded area and to identify white dwarfs and low luminosity stars; in the Virgo cluster galaxies it will be possible to verify and extend the identification of globular clusters carried out by Hanes (1977), to study the chemical composition of individual objects and to discover classical Cepheids; the diameters of HII regions will be measured in galaxies farther than Coma cluster. In the following we review in more details the impact of space observations on the study of ionized nebulae in nearby galaxies. There are three types of emission line nebulae: HII regions associated to newly formed stars, planetary nebulae and supernova remnants.

a) HII regions. Our present knowledge of the chemical abundances in the interstellar gas of nearby galaxies rests on the observations of emission line spectra of giant HII regions associated to the spirarl arm structure. Our understanding of these gigantic complexes of star formation is limited because there is no visible galactic counterpart

TABLE 2 - RADIO POWERS OF HII REGIONS IN NEARBY GALAXIES

Name	Distance Kpc	HII Region	P1415 10^{18} W Hz^{-1}	O5[+]
Galaxy	0.5	Orion Nebula	10^{-8}	<1
	6.5	W51	3.6	33
LMC	48	30 Dor	10	100
M33	720	NGC 595	1.2	11
		NGC 604	3.6	33
NGC 2403	3300	VS 42	10	100
		VS 44	31	30
M101	7000	NGC 5461	112	1000
		NGC 5462	71	630

[+]Number of O5 stars needed to produce the UV photons which ionize the nebula.

of them. In Table 2 the radio fluxes at 1415 MHz for HII regions in nearby galaxies are given together with the estimated number of O5 stars required to produce the ionization and, hence, the observed radio emission. The data are after Israel et al. (1975) and Israel (1976) and form a more homogeneous sample than optical data. The best known galactic HII region, the Orion nebula, is also included in the Table, but its parameters show that it belongs to an entirely different class and, hence, the detailed knowledge of this particular object cannot be extrapolated to the extragalactic HII regions. W 51,   a giant galactic ionized region, is detected only at radio wavelength. With the ST or with an instrument of similar performances the color-magnitude diagram of the ionizing stars will be derived in HII regions as far as M33. This program cannot be carried from the ground due to the crowding of stellar images. One might also hope to estimate the birth-rate of early type stars and, in general, the star luminosity distribution in these peculiar regions. Gigantic arcs are observed at the periphery of these large HII region complexes. Figure 8 is a deep photograph of the brightest HII region in M33, NGC 604, which shows the presence of loops with diameters up to 50 pc. In 30 Doradus, the giant complex in the LMC, these loops have dimensions of a few hundredth parsecs (see, e.g., plate XIV in Davies et al. 1976). Very little can be inferred on their origin; as possible mechanisms they have been proposed stellar winds from O stars, shocks driven by supernova explosions or ionization fronts of an expanding HII region. From the ST observations of UV spectra of OB stars associated to the HII regions, the stellar wind hypothesis should be efficiently tested. Spectroscopic and photographic studies at high angular resolution

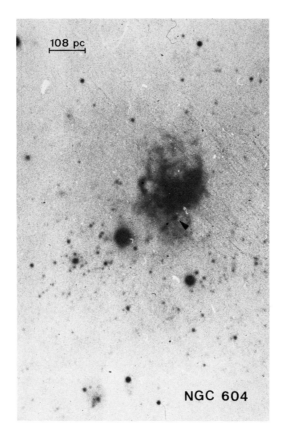

Fig. 8   A deep image-tube photograph of the giant HII region NGC

604 in M33 in the light of the $\lambda 6724\, \text{Å}$ [SII] doublet,

obtained with the 72-inch Asiago telescope. Large arcs of

emission surround the core of this HII complex and the arrow

marks the position of a supernova remnant.

should also clarify the structure and the dynamics of these regions.

b) Planetary nebulae. Unlike the HII regions, the planetary nebulae
are related to the last stages of stellar evolution. The emission line
spectra of their gaseous envelope have been studied in details (see,
e.g., Torres-Peimbert and Peimbert 1978). The abundances derived
from the emission line intensities derived in planetary nebulae

represent the chemical composition of an old population I possibly modified by the nuclear processes such as, for instance, the CNO cycle, which went on in the parental star during its lifetime. The ionized shell eventually goes back to the interstellar medium and contributes to change the primeval abundances. Other competitive processes are violent ejection of star fragments in a supernova explosion and the mass loss by stars, of which the planetary nebulae are likely to be a particular case. Counts and the luminosity function of planetary nebulae in nearby galaxies provide an estimate of the rate of this particular mechanism of mass return to interstellar medium. These parameters are difficult to obtain in our own Galaxy because the observations are hampered by absorption and distance uncertainties. Counts of planetary nebulae in M31 and its companions have been obtained from ground-based observations (Ford and Jacoby 1978, and references therein). From space, this work should be extended to galaxies as far as the Virgo cluster, where a bright PN would show $m_{ph} = 29$ mag. Spectroscopy of planetaries in the Local Group, particularly in M31, would be easier at the high spatial resolution of the Space Telescope, since the galaxy background entering the slit aperture will be greatly reduced. These observations should be particularly rewarding for planetaries in elliptical galaxies where the absence of conspicuous HII regions has so far precluded the determination of the chemical abundances from the study of ionized gas.

c) Supernova remnants. The remnant of a supernova is a shell-like nebula made up by interstellar material swept up and ionized by the

expanding shock following the supernova outburst. It emits the
characteristic optical line spectrum and its radio emission is non-
thermal. The shells have diameters up to 50 pc. The identifications
are based on radio and optical techniques for galactic objects, but the
optical sample is severely limited by absorption. Supernova remnants
have been also identified in the Magellanic Clouds and, recently, in
M33 (D'Odorico et al. 1978). In these galaxies the identification
technique is based on the comparison of narrow filter photographs
taken in the light of Hα and λ6724 Å of [SII] . The supernova
remnants stand out in the [SII] pictures due to their characteristic
emission line spectra. This identification work is rewarding for at
least two reasons: the mapping of the supernova remnants in an
external galaxy will eventually point out the parent population of
their progenitors and lead to a value of the frequency of supernova
explosions. These information is hardly extracted from the galactic
radio sample since the distances are known only on a statistical basis.
Second, the chemical abundances derived from the analysis of the
SNR emission line spectra are independent, unlike than in HII regions,
from the spectra of the ionizing stars and from the dust content.
Thus, they provide an independent check of the observed differences
and gradients in chemical composition in galaxies (Danziger et al.
1978, and references therein).

The identifications of supernova remnants in M33 are made
at the limiting resolution of ground-based telescopes. The situation is
summarized in Fig. 9, where a galactic SNR is shown together with
the SNR 13A in M33 (D'Odorico et al. 1978). This object is only 2."5

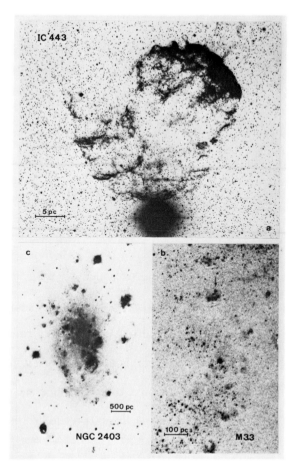

Fig. 9   Comparative views of the angular sizes of supernova
remnants in different galaxies. a) Red band photograph of the
typical galactic SNR IC 443 (Asiago 60/90 Schmidt telescope).
b) Red photograph of a section of M33, a Sc galaxy of the
Local Group. The arrow marks the position of a supernova
remnant identified by D'Odorico et al. (1978) (Asiago 72-inch
telescope). c) Hα photograph of NGC 2403, a Sc galaxy where
supernova remnants appear stellar-like in ground-based observations, but should be easily resolved with the Space Telescope
(Asiago 72-inch telescope).

in size and would appear stellar in NGC 2403, a Sc galaxy four times more distant than M33 (Fig. 9c). The identification of stellar-like remnants or of remnants in crowded HII complexes (see Fig. 8) is difficult if not impossible. Table 3 shows the expected angular sizes and the predicted radio fluxes for typical supernova remnants (the Crab nebula and IC 443) when located in some nearby spiral and irregular galaxies. With an angular resolution of 0."1, the optical technique will be effective as far as M101. The radio fluxes are below present capabilities for distances larger than 1 Mpc. An additional advantage of observations above the atmosphere is the possibility to observe the UV region of the spectrum. As discussed in paragraph 3.1, supernova remnants are expected to have strong UV emission lines. This peculiarity could be used both for identification purposes and for the chemical composition analysis.

TABLE 3 - EXPECTED ANGULAR SIZES AND RADIO FLUXES OF TYPICAL SNR

Galaxy	Distance Kpc	Angular diameters arc-sec		Radio Flux[+] f.u.	
		Crab	IC443	Crab	IC443
LMC	50	9	130	1.7	0.4
NGC 6822	500	0.9	13	1.5×10^{-2}	3.3×10^{-3}
M33-M31	700	0.7	10	10^{-2}	2×10^{-3}
NGC 2403	3000	0.1	2	4.2×10^{-4}	9×10^{-5}
M101	7000	0.06	0.9	8×10^{-5}	2×10^{-5}

[+]At 1 GHz; 1 f.u. = $10^{-26} W\ m^{-2}\ Hz^{-1}$

3.3   NUCLEI OF 'NORMAL' GALAXIES

The central condensation of light in 'normal' galaxies exhibits a large variety of possibilities as far as the surface brightness is concerned. Galaxies having basically the same integrated magnitude such as the dumbbell system NGC 4782/4783 (de Vaucouleurs et al. 1976) have very different nuclei (Burbidge et al. 1964). As an extreme case, nuclei can be exceptionally bright in the optical band. This has been proven to be the case for the elliptical core of the spiral galaxy M31 (Light et al. 1974; see Fig. 10), for the dwarf elliptical M32 and for the two giant ellipticals NGC 3379 (de Vaucouleurs and Capaccioli 1978a, Kormendy 1978) and M87 (Young et al. 1978). This fact is by itself of enormous importance since it seems irreconciliable with the quasi-isothermal models (with constant mass-to-light ratio) which represent very well the observed luminosity profiles in the intermediate regions of elliptical galaxies and the elliptical bulges of spirals (King 1966, Larson 1974, Wilson 1975). The observed cusp of light might be interpreted as due to an anomalously low value of the mass-to-light ratio, as the one produced by a violent burst of star formation or by the presence of a non-thermal source contributing a strong continuum in the optical region.   This hypothesis, while not yet completely ruled out, has been weakened by recent spectroscopic observations of the nucleus of M87 (Sargent et al. 1978) which are consistent with a late type (G5) stellar spectrum. A second interpretation, raised theoretically by Lynden-Bell (1969), Wolfe and Burbidge (1970), Peebles (1972), Hills (1975), Frank and Rees (1976), Young et al. (1977), assumes the presence of a strong

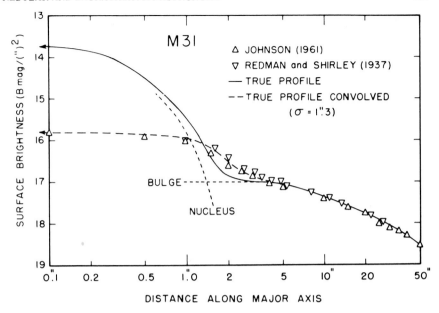

Fig. 10   Photometric profile of M31 along the major axis from a photograph at a resolution of 0."2 taken with the balloon-borne Stratoscope II telescope. The nucleus was observed to be elliptical (1."0x1."6) and appears as a separate feature from the bulge of the galaxy (Light et al. 1974).

potential well generated by a supermassive point source (black hole?) sitting in the center of the galaxy. In the case of NGC 3379 the mass required to account for the observed strong excess of light over the best-fitting isothermal core model with M/L = constant is about $M = 10^7 M_\odot$ (de Vaucouleurs et al. 1978). Its potential is dominant over that of the galaxy at $r < 0."25 \simeq 15$ pc, where the excess is already 0.75 mag. For M87 the requirement is a mass of $6.5 \times 10^9 M_\odot$, acting over a distance of 1."7 $\simeq 160$ pc (Young et al. 1978). The dynamical effect caused by this huge amount of condensed material has been positively tested by Sargent et al. (1978).   They find a sharp increase (50%) of the velocity dispersion inside the core $(r_c = 9\overset{''}{.}6)$, as predicted by the model.

Therefore, in spite of the paucity of presently available data, both theories and observations on the light (and matter) distribution in cores of 'normal' galaxies raise new exciting questions. How may 'normal' galaxies exhibit this 'peculiarity' ? Is there any correlation between the presence of a nuclear light cusp and some morphological, optically are those of cD galaxies. According to Morgan et al. (1975) these objects are supergiants 'having an elliptical-like nucleus surrounded by an extensive envelope'. They appear in clusters of any richness (Morgan et al. 1975) as the brightest members, often associated with powerful radio sources (Matthews et al. 1964). Envelopes of cD galaxies , in excess to de Vaucouleurs' $r^{1/4}$ law, extend over enormous distances (Oemler 1976, Table 3). For instance, the brightest member of the cluster A2670 is followed out to r = 1 Mpc ! The existence of these large coronas poses the problems of their formation and content. According to the dynamical friction picture, cD's are cannibal galaxies (Ostriker and Tremaine 1975, White 1976), growing in mass and size by eating the smaller companions. Therefore they should consist of normal stars, possibly subluminous dwarfs. Another suggestive hypothesis, recently reviewed by Rees (1977a), explains haloes by means of the so-called population III stars, i.e. stars which were formed at z = 100 before protogalaxies collapsed.

Besides stimulating these fashinating speculations, galaxian haloes may also shed light on one of the most intrigued problems of modern cosmology, that of the <u>missing mass</u>. It is well known that the very large values of the mass-to-light ratio given by the Virial Theorem applied to clusters of galaxies are apparently irreconciliable with those measured in the bright regions of individual galaxies. If clusters are bound systems, a good deal of unseen mass placed outside the conventional boundaries of galaxies is required to explain the of the distance $\bar{r}$ from the center, we immediately realize why extended outskirts of galaxies are poorly known. Actually, since more than fifty years ago astronomers have been aware that galaxies might be larger than they appeared. Commenting his classification scheme, Hubble (1926) writes that elliptical galaxies have luminosities fading <u>'smoothly from bright nuclei to indefinite edges'</u> and diameters which are <u>'functions of the exposure times'</u>. The present state-of-the-art of ground-based photometric techniques allows to measure fluxes down to a surface brightness μ_t = 28/29 Bmag arc-sec^{-2} (which is a distance-independent limit provided that the luminosity gradient is not too steep). Within this range several empirical photometric laws have been so far proposed to represent the smoothed observed light distribution in galaxies of different morphological types. Dwarf elliptical galaxies as well as some giants are beautifully fitted by King's (1966) tidally truncated model (King 1978). There are, however, some giant elliptical galaxies that exhibit no cut-off at any distance within the observable range. They follow the $r^{1/4}$ law proposed by de Vaucouleurs (1948), which provides a slow decrease of the luminosity and is consistent with the presence of a faint outer <u>halo</u>. The existence of this latter structure has been directly demonstrated for the first time by the isophotal tracings of deep IIIa-J plates taken by Arp and Bertola (1969, 1971) with the 48-inch Palomar Schmidt. Due to these extended haloes, galaxies should overlap in dense cores of clusters, as shown by de Vaucouleurs and de Vaucouleurs (1970) for the case of Coma. But the most prominent envelopes so far observed pursue a complete coverage of all the types. This is much too large effort for a single telescope and in any case the sample will be poorer than the radio one due to the exclusion of systems obscured by dust and to the limited resolving capabilities. We will make a more effective use of our observing time by concentrating our efforts on the study of one class of objects, e.g. the Seyferts. If a satisfactory understanding of the Seyfert phenomena is reached, we have a good chance to be able to interpret also the general frame and the existing correlations.

3.5   OUTSKIRTS OF GALAXIES

According to a metaphor devised by Disney (1976), galaxies behave alike icebergs in what only a small fraction of their surface is bright enough to rise above the sky background level. There are some galaxies which, due to intrinsic faintness (e.g., the Sculptor dwarf galaxy in the Local Group) and/or galactic absorption (e.g., NGC 6946, Ables 1971), are almost totally overcome by the background. More than 50% of the total blue light of giant elliptical galaxies comes from regions fainter than the night sky (hereafter set at $\mu_s = 22$ Bmag arc-sec^{-2}, which corresponds to good astronomical sites, free from artificial illumination). This appears even more striking in terms of area; the region brighter than μ_s takes only 2% of the area enclosed by the $\mu = 27$ Bmag arc-sec^{-2} isophote. Adding to the above considerations the fact that, with the exception of local deviations, the general trend of a galaxian luminosity profile: $I(\bar{r})$ $= \text{dex}(-0.4\mu(\bar{r}))$, is described by a monotonically decreasing function conditions closer to the values present in the real clouds and not averaged over non-homogeneous extended regions. Our interpretation of the physical processes in the nuclei will be in this way less approximate.

Finally, the improved spatial resolution and the deeper imaging capability can be used to assign a Hubble type to the unclassified Seyfert galaxies. Adams (1977) has studied a sample of 60 Seyferts and has pointed out the difficulty in the classification of the faint and often unresolved features around bright Seyfert nuclei. At the present time, there seem to be very few, if any, Seyfert nuclei in early type galaxies. NGC 1275 is the noticeable exception, but it is a very peculiar system. If this rule is confirmed, it may indicate that a condition for the Seyfert phenomena to develop is the presence of a disc, as suggested by Ekers (1978). It is highly useful to test this hypothesis to increase the number of systems with a clear cut classification.

As we have done for Seyferts, we could draw ad hoc observing programs to investigate from space the nature of active nuclei in other morphological types and in correlated objects. In the past this has been done in particular for the QSO's, protagonists of any comprehensive space observing program. The reasons which make space observations fundamental to our understanding of QSO's are outlined in the review paper by Burbidge (1976).

A consideration can be made on the approach to follow in the space observations of these intercorrelated objects. We should not region, but it is in the UV region that it can be successfully verified. The study of the UV emission lines will help also with two other problems. The amount of dust associated to the nucleus can be derived from the ratio of auroral and trans-aurolat lines over a wavelength baseline which will include emissions from the near infrared to the UV. These values, as well as the extinction measurements from the UV continuum, can be compared with the observed infrared excess, which has possibly the same origin. The shape of the profiles of the UV lines may be important to verify whether the absence of wings in the forbidden lines of the type I Seyferts is indeed due to collisional de-excitation in high density regions. The critical density at which this effect is working (Osterbrock 1974) varies over a large interval for forbidden lines from different ions.   This suggests that forbidden lines with critical densities higher than those characteristic of   OII lines might show broad profiles. Harms et al. (1977) indicate that the best candidate fall in the UV range with $\lambda 1575$ Å of NV ($n_c = 2 \times 10^8$ cm^{-3}) as prime choice.

We have remarked earlier that the 2.4 m Space Telescope will operate with an angular resolution ten times that obtainable with a ground-based telescope. This improvement is not marginal for the study of Seyferts because ground-based optical and radio observations

have indicated the presence of a cloud structure in these nuclei. Even
if space observations will not fully resolve these structures, from the
emission line study we will derive kinematical data and physical
that there is no planned space telescope which can match in the
optical band the resolution of the VLBI technique. Among instruments
that will be flown in the next ten years, the ST should give the best
resolution with 0."1 (0."05 with modest deconvolution). Far from the
radio capabilities, this represents however an improvement by a
factor of ten in comparison with ground observations. In nearby
galaxies, say within 10 Mpc, we will be able to resolve components
with size comparable to most of the resolved radio features. At this
resolution we will study the radial motions of the gas in the nuclei
and we will observe the UV part of the spectrum, a crucial region for
the discrimination among models of primary energy sources. In the
following we will discuss the impact of these new observing
capabilities on our understanding of Seyfert phenomena.

Let us first consider the mechanisms by which energy is
produced in the nuclei. A central non-thermal source, a cluster of hot
stars and collisional processes in a shock wave have been proposed in
the past as possible sources of ionization of the gas of these nuclei.
By extending our knowledge of spectral continuum distribution to the
UV we can make a choice between these models, unless the source
has an extremely steep spectrum. The UV emission line intensities
also represent a crucial test. As discussed in paragraph 3.1, cooling
gas behind a shock wave is expected to have a characteristic
spectrum with strong UV resonance lines as compared to a photo-

ionized gas. A similarity between SN remnants and Seyfert nuclei
spectra was noted by Osterbrock and Dufour (1973) in the visual
Megaparsecs. The high resolution data can be combined with the
observations of extended sources such as the disc emission of spirals
and the radio lobes associated to radio galaxies and QSO's. A power
versus size diagram for all these sources is given in Fig. 11 after
Ekers (1978). This author compares the fundamental parameters (size,
power and spectral index of the sources) observed in the nuclei of
normal spirals, of Seyfert galaxies, of radio galaxies and in QSO's.
The sources associated with Seyferts and normal spirals are similar
but for a factor of ten in power. The sizes can vary up to a few
hundredths of parsecs and the spectrum is non-thermal with $\alpha = -0.7$.
Ellipticals, radio galaxies and quasars have similar central sources,
with powers ranging from a typical Seyfert value upwards and spectra
which are, on the average, flat. Ekers (1978) concludes that the
properties of nuclear sources are dependent on the morphology of the
galaxy, in particular on the presence of the disc. Alternatively, these
properties could have affected the morphology[+].

The radio data do not yet give us a breakthrough on the
nature of the central energy source, but at least they suggest us a
frame within which to pursue the solution of the problem. If we now
consider how space observations can help us, we immediately realize

---

[+]Bertola and Capaccioli (1978), through the analysis of several classes
of optical data, have independently shown that the disc subsystem is a
discriminant feature for the overall dynamics of galaxies.

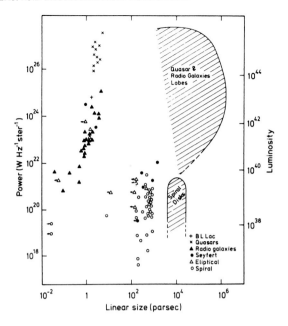

Fig. 11   Linear diameter plotted against radio power at 1.4 GHz for

all the known classes of extragalactic objects (after Ekers

1978).

be much fainter ($M_B \cong -8$ mag), roughly as bright as a nova at

maximum.   It will be therefore possible to study nearby galaxies

($\Delta < 20$ Mpc) with high photometric accuracy over projected surface

elements not larger than about 20 pc in diameter and galaxies of the

Coma cluster at the same level of precision obtainable for nearby

ones using ground-based facilities. Moreover, the access to the UV

band will add information on the nature of the phenomena causing the

light cusp. Finally, let us note that, since the regions to be studied

have a small angular size, it is eliminated the conflict between

resolution and field of view mentioned in paragraph 1.5.

3.4   ACTIVE NUCLEI

The fundamental problem of the nature of nuclei of galaxies and of the relation with their more powerful relatives, the QSO's, presents several facets and, therefore, can be looked at from different points of view. A possible approach is the study of the light distribution as discussed in the previous paragraph. Here, we start by referring to the radio data which form an homogeneous sample of results for objects belonging to different classes.

The radio emission represents only a small fraction of the total energy emitted by the central sources and, in general, it is interpreted as syncrotron emission from the non-thermal nature of the spectrum. In the recent past the Very Long Baseline Interferometer technique (VLBI) has been used to map radio features at a resolution of 0."001. The observations revealed structures less than one parsec in size in galaxies out to distances of hundredths of physical or environmental characteristics of the galaxy ? What is the nature of the light cusp (black hole, compact star cluster or others) ? When does it take place in the life of the galaxy and why ? Are there other phenomena (for instance, active nuclei and QSO's discussed in paragraph 3.4) that can be set within this frame ? Obviously, answers to these questions involve many aspects of formation and evolution of galaxies. As just one example, the light cusps might be related to the mass loss from stars, whether this is a mechanism to fuel the central singularity.

To refine individual observations and to increase the sample we need to improve the spatial resolution of photometric and spectroscopic observations. This is naturally achieved from space. For instance, with the Faint Object Camera of ST we will <u>directly</u> measure optical and UV fluxes at a resolution level two times better than obtainable from Earth through long, painful and risky reductions of the original, seeing-degraded observations. To better understand the achievable gain, let us consider one example. By convolving the luminosity profile model of NGC 3379 with a Gaussian point spread function, de Vaucouleurs and Capaccioli (1978a,b) have computed the magnitude of the source contributing 0.05 mag to the integrated luminosity of the galaxy within centered circular apertures. From this numerical experiment it results that this source corresponds to an exceptionally bright globular cluster ($M_B \simeq -12$ mag) for a PSF = 1", i.e. for typical photometric nights in good astronomical sites on Earth. With the global resolution of the FOC, the source turns out to discrepancy. May galaxian haloes account for the missing mass ? Before commenting on that, we need to discuss spiral galaxies.

Otherwise ellipticals, whose outskirts exhibit a variety of situations, spiral galaxies are characterized by a unique photometric law (de Vaucouleurs 1959, Freeman 1970), which describes the behaviour of the sharply bounded disk subsystem in the classical bands of the optical spectrum. Detection of faint outer extensions has been reported by Kormendy and Bahcall (1974) and by Freeman <u>et al.</u> (1975), who made use of a photographic technique similar to that of Arp and Bertola (1969, 1971). Bertola and di Tullio (1976) have however shown that these haloes cannot rise to the same rank of those of ellipticals, at least as far as the luminosity is concerned.

Nevertheless, there are some gravitational rather than photometric evidences in support to the existence of large massive haloes around spirals. The first suggestion comes from the need to stabilize cold disks of galaxies against large-scale barlike modes. Ostriker and Peebles (1973) find that this condition is simply achieved whether an inner ($r < 10$ Kpc) halo with an high value of M/L is assumed. Extrapolation of this structure to large radial distances would provide a massive but barely detectable corona. The second piece of evidence, while not generally accepted at the time (Burbidge 1975, Schmidt 1975), arises from the analysis of the results of different mass determination methods for spiral galaxies (Ostriker et al. 1974, Einasto et al. 1974). It results that the mass of spirals seems to increase linearly with radius out to 1 Mpc. Additional support to this conclusion comes from the very extended 21-cm rotation curves which are observed to run constant in the outer regions of spirals (Roberts and Whitehurst 1975). In conclusion, hereagain we are led to postulate that spiral galaxies are surrounded by massive undetected haloes with very high mass-to-light ratio.

If this overall picture is correct, predicted massive haloes provide enough matter (no more missed) to bound gravitationally clusters of galaxies. As a consequence, the mean local density approaches the critical density needed to just close the universe (see paragraph 3.6). These non-negligible conclusions are founded on the belief that more than 90% of the matter of the universe is invisible, at least to ground-based eyes. Large space telescopes will give us a chance to get a better insight into it. First of all, taking advantage of the strongly reduced red and infrared sky background (Figure 7), it will be possible to attempt direct detection of presently invisible haloes which, as commonly believed, should be made up of rather evolved cold matter and, therefore, should be particularly efficient in this spectral region. Whenever detection does not succeed, it will be however an important result to set up an upper limit to the luminosity. This information can be used to discriminate among different types of material which have been proposed to populate galaxian haloes. It could be completed with measurements of colors in those outer regions which are at the the limit of detection for ground-based telescopes. The reduced atmospheric scattering and diffusion, which spreads light of bright central regions of galaxies unto the faint outer parts, will allow to detect unambiguously haloes around distant (angularly small) galaxies. This, in addition to the possibility given by the improved resolution to obtain a more complete luminosity function for nearby clusters and to derive the Bautz-Morgan types for distant ones (see paragraph 3.7), will provide a way to test the hypothesis of cannibalism.

3.6   THE EXTRAGALACTIC DISTANCE SCALE

'When the 200-inch telescope went into operation just over a quarter of a century ago it was planned to use the vast light-gathering power of this new telescope to study the size and structure of the universe. As a first step in this process Walter Baade wanted to study Cepheids and other distance indicators in nearby galaxies. With the exciting discovery of radio galaxies and then QSO's Baade's plan for systematic exploration of the nearer

galaxies, which would have required vast amount of large
telescope time, was quietly shelved in favour of what appeared
to be more exciting work on exploding galaxies and quasars'
With this consideration Sidney van den Bergh opens his preface to the
Proceedings of the I.A.U. Colloquium No. 37 'Décalages vers le rouge
et expansion de l'Univers'. Nowadays astronomers are well aware of
the need of an improvement in distance scale calibration. Space will
offer a new chance; it is desirable that the space astronomical
facilities to come will be primarily devoted to this problem which is
at least as fundamental as the study of the numerous puzzles
scattered in the universe. Actually, to improve the distance scale
calibration, we will take advantage of almost any observational gain
obtainable with telescopes working from outer space.

According to de Vaucouleurs (1977a), our way toward the
mapping of the universe is well represented by a tower-like
organization (Fig. 12). Strong pillars should hold the elegant structure, thinnering and weakening toward the top, where the fundamental
distance gauge on cosmological scale, the Hubble constant H_o, is
sitting. Since the time of Hubble (1936b), due to improved observations the constant H_o has passed from a value of 530 km s^{-1} Mpc^{-1} to
the present estimates ranging from 100 to 50 km s^{-1} Mpc^{-1} according
to the different authorities (see Fig. 13). It appears therefore a
clear need to strengthen the structure of our distance-scale tower,
starting from the pillars.

The first rank method of stellar distance measurements is the
trigonometric parallax, since it is straightforwardly related to a

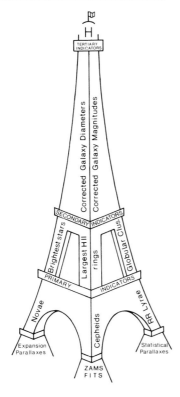

Fig. 12   The tower-like structure of the extragalactic distance scale

indicators (after de Vaucouleurs 1977a).

precisely known quantity, the Astronomical Unit. Unfortunately, this

direct distance indicator requires an astrometric accuracy which, as

near as 100 pc, is already beyond the limits of ground-based

telescopes. At the distance of the Hyades cluster (Δ = 43 pc) the

precision of this technique is about 20% (van Altena et al. 1974; for a

discussion of distance uncertainty in the Hyades see van Altena 1974

and Hanson 1975). The ST in its astrometric mode (Announcement

Opportunity for Space Telescope, March 1977) and the astrometric

satellite (ESA Phase A Report), being able to measure positions of

stars within $\pm 0.''001$, will give an almost tenfold increase of the range

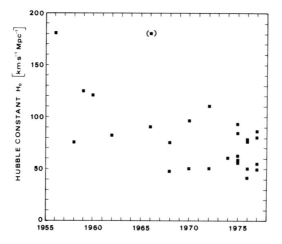

Fig. 13 Since Hubble (1936b) estimates of the Hubble constant H$_o$

have systematically lowered. From the plot of the values of

H$_o$ given by different authors in the last two decades, it is

apparent that two schools exist, giving well distinct mean

estimates of the Hubble constant.

of application of the trigonometric parallax method, i.e. a three-

order of magnitude increase in the number of possible targets. We

will be able to reach several open clusters such as Hyades

(Δ = 43 pc), Praesepe (Δ = 158 pc), Pleiades (Δ = 125 pc) and Coma

(Δ = 80 pc) as well as numerous OB associations, which form the

basis of population I distance scale (see Fig. 14). Accurate photome-

tric measurements in several spectral bands, combined with the

improved distances, will give a more reliable Zero Age Mean

Sequence, possibly extending to much fainter stars. With them, the

distances of farther clusters and associations will be calibrated. Since

a few galactic Cepheids (still too far for direct measurements)

occasionally occur in clusters or associations (van den Bergh 1977),

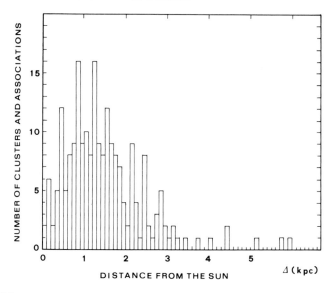

Fig. 14   Number of galactic open clusters and OB associations versus

distance from the Sun (after Becker and Fenkart 1971,

Table 2). Ground-based astrometry allows to measure trigono-

metric parallaxes for objects not farther than 0.1 Kpc, where

only two open clusters are found, namely Hyades and Coma.

With the tenfold increase in resolution provided by the space

astronomical facilities, it will be possible to map the distances

of as much as 74 clusters and associations.

using parent cluster distances (ZAMS fit) we will derive better values

of both zero point and slope of their Period-Luminosity-Color

relations, which presently give distance estimates within 30% (van

den Bergh 1977). Unfortunately, as recently discussed by Tammann

(1978), the sample of available Cepheids is not the most suited for

this type of work.

Cepheids are primary indicators (see Fig. 12). Their mean

blue absolute magnitude ranges conservatively from $\langle B \rangle \sim -2$ mag to $\langle B \rangle \sim$ -5 mag. An easy computation shows that they cannot be reached by ground-based telescopes when farther than, say, 1 Mpc, i.e. outside the Local Group (for the remarkable exception of NGC 2403 see Tammann and Sandage 1968). The ability of space telescopes to detect stellar sources some ten times fainter than any telescope on Earth might provide a larger recognition of these variable stars in the Local Group, whether we will have enough observing time and telescope field of view to map the rather large apparent images of the nearby galaxies (the map of M31 with the WFC of ST down to the isophote μ_B = 22 mag arc-sec^{-2} will require more than 300 exposures !). However it will be possible to discover and study Cepheids in galaxies of the Virgo cluster (here the distance is helping us since it reduces drastically the angular sizes of galaxies). Thus, in two steps only we will obtain the distance of a cluster of galaxies. Our exposition has been oversimplified, of course. The way is not so easy since Cepheid stars are critically sensible to differences in stellar population. In this respect we may agree with de Vaucouleurs (1977c) in saying that since the time of Hubble their importance as distance indicators has been overestimated. Their dependence on stellar population is a double-edge weapon because, while being an additional source of astrophysical information, it complicates the use of Cepheids as distance indicators.

Fortunately, other primary indicators are available as, for instance, novae and RR Lyrae. Without insisting on this subject (a complete review has been given by van den Bergh 1977) we simply note that the same arguments used above for Cepheids can be used for these population II distance indicators to justify the assessement that extra-atmospheric observations will greatly help their finer calibration. As just one example we note that space observations will give us the chance to discover and study RRLyrae in M31. With the improved set of primary distance indicators working out to the distance of the Virgo cluster we can now look with more confidence at the secondary ones (Fig. 12).

Hereagain, from space we will be able to improve their calibration first, as a consequence of the larger number of distant, more precise primary indicators, and then use them over larger distances, due to the higher detection capability available from space. For instance, we will be able to set more properly the relations:

a) $B^o(3) = f(L_c)$, between the mean blue magnitude of the first three brightest stars and the luminosity class L_c of the parent galaxy (Sandage and Tammann 1974b). From space it will work out to the distance of the Coma cluster because -7.5 mag $\lesssim B^o(3) \lesssim$ -9.5 mag (Tammann 1977). Similar results can be achieved using very red stars ($V^o_{max} = (-7.1 \pm 0.1)$ mag, B-V > 2.0 mag), which are possibly all variables and therefore easily identifiable. Moreover, the luminosity of the brightest representative of this class at maximum seems to be independent on parent galaxy luminosity (van den Bergh 1973, Sandage and Tammann 1974b);

b) $D_3(HII) = f(L_c)$, between the mean linear diameter of the three largest HII regions and the luminosity class of the parent galaxy (Sandage and Tammann 1974a). According to de Vaucouleurs (1977b) one should use only ring HII regions to remove the systematic errors arising by photo-optical and seeing effects. This is no more the case out to a distance of 20 Mpc when high resolution photographs are taken from space. This geometrical distance indicator has a remarkable property; it is almost independent on galactic absorption which still hampers extra-atmospheric photometric measurements. Another interesting correlation, deserving attention by space telescopes, has been found by Melnick (1976) between linear diameters of HII regions and internal velocity dispersions measured from the width of global H_α emission.

c) $M^o(0) = f(M_p)$, between the magnitude of the brightest globular cluster and that of the parent galaxy (Sandage 1968, de Vaucouleurs 1970a, Hodge 1974, Sandage and Tammann 1976). The absolute magnitude of galactic globular clusters is ultimately tied to sub-dwarfs with trigonometric parallaxes. The sample of presently available distances of subdwarfs can be blown up by the astronomical facilities from space. Improvement in primary distance indicators' calibration will be useful to get better values of the magnitudes of globular clusters in external galaxies of different morphological types and luminosity classes. Since the most luminous globular clusters of the brightest galaxies have $V \sim -11.0$ mag (Hodge 1974), from space we will detect them out to distances larger than 200 Mpc, i.e. in galaxies with $cz \sim 10000$ km s^{-1}.

The relations listed above as well as others that can be found in current literature are presently rather disperse, since they are based on a small number of calibrators, by themselves not very

precise. After proper revision possible with the major space telescopes they will allow to carefully map distances of galaxies beyond
Coma cluster. This might be enough to calibrate our standard candles,
the first ranked galaxies (see following paragraph).

But still we have one more hand to play. As discussed by
de Vaucouleurs (1977d), the accuracy in morphological classification
of galaxies depends on the degree of resolution of available images. In
this respect, the superiority of space telescopes on the largest
ground-based ones is evident. With them we will be able not only to
classify very distant galaxies but also to revise that of nearby ones,
adding information to the fainter tail of the luminosity function. This
will remove a good deal of selection effects which limit the use of
van den Bergh's classification of galaxies into luminosity classes
(van den Bergh 1960a,b).

As a summary of this adventure toward a better knowledge of
cosmological distance scale we may say that the space facilities to be
flown in the future will certainly provide a more definite answer to
the numerous questions left open by ground-based telescopes. The
scale length of our universe has been raised by a factor 10^6 in the last
three centuries (de Vaucouleurs 1970b). Our present belief gives a
value with an uncertainty of about 50%. It is realistic to assume that,
if space telescopes are not diverted from this subject too much, we
will soon achieve a level of accuracy of about 5%.

3.7   THE HUBBLE DIAGRAM

The ultimate aim of cosmology is the description of the history of the universe. One main problem along this way is the nature of large redshifts of galaxies. Several quite different approaches have been so far considered. In the dominant ortodoxy redshifts (except perhaps quasar absorption lines) are interpreted as due to the expansion of the universe. On the other side, unconventional views exhibit a manifold of ideas going from modest requirements of a non-cosmological explanation for some anomalies (quasars etc.) to radically new reconsideration of the entire expanding universe concept. This matter, recently reviewed by Burbidge (1977) and Rees (1977b), provides a mine for crucial fundamental experiments. Observational cosmology, after stretching capabilities of ground-based telescopes to the limit, still demands more. To illustrate the extra-power provided by future space facilities in performing observations pertinent to this matter, we have taken one example in the framework of 'conventional' cosmology. Other arguments can be extracted by a classical paper of Sandage (1961), entitled 'The ability of the 200-inch telescope to discriminate between selected world models', just updating the observational capabilities to the standard of large space telescopes.

In Friedmann cosmological models the scale factor $R(t)$ describes the history of the universe through its first and second derivatives, the Hubble parameter $h = \dot{R}/R$ and the acceleration factor $q = -R\ddot{R}/\dot{R}^2$. The behaviour of the scale factor for models with different evolution is shown in Fig. 15. Expansion starts from a

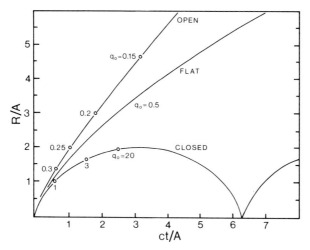

Fig. 15 Behaviour of the scale factor R(t) for Friedmann models
with different evolution.

singularity, possibly in the form of a <u>Big Bang</u>. At any epoch the
density is given by:

$$\rho(t) = 3qh^2/4\pi G \qquad\qquad (4)$$

While always decelerated $(q > 0)$, expansion never reverses in models
with $q < 1/2$ (upper curve in Fig. 15) which describe an open universe
with hyperbolic geodesics. On the contrary, if $q > 1/2$ (lower curve),
the formal solution for R(t) is a periodic function of the cosmic time
and the universe turns out to be closed with elliptical geodesics. A
third possible solution is given by the flat model where the
acceleration factor is always equal to 1/2. This universe is just close;
the corresponding <u>critical</u> density results from (4): $\rho_c(t) = 3h^2/8\pi G$.

Which one of these world models gives the best representation of the real physical universe ? To attempt answering this
question we need to derive the present epoch values of h and q, i.e.
the Hubble constant $H_o = h(t_o)$ and the acceleration parameter

$q_0 = q(t_0)$. The classical (if not unique) procedure is founded on the so-called Hubble diagram. As shown in any textbook of cosmology, the relation between the absolute total bolometric magnitude M_b and the corresponding apparent magnitude m_b of a source seen at a purely cosmological redshift z is given by the power-series expansion:

$$m_b = M_b - 5 \log H_0 + 5 \log z + 1.086 (1 - q_0) z + O^2(z) + const. \qquad (5)$$

which is practically usable only for z 1. Consequently, a family of standard candles, i.e. sources with fixed luminosity and no evolution, will define one of the curves of the Hubble diagram of Fig. 16, according to the value of q_0 for the shape and to the values of H_0 and M_b for the zero point. Thus, all we have to do conceptually in order to obtain H_0 and q_0 is to identify in the zoo of extragalactic sources a family of standard candles with known absolute magnitude and measure their apparent total bolometric magnitudes and purely cosmological redshifts. The Hubble constant will be given by the linear part of equation (5):

$$m_b = M_b \quad 5 \log H_0 + 5 \log z + const. \qquad (6)$$

which holds for small values of z ($z < 0.1$ for $q_0 < 3$) and is equivalent to the Hubble law: $z = H_0 \Delta(z)/c$, where $\Delta(z)$ is the distance of a source having redshift z. The formal value of q_0 is provided by best fitting the experimental distribution in the Hubble plane to the theoretical curves of Fig. 16 for large values of z, where departure from linearity is effective.

This is not a simple task ! First, we have not any a priori solid argument to identify the standard candle families and not enough knowledge of the evolutionary effects which, expecially for large

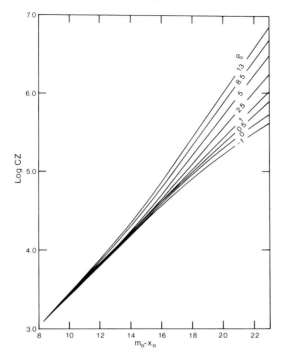

Fig. 16 Hubble diagram for various values of the deceleration

parameter q_0 (after Sandage 1961).

values of z, may spoil measurements of q_0 of their meaning. Let us

first tell our story neglecting evolution which is certainly not

dominant for z $\gtrsim$ 0.4 (Spinrad 1978). Since the time of Hubble (1936b)

the first ranked cluster galaxies were selected as standard candles on

the basis of a statistical justification. In spite of their non-uniform

morphology, these galaxies have proven a posteriori to be good

enough candidates (Humason et al. 1956, Sandage 1975 and references

therein, Sandage et al. 1976, Kristian et al. 1978). Actually, the

Hubble diagram of Fig. 17 for the first ranked galaxies has a very

small dispersion ($\sigma \simeq$ 0.276 mag, Sandage et al. 1976). It is apparent

from equation (5) that this dispersion already causes a 13%

uncertainty on the Hubble constant.   But the main source of

uncertainty acting on the determination of H_0 is due to the fact that

we have no <u>direct</u> method to calibrate the mean value of the absolute

magnitude M_b for our selected standard candle sample. Indirect

calibration via the distance indicators mentioned in paragraph 3.6

produces the scattering of results shown in Fig. 13. Moreover, were

the Hubble flow uniform (as assumed in equation (4), it would be

enough to calibrate galaxies of just the closest cluster (namely Virgo)

to obtain a <u>bona fide</u> value of H_0. Since this seems not the case (at

least at a 5% level, Rubin 1977; see also Sandage 1975), it appears

fundamental to extend calibration to more distant clusters, located in

different regions of the sky. This would improve the determination of

the mean value of H_0 on one side and, on the other side, clarify the

question of redshift anisotropy. The role of observations from outer

space to pursue this goal has already been stressed in the previous

paragraph.

For what concernes the determination of the acceleration

parameter, we have to push observations out to very large distances,

as we said. Progress along this way is indicated in Fig. 17. Presently,

the Hubble diagram extends to galaxies with $z \stackrel{\sim}{<} 0.75$, where the

formal effect of q_0 is still mild and practically masked by

observational errors (Kristian <u>et al.</u> 1978). Moreover, as far as the

redshift increases, heavier and heavier systematic corrections have to

be brought to observed magnitudes, particularly to account for

evolution. Let us first consider geometrical corrections.

First, since magnitude determinations depend on aperture

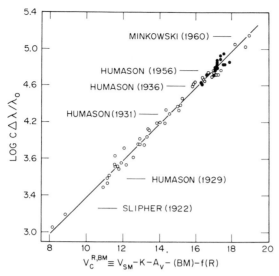

Fig. 17 The Hubble diagram using fully corrected magnitudes (aper-

ture effect, K-term, absorption, Bautz-Morgan type and

richness). The closed circles refer to clusters with new

photometric data by Sandage et al. (1976). The line has the

equation: $V_c^{R,BM}$ = 5 log cz - 6.865 (after Sandage et al. 1976).

used in the photometry, an aperture correction is needed to reduce

measurements to a standard isophote. The procedure outlined by

Humason et al. (1956) rests on the assumption of a unique photome-

tric law for the spatial distribution of light in standard candles. This

assumption seems to hold for the first ranked ellipticals. They exhibit

the correlation: $\vartheta_s \propto z^{-1}$, between apparent angular diameter and

redshift (Sandage 1972), suggesting that the mean surface brightness

is constant all over the class. However, at $z \simeq 0.5$, apparent

diameters are already as small as 1", i.e. of the order of the seeing

disk. The stable PSF of large space telescopes will allow to extend

the available range of isophotal diameters to galaxies with $z > 1$. The new measurements will also provide an independent way to estimate

q_o (Sandage 1972) and they will give us a chance to perform the

Tolman test (Sandage 1974) to ascertain the nature of redshift.

The second correction (the so-called K-correction) to the

observed magnitudes of the first ranked galaxies is required by the

fact that we do not observe bolometric magnitudes but only

magnitudes m_λ over some selected bandpass (the V band centered at

5500 Å in the case of Fig. 17). While in the rest frame the quantity

$m_b - m_\lambda$ depends only upon the sensitivity function for the sources

with fixed energy spectrum, for $z \neq 0$ it is easily shown that the

magnitude difference is also a function of the redshift. In fact, the

rest spectral window is reduced by a factor $1/(1+z)$ and blueshifted by

the same amount. Consequently, to compute the correction tables for

non-evolving galaxies we need to know the UV side of the energy

spectrum of first ranked ellipticals with small redshift. This is

obviously behyond the limits of ground-based telescopes. The prelimi-

nary results provided by the first space experiments show that,

surprisingly, UV continuum of early type galaxies seems to increase

below 2500 Å (Code and Welch 1977). Moreover, it seems that the

observed scatter in the ultraviolet intensity distribution is intrinsic

rather than due to observational errors. It is then clear that more

numerous and precise observations are needed. IUE is now tackling

the problem.

Finally, it is now clear that the sample of first ranked

galaxies is biased by the presence of evolutionary effects, as those

discussed by Sandage and Hardy (1973) and, more recently, by King

(1977) and Spinrad (1977). For instance, the study of magnitude residuals from the Hubble diagram has shown that, while the absolute magnitude of the first ranked galaxies decreases with the Bautz-Morgan (1970) cluster class, the contrast with the second and third ranked increases. This effect seems consistent with the cannibalism picture given by Ostriker and Tremaine (1975). In addition, a weak correlation is found between magnitude of the first ranked galaxy and cluster richness. These are important results, sheding light in the dynamical evolution of galaxies. In order to correct the observed apparent magnitudes and reconstruct a uniform class of sources, we need to see deep into the luminosity function of very distant clusters. We are presently limited to distances not larger than $z \simeq 0.75$, i.e. to the region of the Hubble diagram where the effect of q_0 seems to begin (Kristian et al. 1978). Large space telescopes will provide a spectacular enlargment of our orizon, due to resolution and reduced background luminosity in the red where recession velocity shifts the peak of the energy spectrum of ellipticals.

But the most important source of uncertainty in future determinations of q_0 comes from stellar evolution in galaxies. The problem has been reviewed by Spinrad (1977). To account for that, schematically we need a) to calculate evolutionary changes in magnitudes and colors for galaxies of various types. For that we have b) to study nearby galaxies in the UV to determine their population content. Finally, we can c) make the Hubble diagram observations in the near IR, where evolutionary effects are less dangerous (Spinrad 1977). Again the work is mainly demanded to future large space telescopes. Will they solve the questions on nature and evolution of our universe ? This is clearly hard to say. It is however undoubtful that they will give us many more chances than we have now.

ACKNOWLEDGEMENTS

We are indebted with Prof. I. King for his careful reading of the manuscript and for most valuable criticism.

REFERENCES

Ables, H.D. 1971, Publ. U.S. Naval Obs., XX, Part IV.

Adams, T.F. 1977, Ap.J. Suppl. Ser. 33, 19.

Arp, H.C. and Bertola, F. 1969, Astrophys. Letters 4, 23.

Arp, H.C. and Bertola, F. 1971, Ap.J. 163, 195.

Arp. H.C. and Lorre, J. 1976, Ap.J. 210, 58.

Bautz, L.P. and Morgan, W.W. 1970, Ap.J. 162, L149.

Becker, W. and Fenkart, R. 1971, Astron.Astrophys. Suppl. 4, 241.

Bertola, F. and di Tullio, G. 1976, Proceedings of the Third European
    Astronomical Meeting, Tbilisi, p. 423.

Bertola, F. and Capaccioli, M. 1978, Ap.J. 219, L95.

Blazit, A., Bonneau, D., Koechlin, L. and Labeyrie, A. 1977, Ap.J.
    214, L79.

Bohln, R.C., Marianni, P.A. and Stecher, T.P. 1975, Ap.J. 202, 415.

Boksenberg, A., Carnochan, D., Cahn, J. and Wyatt, S.P. 1975, Mon.Not.R.astr.Soc. 172, 395.

Borgman, J., van Duinen, R.J. and Koorneef, J. 1975, Astron. Astrophys. 40, 461.

Bracewell, N.R. 1955, Austr.J. of Phys. 8, 54.

Bracewell, R. 1965, The Fourier Transform and Its Applications, McGraw Hill, p. 189.

Brault, J.W. and White, O.R. 1971, Astron.Astrophys. 13, 169.

Burbidge, G. 1975, Ap.J. 196, L7.

Burbidge, G. 1977, Décalages vers le rouge et expansion de l'Universe,

I.A.U. Colloquium No. 37, Edit. CNRS, Paris, p. 555.

Burbidge, E.M. 1976, Scientific Uses of the Space Telescope, NASA.

Burbidge, E.M., Burbidge, G.R. and Crampin, D.J. 1964, Ap.J. 140, 1462.

Burr, E.J. 1965, Austr.J. of Phys. 8, 30.

Carnochan, D.J., Navach, C. and Wilson, R. 1975, Mon.Not.R.astr. Soc. 172, 27.

Carruthers, G.R. and Page, T. 1976, Ap.J. 205, 397.

Carruthers, G.R. and Opal, C.B. 1977a, Ap.J. 212, L27.

Carruthers, G.R. and Opal, C.B. 1977b, Ap.J. 217, 95.

Chevalier, R.A. 1977, Ann.Rev.Astron.Astrophys. 15, 195.

Code, A.D. 1976, Scientific Uses of the Space Telescope, NASA.

Code, A.D., Welch, G.A. and Page, T.L. 1971, NASA SP-310.

Code, A.D. and Welch, G.A. 1977, Wisconsin Astrophysics No. 43.

Cox, D.P. 1972, Ap.J. 178, 143.

Dainty, J.C. and Shaw, R. 1974, Image Science, Academic Press, p. 152.

Danziger, I.J, Murdin, P.G., Clark, D.H. and D'Odorico, S. 1978, Mon.Not.R.astr.Soc., in press.

Davidsen, A.F., Hartig, G.F. and Fastie, W.G. 1977, Nature 269, 203.

Davies, R.D., Elliot, K.H. and Meaburn, J. 1976, Mem.R.astr.Soc. 81, 89.

de Vaucouleurs, G. 1948, Ann. d'Ap. 11, 247.

de Vaucouleurs, G. 1959, Hdb.d.Phys. 53, Springer-Verlag, Berlin, p. 275.

de Vaucouleurs, G. 1970a, Ap.J. 159, 435.

de Vaucouleurs, G 1970b, Pub.Astron.Dept.Univ. of Texas, Series II, Vol. III, No. 1.

de Vaucouleurs, G. 1977a, Décalages vers le rouge et expansion de l'Universe, I.A.U. Colloquium No. 37, Edit. CNRS, Paris, p. 7.

de Vaucouleurs, G, 1977b, - ibidem -, p. 69.

de Vaucouleurs, G. 1977c, Le monde des galaxies, edited by the Observatory of Besançon.

de Vaucouleurs, G. 1977d, The evolution of galaxies and stellar populations, B.Tinsley and R.B.Larson editors, Yale University Observatory, p. 43.

de Vaucouleurs, G. and Capaccioli, M. 1978a, preprint.

de Vaucouleurs, G. and Capaccioli, M. 1978b, in preparation.

de Vaucouleurs, G., Capaccioli, M. and Young, P.J. 1978, preprint.

de Vaucouleurs, G. and de Vaucouleurs, A. 1970, Astrophys.Letters 5, 219.

de Vaucouleurs, G., de Vaucouleurs, A. and Corwin, H.G. 1976, Second Reference Catalogue of Bright Galaxies, Austin, University of Texas Press.

Disney, M.J. 1976, Nature 263, 573.

D'Odorico, S., Benvenuti, P. and Sabbadin, F. 1978, Astron.Astrophys. 63, 63.

Dopita, M.A. 1977a, Ap.J. Suppl. Ser. 33, 437.

Dopita, M.A. 1977b, private communication.

Ekers, R.D. 1978, Phys. Scr. Vol. 17, No. 3, 'Quasars and Active Nuclei of Galaxies'.

Ford, H.C. and Jacoby, G.H. 1978, Ap.J. 437, 444.

Frank, J. and Rees, M.J. 1976, Mon.Not.R.astr.Soc. 176, 633.

Freeman, K.C. 1970, Ap.J. 160, 811.

Freeman, K.C., Carrick, D.W. and Craft, J.C. 1975, Ap.J. 198, L95.

Gurzadyan, G,A. 1975, Mon.Not.R.astr.Soc. 172, 249.

Hanes, D.A. 1977, Mem.R.astron.Soc. 84, 45.

Hanson, R.B. 1975, A.J. 80, 379.

Harms, R.J. 1977, Proposal to NASA for the Faint Object Spectrograph of the Space Telescope.

Harwit, M. 1975, Quart.J.R.Astron.Soc. 16, 378.

Hills, J.G. 1975, Nature 254, 295.

Hodge, P.W. 1974, P.A.S.P. 86, 289.

Hubble, E. 1926, Ap.J. 64, 321.

Hubble, E. 1936a, The Realm of the Nebulae, Yale University Press, New Haven.

Hubble, E. 1936b, Ap.J. 84, 270.

Humason, M.L., Mayall, N.U. and Sandage, A.R. 1956, A.J. 61, 97.

Israel, F.P. 1975, Proceedings of the Symposium 'HII regions and related topics', Mittelberg, T.L.Wilson and D.Downes editors, Springer-Verlag.

Israel, F.P., Goss, W.M. and Allen, R.J. 1975, Astron.Astrophys. 40, 421.

King, I. 1966, A.J. 71, 64.

King, I. 1971, P.A.S.P. 83, 199.

King, I. 1977, The evolution of galaxies and stellar populations, B.Tinsley and R.B.Larson editors, Yale University Observatory, p. 1.

King, I. 1978, Ap.J. 222, 1.

Kormendy, J. 1973, A.J. 78, 255.

Kormendy, J. 1978, preprint.

Kormendy, J. and Bachall, J.N. 1974, A.J. 79, 671.

Kristian, J., Sandage, A. and Westphal, J.A. 1978, Ap.J. 221, 383.

Larson, R.B. 1974, Mon.Not.R.astr.Soc. 166, 585.

Light, E.S., Danielson, R.E. and Schwarzshild, M. 1974, Ap.J. 194, 257.

Lynden-Bell, D. 1969, Nature 223, 690.

Matthews, T.A., Morgan, W.W. and Schmidt, M. 1964, Ap.J. 140, 35.

Melnick, J. 1976, Ap.J. 213, 15.

Morgan, W.W., Kayser, S. and White, R.A. 1975, Ap.J. 199, 545.

Oemler, A. 1976, Ap.J. 209, 693.

Osterbrock, D.E. 1974, Astrophysics of Gaseous Nebulae, W.H.Freman and Company, p. 53.

Osterbrock, D.E. and Dufour, R.J. 1973, Ap.J. 185, 441.

Ostriker, J.P. and Peebles, P.J.E. 1973, Ap.J. 186, 467.

Ostriker, J.P., Peebles, P.J.E. and Yahil, A. 1974, Ap.J. 193, L1.

Ostriker, J.P. and Tremaine, S.D. 1975, Ap.J. 202, L113.

Peebles, P.J.E. 1972, Ap.J. 178, 371.

Pottash, S.R., Wesselius, P.R., Wu, C.C., Fieten, H. and van Duinen, R.J. 1978, Astron.Astrophys. 62, 95.

Raymond, J.C. 1976, Ph.D. Thesis, University of Wisconsin.

Rees, M.J. 1977a, Décalages vers le rouge et expansion de l'Universe, I.A.U. Colloquium No. 37, Edit. CNRS, Paris, p. 563.

Rees, M.J. 1977b, The evolution of galaxies and stellar populations, B.Tinsley and R.B.Larson editors, Yale University Observatory, p. 339.

Roberts, M.S. and Whitehurst, R.N. 1975, Ap.J. 201, 327.

Rubin, V.C. 1977, Décalages vers le rouge et expansion de l'Universe, I.A.U. Colloquium No. 37, Edit. CNRS, Paris, p. 37.

Sandage, A. 1961, Ap.J. 133, 355.

Sandage, A. 1968, Ap.J. 152, L149.

Sandage, A. 1972, Ap.J. 173, 485.

Sandage, A. 1974, Large Space Telescope, a new tool for science, 12th Aerospace Sciences Meeting, Washington D.C., p. 19.

Sandage, A. 1975, Ap.J. 202, 563.

Sandage, A. and Hardy, E. 1973, Ap.J. 183, 743.

Sandage, A., Kristian, J. and Westphal, J.A. 1976, Ap.J. 205, 688.

Sandage, A. and Tammann, G.A. 1974a, Ap.J. 190, 525.

Sandage, A. and Tammann, G.A. 1974b, Ap.J. 191, 603.

Sandage, A. and Tammann, G.A. 1976, Ap.J. 210, 7.

Sargent,W.L., Kowal, C.T., Hartwick, F.D. and van den Bergh, S. 1977, A.J. 82, 947.

Sargent, W.L.W., Young, P.J., Boksemberg, A., Shortridge, K., Lynds, C.R. and Hartwick, F.D.A. 1978, Ap.J. 221, 731.

Schmidt, M. 1975, Ap.J. 202, 22.

Spinrad, H. 1977, The evolution of galaxies and stellar poupulations, B.Tinsley and R.B.Larson editors, Yale University Observatory, p. 301.

Tammann, G.A. 1977, Décalages vers le rouge et expansion de l'Universe, I.A.U. Colloquium No. 37, Edit. CNRS, Paris, p. 43.

Tammann, G.A. 1979, Colloquium on European Satellite Astronomy, C. Barbieri and P.L. Bernacca editors, Padova, Tip Antoniana.

Tammann, G.A. and Sandage, A. 1968, Ap.J. 151, 825.

Talbott, R.J.Jr. and Arnett, W.D. 1973, Ap.J. 186, 51.

Torres-Peimbert, S. and Peimbert, M. 1978, Rev.Mex.Astron.Astrof., in press.

van Altena, W.F. 1974, P.A.S.P. 86, 217.

van Altena, W.F., Franz, O.G. and Fredrick, L.W. 1974, New Problems in Astrometry, I.A.U. Symposium No. 61, W.Gliese, C.A.Murray and R.H.Tucker editors, p. 283.

van den Bergh, S. 1960a, Ap.J. 131, 215.

van den Bergh, S. 1960b, Ap.J. 131, 558.

van den Bergh, S. 1973, Ap.J. 183, L123.

van den Bergh, S. 1977, Décalages vers le rouge et expansion de l'Universe, I.A.U.Colloquium No. 37, Edit. CNRS, Paris, p. 13.

van der Kruit, P.C. and de Bruyn, A.G. 1976, Astron.Astrophys. 48, 373.

van Duinen, R.J. Aalders, J.W.G., Wesselius, P.R., Wildeman, K.J., Wu, C.C., Luinge, W. and Snel, D. 1975, Astron.Astrophys. 39, 159.

White, S.D.M. 1976, Mon.Not.R.astr.Soc. <u>174</u>, 19.

Wilson, C.P. 1975, A.J. <u>80</u>, 175.

Wolfe, A.M. and Burbidge, G.R. 1970, Ap.J. <u>161</u>, 419.

Wu, C.C. 1976, <u>Liege Colloquium on the observations from TD1 and ANS</u>, unpublished.

Wu, C.C. 1977, Bull.A.A.S., Vol. <u>9</u>, 2.

Young, P.J., Shields, G.A. and Wheeler, J.C. 1977, Ap.J. <u>212</u>, 367.

Young, P.J., Westphal, J.A., Kristian, J. and Wilson, C.P. 1978, Ap.J. <u>221</u>, 721.

INFRARED ASTRONOMY

Ezio Bussoletti
Gruppo Infrarosso, Istituto di Fisica, Lecce, Italy

Roberto Fabbri
Istituto di Fisica Superiore, Firenze, Italy

Francesco Melchiorri
Gruppo Infrarosso, IROE - CNR, Firenze, Italy

Abstract

We present hereafter an updated review of the most important
results in Infrared Astronomy obtained both by means of ground
based observations and by flying platforms.
Due to the large amount of available information we have
restricted our discussion to galactic astronomy and to cosmic
background radiation for which concerns extragalactic astronomy.

1. INTRODUCTION

Infrared Astronomy is conventionally confined in the wavelength range 1μ - 1000μ. Usually astronomers define as near-
infrared the wavelength region up to 5μ, as middle-infrared
that from 5μ to around 35μ and finally as far-infrared the
remaining wavelength range.
Although some near-infrared observations of cool stars have
been performed earlier, this branch of Astronomy become established only at the end of 1960s., when Neugebauer and Leighton
(1969) performed a 2.2μ Survey of the Northern sky. Their
observations proved, for the first time, the existence of
several objects exhibiting an emission (the so-called "infrared excess") much larger that that expected from their extrapolated visible spectra.
Before the discovery, the revolutioning liquid-helium cooled
germanium bolometer developed by Low opened definitely the
sky to the astronomers allowing observations beyond the limit
of PbS detectors at 4μ.
At the time of the first review (Neugebauer et al., 1971) many
classes of celestial objects showed well defined infrared
excesses at wavelengths shorter than 24.4μ. Since, the avai-

395

P. L. Bernacca and R. Ruffini (eds.), Astrophysics from Spacelab, 395–491.

lability of aircraft and balloon-borne observatories have
revealed the extremely large far-infrared luminosities associated with objects such as H II regions, dark clouds and the
central part of galaxies.
Presently many review papers are available in literature on
this subject, so that we have preferred to avoid any historical description or systematics about this branch of Astronomy.
We have then restricted our discussion mainly to the most
recent findings in galactic astronomy in order to point out
the subjects more interesting for future observations by means
of the new generation instruments able to work outside the
Earth atmosphere.
      In recent years IR Astronomy has acquired also an
enormous relevance for cosmology due to the discovery of the
diffuse microwave radiation. Since this entire field has
enormously expanded we will limit our discussion only to
galactic sources and to the observations of cosmological
significance, neglecting the field of discrete extragalactic
sources and the related problems. As far as galactic astronomy is concerned, we shall assume the general physics of
emission mechanisms familiar to the reader, and we shall
emphasize the recent observations. On the other hand, we
shall discuss in some detail the theoretical predictions
concerning the microwave background, since they originate
from a discipline remote from Infrared Astronomy.
The length of this paper does not allow us to present, in
addition, a short review about presently available infrared
detectors, that we had in mind to write.

2. GALACTIC INFRARED ASTRONOMY

2.1 Infrared Stars

      Many galactic IR sources are stars which exhibit an in-
frared excess over the radiation which would be expected if
the star were emitting like a black body at the effective
temperature appropriate to its spectral type. This excess is
commonly attributed to the emission of a dust shell, or a
dust clumpiness surrounding the central star. The particles
absorb the UV and optical radiation produced by the star,
are heated and then reemit energy in form of IR radiation.
Mass loss is generally indicated as the agent of this shell
production in late-type objects. For young stars (or protostars), the presence of dust around the central core is
instead essentially due to the remnant of the parent cloud

which has created the star itself after collapse and occurrence of the nuclear reactions. Finally, in few classes of
objects where there is a clear absence of dust or where its
number density is reduced, the IR excesses are due to other
causes (for example, free-free radiation).

2.2 IR Surveys

A complete survey of the northern sky at declinations
between -33° and $+80^{\circ}$ was carried out at Caltech to provide
information of celestial 2.2μ emitters (Neugebauer and
Leighton, 1969). Approximately 5600 sources were found to
emit more than $4 \times 10^{-25} Wm^{-2} Hz^{-1}$. In this sample, the large
majority of the identifications are late type giant stars.
Following this effort, in 1971-72, the Air Force Cambridge
Res.Lab. carried out the first survey of the sky at 4μ, 11μ,
and 20μ by using a small criogenically cooled telescope
mounted on a rocket. A total of 79% of the celestial sphere
was surveyed in seven flights from -46° up to $+90^{\circ}$.As much as
2507 sources were detected at 4μ, 1441 at 11μ, 873 at 20μ,
for a total of 3198 sources (Walker and Price, 1975). Two
extra flights performed in 1974 carried a broad band filter
centered at 27.4μ instead of that at 4μ. A new updated and
extended catalog has been produced which covers 90% of the
celestial sphere, i.e. 37000 squared degrees in the 11μ survey (Walker and Price, 1976) providing a quite complete IR
picture of the sky. In this work 2363 objects have been
detected in one or more colours after all the necessary corrections to eliminate spourious sources. The analysis of selected samples of these sources is at present in course among
various research groups at the ground. Following the first
systematic study of the 235 anonymous sources in the 2μ
Catalog by Strecker and Ney (1974) several other investigations have been performed showing that many of these objects
are stars of the conventional spectral sequence sourrounded
by optically thick circumstellar dust shells, while others
exhibit no evidence of classical cool stellar objects and
may represent various transitions stages of stellar evolution.

2.3 Infrared Excesses in Stars

As we have already mentioned, almost all IR excesses in
stars are believed to arise from thermal radiation of a dust
shell sourrounding the central source of energy. In addition,

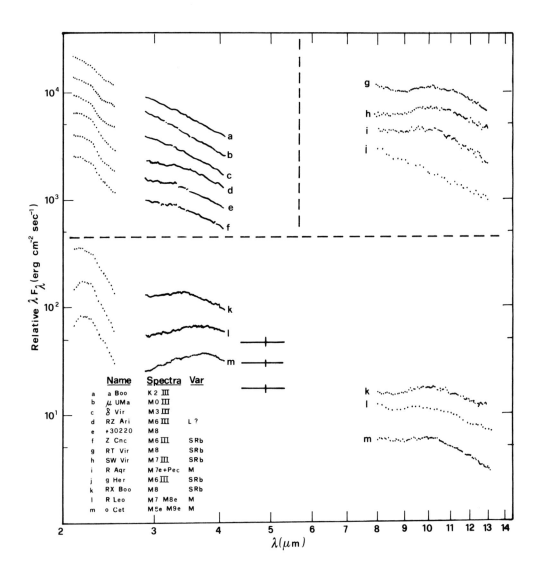

Figure 1 - Spectra of oxygen-rich giants. The variability
        class and approximate spectral type are also indicated.
        (Adapted from Merrill and Stein, 1976a)

the observed spectrum exhibits a short wavelength component
typical of a normal star and a relatively smooth long wavelength component. It is our aim here to give an updated review of some of the most studied classes of objects neglecting any systematics of the star properties, for which we recall specific reviews (Johnson, 1966 (a,b) Neugebauer et al.
1971 Allen 1975).

(a) Normal cool stars.

   We assume under this group objects with T≤4000 K° for
which many IR observations are available in the literature.
In particular, spectroscopy (up to $\lambda/\Delta\lambda \sim 400$) in the near
and mean IR has been performed by several authors (McCammon
et al.1967, Frogel and Hyland 1972, Thompson et al. 1969,
Gammon et al. 1972, Treffers and Cohen 1974, Forrest et al.
1975). An emission peak at 9.7μ has been found and attributed to silicates while the narrower emission at 11.5μ in the
spectra of C stars was identified with SiC grains.

(a.1) Oxygen rich giants and supergiant stars.

   Very recently Merrill and Stein (1976a) have performed
spectroscopic observations ($\Delta\lambda/\lambda \sim 0.015$) in the range 2-14μ
for these stars as well as for C stars. Fig.1 reports the
spectra of several oxygen rich cool giants with their variability class and approximate spectral type. The 10μ excess
for these stars is clearly small. Broad-band observations
suggest that the onset of the silicate excess is dependent
both on temperature and luminosity. In addition, it is worth
noting that the circumstellar IR excess is clearly seen in
every star later than M6III, M5II, M1ab, K5Ia and G0 Ia-0.
From these observations it comes out that, as the temperature
decreases, the molecular species become more and more important
sources of opacity. The first overtone of CO at 2.3μ is
weakly present in the K2IIIαBoo star, while it becomes
stronger for spectral type stars M0 and M3. The 2.7μ band of
water dominates in all these stars and appears to get stronger
as temperature decreases from class M0(T=3800°K) to class
M8(T=2800°K) and later. In general, the entire spectral range
is marked by discrete molecular and particulate absorption
and emission bands so that the continuum level is hard to
define. The spectra of several oxygen rich M supergiants observed by Merrill and Stein (1976) are displayed in Fig.2.
All but VY Ca Maj are semiregular variables. The 9.7μ silicate
band emission appears to have a large variation which is not
correlated with the spectral type but only roughly with the

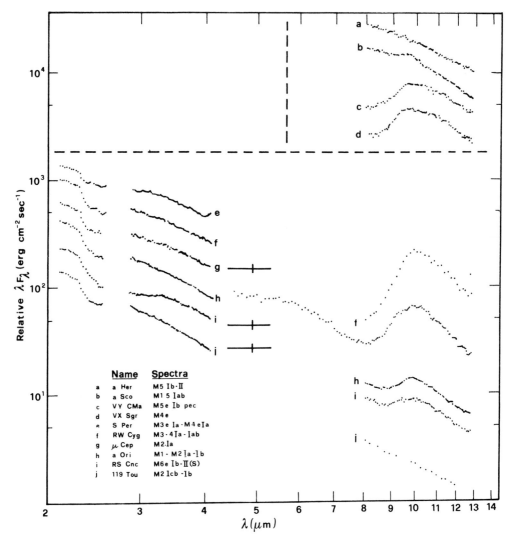

Figure 2 - Spectra of oxygen-rich M supergiants. Spectral classes
are also indicated as well as variability classes.
(Adapted from Merrill and Stein, 1976a)

luminosity class. The largest peaks appear in the spectra of
the most luminous members of the class (IA). Two cases, where
no obvious silicate emission appears, have anyway a 10μ
continuum flatter than that of a black body at the photospheric temperature of the star. In this case too, strong
CO absorption at 2.3μ is clearly seen in all objects while
in μCep the first overtone at 4.6μ is also present. The water
vapour absorption band is rougly hinted in all the stars but
RS Can where it appears clearly. Other minor differences are
evident in the continua of the other stars.

(a.2) Carbon rich stars.

They are usualy indicated as stars with $N_c/N_o \geq 1$. Fig.
3 shows some spectra measured by Merrill and Stein (1976a)
as examples of their properties. A discrete emission feature
at 11.5μ is seen, as in all the stars surveyed up to now.
This feature has been attributed to SiC following the indications given by Gilra (1972) and supported furtherly by
Treffers and Cohen (1974). The continuum in the 8-13μ range
appears smooth for Mira variables, roughly that of a gray-
body (Forrest et al.1975), and it is likely produced by
graphite particles condensed in their atmosphere (Hoyle and
Wickramasinghe 1962, Gilman 1969, Salpeter 1974a). In addition, several molecular features appear in the 2-4μ spectra.
CO has been identified at 2.3μ while a band falling at 2.4μ
cannot be easily identified. Again, a sharp absorption band
occurs in all C stars surveyed to date, centered at 3.7μ. A
further absorption band at 3.9μ is also definitely seen in
the Mira variables. Many molecules are know to exist in
abundance within the atmospheres of C stars. Among others C_3,
SiC_2, HCN, H_2S, C_2H_2 and C_2N_2 are possible candidates. Since
many of their bands overlap, even high resolution studies
may produce ambigous identifications and not much can be said
at present. In conclusion some statements can be done for
the stars considered here:

i) the broad 9.7μ emission is unique to oxygen rich stars
   which show also the 2.7μ water vapour band;

ii) the narrow SiC 11.5μ emission is unique to C rich stars;

iii) the 2.3μ CO absorption band is present in all photospheres of M and C stars;

iv) the 3.07μ absorption band is present in all C stars surveyed and, at moderate strenght, it is clearly distinct
   from the water ice absorption band. In most cases water
   ice absorption is accompained by clearly distinct 9.7μ
   silicate absorption;

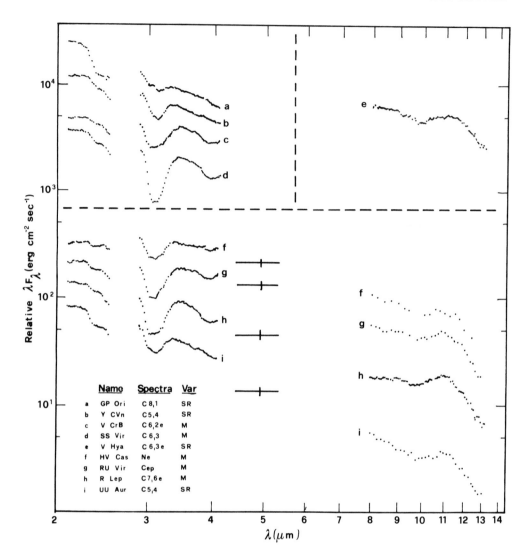

Figure 3 - Spectra of carbon rich stars. Variability class
and spectral type are also indicated.
(Adapted from Merrill and Stein, 1976a)

v) C stars present 3.07μ and 3.9μ absorption bands whose
identification (molecular) is not yet sure.

(b) Selected stars from sky survey catalogues.

Following a comparison with various catalogs of faint
red stars performed by Grasdalen and Gaustad (1971), about
200 objects have remained unidentified in the Northern hemisphere. All these objects are very red, many have apparent
colour temperatures T=1000°K and occur often in regions of
high visual obscuration.

Intensive studies for the Northern (Vogt, 1973) and the
Southern emisphere (Hansen and Blanco, 1973, 1975) have led
to classical spectral classification for most of the anonymous sources. Flux distributions have been established for
a large number of these objects (Hyland et al.1972, Dick et
al. 1974, Treffers and Cohen 1974, Gaustad et al. 1969) while
systematic studies of the optically unidentified portion of
the catalog has been reported by Strecker and Ney (1974a).
They examined 232 of the 235 anonymous northern sources between 2μ and 18μ, looking for the 9.7μ signature often seen
in oxygen rich stars. Very recently, Merrill and Stein
(1976b) have surveyed 55 objects in the 2-14μ range both
using broad band photometry and intermediate resolution spectrophotometry at a resolution of $\Delta\lambda/\lambda$ 0.015. Fig.4 (a,b,c)
reports the broad band observations of those objects. Fig.4
shows the results for oxygen rich stars. The usual 9.7μ silicate emission is clearly seen in all spectra while the overall flux distribution shows a clear progression from the top
to the bottom of the figure. For normal cool M type stars
(top) only a small fraction of the total flux is in the silicate emission band superposed on a continuum due to a cool
blackbody (T=1700-2000°K). At the bottom (see NML Tau for
comparison) the level of the 9.7μ emission band is comparable
to the emission maximum near 1.5μ . This last seems to be
due to some combination of short wavelength extinction in
conjunction with excess emission at $\lambda=3\mu$. The flux distribution of C rich stars is reported in Fig.4b. In this case, a
clear progression appears from overall fluxes following
closely blackbodies at T=2200°K (cool stellar photosphere)
to extreme cases of very cold blackbodies (T=600°K). Oxygen
rich stars with overall flux distributions broad such NML
Cyg are reported in Fig.4c. Optically thick circumstellar

INFRARED STELLAR SPECTROPHOTOMETRY

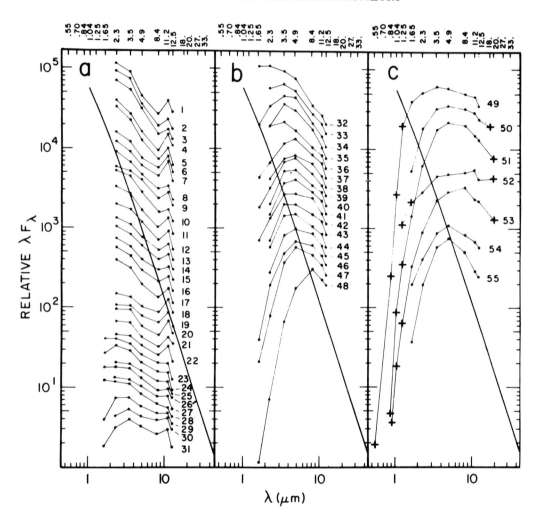

Fig.4(a,b,c) - Flux distribution for stars in the 2μ sky
          survey.
          a) Oxygen rich M stars
          b) Carbon rich stars
          c) Oxygen rich stars with optically thick cir-
            cumstellar dust envelopes.
          (Adapted from Merrill and Stein, 1976b)

dust shells are strongly evidenced in these objects. The
observations have shown that all the C stars present the
11.5μ SiC emission band as well as the 2.3μ CO band. In one
case, +10216, it seems that the thermal reemission by dust
has contributed to wash out this band. Again the 3.07μ and
the 3.9μ absorption bands appeared to be present in most
of these objects. The oxygen rich stars observed appear, all
but   two, to be late M spectral type. As usual, the 9.7μ
band is evident in all cases with a large height compared to
the continuum. Most of these stars present also 1.9μ and
2.7μ water absorption bands. In summary, evidence for stellar
photospheres similar to those in oxygen rich stars and carbon
rich stars exists in all but one case. These objects seem
to be essentially an extension of the normal spectral classification where also cooler stars are present. Large scale
mass losses with circumstellar condensations are seen to
exist in these objects due to their apparent high luminosity
and large diameter. The circumstellar envelopes of these
sources are generally optically thicker than those found in
other infrared stars.

(c) Selected stars from AFCRL catalog.

The AFCRL rocket-borne survey has resulted in a catalog
of 3198 objects, some of which were previously known to be
IR sources. Following these observations, ground-based
observational programs to produce a confirmation list have
been performed at 2.2μ (Allen et al.1976), at 11μ(Low et al.
1976) and in the band 9.5-12.5μ(Lebofsky et al. 1976). Over
1350 square degrees of the sky in the northern emisphere
have been surveyed by the University of Arizona team (Low
et al. 1976) finding only 27 over 162 sources described in
the preliminary Catalog (Walker and Price 1972). A first
result is that these sources are mainly concentrated along
the galactic plane (bll=20°) while the unconfirmed ones are
widely distributed in latitude. The improved verification
program of Lebofsky et al. (1976) has found only 34 sources
of 281 listed in the final catalog finding the same spatial
distribution. In addition, narrow band photometry of these
sources has evidenced structures in the 8-14μ range suggesting
the presence of silicates. In parallel, observations at 2.2μ
of anonymous sources in the southern emisphere lead to the
same distribution of galactic latitudes: in this case only
4 sources over 58 were clearly seen and identified as M type
bright stars. The reason of this common failure is at present

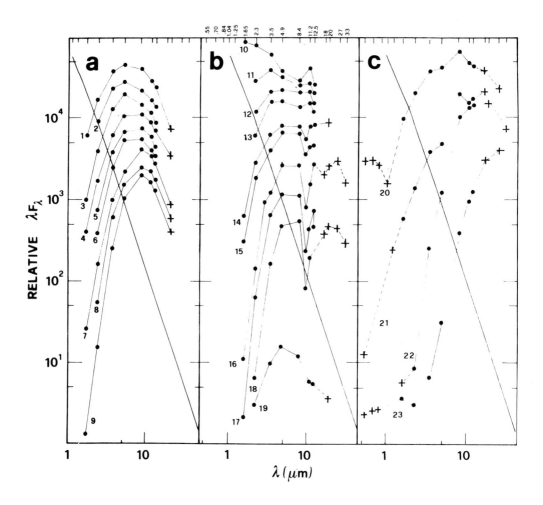

Fig. 5.(a,b,c). Flux distributions of objects in the AFCRL catalog.
Group (a) contains featureless sources (at the phto-
metric resolution) which are essentially C stars.
(Oxygen rich stars may also be present). Group(b)
contains oxygen rich stars with silicate emission or
absorption bands. Group (c) is a heterogeneous group
of compact bright sources which show evidence of circumstellar dust envelops.(Adapted from Merrill and
Stein, 1976c)

puzzling. Possible explanations may be attributed to several
reasons: a) not correct positional accuracy in the catalog,
b) moving sources, c) variability (the sources have been
observed at their maximum by the rocket), d) very extended
sources not detectable because of the low ground base field
of view and spatial modulation, e) most of the anonymous
sources are spurious. This problem is still open, however,
and further observational campaigns are in progress to solve
it. A large sample of these anonymous sources has been observed again by Merrill and Stein (1976c) between 2μ and
14μ. Fig.5 (a,b,c,) reports their broad band spectral distribution in some cases. Group (a) contains featureless sources
(at the photometric resolution) which are essentially C rich
stars (oxygen rich stars may also be present) which have
spectra peaking at 5-8μ well fitted by a blcakbody at a single
temperature. The broader flux distributions resemble that
of NML Cyg, while the narrower that of IRC+10216. Group (b)
presents the usual oxygen rich star 9.7μ silicate emission
or absorption bands. These objects are not associated with
bright HII regions, are compact (≤5") and are apparently
isolated from other IR sources. Some are associated with
compact molecular clouds and hence are presumed to be young
objects (Merrill et al. 1976). Finally Fig.5c shows a heterogeneous group of compact bright sources, some of which have
been already observed by Ney et al. (1975), Westbrock et al.
(1975), Cohen et al. (1975), with fluxes sharply peaking at
λ=8μ. These sources are thought to be sourrounded by flattened
disk-like (or not spherically symmetric) optically thick
dust clouds. They are visible only in part by direct transmission of the unocculted portion of the central star, or
indirectly, by reflection from material along an axis perpendicular to the disk. Their IR spectra between 2-14μ are
shown in Fig.6. Some of these sources (618 and 2688) are
highly polarized in the radial direction suggesting an efficient mechanism of scattering. Actually it appears that the
circumstellar dust envelopes of these objects are more optically thick than their counterparts of equivalent spectral
type found in the 2.2μ Catalog. In summary, these sources
(AFCRL Catalog) may be grouped as follows:
i) stars which, apart from large IR excesses, otherwise
    resemble normal cool M,S or C stars;
ii) sources embedded in optically thick circumstellar dust
    envelopes whose IR spectra resemble the envelopes found

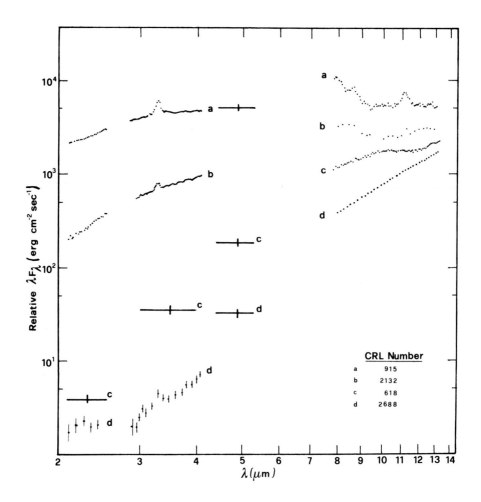

Figure 6 - Spectra of some objects with disk-like circumstellar dust envelopes in the AFCR catalog. (Adapted from Merrill and Stein 1976c)

for normal cool stars. The 9.7μ band may be present but
the water ice 3.1 μ band is absent. These stars are more
likely evolved stars undergoing extensive mass losses;
  iii) objects embedded in optically thick circumstellar dust
envelopes where both the 9.7μ and the 3.1μ bands are seen,
while the characteristic absorption bands seen in cool
atmospheres are lacking. These may be young objects due
to their association with molecular clouds;
  iv) objects embedded in distinctive flattened disk-like cir-
cum-stellar envelopes. Their evolutionary state remains
uncertain: apparently these stars represent some sort of
transition state in stellar evolution.
Additional ten sources have been observed at the ground with
a narrow band multifilter photometer by Gehrz and Hackwell
(1976) between 2μ and 23μ , see Fig.7(a,b). These objects,
two of which show the 11.5μ SiC band, may represent different
evolutionary stages in a single class of stars. Actually
they lie in the direction of the Perseus arm. Their galactic
latitude distribution is similar to that of the disk population cool luminous giants and planetary nebulae. In the first
case this would explain, by their typical high rate of mass
loss, the large circumstellar dust envelopes eventually
sufficient to extinguish the exciting star (optically thick
envelopes). Alternatively, these sources may be a class of
dust enshrouded objects which are evolving towards planetary
nebulae. Presumably the dust would have condensed as the
outer layers of the star expanded during the final stages of
the nuclear burning. As the hot inner layers of the embedded
star become exposed, an ionization front moved outward through
the thermosphere eventually vaporizing much of the dust
(optically thin sources). Indications in this sense come out
also from recent observations of CRL 618, a source in Perseus,
which seems to be some kind of young planetary nebula, perhaps still in the process of formation (Westbrook et al. 1976).
Very recently other two sources have been observed: CRL 2789
(Cohen 1977) and GL 437 (Kleinmann et al.1977). The first
one is a young object and coincides with a nebulous object
containing a starlike condensation known as V645 Cyg. It is
a bipolar nebula, with strong optical polarization, whose
central star is directly invisible. The IR source is supposed
to be a toroid of dust grains around the star with a characteristic temperature of 450°K. GL 437, one of the faintest
objects discovered (m (11μ)=+0.1), presents a complicate

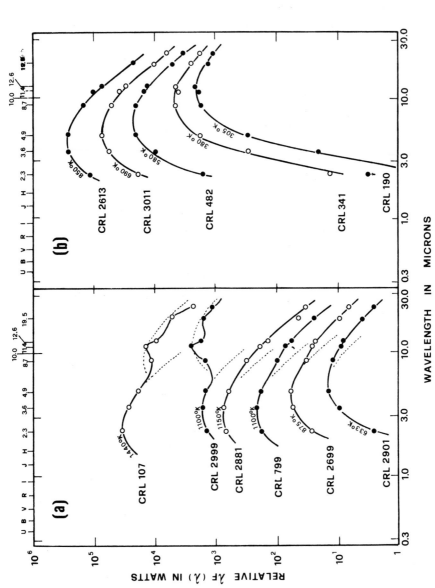

Fig.7(a,b). Spectra of CRL objects which seem to represent different evolutionary stages in a single class of stars. Group (a) shows evidence of optically thin circumstellar shells, while in Group (b) the objects appear to radiate like blackbodies from 2.3 μ up to 23μ. (Adapted from Gehrz and Hackwell, 1976)

spatial structure and it is found associated with a massive
molecular cloud (Barrett et al. 1976). A 3.5μ mapping shows
clearly three members in a cluster. Narrow emission features
at 3.28, 8.7 and 11.3μ are also seen while the presence of
water ice and silicate absorption bands is unclear. Its structure and its association with molecular clouds seem to indicate young stars also if not of very early spectral class
(G or earlier due to Hα emission from the southernmost optical condensation in the cluster). Indications of a probable
location above the zero age main sequence comes out from the
lower limit of the luminosity, $L=4000 (d/3Kpc)^3 L_0$, and from
the upper limit to the surface brightness. It is worthwhile
to note, in addition, that the features above mentioned
appear, for the first time, to be originated outside the
ionized region due to the low ionizing mass sourrounding an
early type star.

(d) Novae.

      One further prediction of the hypothesis that stellar
material ejected during mass loss can condense into a circumstellar shell comes from the observations that galactic novae
have large IR excesses as a consequence of the matter ejected
during their outbursts. These excesses are expected to become
apparent after several months following the optical out-burst
when the ejected material encounters suitable conditions for
condensation to occur. Detailed studies of the IR development
from novae have been reported only for two objects: Nova Ser
(Hyland and Neugebauer 1970, Geisel et al. 1970) and Nova
Cyg (Gallagher and Ney 1975, Ennis et al. 1977). In the former
case the observations started 19 days after discovery and
therefore nothing was known of the behaviour near maximum
light. Recently 1-10μ and 1-3μ photometry has been reported
respectively by Gallagher and Ney (1975) and Kawara et al.
(1976) of the early behaviour of Nova Cyg 1975 as well as
limited 2-4μ spectroscopy by Grasdalen and Joyce (1976). The
only extended work in time is that of Ennis et al. (1976)
who report IR photometry from 1μ to 20μ (and spectroscopy
at 2μ) for two days before to one year after the maximum
light. Fig.8 reports the energy distribution of the nova on
each of six days around maximum and at 14, 31, 43 days after
maximum. It appears that the IR spectra fall into two distinct classes: a blackbody spectrum up to the time of maximum
and a free-free spectrum, similar to that of the old nova RR
Tel (Glass and Webster, 1973), after that time. The data led
actually to a simple model in which the nova consists throug-

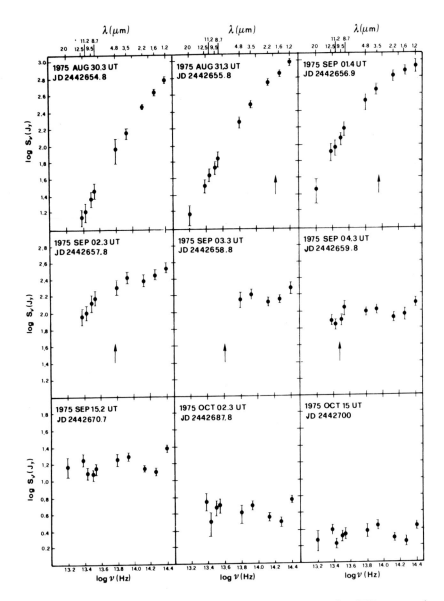

Fig.8. 1 - 20 μ energy distribution of Nova Cygni 1975 on each of
        six days around the maximum, and at 14, 31, 43 days
        after the maximum. The Julian days are also reported.
        (Adapted from Ennis et al. 1976)

hout its lifetime of an expanding mass of ionized gas. Initially the gas is optically thick at IR wavelengths and absorption lines are produced at the front edge of the cloud. Later, as the gas continues to expand, it becomes optically thin resulting in a bremsstrahlung spectrum with hydrogen and helium lines in emission. Its temperature may be assumed $T=10^4$ $^{\circ}K$. The observed increase in total flux of the nova during the optically thick phase is caused by the change in size of the object. Its expansion velocity has been estimated to range in 1300-2500 kms^{-1} while its distance ranges between $1.3 \leq D \leq 2.5$ Kpc according to similar results obtained by means of other observing techinques. The optically thin stage, whose onset occurs four days after the nova onset, remains stable during the next period of observation. The resultant ejected mass can be estimated to be $\sim 10^{-4}$ M_0 and it is typical of other novae studied at optical wavelengths (Payne-Gaposkin, 1957). The temporal dependence of the flux during the optically thin phase gives an indication that the spatial distribution of the expanding gas is in form of a shell. We note also that the time dependence of the 10μ flux, shows a new increase after 300 days indicating the beginning of thermal emission by dust grains similar to that found for other novae. Studies of this phase are still in course and are not available at present.

(e) Young stars.

IR observations of young stars have been performed in the past by several authors. A first review has been published by Neugebauer et al. (1971). Since the observations proceeded both photometrically and spectroscopically. Some very recent results about particular groups of objects have been obtained by Merrill and Stein (1976c) and already reported here. One of the most complete work has been that of Cohen (1973a,b,c) to which we recall. This author has studied between 2μ and 22μ a very large sample of young objects systematically: stars in young clusters, T Tau stars and the Orion population,and finally nebulous emission-line stars. Examples of these observations are reported in Fig.9(a,b,c) where the energy distributions of stars are compared to the slope of a blckbody and that of the optically thin regime of free-free radiation. In general it is found that the energy distribution follows the slope of free-free emission up to 5μ such that found in several Be stars (Woolf et al. 1970). Few stars however extend this trend up to 11μ (for example

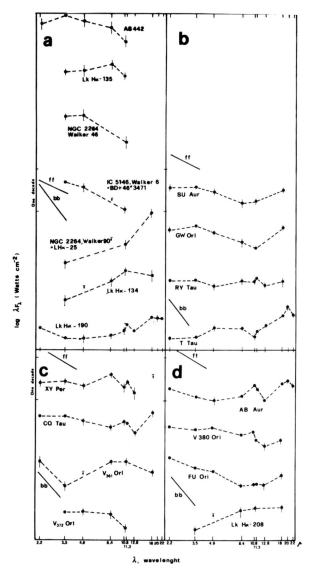

Fig. 9 (a,b,c,d). Examples of energy distributions of young stars.
(a) Stars in young clusters, (b) T-Tau stars, (c) Orion
variables, (d) nebulous early-type stars. The lines marked
"bb" and "ff" give respectively the slope of the Rayleigh-
Jeans tail of a blackbody, and that in the optically thin
regime of free-free radiation. (Adapted from Cohen 1973a,b,c)

Walker 46 in NGC 2264 and BD+46°3971 in IC 5146). Beyond
the 5-8μ region, the energy for nearly all the stars presents
an excess while few nebulous early-type stars show clearly
a 10μ silicate feature too. At these wavelengths again circumstellar dust environments are believed to be responsible for
the IR emission of young objects. These observations are confirmed by the results of many faint members of the Orion
population stars which show flat energy curves, featureless
at all, for T Tau stars (Cohen, 1974). The trend between
photographic properties of variables at maximum and their
present 10μ and 18μ mag argues definitely that circumstellar
clouds, in some way, influence the long term visible variations of these stars, perhaps by simple obscurations and
clearings (Cohen 1973d, 1974). Narrowband spectrophotometry
(Δλ/λ=0.015) between 2μ and 4μ has been performed for several
of these stars (Cohen 1975). Virtually all these objects are
characterized by smooth featureless continua. Some of these
appear to have hot blackbody continua ranging from T=700°K
up to T=1000°K. These stars are either of early type or so
heavily veiled optically that there is no hint for their
spectral type. HL Tau, a very young object, shows clearly an
absorption feature attributable to ice particles.Two older
stars, FU Ori and V 1057 Cyg show instead absorption due to
water vapour near 2μ. From these last observations some
conjectures about the star structure may be made:
i) only in the very earliest phases of star formation interstellar ice is present in abundance;
ii)as the star evolves ice evaporates due to the increasing
   UV field;
iii)in the case of some higher mass stars which undergo FU
   Ori-type flares, the resulting water vapour persists and
   remains visible.
In conclusion, while near IR emission in young stars appears
to be not uniquely due to free-free radiation,thermal dust
emission is again the only effective radiation mechanism at
longer wavelengths.

2.4 HII Regions

At the time of the first review of infrared astronomy
(Neugebauer et al. 1971) only three HII regions were described in the literature. Since, a lot of work has been done
and more than 50 regions have been detected at IR wavelengths.
Recent reviews and discussions of their properties can be

found in Wynn-Williams and Becklin (1974), in the Proceedings
of the 8th ESLAB Symposium (1975), the Symposium HII regions
& related topics (1975), while the most complete one concerning also theoretical aspects of the problem has been produced
by Panagia (1976) that we follow here.

(a) Observations.

The IR spectrum of HII regions is essentially the same
for all but few exceptions. The spectrum is due to a real
continuum; its shape is typical of thermal emission at low
temperature, largely in excess that expected from the free-
free emission of the gas, see fig.10. Dust associated with
individual objects is generally suggested as the cause of the
IR excess. This is also confirmed by the fact that the absorption for wavelengths shorter than 2.2μ varies strongly from
source to source due to the gas emission. Generally, the
spectra are broader than those of an isothermal nebula and
peak at λ≈50-100μ. In the near and mean IR (2-20μ) the spectra
can be fitted by a power law such as

$$F(\lambda) \sim \lambda^{+\gamma} \qquad\qquad \gamma \simeq 2.5 \div 4.0$$

At these wavelengths the nebulae are generally optically thin
because their brightness temperatures are significantly lower
than blackbody colour temperatures. At $\lambda > \lambda_{peak}$ the spectra
are steeper than a blackbody curve, implying that the
sources are optically thin and the grain emissivities must
be decreasing functions of λ, so that the grain dimensions
must be smaller than 50μ. Some peculiar features appear in
either emission, at 9.7μ, or in absorption at 3.1μ and 9.7μ,
superposed to the continuum. This is interpreted as due to
the presence of peculiar components in the dust grains sitting
within or immediately outside the HII regions. (Soifer et al.
1976, Gillett et al. 1975, Person et al. 1976, Forrest and
Soifer 1976). Very often molecular clouds have been observed
associated to these objects emitting at radio wavelengths.
They are rich of molecules such as CO, H_2S, HCN, H_2CO etc.
and, in particular, of molecular hydrogen whose mass has been
estimated to be as long as $10^4 M_0$.

(a1) The brightness distributions.

Many HII regions have been mapped up to 30μ with angular
resolutions of 1"-10". Their brightness distributions are
similar to each other, with few exceptions, so that the temperature range implied by the spectra cannot simply be assumed
as a dilution of radiation from the central star. In addition,

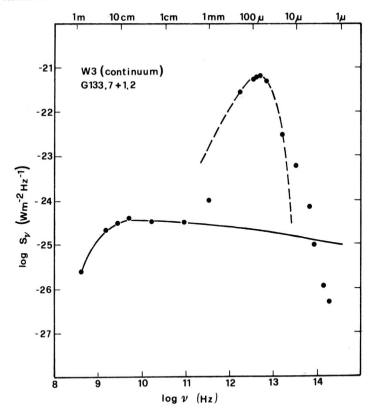

Figure 10 - IR and radio emission of the HII region W3.The lower
          curve which fits the radio measurements represents the
          free-free emission.The upper curve is a diluted 70 K
          Planck function.(from Wynn-williams and Becklin 1974)

these maps coincide quite well with the corresponding radio
maps, indicating that dust must be mixed with ionized gas.
The Far Infrared maps (30-300μ) have been obtained at a poorer
angular resolution (1'-4') and again they well coincide, both
in position and shape, with the radio counterpart within the
limits of the angular resolution. These maps are, in general,
somewhat more extended than those at near and intermediate
IR indicating that the dust is sitting also outside the ionized regions and that it is heated by the same sources which
ionize the gas. Due to the difficulty of observing at long
wavelengths, (300μ-1mm) only few sources have been analized.
At present, there is a tendency to indicate that this radiation is emitted by the grains in the molecular clouds associated to the HII regions. Protostars may be the energy input
for dust and molecules present in these objects.

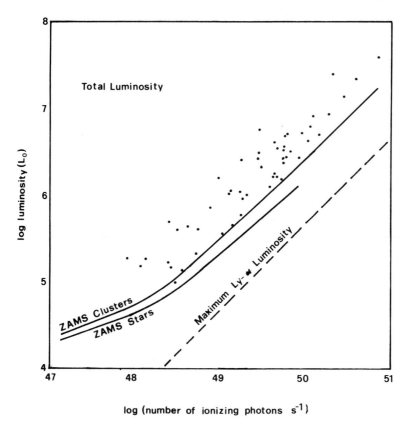

Fig.11 - The total infrared luminosity of 46 HII regions plotted
         against the Ly-continuum photon flux.The curves of ZAMS
         cluster,ZAMS stars and the line $L_{IR} = L_{max}$ (Ly- α)are also
         reported in the figure.(from Panagia 1976)

(a2) Infrared and radio data.
        The far IR fluxes of HII regions have been found to be
linearly related to the corresponding radio fluxes. This can
be interpreted as an evidence for absorption of Lyman continuum photons by the dust in competition with the sourrounding
gas. Fig.11 reports the total IR luminosity, L_{IR}, against
the observed flux Ly-cont photons, N_L, for 46 HII regions
(Felli and Panagia, 1976). From this relationship one can
note:
i) all the sources lie well above the curves difining the
   loci of ZAMS stars and clusters,
ii)the IR excess

$$IRE = L_{IR} / \left[N_L \ h\nu \ (Ly\alpha) \right]$$

is essentially independent of the luminostiy with a mean
value of 14 and a dispersion of 1÷7. The high values of IR
excesses indicate that a large amount of dust is mixed with
the ionized gas and that it absorbs a considerable fraction
of Ly-continuum photons. It comes out therefore that the N_L
values estimated from radio observations are smaller than
the Lyc- flux emitted by the exciting stars by a factor 2-3.
The fact that the IR excesses are the same at all luminosities
seems to imply that the most powerful sources consist of several smaller components, still unresolved at far IR. This
seems to be confirmed from the fact that the largest radio
components found in well resolved HII regions have an excitation parameter of only u $\sim$ 85 pc cm^{-2} (Felli and Panagia,1974;
Wink et al. 1975). Again, if we consider Fig.12, we may analize
the correlation between the intermediate IR luminosity against
the expected Ly-α luminosity of several HII regions (Felli
and Panagia, 1976). All but few sources emit significantly
more than Ly-α luminosity. As the intermediate IR flux comes
from within the HII region where dust and ionized gas coexist,
it comes out that these grains must absorb directly a considerable fraction of stellar radiation and that absorption of
Ly-α photons is not the dominant heating mechanism as it is
usually accepted in the literature.

(b) Discussion of observations.

     The dust opacity in the Ly-c can be derived from a compa-
rison L_{IR}-N_L. Adopting a spherical, constant density model
to describe an HII region, the data presented in Fig.12 yield
to an average opacity in the whole Ly-c (i.e. λ < 912 Å) of
k=5.7x10$\bar{n}_H^{22}$;n_H being the number density of hydrogen atoms. If
we assume that k(He) = 4 k(H), then k(H) = 3.5x10^{-22} and k(He)=
=1.5x10$\bar{n}_H^{21}$. These values however represent lower^{nH} limits to
the actual opacities because any deviation from a uniform
spherical model would raise them, as it is clear from geometrical considerations. From the far IR fluxes it is also
possible to estimate the corresponding dust optical depth
and, by comparing it with the Ly-c optical depth, the absorption efficiency at $\lambda\sim50\mu$m can be inferred (Panagia,1974;
Natta and Panagia, 1976). It has been found that the absorption efficiency of dust at 50μ must be $Q_{50\mu}$ $\sim$ 0.01. This
implies for the grains, responsible for the bulk of far IR
emission as well as the bulk of the UV absorption (500-1500Å),
an average size greater than 0.05μ and a dust to gas ratio by
mass M_d/M_g >4x10^{-3}. Since in order to account for selective
absorption of He-ionizing photons grains with size $\leq$10 Å are

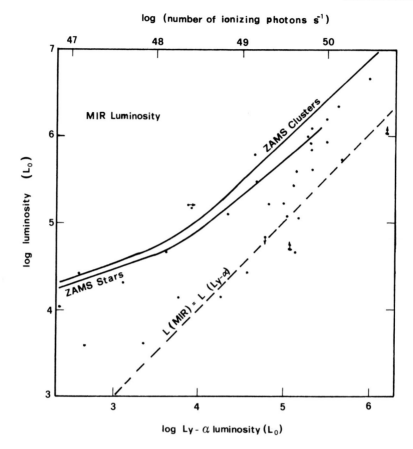

Fig.12- The mean infrared luminosity of 36 HII regions is plotted
        against the Ly-α luminosities of these objects.The curves
        of ZAMS cluster and ZAMS stars are also reported.In addi-
        tion the line L(MIR)= L(Ly-α) is also shown in the figu-
        re.(from Panagia 1976)

required, this lower limit to the size of far IR emitting
grains gives a first indication that more than one dust components must be present in HII regions. Further information
about the dust properties in the far IR and in the submillimeter
range is provided by some recent studies at 350-400μ and 1mm.
It has been found that, at around 400μ, the dust absorptivity
is approximately 20 times greater than what it could be for
silicate and water ice grains but still about 10 times less
than the maximum allowed by the Kramers-Kröning relations
(Hudson and Soifer, 1976). In addition, it has been found

that $Q(\lambda)$ decreases as fast as $\lambda^{-1} \div \lambda^{-2}$ in the submillimeter range (Righini et al. 1976). Since these slopes for $Q(\lambda)$ have been determined by assuming that the spectrum is produced by an isothermal nebula, they should be regarded as lower limits to the actual values. All the above results imply then the picture of a dust component which dominates the UV absorption and the far IR emission, which has a high far IR and submillimeter absorption efficiency, and which has a relatively high abundance and large size. The observations show that it cannot consist of either silicate or water ice. Its nature is still puzzling at present: frozen molecules may be perhaps considered but are still object of speculation. On the other hand, evidence for the presence of water ice grains near, or even inside, HII regions is provided by the absorption band centered at 3.1μ (e.g. Gillett and Forrest, 1973; Soifer et al., 1976). Silicate grains must also be present, both within and around HII regions, because several sources display a spectrum with an emission band centered at $\sim9.7\mu$ attenuated by an absorption band with the same profile (Aitken and Jones, 1973; Gillet et al., 1975; Persson et al., 1976). Furthermore, recent measurements by Forrest and Soifer (1976) have revealed in the Orion nebula spectrum the presence of a band emission around 20μ which supports the silicate identification. However, even for these silicate bands, the observed profile is broader than that of any known silicate material (Bussoletti and Zambetta, 1976). Looking at Fig.13 we see that the intensity of the band observed in the Trapezium region (Forrest et al. 1975) is significantly in excess of that of the lunar rock No. 14321 (Bussoletti and Zambetta, 1976) expecially around $\sim8.5\mu$ and $\sim11.8\mu$. Such excesses may be explained by the presence of grains with a suitable band structure or, by some unknown extra source of continous opacity. A mixture of different silicates does not seem to ba able to reproduce the observed band profile, and some other constituent is needed. This is also evidenced by the observations of Becklin et al. (1976) in a region about 2' south-east of the Trapezium where spectra with local peaks at $\sim8.7\mu$ and 11.2μ have been detected. Also, in the near IR range, there is an indication that dust other than silicate contributes to the absorption. In fact, it has been found that $\tau(2.2\mu) \overset{\sim}{\sim} \tau(9.7\mu)$ (Gillett et al. 1975; Soifer et al. 1976), whereas for any silicate the effective optical depth at 2.2μ is from two to ten times lower than at 9.7μ. The possibility of two coexisting dust components was

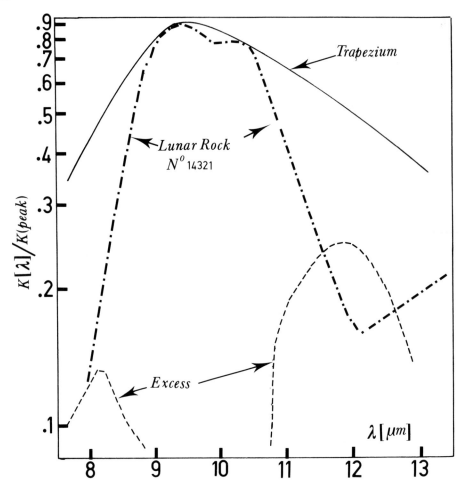

Fig.13- Comparison between the observed emission of the Trapezium
        in the range 8 - 13 μ and that computed by using dust gra-
        ins formed by a lunar silicate(the sample chosen is that
        labeled 14321).The opacities are normalized to the peak va-
        lue.Possible excesses around 8.5 μ and 11.5 μ are indicated
        bu the dotted lines.(from Panagia 1976)

first suggested by Lemke and Low (1972) to explain the extremely broad spectrum of M17. This suggestion has recently been
repeated and reinforced by Harper et al. (1976) on the basis
of more detailed data. Other direct and clear evidence for
the presence of at least two components is provided by the
observations of the bar in the Orion Nebula, approximately

2' south-east of the Trapezium (Becklin et al. 1976). It has
been found that the spectrum requires a range of temperatures
which cannot be due to dilution of radiation but which can
be easily explained in terms of coexisting "hot" and "cold"
dust. Natta and Panagia (1976) have demonstrated that a model
with two, or more, dust components can naturally account for
all the observed features of HII regions as infrared sources.
In particular, they have found that it is possible for dust
to absorb a considerable fraction of the Ly-c flux and for
the 10-20μ color temperature to remain almost constant across
the HII region only if the presence of two components of dust
is allowed.

(c) The dust in HII regions.

The above discussion can be unified in a picture which
contains four different components for the dust associated
to HII regions:

1) a first component which dominates the UV absorption (500-
   1500 Å), the far IR and the submillimeter emission. It
   must have a size of about 0.1μ, a Ly-c absorption opacity
   of about $k \sim 9 \times 10 \bar{n}_H^{22}$ a high absorptivity in the IR $\sim (Q \sim 2\pi a/\lambda$
   up to $\sim 100\mu$) and an abundance relative to the gas of M_D/M_{gas}
   $\approx 5.5 \times 10^{-3}$. Its composition is unknown, but it may
   consist of frozen molecules, so that it should have a sublimation temperature of the order of 100K. Thus it can
   exist only in the outer parts of, or outside, a compact
   HII region. Therefore its apparent UV opacity is two to
   three times lower than the actual one. Furthermore, the
   high abundance and relatively large size make it an important contributor to the visual and near infrared extinction.
   Scaling the absorption and scattering cross section by
   λ^{-1} and λ^{-4}, respectively, for $\lambda > 3000$Å, it is possible to
   estimate the extinction coefficients k_{ext} (5500 Å) $\approx 5 \times 10 \bar{n}_H^{22}$
   and k_{ext} (2.2μm) $\approx 1 \times 10 \bar{n}_H^{22}$. Then the deficiency of dust
   found in the center of the Orion Nebula (Schiffer and
   Mathis, 1974) may be the effect of evaporation of this
   kind of dust. In the most compact HII regions this type
   of dust should be completely absent and only more resistent
   grains could survive. This may explain why regions such
   as K3-50 and DR21 have relatively small infrared excesses
   (<6) but lower than normal He^+/H^+ ratios, 0.07 and 0.05
   respectively (Churchwell et al., 1974). The fact that this
   dust exists only at low temperatures tends to tie the peak

of the spectrum to some constant far IR wavelength somewhat longer than the peak wavelength which corresponds to
the sublimation temperature (Bollea et al. 1976). However,
it is worth noting that even without invoking evaporation
the peak wavelength of the IR spectrum has been found to
be approximately the same for very different conditions
(cfr. Natta and Panagia, 1976).

2) A second component is responsible for the increased opacity
in the He-ionizing continuum. It may be identified with
graphite which seems to have a "resonance" at around 25eV
(Taft and Philipp, 1965). Its typical size is 0.01µ and
the opacity at 500 Å is about $k(He) \simeq 1.5 \times 10 \bar{n}_H^{21}$. This corresponds to a relative abundance by mass of 2×10^{-3}, which is
about 40% of the cosmic abundance of carbon. This component
is an efficient absorber only in the He-ionizing continuum,
it can absorb about 10-15% of the total stellar energy which
is re-emitted mostly in the range 3-30µ. It is interesting
to note that these parameters for graphite are just those
required for explaining the interstellar absorption bump
peaking at 2175 Å (Savage, 1975).

3) Silicate dust is undoubtedly present in HII regions. In
the ionized region, the abundance by mass is about 1.3×10^{-3}.
This implies that ∿50% of the silicon is tied up in grains.
This abundance is consistent with the data of Frogel and
Persson (1974), Persson et al. (1976), and Gullett et al.
(1975) when interpreted according to the prescriptions of
Panagia (1975, 1976) and Natta and Panagia (1976). The
typical size is about 0.15µ and the absorption opacity in
the Ly-c ∿$8 \times 10 \bar{n}_H^{23}$. Thus, silicate dust absorbs and re-
emits in the intermediate IR (say 5-50µ) about 20% of the
total stellar energy as it is observed. At optical wave-
lengths, the expected extinction is about $1.3 \times 10 \bar{n}_H^{22}$ corresponding to nearly pure scattering.

4) Water ice is also present, mainly outside the HII regions,
where the grain temperature is lower than 100K. Its abundance is about 1/4 that of silicate, as deduced from the
optical depths in the 3.1µ and 9.7µ bands. If the same
abundance is taken for silicate in the cold regions of a
nebula as is taken in the ionized region, the ice abundance
turns out to be M(ice)/M(gas) ∿ 4×10^{-4}. Since ice should
be present in the form of a coating of other grains, the
size may be about 0.2µ. Finally, other constituents may
also be present (e.g. silicon carbide, carbonates, etc.),
but they would noy be important in determining any gross

property of the IR emission or of the UV and IR absorption.
Actually we may summarize that the total abundance by mass
of the four listed components is M(dust)/M(gas) = 9.2×10^{-3};
the extinction expected in the visual is k_{ext} (V)$\simeq 6.6 \times 10 n_H^{-22}$
(with an albedo $\overset{o}{A}$ 0.3), mainly slightly larger than
those appropriate for dust in interstellar space.

2.5 Planetary Nebulae

The detection of unexpected and strong radiation from
the planetary nebula NGC 7027 much exceeding the emission
expected from free-free (Gillett et al. 1967) led investigators to look for IR radiation from other planetaries. Observations in this sense have been performed, among others, by
Gillett and Stein (1970), Neugebauer and Garmire (1970),
Gillett et al. (1973, 1975), Jameson et al. (1974), Telesco
et al. (1976) and Dyck and Simon (1976). Particularly interesting, because of its completeness, is the survery of 113
objects studied photometrically between 2.2μ and 24μ by Cohen
and Barlow (1974) who provided a large homogeneous set of
observations covering a wide variety of types. Examples of
these results are reported in Fig. 14(a,b,c,d) where the
spectra have been grouped according to the overall shapes of
the distributions. Each figure includes, for comparison, the
slopes of the Rayleigh-Jeans region of a blackbody (bb) and
of optically thin free-free emission (ff). In nine nebulae,
only several broad spectral features appear but none clearly
resembles the usual silicate emission signature. In general,
from the above observations and from theoretical computations
some conclusions may be argued:
i)   many nebulae are detected at long wavelengths while they
     are either undetected or very faint at near IR;
ii) the long wavelength radiation consists almost entirely
     of a continuum emission;
iii) the fluxes at 10μ and 18μ are substantially greater than
     those to be expected from an extrapolation of free-free
     emission from the radio region, though this mechanism may
     account for the radiation at shorter wavelengths;
iv) the excesses at 10μ and 18μ are interpreted as being due
     to thermal emission by cool dust grains;
v)   the dust seems to coexist with the ionized gas within the
     nebula. This is confirmed by the maps of NGC 7027 per -
     formed by Becklin et al. (1973) which showed that the
     10μ brightness distribution is essentially identical to

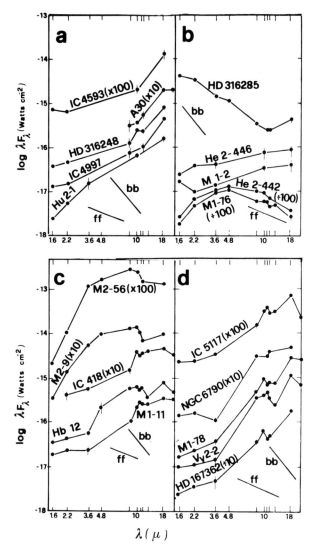

Fig.14(a,b,c,d) -(a)Photometric energy distributions of planetary
nebulae;(b)Emission line objects,including a possibly symbiotic star(HD 316285),the peculiar object M1-76 and the pro-
toplanetary nebula M1-2;(c)Planetary nebulae which are flat
at long wavelengths;(d)Steeply rising spectra of planetary
nebuale.The curves labeled bb and ff reproduce respectively
the slope of the Rayleigh-Jeans tail of a black-body and that
in the optically thin regime of free-free radiation.(from
Cohen and Barlow 1974)

        that at 5 GHz;
vi)   the grains are heated by Lyα photons resonantly trapped
       in the ionized region (O'Dell and Krishna Swamy, 1968).
       In several nebulae however,non-ionizing continuum UV
       longward of the Lyman edge is needed to justify the total
       dust heating;
vii)   low values of dust to gas ratio are found, $M_d/M_g \sim 10^{-4} \div 10^{-3}$.
       (Bussoletti et al. 1974). These values are reduced down
       by a factor 10 when IR radiation up to 300μ is considered
       (Telesco et al. 1976);
viii) evidence is given, principally from the observations of
       two very large low surface brightness nebulae (A30, A78),
       that the formation of dust occurs in the mass-loss out-
       flows from the emission-like nuclei of several sources.
       The most complete information presently known concerns
       the nebula NGC 7027 which has been extensively observed
       between 0.9μ and 300μ(Treffers et al. 1976, [0.9-2.7μ],
       Merrill et al. 1976,[2.1-4.1μ], Gillett et al. 1973,[8-14μ]
       Jameson et al. 1974, [5-27μ], Dyck and Simon, 1976
       [20-33μ] and Telesco et al. 1975 [28-320μ]).
In the first spectral range, emission lines of the Brackett
and Paschen series of hydrogen and of ionized helium have
been detected as well as three lines of the I-0 quadrupole
spectrum of molecular hydrogen. Four lines of HeI have also
been identified. Spectrophotometry from 2.1 and 4.1μ has
revealed a rich spectrum including three hydrogen recombination lines and several unidentified features respectively at
2.43μ, 3.09μ and 3.27μ plus a broad feature peaking at 3.4μ
which are presently object of speculation. This portion of
the spectrum is reported in Fig.15 (a,b). In the figure we
report also the middle and far IR spectrum of NGC 7027 according to Telesco et al. (1975). All the inherent narrow band
and spectrophotometric observations are here summarized. It
clearly appears that a significant turnover occurs at λ=36μ.
The total IR flux has been estimated to be rougly 2.4×10^{-10} W m^{-2} with a
luminosity of $L=2.3 \times 10^4 L_0$ for an assumed distance of 1.77Kpc. The
total IR flux exceeds the Lyman flux by a factor 3.6 so that
Lyman stellar continuum as well as optical and UV lines are
requested to justify the dust heating. The presence of a
neutral shell may also contribute by a certain percent to the
total IR excess, as suggested theoretically by Bussoletti et
al. (1974) and confirmed experimentally by Dick and Simon
(1976). The existence of a large CO emission around NGC 7027

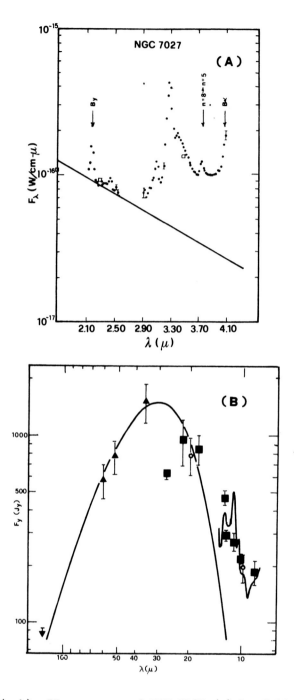

Fig.15(a,b)- IR spectrum of NGC 7027:(a)the full line represents the
free-free emission;(b)Middle and FIR spectrum of the
nebula.See text for the interpretation of the observations.(from Merrill et al.1976(a),Telesco et al.1975(b))

discovered by Mufson et al. (1975) supports the indication
in this sense. Pure graphite grains, plus some carbonated
such as $MgCO_3$, are well suitable to reproduce the observed
spectra in the middle and far infrared.

2.6 IR Lines

The importance of fine structure lines emitted in the
ground state by some ions has been noticed for the first time
by Burbidge et al. (1963) discussing the cooling mechanisms
of the interstellar medium. In this case, collisions play
an important role and two different configurations may occur.
The first one is the excitation of the low levels of the
ground state. The deexcitation energy ranges usually between
I and 5 eV so that forbidden lines, mainly in the visible,
are produced. In the second configuration where only the highest levels of the ground level are excited, the deexcitation
energy ranges between 0.01 and 0.2 eV so that IR emission
lines are expected to occur.

(a) Fine structure lines.

These lines have been studied theoretically, among others
by Pottasch (1968), Bachall and Woolf (1968), Petrosian (1970)
and Simpson (1975). They arise from fine structure transitions
between levels of the $2p^n$ and $3p^n$ configuration, where n=1,2,
4, 5. These lines are quite insensitive to the electron temperature because their excitation energy is very small. Consequently, their detection represents a powerful instrument
for the determination of chemical abundances and temperature
both in planetary nebulae and HII regions. In particular,
ions in the configuration p^1 and p^5 do not have forbidden
optical lines so their abundances can be only determined in
the IR. It is worthwhile to note that these lines fall in a
range from few up to hundred microns, that is, usually outside the normal ground atmospherical windows. This explains
the scarcity of observations appeared in the literature outside the range 8-12μ window. Only recently, due to the NASA
facilities, there has been a certain increase of activity in
the field. In addition to these problems, the rest frequency
of many transitions is still known with a large uncertainty
which may go up to I cm^{-1}, as for SIII. (The best determinations are for neutral elements as OI or NI). A good survey
of the present status of observations and their interpretation
has been reported recently by Baluteau (1977) to whom we recall. Since the first observation (Gillett and Stein, 1969)

TABLE I

Transition Element	ν predicted cm^{-1}	ν observed cm^{-1}
SIV($^2P_{3/2} \to {}^2P_{1/2}$)	950.2	951.5 $\pm$ 0.1
ArIII($^3P_1 \to {}^3P_2$)	1112.1	1112.2 $\pm$ 0.1
NeII($^2P_{1/2} \to {}^2P_{3/2}$)	781.3	780.43$\pm$ 0.03
SIII($^3P_2 \to {}^3P_1$)	535.3	534.39$\pm$ 0.01
OIII($^3P_1 \to {}^3P_0$)	113.4	113.18$\pm$ 0.01

of NeII at 11.8µ in the planetary nebula IC 418, several
measurements have been performed both in HII regions and in
planetary nebulae. Only five lines have been detected and
clearly identified, see Table I. In the first column the
element and its transition are reported. The second and third
columns give respectively the expected frequency and the best
rest frequency determined. (It is worth noting that the largest discrepancy occurs for SIV).

(a1) Planetary nebulae.

These objects have been most extensively studied. Table
II reports the updated observations of IR lines to authors'
knowledge. We note that in the Table when the field of view
does not contain the nebula entirely, a correction has been
made according to the object etendue observed in the visible
forbidden OIII line at 5000 Å. We want to note that this
method may be not completely correct when we discuss elements
with a much different ionization energy such as NeII and SIII.

(a2) HII regions.

Very few observations have been performed of compact
and diffuse regions. This is due to the fact that these
objects are much more extended and have a lower surface
brightness compared to planetaries. Table III reports these
observations giving both line identification and fluxes. NeII
has been also observed in three galaxies (M82, NGC 253) by
Gillett et al. (1975), NGC 1068 by Kleinmann et al. (1976)
and in the galactic center (Aitken et al. 1974, 1976 and
Wollman et al. 1976). These last authors mapped several zones
finding that the emission comes from a region larger than
45". The line has been resolved with HPBW of roughly 200km/s
while its central frequency LSR is of +75±20 km/s. This
result has been confirmed recently by hydrogen recombination
lines measurements. As we have already mentioned above, the
element abundances and electron densities are the main information coming from IR line observations also if no unified
pictures can be given at present.

(a3) Ionic abundances.

- Sulfur: its abundance has been derived until now by using
the visible forbidden lines of SII and SIII. Actually, the
total S abundance in planetary nebulae remains not well
known because the ions SV and SVI are expected to be very
abundant, due to the strong excitation in these objects. However they have been not yet observed. Present observations
confirm this feeling as the ratio n(SIV)/n(SIII) ranges between 3 and 10. According to Simpson (1975), in planetary

TABLE II

Planetary Nebulae

Object	ArIII 9.0 μ	SIV 10.5 μ	NeII 12.8 μ	SIII 18.7 μ
NGC 7027	∿10	∿36 ∿35 69±6	<18	∿6.0
IC 418	<5	<1.5	21±5 <7	
BD+30 3639	<5	<1.5	<15 22±3	∿1.7
NGC 6572	4 5.2±1.3	∿9	<15	
NGC 7009	∿44	∿22 ∿16		
NGC 7662	<90	∿9	<180	
NGC 6210	<13	∿8		
IC 4593		∿1.4		
NGC 6309		∿3		
NGC 6369		∿4		
NGC 6537		∿4		
NGC 6790		∿1.4		
NGC 6818		∿4		
NGC 6826		∿1.5		
NGC 6884		∿6		
IC 4997		∿3		
NGC 7026		∿9		
IC 5217		2		
NGC 6629		<1.3		
NGC 6778		<6		
NGC 6891		<2		

(+) in units of $10^{-18} W\ cm^{-2}$

nebulae where SIV is observed, a mean value of n(SII+SIII+SIV)/
/n(HII) $\sim$ 1.1x10^{-5} can be estimated which is slightly lower
than previous predictions. Variations up a factor 10 are seen
in these objects. They are not extremely significant because
of the lack of information about SV and SVI. Greenberg et al.
(1976) find in NGC 7027 a density well in agreement with
Simpson predictions allowing the estimation of SIII column
density to be 3.2x10^{15} cm^{-2} and of an abundance of n(SIII)/
/n(HII) $\sim$ 1.6x10^{-6} under the hypothesis of an uniform nebula
with density n_e=1.6x10^5 and a temperature T_e=12500°K. The SIV
observations of Holtz et al. (1971), Gillett et al. (1972)
and Aitken and Jones (1973) give a value n(SIV)/n(sHII)$\sim$1.4-2.8x
x10^{-6}. In this object it is also possible to estimate a total
abundance n(SII+SIII+SIV)/n(HII)$\sim$(4$^{\pm}$1)x10^{-6} when also the
measurements of SII by Peimbert and Torres-Peimbert (1971)
are taken in account. At present however the case of Orion
is still open: Baluteau et al. (1976) find an intensity of
SIII twice lower than predicted while they have not been able
to find SIV. More refined measurements to solve the problem
are in course.
- Neon: this element presents a p^5 configuration and therefore
only IR transitions allow an abundance determination. Only
IC 418 and BD+30°3639 show this line among planetary nebulae.
In the first one it has been found that n(NeII)/n(Ne)$\sim$0.7
with an abundance n(He)/n(H)$\sim$8x10^{-5} which is quite near the
expected value. In the second object, Ne appears to be overabundant by a factor 3 without any significat justification.
For compact HII regions, Gillett et al. (1975) find an under-
abundance in G29.9-0.0 (n(NeII)/n(HII)$\sim$2.5x10^{-4}), a value
slightly lower than that expected in G333.6-0.2 ($\sim$5x10^{-5}) and
very low values in W49 and W51 respectively (3x10^{-5} and
1.5x10^{-5}). We must stress however that silicate opacity surely
affects these estimations as well as the excitation of NeIII
at 15, 4μ and 36.1μ which may occur. Galactic centre observations give an abundance n(NeII)/n(HII)$\sim$1.7x10^{-4} which is
slightly higher than the expected value. These observations
are easily fitted by a situation in which heavy elements are
overabundant by a factor 3 compared with solar values and
ionization is provided by stars with effective temperature
of 40000°K (Aitken et al. 1976).
- Argon: it presents also a forbidden transition in the visible (λ=7500Å) whose cross section is not well known. In addition, its intensity is much more sensible to the electron

TABLE III

HII REGIONS

Object	ArIII 9.0 μ	SIV 10.5μ	NeII 12.8μ	SIII 18.7μ	OIII 88μ
G 333.6-0.2	<13	<9	163±12	39±2	
G 29.9-0.0	~0.9	<0.5	~9.2		
W 51-IRS1			~3.3		
W 51-IRS2			≤3.3		
W 49-OH			~1.1		
K 3-50			≤1.3		
W3-IRS1	~3.5	~5.3	~7.1		
N 17					0.7 / 1.6±0.5
M 42		<6		2.8±0.3	2.0±0.4

(+) in units of 10^{-9} W cm^{-2}sr^{-1}

temperature so that the IR lines respectively at 9.0 and at
21.84μ are much more suitable for the abundance determination.
The 9.0μ line has been observed by Geballe and Rank (1973)
in NGC 7027 and NGC 6572. The abundances have been found respectively 7×10^{-7} and 9×10^{-7}, which correspond to fractional
ionizations of $n(ArIII)/n(Ar)$ 7% and 9% if we assume for planetary nebulae a normal abundance $n(Ar)/n(H) \sim 10^{-5}$. The same
line has been observed also in the HII region G 29.9-00. A
large fraction of uncertainty arises, as it is for the Ne lines, from the proximity of the silicate band in the determination of its intensity and then of its abundance. The values
found, $n(ArIII)/n(H^+) \sim (1.5-5.7) \times 10^{-6}$, are lower than the expected Ar abundance $n(Ar)/n(H) \approx 7.6 \times 10^{-6}$ according to Brown
and Gould (1970).
- Oxygen: only the OIII line at 88.3μ has been so far observed in Orion and M17 respectively by Baluteau et al. (1976)
and by Ward et al. (1975). Their intensities agree well with
Simpson's predictions based on an abundance of $n(O)/n(H) \sim$
$\sim 2.8 \times 10^{-4}$. The other line at 51.7μ falls in a spectral region
where the atmospheric opacity is very high and no observations
have been possible until now.
(a4) Electron density.
      The most precise determination of the electron density
of a ionized source can be done by means of the observation
of the intensity ratio of two lines of the p^2 and p^4 electronic configurations but unfortunately no such observations
are presently available. A second method can be derived from
the ratio of the intensities of an IR fine structure line
and of a forbidden optical line of the same element. The
latter method is actually less precise than the former due
to the indetermination of the measurement of the optical line,
and it is restricted obviously only to visible objects.
Greenberg et al. (1976), have determined the mean density of
NGC 7027 and of BD+30° 3639 by using the SIII line at 18.7μ
and those at 9069Å and at 6312Å. They find in the first case a
value $n_e \approx 1.6 \times 10^5$ cm^{-3} which agrees with previous optical determinations (Kaler et al. 1976). On the contrary, in the lower-
second object they find a value $n_e \approx (1-3) \times 10^5$cm^{-3} lower by a
factor 4 respect to the optical determinations. This result
implies then that the fine structure levels of the SIII ground
level are strongly saturated in the planetary BD+30° 3639.
(b) Recombination lines.
      Petrosian (1970) has discussed their emission in IR. Only
hydrogen can be expected to be observed at these wavelengths

due to its very large abundance. Some of these lines have
been detected recently in the near IR, at $\lambda=2\mu$, by Gautier
et al. (1976) while others are presently looked for in several other IR sources.

(c) Molecular lines.

Molecular hydrogen and HD have been observed in absorption in the spectra of several stars (Carruthers 1970, Smith
1973, Spitzer et al. 1973) providing a first evidence of
their presence in interstellar clouds. Up to now, however,
no direct measurements are available in IR in dense clouds
where their presence should be of fundamental importance for
physical and chemical processes involved. The significance
and feasibility of these observations have been pointed out
by various authors (Field et al. 1968, Dalgarno and Wright
1972, Bussoletti 1973, Drapatz and Michel 1974, Bussoletti
and Stasinska 1975) while in the last two papers the effects
of absorption by dust has also been taken in consideration.
In dense clouds, where these lines are expected to be easily
detectable, collisions between H_2 molecules prevail in ex -
citing and deexciting the molecular levels. Transitions between as closely adiacent levels as possible ($\Delta J=2$ for H_2 and
$\Delta J = 1$ for HD) prevail because multiquantum effects are at
least one order of magnitude less efficient. Detailed computations in several objects such as the Orion HCN cloud, Sgr A,
Sgr B2, globules and protostars have been performed by Bussoletti and Stasinska (1975) looking for the lines detectable
with present techniques. It comes out that the most promising
objects are SgrB2 and Orion where the S(0) line of H_2 at
28.2μ is expected to radiate respectively 1.7×10^{-10}erg cm^2/s
and 3.7×10^{-9}erg cm^2/s taking account of the dust absorption.
More favourable are the HD lines R(0), R(1) and R(2) at 112μ,
56.2μ and 37.7μ which are expected to radiate respectively
3×10^{-10} erg cm^2/s, 5×10^{-10}erg cm^2/s and 7×10^{-11}erg cm^2/s due
to the low values of the dust absorption at these wavelengths.
These intensities and the wavelengths of occurrence indicate
that airborne platforms are needed for the observations. An
estimate of the line HPBW obtained by molecular radio observations in the same objects indicates that high spectral reso-
lution, $\lambda/\Delta\lambda \sim 10^4$, is necessary to perform these measurements.

2.7 The Orion Complex

The Orion nebula has been the most extensively object
of our Galaxy observed in IR due to its vicinity to the earth

(d=450-500 pc) and to the fact that it is a region of recent
star formation. The main first discoveries were those of
Becklin and Neugebauer (1967) who detected at 2.2μ a point-
like source 1'NW of θ'C of Trapezium and of Kleinmann and
Low (1967) who observed a complex diffuse nebula at 22μ which
contained the previous BN object. In addition, Ney and Allen
(1969) discovered a diffuse source, at 11.6μ and at 20μ, centered on θ'C of Trapezium. Later on Gatley et al. (1974)
discovered a large complex of sources, 12"NE of the Trapezium,
which had a total dimension of 1'. This complex, named Orion
Molecular Cloud 2 (OMC2), is associated with a large CO molecular cloud with 6' of dimension. The complexity of these regions and their richness in different kinds of sources has
represented one of the major scientific targets for many
observations since their discovery. For example, Ricke et al.
(1973) have mapped the KL nebula at 5μ, 10μ and 21μ finding
four compact sources south of the BN object which still remains the brightest object. Some of these sources seem to
coincide with OH masers detected by Hills et al. (1972),
i.e. IRCI and IRC3. In addition no compact radio continuum
sources have been detected in this region of space. Again,
maps of KL and NA in the 8-13μ range have been performed by
Gehrtz et al. (1975) who found the general shapes, the locations, and surface brightnesses of the prominent IR features
in agreement with previous observations. In addition, they
found an extended region of moderate surface brightness extending to the east of the main body of BNKL which not appeared in the data of Rieke et al. (1973). A silicate emission
appears clearly over the entire region excluding BNKL with
maxima centered at NAI and NAII sources position and coincides with the spatial distribution of hot ionized gas (Martin
and Gull, 1975). On the contrary, absorption by silicate is
evident from the BNKL region and from the cool eastern source
in a less stringent way. The maximum grain temperature of
$T=250^{\circ}K$ is centered on θ'Ori D while the temperature drops
down to $160-180^{\circ}K$ 10"-15" of θ'D remaining stable over the
entire extent of NAI and NAII sources to distances as large
as 50". This stability rules out direct heating from the
Trapezium stars as well as stellar photons, suggesting instead
the interpretation that the primary heating mechanism is the
absorption by dust of Lyman-α resonantly trapped photons.
This seems to be confirmed by the coincidence between dust
and hot gas. The heating in the immediate vicinity of the
Trapezium region must also take in account the previous other

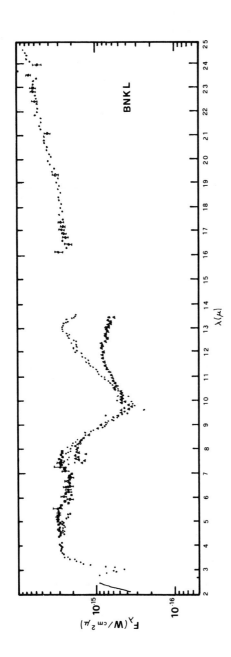

Fig.16 - The composite spectrum of BNKL in the wavelength range 2-25 μ .The different measurements are labelled as follows: ● 2.8-5 μ Gillett and Forrest(1973), (——)2.1-2.5 μ Gillett et al.1975, ◨ 4.5-8 μ Russel et al.(1977), ▲ 7.5-13.5 μ Gillett and Forrest (1973), o 16-25 μ Forrest and Soifer(1976). All the observations have been performed at different apertures.(Adapted from the above authors)

two mechanisms to justify the observed enhanced temperature.
Spectrophotometric observations of BNKL have been performed,
in the range 2.1-25μ by several authors: $\lfloor 2.1$-$2.5\mu \rfloor$ (Gillett
et al. 1975), $\lfloor 2.8$-$5\mu \rfloor$ (Gillett and Forrest, 1973), $\lfloor 4.5$-$8\mu \rfloor$
(Russell et al. 1977), $\lfloor 7.5$-$13.5\mu \rfloor$ (Gillett and Forrest,1973),
$\lfloor 16$-$25\mu \rfloor$ (Forrest and Soifer, 1976). Figure 16 reports the
entire spectrum in this range. According to Becklin et al.
(1973) the spectrum may be interpreted as due to a hot dust
cloud (T≈530°K), maybe in a prestellar stage of its evolutionary life. Absorption features appear clearly at 3.1μ and at
9.8μ and are currently attributed to water ice and silicates
respectively. Attempts to fit these features by using ice-
mantle particles with silicate cores (olivine, quartz or
lunar silicates) have been performed by Aannestad (1975) but
the results are quite poor. No single type of these particles
can adequately account for the 9.8μ feature and at least a
mixture of different types of silicates must be required.
Russell et al. (1977) have also claimed the existence of an
extra source of emission to give account of the smooth continuum at λ=8μ. Graphite, which has an IR featureless spectrum, is one likely candidate. M42 has been also extensively
studied in the far IR by various authors who have attempted
to define its spectral behaviour. Comparing all the available
observations at these wavelenghts, it comes out that the
source is well fitted by a 100°K blackbody. The only departure from a smooth spectrum occurs at the short wavelengths
where two observed points respectively at 43μ and at 47μ (Ward
et al. 1976) appear higher than predicted. They fall however quite
near the water-ice band at 43μ and the signal levels are quite
low in this region so that systematic errors due to instrumental scatter may be the actual cause of these minor discrepancies. Maps of the central 5' of the nebula have been obtained with high resolution, 1' at 20μ, 50μ and 100μ by
Werner et al.(1976) and are reported in Fig.17 (a,b,c). These results
are of extreme interest because provide a good understanding
of the interactions between the various components of the high
density central regions where star formation seems to proceed
most rapidly. Within the mapped region, one can discern heating effects due to both the Trapezium stars and the IR cluster
resolved firstly by Rieke et al. (1973). The principal features
seen both at 50μ and 100μ are a sharp peak in emission near
the position of the IR cluster about 1' NW of the Trapezium,
a ridge of emission elongated roughly NS abour this peak and
a barlike feature running NW to SW about 2' to the SE of the

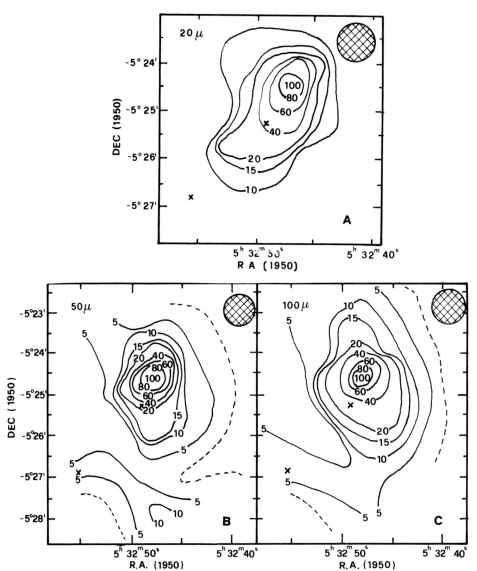

Fig.17(a,b,c)- Maps of the Orion Nebula respectively at $20\,\mu$, $50\,\mu$
and $100\,\mu$ .In each map the contours are normalized
relative to the peak flux density(see text).The cross
near the center of each map marks the position of
ϑ^1 Ori C,while that at the lower left gives the position of ϑ^2 Ori A.(from Werner et al.1976)

Trapezium. The 20µ map, similar to that obtained by Lemke et al. (1974), shows an emission which also peaks at the IR cluster but with an intensity not so pronounced as that at the longer wavelengths. The 2'x4' ridge of 50µ and 100µ radiation is similar, in extent and orientation, to the central ridge of the molecular cloud as observed in emission from both dust and gas by Harvey et al. (1974). On the contrary, the 20µ map is more similar to the distribution of free-free emission from the ionized gas. As the position of maximum 50µ and 100µ surface brightness coincide, the total IR luminosity is considerably greater than the luminosity from heating by the Trapezium stars for any source geometry, so that the IR cluster heating is important. In addition, at these wavelengths the gradients of surface brightnesses and the color temperature seen in the ridge of IR emission surrounding the peak are consistent with those expected if the emission arises from material sitting within the molecular cloud heated by the IR cluster (Gezari et al. 1974, Scoville and Kwan, 1976). From these measuremnts the total 10-1000µ luminosity of the cluster is estimated to be in excess of $1.2 \times 10^5 L_0$ in agreement with the values found by Fazio et al. (1974) and by Harper (1974). Heating of the Trapezium stars appears to be important in the southern portion of the ridge implying that the Trapezium lies at 0.1 pc from the molecular cloud. Finally, far IR radiation is seen from an optically prominent ionization front to the SE of the Trapezium. The emission seems to be produced by material at, or near, the HI-HII boundary which is heated directly by the Trapezium stars. In contrast with the observations above mentioned, the complex OMC-2 has been, up to now, observed only at the ground photometrically from 1.6µ up to 20µ. A map at 2.2µ of this region is reported in Fig.18 according to Gatley et al. (1974). Five components, identified as IRS1-5, appear clearly and they have not any visible counterpart. In general OMC-2 is less bright than BNKL in infrared. The brightest source of the complex is IRS3. Its spectral trend resembles the energy distribution of the BN object except for the absence of the absorption features at 3.1µ and at 10µ. IRS4 and KL, which is 10^3 times brighter, have a spectral analogy, but not very stringent. Actually, due to the small number of observations available in the literature, not much can be said about the nature of the sources.

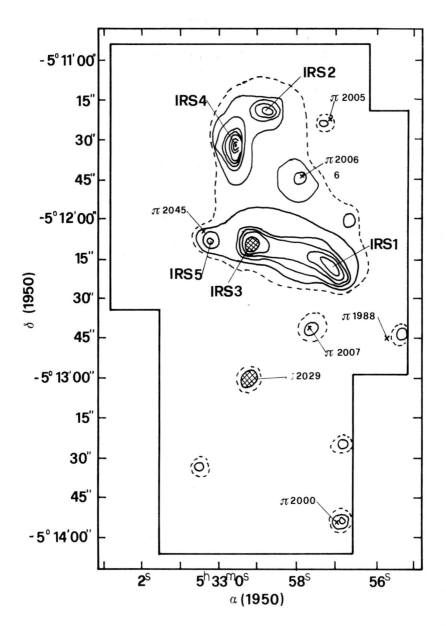

Fig.18- The 2.2 map of the Orion Molecular Cloud 2.Five components
        are clearly identified (IRS 1-5)which have not a visible
        counterpart.The objects labelled by indicate visible stars
        in the Parenago Catalog.(from Gatley et al.1974)

2.8 Galactic Center

A wide variety of interesting physical phenomena take place in the region of the center of our Galaxy, where complex structures appear in IR. Many observations have been performed in the past and have been reviewed by Neugebauer et al. (1971). Further studies have been performed by Rieke and Low (1973) and by Borgman et al. (1974) revealing the unicity of this region of space. Recently high resolution mapping ($\theta \approx 2.5$) has been performed of the central I' region by Becklin and Neugebauer (1975) respectively at 2.2μ and 10μ. Fig. 19 (a,b) reports these results as well as the polarization of the sources measured respectively at 2.2μ, 3.5μ, and 11.5μ by Capps and Knake (1976) and by Knake and Capps (1977). The 2.2μ map, where 19 sources have been identified, shows that the radiation from the central 30" of the galactic center results from three main components: about one third comes from a point source (N°7 in the map) already observed by Becklin and Neugebauer (1968) and conventionally assumed as the center of the Galaxy. About one third comes from discrete sources and the remaining from a general background. At 10μ there are 9 discrete sources as well as a curved ridge of extended emission, probably optically thin. In this case, one quarter of radiation comes from the discrete sources, one half from the emission ridge and the remainder from low surface brightness radiation. These maps agree well, within positional errors, with those obtained by Rieke and Low (1973) at lower resolution at 3.5μ, 5.0μ, 10μ, and 20μ and with that of Borgman et al. (1974). In addition, some of the sources observed by Rieke and Low have been also resolved in discrete components. Recent spectrophotometry between 0.83μ and 4μ has been performed respectively by Treffers et al. (1976) $\lfloor 0.83-2.7\mu \rfloor$, Neugebauer et al. (1976) $\lfloor 2.10-2.45\mu \rfloor$ and by Solfer et al. (1976) $\lfloor 2.1-4.1\mu \rfloor$. An absorption feature centered at 2.3μ has been detected in the point-like source as well as in some IR sources within 20" of the galactic center. This feature has been identified as the v=2 CO band head. Its presence, in some but not all sources, indicates that CO molecules are local to each source rather than being a property of the intervening material. The general spectrum and this band depth in the point-like source (definitely proved to be a cool star) indicate that it is a supergiant while the lack of C$_2$ band shows that it is an oxygen rich star. Because M supergiants are generally assumed to have evolved from O and

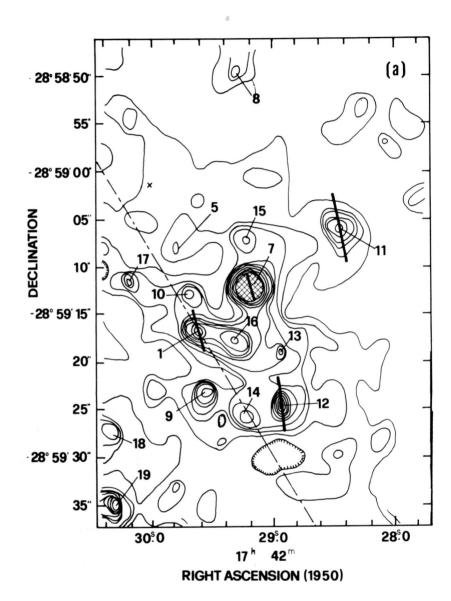

Fig.19(a,b)- Maps of the central 1'region of the galactic centre performed at high resolution($\vartheta \cong 2.5"$)by Beclin and Neugebauer
(1975) respectively at 2.5μ (a) and at 10μ(b).The maps contain
(a)the polarization measurements by Knake and Capps(1977)at
2.2μ and those at(b)3.5μ (- - -)and at 11.5μ (——).

B stars it seems confirmed that this region of the galaxy
contains a rapidly evolving cluster of young objects. No 3.1μ
absorption bands due to water ice are seen at all while a
quite significant feature appears at 3.3/3.4μ. Whether it is
an emission at 3.3μ or an absorption at 3.4μ is presently unclear as well as its identification. It is worthwhile to note
however that no similar stellar feature has been seen in the
spectra of late type carbon or oxygen rich stars by Merrill
and Stein (1976). Recently, many far IR observations of both
SgrA and SgrB2 have been performed (Harvey et al.1976, Righini et al. 1976, Ward et al. 1976, Low et al. 1977, Ericson
et al. 1977, Thronson and Harper 1977, Harvey et al. 1977,
Gatley et al. 1977). SgrB2, the weakest of the two sources,
has been observed up to 3.2mm and mapped at these wavelengths.
As usual, the far IR emission is definitely attributed to
dust emission. It is also interesting to note that, in general, the maxima of the ^{13}CO and H_3CN emissions correspond,
within positional uncertainties, to the peaks in the 3.2mm
and 350μ (Righini et al. 1976), the 1mm map of Westbrook et
al. (1977), and the 53μ, 100μ and 175μ maps of Harvey et al.
(1977). The morphology of these maps is quite similar although
the peak widths reduce from 2' down to 0.4' for λ=3.2mm to
λ=53μ indicating a general temperature gradient in the source.
Low et al. (1977) have shown that the central core is broader
in IR than at 5Ghz indicating that the distribution of cool
stars is broader than that of hot ionizing stars. This is
confirmed by the 350μ map which seems to indicate that the
dust radiation arises from grains sitting within the molecular cloud. A lower limit to the total luminosity may be deduced from the observations: it is equal to $L=10^9 L_0$. All the
observations indicate a low temperature for the source, $T=32-33^o K$
(except Ward et al. who find $T=40^o K$) while the CO temperature
is $T=20^o K$. In conclusion it appears that while thermal coupling between dust and gas is good it is however inadequate
to establish equilibrium between the two. This result is consistent with the calculations of Goldreich and Kwan(1974)
and Scoville et al. (1974) provided that the H_2 density is
greater than $2 \times 10^4 cm^{-3}$. For SgrA, several complementary observations have been performed (Harvey et al. 1976, Gatley et
al. 1977, Rieke et al. 1977) and the source has been mapped
to a very high resolution (∿20") in the central few minutes
of its core. Several similar results have been obtained by
many authors so that we prefer to report here the more complete ones by Gatley et al. (1977) who have mapped the region

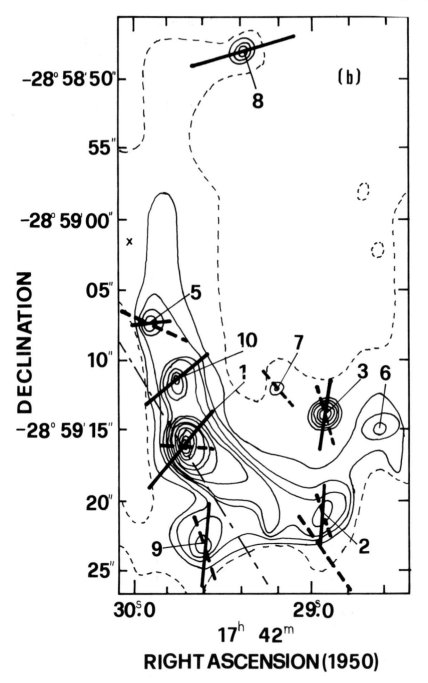

Fig19(b)- See the caption reported in Fig.19(a)

simoultaneously at 30μ, 50μ, and 100μ and in addition also
at 1mm. Fig.20 (a,b,c) reports these IR maps while the parts
(e) and (f) of the figure report the 10.7 Ghz map of Pauls
et al. (1976) and the ^{12}CO emission detected by Solomon et
al. (1972). All the far IR maps consist of a dominant bright
source, coincident with the galactic center, plus a variety
of fainter sources and regions of extended emission. The
position of the highest surface brightness is coincident with
the peak of stellar distribution (2.2μ source) and with the
HII region SgrA West. No far IR features appear at the position of the supernova remnant SgrA East. The total FIR luminosity of the central 1' of the galaxy can be estimated, from
these observations, to be $2.3 \times 10^6 L_0$. The two sources appearing NW of the peak brightness are identified with the known
HII regions G0.01+0.02 and G0.07+0.04. A third HII region,
G0.01+0.12, is seen also SE. Gradients of the dust temperature
are evidenced by the increase of the diameter of the SgrA
West source with increasing wavelength. These data, integrated by those previously mentioned, provide very strong support
for the idea that the far IR radiation is due to thermal
emission from dust heated by a centrally concentrated energy
source. Both, late-type stars and ionizing sources are able
to produce this emission. Because both sources of luminosity
peak at the same position and can supply comparable amounts
of energy, it is not possible to distinguish effects due to
one or other within the central few parsecs. On the contrary,
stellar heating may be more important in the outer regions
because the radio flux is more circularly symmetric while
the far IR emission and the stellar distribution are elongated
along the galactic plane. It is also found that the visual
extinction in the central 10pc' is only about 3 mag while
the dust density is fairly low but in the molecular clouds
where it may become one order of magnitude greater than the
elsewhere. By fitting the emission with appropriate, also if
not unique, models one obtains a temperature for the dust
of $80°K$ at the peak of brightness which reduces smoothly to
$50°K$ 10pc away. These values agree with those found by other
authors when the experimental uncertainties (up to $\pm 30\%$) are
considered. It is finally interesting to note from Fig.20
(e,f) that there is no coincidence between IR peaks and the
CO strongest emission which are, on the contrary, respectively
the hottest and the coldest points on the maps. The IR polarization of SgrA, firstly measured by Low et al. (1969) and
by Dick et al. (1974), has been measured recently by Capps

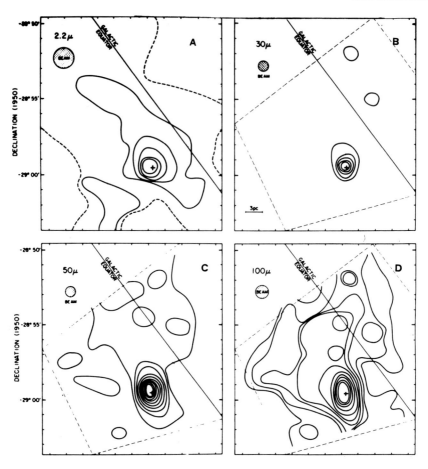

Fig.20(a,b,c,d)- (a)Map of the galactic centre region at 2.2 μ .The
   contour intervals are equivalent to 3Jy.The dashed contour represents half level.(b)Map at 30 μ .The diagonal dashed lines
   indicate the mapped area.The contour interval is equivalent
   to 1.2 x 10^3 Jy.(c)Map at 50 μ .The contour interval is equivalent to 1.1 x10^3 Jy.(d) Map at 100 μ .The contour interval
   is equivalent to 600 Jy.
   The cross in each figure indicates the position of SgrA West.
   (from Gatley et al.1977)

and Knake (1976) and by Knake and Capps (1977) between 2.2μ
and 12μ with a 3.5" beam. At 2.2μ and 3.5μ , see Fig.19(a,b),
the polarization orientation has been found nearly parallel
to the galactic plane and therefore coincident with what is
observed for visible stars. Values from 1.8±0.5% up to 6±0.3%
have been found in the sources at 2.2μ . At 3.5μ values between 2.7±0.3% and 5.4±0.8% are seen, while at 11.5μ these
values range from 2.4±0.4% up to 5.9±0.5 and spatial variations across the sources are present. The observations suggest
that the near IR radiation is polarized by grains in the
spiral arms and the 11.5μ flux is polarized by dust sitting
whithin or near the IR sources in SgrA. The general spectral
dependence of the polarization suggests that this dust must
contain silicates. Magnetic fields strengths of 10-20 micro-
gauss are required in the plane and 10^{-3} gauss in the sources
to accept the Davies and Greenstein (1951) type of alignement.
Since these values are generally greater than those found
with other observations a more efficient alignement seems
to be required but at present the question is still open.

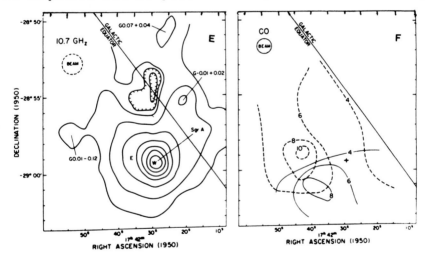

Fig.20(e,f)-Map of the 10.7GHz radio continuum emission from the galactic centre.The contour intervals are at 5,10,20,40,60,80,
   100 times the unit of 0.47 main beam brightness temperature.
   The letters W and E indicate the positions of Sgr A West and
   East respectively.(f)Map of the ^{12}CO emission near the galactic centre.The contours are labeled in units of the peak antenna temperature.(from Gatley et al.1977 who report data from
   Pauls et al.1976 and from Solomon et al.1972,see text)

3. BACKGROUND FAR INFRARED ASTRONOMY

Three different processes produce a far infrared (FIR) background in
the sky:
(a) Emission from galactic clouds
(b) Emission from extragalactic non-resolved objects
(c) The 3 °K cosmic background
However, other backgrounds may be given both by local sources (like the
interplanetary dust) and by unresolved galactic objects, especially if
low spatial resolution telescopes are coupled to high sensitivity detectors presently available in the middle infrared. The problem may be so
severe as to limit the possibilities of the medium-size space telescopes
cooled at low temperatures, like IRAS or SIRTF.

In the following we present a few remarks on points (a) and (b), and then
we will concentrate our attention on the third point which represents a
fundamental topic of research in experimental and theoretical cosmology.

3.1 Galactic background.

Stein (1966) has proposed the existence of a FIR background due to emission from galactic dust: The question has been discussed later by Pipher
(1973). Recent observations of millimeter emission by CO in galactic
molecular clouds have allowed a more detailed study of this guess. The
basic idea is that FIR emission originates from dust heated by early
type stars or young stellar associations located near molecular clouds
or H II regions. Intercloud dust might also contribute to this background
but the temperature of such grains is unknown due to the uncertainties
of the spectrum and intensity of the stellar radiation field. Some estimates (Puget, 1977; Metzger, 1977) give percentages ranging from 20%
up to 100% of molecular cloud emission. Fazio and Stecker (1976) have
evaluated the galactic background at small latitude on the basis of CO
surveys in molecular clouds. Their results are reported in Figure 21.
The total infrared luminosity has been evaluated by Serra et al. (1978)
on the basis of balloon measurements in the 100-200 μ bandwidth. They
found

$$I_{ir} \simeq 7 \times 10^{-5} \ W \ m^{-2} sr^{-1}$$

Figure 22 reports the longitude profile of the far infrared diffuse emission from the galactic plane. The flux is shown to have a maximum near
$l = 30°$. Serra et al. (1978) have pointed out that this maximum is likely
associated with the so-called "5 Kpc ring" which has a high density of
molecular clouds, a high gamma-ray flux and a high star formation rate.
The latitude profile of the galactic emission is unknown: wide-beam
observations carried out by Weiss (1977) give some evidence for large
dust clouds not associated to molecular clouds, outside the galactic
plane. More recently, an airborne experiment (Fabbri et al. 1979) has
shown that the far infrared flux is negligible outside the galactic
plane , at least for longitudes close to the galactic center. Not much

more can be said at present due to the lack of information.

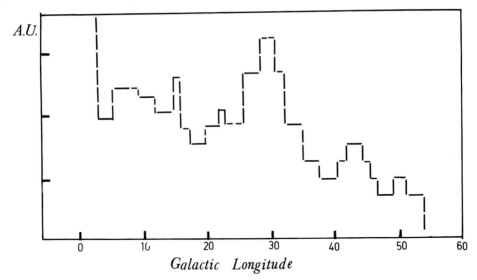

Figure 21. Galactic background at zero latitude as evaluated
on the basis of CO surveys in the 100-500 μ region.
(Adapted from Fazio and Stecker, 1976).

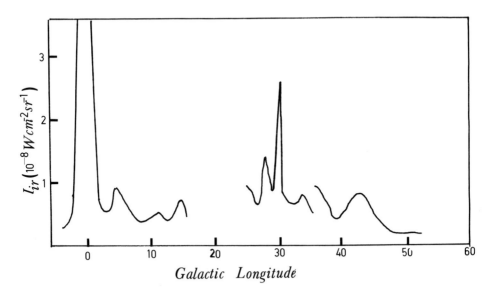

Figure 22. Galactic background at zero latitude as measured
by various authors.
(Adapted from Serra et al. 1978).

3.2 Extragalactic background.

At high galactic latitude it may become possible to detect the extragalactic background. FIR background from external galaxies was investigated theoretically in the past with the aim of explaining the radio
and millimeter excess of radiation currently attributed to the 3 °K
cosmic background radiation. However, a conspicous extragalactic background is today believed to exist in the infrared region, between 1 and
500 μ (see Figure 23). The detection of such a radiation is considered
one of the main tasks of infrared astronomy in the Eighties, because
the investigation of its properties will provide precious information
of astrophysical and cosmological interest (Peebles, 1971; Mather, 1977;
Bussoletti, Fabbri and Melchiorri, 1979). The background arising from
the integrated emission of distant, unresolved sources may be the only
way for the study of the primeval evolution of normal galaxies (Partridge
and Peebles, 1967; Tinsley 1973; Puget et al. 1976; Stecker et al. 1977),
infrared galaxies (Low and Tucker, 1968; Stecker et al. 1977) and possibly of more exotic objects such as pregalactic stars (Thorstensen
and Partridge, 1975). However, difficult problems are represented by
the galactic background and the planetary background, which are expected
to mask the extragalactic background almost at all frequencies. A new
technique has been proposed recently by Fabbri and Melchiorri (1979)
which automatically cancels the galactic and zodiacal background. This
method consists in measuring the spectral properties of the anisotropy
of the extragalactic background, arising from the inverse Compton scattering of the background photons on a very hot gas in cluster of galaxies. This anisotropy is known as Sunyaev- Zeldovich effect in the case
of the 3 °K cosmic background (see Section 4.2). The proposed technique
combines the advantages of the differential (double-beam or beam switching)
techniques with the special advantage of dealing with a signal with a
very peculiar spectral shape.

Only few upper limits are available up to now for the infrared sky
background: as pointed out by Peebles (1971) , the experiments were
designed for quite other purposes and the cosmic limit was always a
byproduct. Hopefully we shall be able in the future to get significant
information on the source evolution at redshifts z = 3-30. While only
very powerful sources are observed at z ∿ 3, and the 3 °K cosmic background radiation measured in the microwave and millimetric region provides information mainly on very early epochs, the intermediate redshifts
might be covered by the infrared background (For the radio background
at metre wavelengths, cfr. Longair 1978). The infrared detectors presently available are certainly adequate to this purpose. In the near
future the entire near and middle infrared region will be covered by
photoconductors with Noise Equivalent Power (NEP) $\lesssim 10^{-17}$ W Hz$^{-\frac{1}{2}}$. Such
detectors are certainly background limited at balloon or airborne al-
titude: However, space experiments should allow us to fully employ the
intrinsic high sensitivity of these detectors.

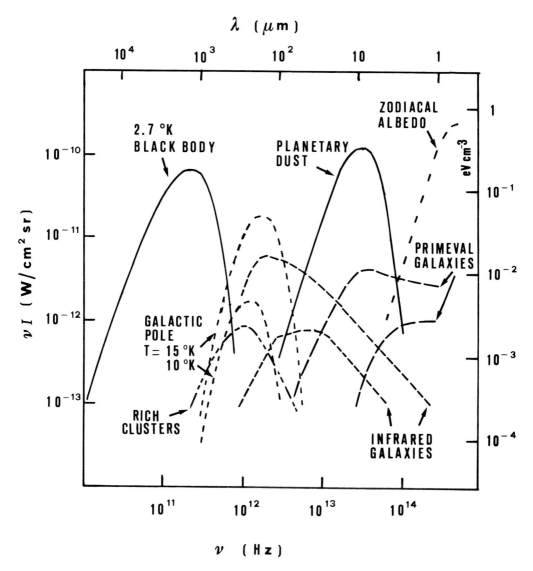

Figure 23. Anticipated extragalactic background (due to infrared galaxies, primeval galaxies, rich clusters) compared to
the "local background" due to planetary dust, zodiacal albedo
and galactic pole emission.
(Adapted from Mather, 1977).

4. THE COSMIC BACKGROUND RADIATION

4.1 The cosmic background in the Standard Hot-Big-Bang Theory

The origin of the cosmic background radiation can be understood by considering the early history of our Universe. The description of the initial stage (the big bang) is a challenging matter, but to the purposes of the present review it is enough to give a rough sketch starting at about 1 second after the big bang. For a detailed description of the cosmic evolution we refer to Sciama (1971), Peebles (1971) and Zeldovich and Novikov (1974).

At t $\sim$1 second the universe is filled by a mixture of electron-positron pairs, neutrinos, protons and neutrons, and photons in thermal equili-brium. The temperature is in the range T $\sim 10^{10}$ - 10^{12} °K and nuclei can-not form because of the large energies of particles. As the temperature drops below 10^{10} °K, neutrinos decouple for the electron-positron an-nihilation. Between 3 sec and 10^3 sec light elements are formed; however in the following we shall neglect the details of the chemical composition of cosmological matter in spite of its significance (Wagoner 1974). Thus for T < 10^{10} °K we assume a plasma of electrons, protons and photons. The photon spectrum is that of a blackbody having the same temperature as matter. This picture is appropriate untill hydrogen neutralizes (Peebles 1968; Zeldovich et al., 1968). The "recombination" starts at T = 4000 °K (at this temperature the Saha equation predicts 50% of neutral hydrogen). Since neutral atoms do not interact with radiation, except at few frequen-cies, the bulk of radiation is thermally decoupled at T $\sim 10^3$ °K, or t $\sim 10^5$ years. In an isotropic universe the cosmic expansion is suitably described by a cosmic scale factor R(t) (see for instance Weinberg, 1972): The volume of the universe V increases like R^3. Now assume that at some time the photon distribution is Planckian:

$$dN = \frac{8\pi V}{c^3} \frac{\nu^2 d\nu}{\exp(h\nu/kT)-1} \qquad (4.1)$$

As the universe expands and the volume becomes V' = V $(R'/R)^3$, the number of photons dN remains constant if the process is adiabatic (no overall heat exchange) but the frequency is shifted to $\nu' = \nu R/R'$ (this is the well known cosmological redshift). Therefore

$$dN' = \frac{8\pi V'}{c^3} \frac{\nu'^2 d\nu'}{\exp(h\nu'/kT')-1} \qquad (4.2)$$

where T' = T R/R'. Eq. (4.2) shows that the blackbody spectrum is conser-ved, but the temperature is adiabatically cooled. These considerations clearly apply to the epoch when photons propagate freely. But they apply also to the epoch of thermal equilibrium, because no large heat transfer is found to occur in the temperature range 10^9 °K $\gtrsim$ T $\gtrsim$ 4000 °K in a strictly homogeneous and isotropic universe. Therefore the temperature of the background radiation is simply related to the cosmic scale factor, or more precisely to the redshift $Z = R(t_o)/R(t) - 1$ which is a useful time parameter in cosmology:

$$T(Z) \simeq 3(1+Z) \ ^\circ K \qquad\qquad\qquad\qquad (4.3)$$

To be precise, the recombination epoch should be treated more carefully. Since the recombinig atoms emit radiation, the process is not strictly adiabatic. Let us assume for simplicity that each hydrogen atom emits one photon while going to the ground state. The recombination photons add to the background and produce a radiation excess around the resonant frequency. According to Zeldovich and Sunyaev (1969) the temperature at which recombination proceeds effectively is lower than 4000 $^\circ K$, and is given by $kT \simeq h\nu_\alpha/30$, with ν_α the frequency of a Lyman-α photon. The background photons with frequencies larger than ν_α are 200 times less numerous than the recombining atoms. Thus the spectral distortion is large. However, the wavelength of a Lyman-α photon being redshifted by the cosmological expansion, the present-time wavelength is about 200 μ. In this region the intensity of the background radiation should be much smaller than the infrared emission from interstellar dust. This simple analysis is inadequate, because the Lyman-α photons are easily readsorbed, and for an efficient recombination the two-quanta decays are required (Zeldovich et al. 1968; Peebles, 1968; Zeldovich and Sunyaev, 1969). As a result a broad infrared excess occurs below $\sim$ 200 μ, but the spectral distortion remains unimportant. This fact does not imply that any spectral distortion of the background is undetectable: In the real universe which is not strictly homogeneous and isotropic large amounts of energy may have been released to the background through several mechanisms, and some trace should remain in the spectrum of the radiation.

4.2. Distortions in the spectrum of the cosmic background radiation.

Several processes are known which can result in some injection of energy into the background radiation. At $Z > 1000$ the most realistic mechanism is the dissipation of the energy associated to local motions of the cosmological fluid, such as primeval turbulence or sound waves. As a general rule, significant distortions can be produced even if the energy transfer does not increase the total photon number appreciably; so there is a good chance for gaining insight into the physical processes which preceeded the recombination of hydrogen. In more recent epochs the radiation may have interacted with a very hot intergalactic gas; as a matter of facts, it is generally believed that the neutral hydrogen was reheated at some redshift $Z < 10^3$, reaching a temperature $> 10^4$ $^\circ K$ sufficient to ionize it again. If the density of this gas was large enough at reheating, it might have produced appreciable distortion. Other distortions are expected to arise due to emission from unresolved extragalactic sources and from our galaxy (diffuse non-thermal background and infrared emission from H II regions). Now we shall briefly review the main features of the spectral distortions.

I. Energy injection before the decoupling.
The spectral distortions due to an energy release at early times have
been discussed by several authors (Zeldovich and Sunyaev, 1969; Sunyaev
and Zeldovich, 1970a, Zeldovich et al., 1972; Chan and Jones, 1975a,b,c;
Danese and De Zotti, 1977). Following Sunyaev (1974) we divide the thermal
history of the primeval fireball before decoupling into a few epochs.
If the energy release occurs at a redshift $Z_h \gtrsim 10^8$, the spectral distortion cannot reach the present epoch, because the injected energy is
thermalized by free-free processes and Compton scattering and a Planckian
spectrum is quickly recovered. In order to identify the subsequent epochs
let us introduce the redshifts Z_a and Z_b defined by

$$Z_a = 4 \times 10^4 \ (\bar{\Omega h}^2)^{-\frac{1}{2}}$$

$$Z_b = 3 \times 10^5 \ (\bar{\Omega h}^2)^{-6/5}$$

(4.4)

where $\Omega = \rho/\rho_c$ is the ration of the present-time matter density to the
critical density $\rho_c \simeq 5 \times 10^{-30}$ gr cm^{-3} and $\bar{h} = H_o/(50$ Km sec^{-1}Mpc$^{-1})$
where H_o is the Hubble constant. In the epoch $10^8 \gtrsim Z \gtrsim Z_b$ the bremsstrahlung emission is able to estab lish a radiation field in equilibrium
with electrons at temperature T_e only at low frequencies, namely for

$$X_e \equiv \frac{h\nu}{kT_e} < X_{cb} \simeq 10^2 \ (\bar{\Omega h}^2)^{\frac{1}{2}} \ Z^{-3/4}$$

For $X_e > X_{cb}$ Compton scattering drives the photons toward higher frequencies and produces a Bose-Einstein distribution. The photon occupation number is given by

$$n = \{\exp(X_e + \mu) - 1\}^{-1}$$

(4.5)

where the chemical potential μ depends on frequency and time: for small
distortions

$$\mu(X_e, Z) = \mu_o(Z) \ \exp \ -\left[\frac{2X_{cb}}{X_e}\right]$$

(4.6)

If the heating redshift Z_h is larger than Z_b the function $\mu_o(Z)$ decreases
sharply as soon as the energy input stops. On the other hand, the damping
of the distortion is small if $Z_h < Z_b$. It can also be shown that larger
distortions ($\mu_o > X_{cb}$), for which (4.6) does not hold, are not subject
to a quick damping and can survive up to the present epoch. In all the
cases where the damping is negligible we have

$$\mu_o(Z=0) = 1.4 \int \frac{Q \ dt}{U}$$

(4.7)

where U is the radiation density energy, and Q is the power realesed
in the unit volume. If the energy release occurs in a short time we have

$$\mu_o = 1.4 \, \frac{\Delta U}{U} \qquad\qquad\qquad (4.8)$$

The distribution (4.5) exhibits a dip in the radio region; its size is simply related to the energy input according to (4.7) or (4.8) (see Figure 24).

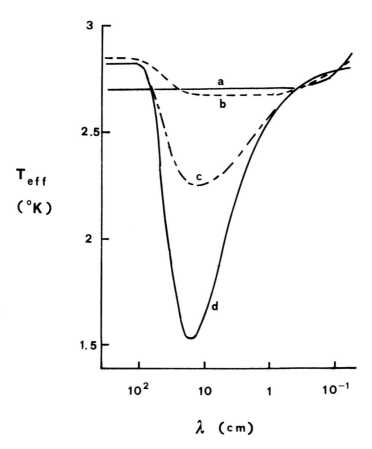

Figure 24. The effective temperature vs wavelength for some distorted spectra. The energy input is $\Delta\varepsilon/\varepsilon = 3.5\%$, and the heating redshift is (a) $Z_h = 1.4 \times 10^3$, (b) $Z_h = 4 \times 10^4$, (c) $Z_h = 1.2 \times 10^5$, (d) $5 \times 10^6 \geq Z_h \geq 2 \times 10^5$.
(Adapted from Sunyaev, 1974).

For $Z_h < Z_a$ the Compton scattering is no more able to establish a Bose-Einstein distribution. A superposition of blackbodies with different temperatures takes place, and the photon occupation number is given by

$$n = \int R(T,T',u) \, p(\nu/T') \, dT', \qquad\qquad (4.9)$$

where $p(\nu,T')$ is a Planckian spectrum at temperature T', and

$$R(T,T',u) = (4\pi u)^{-\frac{1}{2}} T^{-1} \exp\left[4u - \frac{1}{4u} (\ln T'/T + 5u)^2\right],$$

$$u = \int \frac{k(T_e-T)}{mc^2} d\tau \qquad\qquad (4.10)$$

In Eqs. (4.9) and (4.10) T is the unperturbed radiation temperature, m is the electron mass and τ is the optical depth for Thomson scattering (Zeldovich et al., 1972; Chan and Jones, 1975a). As a first approximation we have

$$\frac{T_{eff}(X) - T}{T} = u \ (X \coth X/2 - 4) \qquad\qquad (4.11)$$

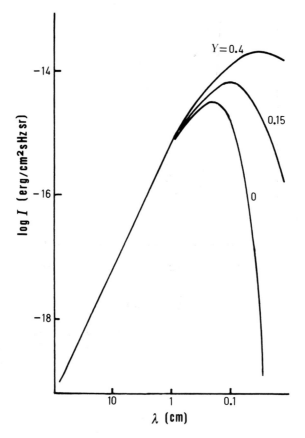

Figure 25. Comptonized spectra corresponding to a few values of y. (We assume that the distortion parameter u is equal to the Comptonization parameter y).

where $T_{eff}(X)$ is the effective temperature at the frequency $\nu = XkT/h$.
The distortion parameter u is related to the energy input by

$$u = \frac{1}{4} \; \frac{\Delta U}{U} \tag{4.12}$$

The Comptonized spectrum (4.9) typically gives an infrared excess (Fig.
25). A superposition spectrum arises also from dissipation of primeval
turbulence although u is no more given by (4.10) (Chan and Jones, 1975a,
b).

II. Energy injections after decoupling.
We have already remarked that a secondary ionization of the cosmological
plasma is believed to occur at low redshifts, although the heating mechanism is not clear. The bremsstrahlung of the hot gas should be observable at low frequencies, whereas an infrared excess should be produced by
Compton scattering in agreement with Eqs. (4.9) - (4.11). In this case
$T_e \gg T$, and the distortion parameter coincides with the well known
Comptonization parameter y:

$$u = y \equiv \int \frac{kT_e}{mc^2} \; d\tau \tag{4.13}$$

An effect analogous to this one may take place in intracluster gases
(Sunyaev and Zeldovich, 1970b). Many clusters are sources of X rays, due
perhaps to bremsstrahlung of free electrons in very hot intergalactic
gas, $T_e \sim 10^8$ °K. Compton scattering on this gas produces a superposition
spectrum in the radiation traversing a rich cluster. The effect is particularly interesting because at relatively low frequencies (X<3.8) a
decrease of the microwave temperature is anticipated in direction of the
cluster, while an increase of the temperature at high frequencies may
be simulated by some different emission. However, measurements around
X = 3.8 are especially desiderable (Fabbri et al., 1978) to exploit the
spectral shape of the signal. A fuller discussion is given in section 6.

III. Extragalactic sources and galactic emission.
Another source of distortions of the spectrum is represented by radiation
emitted by unresolved sources. Their contribution to the background is
not easy to evaluate, and several authors suggested that the entire
microwave background could be generated in this way (Sciama 1966, Gold
and Pacini, 1968; Parijskij, 1968; Wolfe and Burbidge, 1969; Rowan-Robinson, 1974; Narlikar et al., 1975). As a general rule it is rather easy
to mimic the blackbody spectrum in the Rayleigh-Jeans region by means of
suitable populations of discrete sources, but large spectral distortions
arise in the submillimetric region, unless one introduces very artificial
hypothesis on the source luminosities. The work of Ade et al. (1976) supports the view that discrete sources cannot have produced the background
(moreover, see the section devoted to fine-scale anisotropy). Thermali-
zation of stellar light by dust grains has been proposed by some authors
(Layzer 1968; Layzer and Hively, 1973; Rees 1978), but since there is no
evidence for an extragalactic medium with suitable properties we adopt

the view that the background is really cosmological in accordance with
the standard big-bang theory. However discrete sources can induce appreciable distortion. Their contribution to the background depends on
the intrinsic spectrum and on evolutive effects, but in any case it must
be important at very short wavelengths. The radiance of radio and infrared sources is no more than 0.2% of the background at 1 mm (Ade et al.
1976), but should be widely dominant around 0.2 mm.
Galactic emission too, could dominate the short wavelength side of the
spectrum, as already remarked. The basic source of far infrared radiation
is dust heated by starlight, whose maximum emission is expected around
100-300 μ. A map of the galaxy at 1 mm is required, before attributing
any infrared excess to distortions of cosmological origin.

4.3. Review of the experimental results.

After the review of Blair (1974) new and more precise observations have
extended our view deeply into the infrared. Perhaps the most important
innovation in measurements of the cosmic background is the use of infrared
high-resolution spectrometers. However, we cannot assert that the use
of balloon-borne spectroradiometers has completely solved two problems
encountered by earlier investigators, namely the absolute calibration
and the correction for atmospheric effects. The first published measurements of the Queen Mary College and Berkeley groups showed discrepancies
in the contribution of the residual atmosphere, probably depending on
different elaborations of the data (Robson and Clegg, 1977). More recent
spectrometric data have not allowed to give a definite answer to the
problem of the atmosphere so that there is still the possibility of systematic errors larger than the quoted errors. The data obtained in the
1977 flight of Berkeley group are consistent at the 80% confidence level
with a two parameter curve with the shape of a 2.79 °K blackbody, but
an emissivity of 1.27 (Woody and Richards, 1978). Tables IV a,b,c, report
the monochromatic, broad-band and spectrometric measurements, respectively. Data in numerical form for the measurements of Robson et al.(1974)
have been provided by Clegg (1978); for the measurements of Woody et al.
(1975) they are quoted by Danese and De Zotti (1978). The results of
Puzanov et al. (1975) have been lowered according to Kisyakov et al.(1971).
The CN data from Zeta Oph have been revised by Danese and De Zotti (1978).
The data obtained by Woody and Richards in 1977 are not available in
numerical form and are reported in Fig.26. In spite of the present uncertainties, the data collected in tables IV allow us to state:

(a) The measurements in the radio region are consistent with a 2.7 °K
blackbody. More precisely the analysis of sixteen radio data (Robson and
Clegg, 1977) gives

$$<T> = (2.73 \pm 0.08) \ °K \quad \text{with} \ \chi^2 = 9$$

The expected value of χ^2 would be 15±5. The low value found is due to the
fact that the declared errors are not purely statistical, but take into
account the calibration errors.

(b) There is some evidence for a higher brightness temperature in the millimetric region. Probably T_b = 2.9-3.1 °K.

The interpretation of data is not clear. Actually the Comptonized spectrum considered in section 4.2 exhibits an infrared excess and may fit the available data. Field and Perrenod (1977) assumed a spectrum like Eq. (4.9) and calculated the best values for the unperturbed temperature T_r and the distortion parameter u. They found T_r = 2.95 °K and u of the order of 10^{-2}. However it is better to limit ourselves to give only upper limits to the cosmological distortion: As a matter of facts the observed infrared spectrum may be contaminated by local emission from our galaxy, so that attempts to fit the data may be untimely. Thus we shall set

$$u \lesssim 3 \times 10^{-2}$$

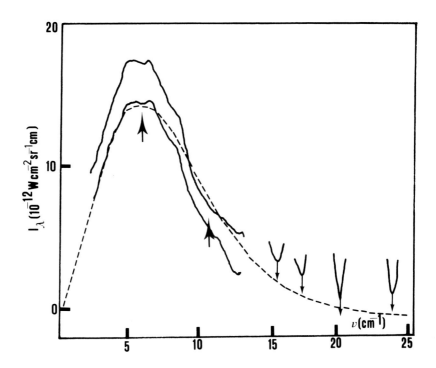

Figure 26. Measurements of the emission spectrum of the night sky in the frequency region from 1.7 to 40 cm^{-1} as obtained by Woody and Richards (1977). Note that the measured intensity results to be about 10% higher than that of the 2.96 °K Blackbody at 6 cm^{-1} and about 20% smaller at 11 cm^{-1}.

Table IV a, Monochromatic Measurements

Wavelength (cm)	Brightness Temperature (°K)		Reference	
73.5	3.7	± 1.2	Howell et al.	1967
49.2	3.7	1.2	Howell et al.	1967
21.0	3.2	1.0	Penzias and Wilson	1967
20.7	2.8	0.6	Howell et al.	1966
7.35	3.5	1.0	Penzias and Wilson	1965
3.2	3.0	0.5	Roll and Wilkinson	1967
3.3	2.69	+ 0.16 − 0.21	Stokes et al.	1967
1.58	2.78	+ 0.12 − 0.17	Stokes et al.	1967
1.50	2.0	0.8	Welch et al.	1967
0.924	3.16	0.26	Ewing et al.	1967
0.856	2.56	+ 0.17 − 0.22	Wilkinson	1967
0.82	2.4	0.7	Puzanov et al.	1967
0.358	2.4	0.7	Kislyakov et al.	1971
0.33	2.46	+ 0.40 − 0.44	Boynton et al.	1968
0.33	2.61	0.25	Millea et al.	1971
0.264	2.3	?	Mc Kellar	1941
0.264	3.22	0.15	Field and Hitchcock	1966
0.264	2.63	0.24	Clauser and Thaddeus	1972
0.264	2.82	0.10	Bortolot et al.	1972
0.264	2.81	0.19	Thaddeus	1972
0.264	2.59	0.15	Hegyi et al.	1972
0.132	2.8	0.4	Hegyi et al.	1974

NOTE: The dotted line separates radio data from optical data derived from measurements of molecular absorption.

Table IV b, Broad Band Measurements

Wavelength (mm)	Brightness Temperature (°K)		References	
0.08 − 6	3.1	+ 0.5 − 2.0	Blair et al.	1971
1.8 − 10	2.7	+ 0.4 − 0.2	Muehlner and Weiss	1973
1.0 − 1.4	2.4	+ 0.3 − 2.4	Dall'Oglio et al.	1974

Table IV c, Spectroscopic measurements

Wavelength (mm)	Brightness Temperature (°K)		Reference	
5.00	3.67	+ 0.79 / − 0.80	Robson et al.	1974
3.33	3.52	0.37		
2.50	3.11	+ 0.24 / − 0.26		
2.00	2.88	+ 0.19 / − 0.20		
1.67	2.86	+ 0.17 / − 0.18		
1.43	2.76	+ 0.16 / − 0.18		
1.25	2.90	+ 0.15 / − 0.16		
1.11	3.10	+ 0.13 / − 0.16		
1.00	3.03	+ 0.15 / − 0.17		
0.909	2.82	+ 0.19 / − 0.17		
0.833	2.78	+ 0.23 / − 0.37		
0.769	2.75	+ 0.28 / − 0.62		
2.50	2.64	+ 0.37 / − 0.38	Woody et al.	1975
2.04	3.43	+ 0.20 / − o.19		
1.72	3.36	+ 0.13 / − 0.13		
1.43	3.21	+ 0.10 / − 0.09		
1.19	2.99	+ 0.08 / − 0.07		
0.990	2.85	+ 0.06 / − 0.07		
0.826	3.01	0.06		

(continued)

Table IV c (continued)

Wavelength (mm)	Brightness Temperature (°K)	Reference	
0.690	3.2 ± 0.08	Woody et al.	1975
0.575	3.23 +0.11 −0.15		

This limit is mainly based on infrared data. Limits for $\mu_\circ$ (see Eq. 4.7) may be given using the radio data. At present we can conclude that data are consistent with an undistorted blackbody within a few percent; the basis of the hot-big-bang theory is confirmed, although the details are still to be settled. The next goal should be the measurement of the spectrum with an accuracy of the order of 10^{-3}. This goal cannot be achieved by means of balloon-borne instruments, due to the residual atmospheric emission. The future of the background spectral observations is bound to space veichles. Two main experiments have been proposed: the CBE (Cosmic Background Explorer) is the European proposal for Spacelab (Queen Mary College, Service d'Aeronomie of CNRS, IROE-CNR, Estec). COBE (Cosmic Background Experiment) is the American proposal for a devoted satellite (JPL, Goddard, Berkeley, Princeton and MIT). The CBE instrument consists of a Michelson interferometer , similar to that employed in balloon experiments, having a resolution of 0.2 cm^{-1} and an expected sensitivity of $\Delta T/T \sim 10^{-3}$. The COBE is a more ambitious program, intended to explore the entire infrared sky background between 10μ and a few centimeters. It is difficult to anticipate the information which is to come out from these experiments. However a possible "scenario" which has to be investigated is represented in Figure 23. It is evident that many sources are able to contaminate the infrared spectrum of the cosmic background.

5. THE LARGE SCALE ANISOTROPY

5.1. Radiation patterns in non-Friedmannian universes.

The large scale anisotropy of the background radiation might be one of the properties most deeply related to the geometry of the cosmological expansion and the large-scale structure of the universe. More precisely we should distinguish between a simple anisotropy due to the local motion of the earth and the other contributions associated to the global properties of the universe.

A dipole pattern is known to arise from the motion of the observer with respect to the frame at rest with radiation: the temperature of a blackbody appears to be angle-dependent to the moving observer (although the spectrum remains Planckian) and is given by

$$T(\theta) = T_{rest}(1 - \frac{v}{c} \cos \theta)$$ (5.1)

where θ is the angle between the observation direction and the earth velocity. Eq. (5.1) describes the well known non-relativistic Doppler effect. The existence of a local velocity is not unexpected; the expected anisotropies due to various sources of motion are listed in Table V.

Table V, Motions of the earthly frame with respect to the radiation frame.

Source of motion	Expected velocity (Km sec^{-1})	Anisotropy (m°K)
Earth rotation	0.46	10^{-2}
Orbit around Sun	29.8	0.53
Solar system around Galaxy	270 $\pm$ 40	4.9 $\pm$ 0.7
Galaxy around Local Group	80 $\pm$ 20	1.4 $\pm$ 0.4

We shall discuss now cosmological anisotropies, which do not arise in an isotropic universe, but are predicted by more complex cosmological models. General relativity (which is the basis for contemporary cosmology) allows a large variety of cosmological models if one relaxes the requirement of isotropy of the Hubble expansion. It is known that the universe is quasi-isotropic today, but the situation may have been quite different at other times, since the standard isotropic models are unstable (Lifshitz et al. 1971; Collins and Hawking, 1973a; Caderni and Fabbri, 1978). This reason (and others) justifies the effort of cosmologists in the investigation of anisotropic models. The simplest of such models may be described by introducing three scale factors $R_i(t)$ corresponding to the x, y, z, axes, and three directional Hubble "constants" (which depend on time):

$$H_i = R_i^{-1} \frac{d}{dt} R_i, \quad i = 1, 2, 3$$ (5.2)

Heckmann and Schücking (1962) give the general solution for anisotropic flat-space models. They find that at early times the universe evolves independently of the matter and radiation content and is highly anisotropic, since it expands in two directions and contracts in the third

one. Later when the radiation is able to affect the cosmic evolution,
Hubble expansion becomes quasi-isotropic; the anisotropy decays according
to the law

$$\frac{H_i - H}{H} \propto (\frac{t_f}{t})^{\frac{1}{2}} \tag{5.3}$$

where H is the average Hubble parameter, and t_f is the time at which
the isotropization starts. Although the anisotropy of the expansion rate
becomes small when $t \gg t_f$ in such models, a trace is left in the angular
distribution of the cosmic background (Thorne, 1967; Jacobs, 1968). Let
$T(\vec{\alpha})$ be the temperature of the radiation in the direction specified by
the unit vector $\vec{\alpha} = (\alpha_1, \alpha_2, \alpha_3)$ and T the average temperature; then
we have

$$\Delta T/T \equiv \frac{T(\vec{\alpha}) - T}{T} \simeq \Sigma \alpha_i^2 \frac{H_i - H}{H} \tag{5.4}$$

where the Hubble parameters are evaluated at the last-scattering time.
The cosmological radiation conserves the memory of a time when the universe may have been more anisotropic than today, according to (5.3).
Eq. (5.4) describes a quadrupole-type anisotropy. Similar patterns are
found in many models but not in all of them. In a class of models (type
V and type-VII spaces) a quadrupole or quasi-quadrupole pattern should
arise in the quasi-isotropic stage of the cosmic expansion (Novikov 1974;
Matzner, 1971a,b; Doroskevich et al., 1975), but it is later replaced
by a different pattern. When this regime takes place, the anisotropy
is squeezed in a small region of the celestial sphere (Novikov, 1968;
Grishchuk et al., 1969; Hawking,1969; Collins and Hawking, 1973b, Matzner
1969a; Mac Callum and Ellis, 1970; Doroskevich et al., 1975). The following equation has been proposed:

$$\Delta T/T \simeq K^2 \frac{\sin^2\theta \, \tanh \, \ln(t/t_d)}{1 + \cos\theta \, \tanh \, \ln(t/t_d)} \tag{5.5}$$

where $K^2 \ll 1$ is a parameter of the world model. Eq. (5.5) describes
a "ring-of-fire pattern". The angular size of the ring of fire is determined by the matter content of the universe, $\Delta\theta \sim 4\Omega$ and is very small
in a low-density universe. Definite predictions about the amount of
anisotropy are difficult. It was hoped that accurate models, where dis-
sipative processes in the early history of the universe are considered,
could give limits on the amplitude of the large-scale anisotropy. If
the universe was highly anisotropic at the beginning, irreversible processes played an important role in damping the expansion anisotropy. Two
damping processes have been investigated, namely the particle production
(Starobinsky and Zeldovich, 1972; Lukash and Starobinsky, 1974; Lukash
et al., 1976) and the neutrino viscosity (Misner,1967, 1968; Matzner
1969; Matzner and Misner, 1972; Caderni and Fabbri, 1978a, b ,c). It has
been shown that neutrino viscosity suffices to damp the primordial anisotropies at $t \sim 1$ sec, but the subsequent evolution may give rise to a

"new" anisotropy if the space is not flat(Collins and Hawking, 1973b, Caderni and Fabbri, 1978b). Equations (5.4) and (5.5) have been found for world models where the matter is at rest in the homogeneity frame. More complications may arise if we allow for a peculiar velocity field u of the cosmological matter as a whole. This has nothing to do with the random motions of relatively small aggregates of matter, such as those considered in Table V; now we consider the overall motion of matter with respect to the homogeneity frame. The existence of a non-zero velocity is not ruled out by the available experimental data (Grishchuck et al., 1969; Grishchuck et al, 1972; Novikov, 1970; Collins and Hawking 1973b). It is also interesting to note that many models predict a vorticity of the cosmological matter of order ω ᷈ (u/c) H. This means that distant masses rotate with respect to any local inertial frame, contrary to Mach's principle. In such models dipole and quadrupole anisotropies may be correlated to each other, as well as to other multipoles.

5.2 Experimental results

After several experiments which produced only upper limits, we have today two positive measurements of the dipole anisotropy of the background. The solar system appears to be in motion with a velocity of 330 ± 50 Km/sec in the direction identified by 12 h in right ascension and 10° ± 30° in declination (see Fig. 27).

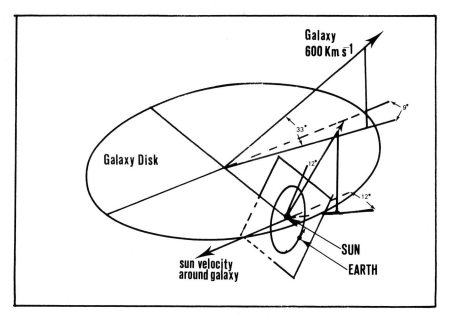

Figure 27. Motions of the earth, Sun, galaxy with respect to the cosmic background radiation. (After Muller, Scientifc American, 1978)

The hottest spot in the sky lies in the costellation of Leo while the
coldest spot lies 180 degrees away in Aquarius. If the temperature
difference between hottest and coldest regions is plotted again distance,
the result is a cosine curve, as shown in Figure 28.

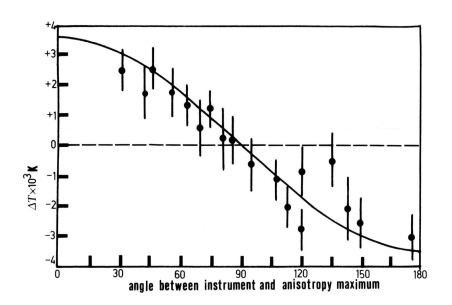

Figure 28. The large scale anisotropy of cosmic background
as measured by Smoot et al. (1977).
Adapted from Smoot et al. (1977).

All the experimental results are collected in Table VI. The first point
to be underlined is that the measured velocity of the sun is different
from that observed by Visvanathan and Sandage (1977) and Rubin et al.
(1976). The obvious interpretation is that the galaxies used as a reference by Sandage and Rubin are not at rest with respect to background
radiation. The unexpected high velocity of our galaxy represents a
challenge to cosmological theories. Since it is known that the non-cosmological motion of our local group of galaxies relative to the Virgo
cluster is small, the entire Virgo cluster must have a high velocity:
A vast volume of space, tens of millions of light-years in radius, is
moving with a velocity of about 600 Km/sec with respect to the matter
from which the cosmic background radiation originated. The second point
is that the search for quadrupole effects has given a null result within
10^{-3} °K. This limit gives upper limits on the anisotropy of the cosmological expansion and on the rotation of the distant matter (Collins and
Hawking, 1973b, Smoot et al. 1977).

Table VI, Measurements of the large-scale anisotropy

Observer	Declination of the scan	Declination of maximum	R.A. of maximum	ΔT (m°K)
Partridge et al. (1967)				
Dipole	−8°			2.2 ∓ 1.8
Quadrupole	−8°			2.7 + 1.8
Boughn et al. (1971)				
Dipole	0°			7.5 ∓ 11
Quadrupole	0°			5.5 + 6.6
Henry (1971)				
Dipole		≈ −30°	≈10 h	3.2 ∓ 0.8
Conklin (1972)				
Dipole	32°		≈ 11 h	2.3 ∓ 0.9
Quadrupole	32°		≈ 6h, 18h	1.3 + 0.9
Muehlner et al. (1976)				
Dipole				≲ 3.2
Corey et al. (1976)				
Dipole	−25° ± 20°		13 ∓ 2h	2.4 ∓ 0.6
Smoot et al.(1977)				
Dipole		6° ± 10°	11.0 ∓ 0.6	3.5 ∓ 0.5
Quadrupole				≲ 1

Future experiments should explore all the sky with better angular resolution and sensitivity. Also, the observations should be done at various
frequencies, although the emission from our galaxy is expected to be not
negligible both in the radio and in the infrared region.

6. THE FINE-SCALE ANISOTROPY

6.1. Perturbations of the Friedmann universe and anisotropies of the background radiation

In the previous sections we discussed deviations from isotropy which might show up in our universe. Here we consider the fact that the universe apparently is not structureless because condensed systems exist at various scales. In the context of the hot-big-bang theory galaxies, clusters and superclusters are thought to be originated by primeval perturbations of the Friedmann background, namely sound waves (Bonnor 1957, Doroskevich and Zeldovich 1963, Peebles 1965, Silk 1968, Zeldovich and Novikov, 1970; Sunyaev and Zeldovich 1972; Field, 1972), or alternatively whirl motions (von Weizsacker, 1951; Gamow, 1952; Nariai, 1956; Ozernoi and Chernin, 1968; Ozernoi and Chibisov, 1971). These perturbations are expected to induce small-scale anisotropies in the cosmic background, as do other kinds of perturbations, such as isothermal fluctuations (where only the matter density oscillates in a constant background of radiation), shear motions, and gravitational waves. More precisely, the anisotropy that may be observed today should be the one impressed at the last scattering of the background photons. Assuming that the last scattering occured at $z = 10^3$ (the recombination of hydrogen), Longair and Sunyaev (1969) and Sunyaev and Zeldovich (1970) find temperature fluctuations of the radiation originated by Doppler shift

$$\Delta T/T \simeq 10^{-5} \left(\frac{\Omega^{\frac{1}{2}}}{10^{15}} \frac{M}{M_\Theta} \right)^{1/3} (1+Z_0) \qquad (6.1)$$

Here M is the mass contained in a sphere whose radius is equal to half wavelength of the density perturbation and M_Θ is the solar mass; Z_0 is the redshift at which the matter density contrast becomes unity. This equation should hold for masses of the order of those of rich clusters; M is related to the angular scale of the anisotropies θ^* by

$$\theta^* \simeq 10' \left(\frac{\Omega^2}{10^{15}} \frac{M}{M_\Theta} \right)^{1/3} \qquad (6.2)$$

At small scales ($\theta^* \ll 10'$) we have rather small fluctuations given by

$$\Delta T/T \simeq 10^{-5} \left(\frac{\Omega^{\frac{1}{2}}}{10^{15}} \frac{M}{M_\Theta} \right)^{5/6} (1+Z_0) \qquad (6.3)$$

At large scales ($\theta^* > 2°\Omega^{\frac{1}{2}}$) one should take gravitational-potential effects into account (see below). Since the peak mass M and the redshift Z_0 are arbitrary parameters, the above formulas do not assign a definite spectrum to the angular anisotropies. Peebles and Yu (1970) assume that the power of each mode of the acoustic waves is proportional to the wavenumber and find the characteristic scale $\theta^* \simeq 7'$ (see Fig.29).
The anisotropies described above, which are induced at the recombination

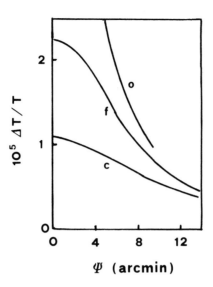

Figure 29. The temperature anisotropy vs. the beamwidth of
the detector according to Peebles and Yu (1970): (o) open
model, $\Omega = 0.03$; (f) flat model; (c) closed model, $\Omega = 5$.

of hydrogen, survive up to the present epoch if the optical depth of the
reheated gas is small (see sect. 4.2). More generally, one should multi-
ply the above expression for $\Delta T/T$ by a damping factor $\exp(-\tau)$. We note
also that, if τ is very large, some new anisotropy may be induced by the
growing perturbations of the reheated matter. We recall also that Silk
(1974) has suggested a different choice for the spectrum of the primor-
dial fluctuations, setting the maximum fluctuations at $\theta^{*} \simeq \Omega^{\frac{1}{2}} 2'$ (see
Fig. 30).

Now let us consider primeval vortex motions. According to the whirl theory
of galaxy formation, the cosmological fluid exhibits subsonic turbulence
prior to the recombination of hydrogen. At the decoupling the matter motions become supersonic and generate density fluctuations, which eventually condense into bound systems. Temperature anisotropies are induced into
the background radiation at $Z_d \sim 10^3$, owing to Thomson scattering. If the
optical depth of reheating is $\tau < 1$ we have at the present epoch

$$\Delta T/T \simeq \frac{\omega_o L_o}{c} (1+Z_d) \tag{6.4}$$

where L_o is the characteristic scale of the whirls and ω_o their vorticity
at $Z = 0$ (Anile et al., 1976; Anile and Motta, 1976a). The length L_o
determines the angular scale of the temperature fluctuations:

$$\theta^{*} \simeq 1' \frac{\Omega L_o}{1 \text{ Mpc}} \tag{6.5}$$

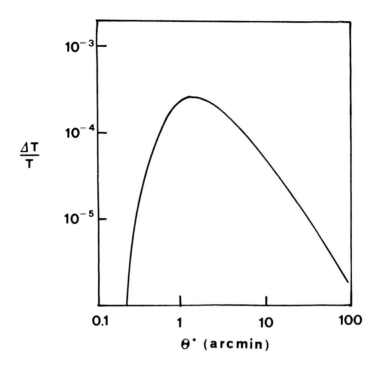

Figure 30. The primordial temperature anisotropy vs the angular scale according to Silk (1974).

If the radiation was last scattered in recent epochs ($\tau \gg 1$) we have instead

$$\Delta T/T \simeq \Omega^{1/4} \left(\frac{H_o L_o}{c} \right)^{3/2} \frac{\omega_o}{H_o} Z_d^{5/4} \qquad (6.6)$$

where it is usually assumed $Z_d \simeq 10 \, \Omega^{-1/3}$.

One should also consider the effect of gravitational fields on the red-shift of the radiation. Such fields may be gravitational waves (which propagate with the speed of light c), or may be associated to the sound waves; in any case the photon frequency along a geodesic path is affected by the gravitational-potential perturbation. For sound waves the perturbation can be expressed in terms of the density contrast, and the temperature fluctuations impressed into the background radiation at $Z \simeq 10^3$ is

$$\Delta T/T \simeq (H_o L_o/c)^2 \left(\frac{\Delta \rho}{\rho} \right)_o \qquad (6.7)$$

where $(\Delta \rho/\rho)_o$ is the density contrast at the present epoch (Sachs and Wolfe, 1967; Anile and Motta, 1976b). For density waves the gravitational

effect is important at large scales ($\theta > 2° \Omega^{\frac{1}{2}}$). For gravitational waves
Dautcourt (1969) found

$$\Delta T / T \simeq 6\times10^{-4} \quad \frac{L_0}{1\ \text{Mpc}} \left(\frac{\rho_g(0)}{\rho_{em}} \right)^{\frac{1}{2}} \tag{6.8}$$

where ρ_{em} is the energy density of the cosmological background radiation
($\rho_{em} \simeq 5^m x\ 10^{-34}$ g cm^{-3}) and $\rho_g(0)$ is the energy density of gravitational
waves with wavelengths $\lambda \sim L_0$ at the present epoch (Sachs and Wolfe,
1967).Anile and Motta (1978) give an expression valid whatever the last-
scattering redshift.
Here we have considered only the anisotropies of cosmological origin.
But unresolved discrete sources and galactic sources may introduce appreciable anisotropies in the spectral regions where the blackbody is
not largely dominant with respect to the other emissions. According to
some models the entire background would be emitted by far-away sources;
however the fine-scale anisotropies are too large in these models unless
one assumes strong evolutionary effects (Smith and Partridge,1970).
Generally speaking, discrete sources can contribute to the anisotropy
more effectively than to the intensity of the bakcground. It is also
interesting to note that their contribution depends on the frequency,
whereas the anisotropies of cosmological origin are all frequency-indepen-
dent. The most interesting non-cosmological anisotropy is the Sunyaev-
Zeldovich effect. Although it is based on a spectral distortion (Comptonization effect, considered in section 4.2) it is detected as a small-
scale anisotropy located in the direction of a rich cluster of galaxies.
Suppose that an isotropometer measures the difference between the radiation intensities in two adiacent sky spots, one of which covered by the
hot gas cloud. Then the detected monochromatic signal is given by

$$I_\lambda = 2y \frac{(kT_0)^5}{h^4 c^3} \frac{x^6 e^x}{(e^x-1)^2} (x\ \coth(x/2) - 4) \tag{6.9}$$

where T_0 is the radiation temperature at zero redshift and $x = hc/(kT_0\lambda)$.
The signal vanishes for $x= 3.83$, that is at $\lambda_0 = 1.4$ mm, and is negative
at longer wavelengths (see Fig. 31). Because of the peculiar spectral
shape this effect should be easily separated by other sources of anisotropy. Intensity and spectral shape do not depend on the cluster redshift.

6.2. Experimental results.

Observations of the fine-scale anisotropies have been carried out at
various wavelenghts by means of telescopes of various diameters. In the
radio region the main limitation is represented by the noise temperature
of the radioreceivers. Since in the best case the noise equivalent temperature is of the order of 10^{-2} °K Hz$^{-\frac{1}{2}}$, we would need a time of the order
of 10^6 seconds in order to reach a sensitivity of 10^{-5} °K. Moreover, the
angular resolution is limited by the diffraction effects at long wavelengths.

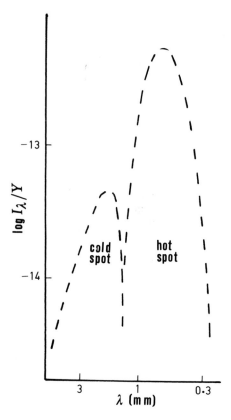

Fig. 31. The Sunyaev-Zeldovich effect as function of wavelength
(After Boynton and Melchiorri, 1977).

Perhaps the ultimate performances of radiosystems has been reached by
Parijskij (1977), who gave the upper limit of 8×10^{-5} (at two standard
deviations) for the rms fluctuations at the angular scale $\theta^{x} \sim 5'$.

On the other hand infrared observations are facilitated by the better
sensitivity of broad-band infrared detectors: the most sensitive systems
are operating at a level of 5×10^{-4} °K $Hz^{-\frac{1}{2}}$. Up to now the most stringent
limit on fine-scale anisotropy in the IR band has been given by Caderni
et al. (1977). We note that ground-based observations have been attempted
only recently, because of the anticipated difficulties arising from
the atmospheric noise. Very dry sites must be chosen (high mountain or
arctic station) as well as cold seasons; it has been proved that the
atmospheric noise is negligible under these conditions (Caderni et al.
1977). In order to compare experiment and theory one should estimate
for instance the minimum perturbations that are required in order to
produce the presently observed structure in the universe. This has been

done by Boynton (1977) using the sound-wave theory of galaxy formation. In Figure 32 we have reported the temperature fluctuations $\Delta T/T$ predicted by this theory versus the angular scale θ^* for several values of Ω and $Z_0 = 1/\Omega$ (see also eqs. 6.1-6.3). Many results given in the literature seem to neglect important corrections, so that the resulting upper limits may be overrated. Partridge (1978) proposed revised values that we quote in Table VII.

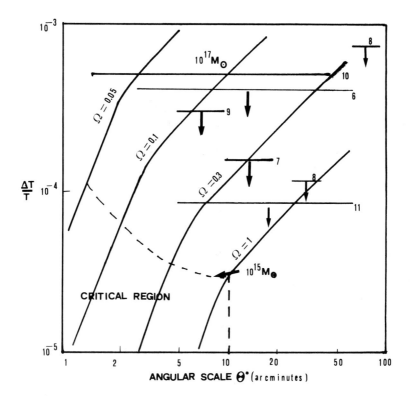

Figure 32. Fine-scale temperature fluctuations as a function of angular scale. Bold curves represent the calculations of Sunyaev and Zeldovich (1970). Horizontal bars indicate current observational limits identified in Table VII. The critical region is discussed in the text.
(Adapted from Boynton and Melchiorri, 1977)

It is clear from Table VII and fig. 32 that the experimental upper limits exclude the existence of bound systems with masses greater than 10^{16} $M_\odot$ for $\Omega > 0.03$. However the critical region indicated in the Figure has

not been violated by the experiment. For analogous calculations concerning
vortex motions see Kurskov and Ozernoi (1978). Infrared observations
appear to be very promising, especially if performed outside the atmosphere , for instance on board of large Spacelab telescopes, like ST
or LIRTS: a sensitivity of few 10^{-5}°K with an angular resolution of 2'
is anticipated in the case of LIRTS observations.

It is worth to discuss separately the observations of the Sunyaev-Zeldovich
effect. In fact, this is the only small-scale anisotropy of the background radiation which seems to have been detected up to now. Radio observations give ambigous results (Parijsky, 1973; Gull and Northover,
1976; Lake and Partridge, 1977).
However a recent experiment of Birkinshaw et al. (1978) seems to give
a definite result for A 2218. This work reports a maximum negative
brightness $\Delta T = -(1.5 \pm 0.2)$ mK and a thermal core radius $\theta = 0.7$ arcmin.
The correlation between radiation dip and the distribution of galaxies
in the cluster supports the view that hot gas has been really observed.
By 1978 all the measurements have been performed in the centimetric
region (3 cm or 0.9 cm). Such measurements require long integration times
(about 200 hours in the experiment of Birkinshaw et al.), so that it is
problematic to reach sensitivities of the order of 0.1 mK. A substantial
improvement might be obtained by using infrared bolometers, both for the
better sensitivity and for higher intensity of the signal in the millimetric region. Infrared detectors may allow the discovery of the S-Z
effect in a large number of objects. This would have several applications
of cosmological interest. First, the detection of this effect in X-ray
sources proves that its emission is due to thermal bremsstrahlung and
the problem of the existence of hot intracluster gases may be clarified
by such an experiment; it would be also interesting to investigate unidentified X-ray sources: Part of them may be young clusters (Cavaliere
et al. 1976). Moreover, one can get information on the cosmic background
itself. As matter of fact, the zero signal wavelength depends on the
unperturbed temperature of the radiation $\lambda_0 = ch/(3.83 \, kT)$. Since the
signal intensity changes very sharply around the zero, it should be not
too difficult to measure λ_0 with a good accuracy using spectrometric
techniques in the millimetric region. Then one is able to calculate T.
This infrared measurement does not present the difficulties of the
absolute calibration which one encounters in the usual experiments (Fabbri,
Melchiorri and Natale, 1978; Bussoletti,Fabbri and Melchiorri 1979).

7. THE LINEAR POLARIZATION OF THE COSMIC BACKGROUND

7.1 Polarization of cosmological origin.
In the hot big-bang picture the cosmological background is unpolarized.
If a temporary mechanism induces some polarization into the background
radiation before the last scattering of photons, the thermalizing processes destroy it in a time comparable to the photon collision time.
After the decoupling the radiation propagates freely and its polarization state is conserved. Thus the polarization degree remains zero up
to the present epoch. This picture is inadequate if a lasting mechanism

Table VII. Observations of fine-scale anisotropy

Observers	Wavelength cm	Angular scale	$\Delta T/T$ x 10^3	corrected $\Delta T/T$ x 10^3
1. Conklin and Bracewell (1967)	2.8	10'	1.8	1.8
2. Penzias et al.(1969)	0.35	2'	9	20
3. Boynton and Partridge (1973)	0.35	1.5	3.7	2
4. Carpenter et al (1973)	3.6	2' – 1°	0.7	0.7
5a. Parijskij (1973a)	2.8	3' – 1°	0.03	0.4
5b. Parijskij (1973b)	4.0	12' – 40'	0.05	0.16
6. Stankevich (1974)	11.1	8' – 20'	0.15	0.3
7. Caderni,Fabbri,Melchiorri Natale (1977)	0.13	20'	0.12	0.12
8. Partridge (1978)	0.9	3'	0.15	0.15
9. Parijskij (1977)	4.0	5' – 3°	0.08	????
10. Pigg (1978)	2.0	1'	0.4	???
11. Fabbri,Melchiorri,Natale (1978)	0.08	2°	0.3	0.3

Note: all the listed values are upper limits: they have been corrected by Partridge (1978).

perturbs the isotropy of the blackbody before the last scattering, provided
some special direction of space may be selected as the polarization direction. This situation can occur in the real universe, which is not - as
already discussed- strictly homogenous and isotropic. It is well known
that when a beam of natural light undergoes Thomson scattering, the
radiation becomes partially polarized in the direction ortogonal to the
scattering plane. The overal effect does not vanish after averaging over
the incidence directions of the beams, if the beam intensity depends on
the direction. This is just what happens for the background radiation
in anisotropic cosmologies. Rees (1968) studied the mechanism here described by considering axisymmetric Bianchi type-I spaces. In such world
models two directional Hubble constants, say H_1 and H_2, can be introduced:
the former describes the cosmological expansion along the polar axis of
a suitable $\{\theta,\varphi\}$ frame, and the latter the expansion along the polar
axis of a suitable plane. One can also introduce two background temperatures, T_1 and T_2, corresponding to radiation polarized along the polar
axis and in the equatorial plane, respectively. These temperatures depend
on the angle θ:

$$T_i = T_o(1 + \varepsilon_i \sin^2\theta) \qquad i = 1, 2 \qquad\qquad (7.1)$$

The linear polarization degree of the radiation is given by

$$P(\theta) = \frac{xe^x}{e^x - 1} \left|\varepsilon_1 - \varepsilon_2\right| \sin^2\theta \qquad\qquad (7.2)$$

where $x = h\nu/kT$ is the time-independent frequency parameter. (The dependence of P on x is due to the fact that P is defined in term of the radiation intensities). Rees solved the transfer equations for a photon
gas interacting with a completely ionized plasma through Thomson scat-
tering, and found that the stationary solution gives $\varepsilon_1 \neq \varepsilon_2$. This solution applies if the reheated plasma scattered the background photons
at redshifts $Z \gtrsim 10 \ \Omega^{-1/3}$. In such a case the polarization degree at
the present epoch is given by

$$P(\theta) = \frac{x \, e^x}{e^x - 1} \frac{\Delta H}{3N_o\sigma c} \sin^2\theta, \qquad\qquad (7.3a)$$

and

$$P(\theta) = \frac{x \, e^x}{e^x - 1} \frac{\Delta T}{T} \qquad\qquad (7.3b)$$

Here N_o is the present particle density and σ the Thomson cross-section.
If photons have propagated freely since the recombination of hydrogen
we have $P \simeq 10^{-2} \ \Delta T/T$. The most favourable situation should occur if
photons where scattered just a few times by the reheated plasma, so that
the stationary solution does not apply although matter is completely
ionized at the last scattering: in this case P may larger than $\Delta T/T$
even in the Rayleigh-Jeans region (Rees, 1968). Thus the order of magnitude of $P/(\Delta T/T)$ seems to depend strongly on the last-scattering redshift.
This might be useful to infer the epoch of the last scattering.

It is not necessary to limit the analysis to large-scale anisotropies
and large-scale polarization. Thomson scattering effectively converts
temperature fluctuations into partial polarization also at smaller scales
if the photon mean free path is smaller than the characteristic length
of the perturbations before the last scattering time (Caderni et al.
1977a). We have already remarked that the redshift anisotropies can be
generated by density waves, turbulent motions and gravitational waves.
All these perturbations should produce a polarization $P \simeq 10^{-2} \Delta T/T$ at
$Z = 10^3$. But more important effects take place if the radiation interacts
with the matter perturbations (or the metric perturbations) at the reheating epoch, because in such a case Eq. (7.3b) still holds. For instance
gravitational waves prop agating in a dense hot medium should give

$$P \simeq \frac{x \, e^x}{e^x - 1} \quad \frac{A}{N_0 L_0 \sigma Z_d} \qquad\qquad (7.4)$$

for very large values of L_0. According to our discussion an appreciable
amount of polarization may indeed be induced into background radiation.
A crucial problem, however, concerns the survival of the primordial
polarization during the prop agation of photons over cosmological distances and within our galaxy. As a matter of facts, the radiation might be
depolarized by random magnetic fields through Faraday rotation. At
present there is no evidence for the existence of intergalactic fields
of sufficient intensity, and the magnetic fields of our galaxy cannot
affect the polarization of the radiation at wavelengths of the order
of 1 mm. It is therefore recommendable to look for a cosmological polarization near the peak of the spectrum. The Faraday rotation of course
requires the existence of both a magnetic field and a ionized medium.
On the other hand, a peculiar general-relativistic effect occurs even
in vacuo (Brans 1975). The transverse polarization of any wave turns out
to change during the free propagation if cosmological expansion is ani-
sotropic.Since the rotation of the polarization vector is not the same
for all the background photons, Brans argued that any primordial polarization should be smeared out. However Caderni et al. (1978b) calculated such rotation angles in many models and found that the smearing
effect is small. Anile and Breuer (1977) showed that wave optics effects
are exceedingly small.

7.2 Experimental results

Radio observations of the linear polarization of the cosmic background
have been carried out by Nanos (1973) by means of a heterodyne receiver
operating at 3.2 cm of wavelength. The data were collected between August
1972 and August 1973 and about 40% of them were used in determining the
Stokes parameters of the radiation. Nanos has fixed an upper limit of
1.6 10^{-3} °K (at 90% of confidence) for the 24 h polarization.
A balloon borne polarimeter was flown in 1972 by the Infrared Group of
IROE with the intent of studying the far-infrared linear polarization.
Preliminary results show that the instrument was able to detect few
celestial sources (Dall'Oglio et al. 1974). In four hours of flight

the measurement covered a sky region ranging from 17h 30m up to 20h 30m
in right ascension and from -10° up to -45° in declination. Upper limits
on the polarization degree have been given at several angular scales
(Caderni et al. 1978a). The result can be expressed as

$$P(\theta) < 8 \ 10^{-4} \ (40°/\theta)^{\frac{1}{2}},$$

where the angular scale θ ranges from 0.5° up to 40°. The relevance of
this result consists in extending the investigation at small and intermediate scales, and in a wavelength region where the Faraday dispersion
is reduced by a factor 10^3 with respect to the centimetric region studied
by Nanos.

7.3 Conclusions

The discovery of the background radiation has opened a new world for
cosmological investigations. The spectral shape of this radiation and
its distortion can be related to highly energetic processes preceding
the recombination of the primeval hydrogen, or to the interaction with
hot gases in more recent epochs. The angular distribution of the radiation
can give insight into the large-scale structure of the universe and
the formation of galaxies and larger systems. Measurements of both the
anisotropy and the polarization degree might tell us when the radiation
was last scattered by the intergalactic gas, and so how strong the secondary ionization has been.
The data available today confirm the cosmological origin of the radiation
and thereby the big-bang picture of our universe; but more accurate
data may give information on many cosmological parameters and phenomena
which are not described by the idealized, isotropic and homogeneous models. Our knowledge of the cosmic history may be strongly increased by
progress in infrared observations of the background.

Aknowledgements

E. Bussoletti is partly supported in his work under Nato grant N° 861.

F.Melchiorri is supported in his work by Servizio Attività Spaziali
of Consiglio Nazionale delle Ricerche.

REFERENCES

Aannestad, P.A., 1975, Astrophys.J., 200, 30

Ade, A.R., Rowan-Robinson, M., Clegg, P.E., 1976, Astron.
        Astrophys., 53, 403

Aitken, D.K., Jones, B., 1973, Astrophys.J., 184, 127

Aitken, D.K., Jones, B., Penmann, J.M., 1974, MNRAS, 169,
        35P

Aitken, D.K., Griffiths, J., Jones, B., Penmann, J.M., 1976,
        MNRAS, 174, 41P

Allen, D.A., Hyland, A.R., Longmor, A.F., 1976, MNRAS, 175,
        61P

Allen, D.A., 1975, "Infrared, the new Astronomy"

Anile, A.M., Motta, S., 1976a, Astron.Astrophys., 49, 205

Anile, A.M., Motta, S., 1976b, Astrophys.J., 207, 685

Anile, A.M., Breuer, R.A., 1977, Astrophys.J., 217, 353

Anile, A.M., Motta, S., 1978, MNRAS, 184, 319

Bachall, J.N., Woolf, R.A., 1968, Astrophys.J., 152, 701

Baluteau, J.P., Bussoletti, E., Anderegg, M., Moorwood, A.F.M.
        Coron, N., 1976, Astrophys.J., 210, L45

Baluteau, J.P., 1977, Thèse Doctorat d'Etat ès Sciences
        (Paris)

Barrett, A.H., Ho, P.T.P., Martin, R.N., Schneps, N.H., 1976,
        Bull AAS, 8, 372

Becklin, E.E., Neugebauer, G., 1967, Astrophys.J., 147, 799

Becklin, E.E., Neugebauer, G., 1968, Astrophys.J., 151, 145

Becklin, E.E., Neugebauer, G., Wynn-Williams, C.G., 1973,
        Astrophys.Lett., 15, 87

Becklin, E.E., Neugebauer, G., 1975, Astrophys.J., 200, L71

Becklin, E.E., Beckwith, S., Gatley, I., Matthews, K.,
        Neugebauer, G., Sarazin, C., Werner, M.W., 1976

Belinsky, V.A., Khalatnikov, I.M., 1969, Sov.Phys.JETP, 29,
        911

Birkinshaw, M., Gull, S.F., Northover, K., 1978, Mon.Not.Roy.
        Astron.Soc., 185, 245

Blair, A.G., 1974, Confrontation of Cosmological Theories
        with Ovservational Data, ed. M.S.Longair (Reidel)

Bollea, D., Cavaliere, A., Panagia, N., Natta, A., 1976

Bondi, H. and Gold, T., 1948, Mon.Not.Roy.Astron.Soc., 108,
        252

Bonnor, W.B., 1957, Mon.Not.Roy.Astron.Soc., 117, 104

Borgman, J., Koorneef, J., de Vries, M., 1974, Proc. 8th
        Eslab. Symp., Frascati

Boynton, P.E., 1977, talk presented at International School

of Relativistic Astrophysics, Erice

Boynton, P.E., Melchiorri, F., 1977, Proposal to NSF and CNR for International Collaboration

Brans, C.H., 1975, Astrophys.J., 197, 1

Brown, R., Gould, R., 1970, Phys.Rev.D, 1, 2252

Burbidge, G.R., Gould, R.J., Pottasch, S.R., 1963, Astrophys. J., 138, 945

Bussoletti, E., 1973, Astron.Astrophys., 23, 125

Bussoletti, E., Baluteau, J.P., Epchtein, N., 1974, Mem.Soc. Astron.Ital., 45, 387

Bussoletti, E., Fabbri, R., Melchiorri, F., 1979, Infrared Astronomy (Pergamon Press, to be published)

Bussoletti, E., Stasinska, G., 1975, Astron.Astrophys., 39, 177

Bussoletti, E., Zambetta, A.M., 1976, Astron.Astrophys.Suppl., 25, 549

Caderni, N., Fabbri, R., 1978a, Nuovo Cimento, 44B, 228

Caderni, N., Fabbri, R., 1978b, Physics Letters,67A, 19

Caderni, N., Fabbri, R., 1978c, Physics Letters, 68A, 144

Caderni, N., Fabbri, R., Melchiorri, B., Melchiorri, F. and Natale, V., 1978a, Phys.Rev.D,17, 1901

Caderni, N., Fabbri, R., Melchiorri, B., Melchiorri, F. and Natale, V., 1978b, Phys.Rev.D, 17, 1908

Caderni, N., De Cosmo, V., Fabbri, R., Melchiorri, B., Melchiorri, F. and Natale, V., 1977, Phys.Rev.D,16,2424

Capps, R.W., Knakz, R.F., 1976, Astrophys.J., 210, 76

Carruthers, G., 1970, Astrophys.J., 161, L85

Cavaliere, A., Danese, L., De Zotti, G., 1977, Astrophys. J. 217, 6

Cavaliere, A., Danese, L., De Zotti, G., 1978, XXI Cospar Meeting, Innsbruck

Chan, K.L. and Jones, B.J.T., 1975a, Astrophys.J., 195, 1

Chan, K.L. and Jones, B.J.T., 1975b, Astrophys.J., 200, 454

Chan, K.L. and Jones, B.J.T., 1975c, Astrophys.J., 200, 461

Churchwell, E., Mezger, P.G., Hutchmeier, W., 1974, Astron. Astrophys., 32, 283

Clegg, P.E., Rowan-Robinson, M. and Ade, A.R., 1976, Astron. Astrophys., 53, 403

Cohen, M., 1973, MNRAS, 161, 85

Cohen, M., 1973, MNRAS, 161, 97

Cohen, M., 1973, MNRAS, 161, 105

Cohen, M., 1973, MNRAS, 164, 395

Cohen, M., 1974, MNRAS, 169, 257

Cohen, M., Barlow, M.J., 1974, Astrophys.J., 193, 401

Cohen, M., 1975, MNRAS, 173, 279

Cohen, M. et al., 1975, Astrophys.J., 196, 179

Collins, C.B. and Hawking, S.W., 1973a, Astrophys.J., 180, 317

Collins, C.B. and Hawking, S.W., 1973b, Mon.Not.Roy.Astron. Soc., 162, 307

Corey, B.E. and Wilkinson, D.T., 1976, Bull.Am.Astron.Soc., 8, 351

Dall'Oglio, G., Fonti, S., Melchiorri, B., Melchiorri, F., Natale, V., Lombardini, P., Trivero, P. and Siversten, S., 1976, Phys.Rev.D, 13, 1187

Dall'Oglio, G., Melchiorri, B., Melchiorri, F., Natale, V., Aiello, S. and Mencaraglia, F., 1974, in Planets, Stars and Nebulae Studied with Photopolarimetry, ed. T. Gerhels (University of Arizona Press., Tucson)

Danese, L., De Zotti, G., 1977, Riv.Nuovo Cimento

Danese, L., De Zotti, G., 1978, Astron.Astrophys., 68, 157

Dautcourt, G., 1969, Mon.Not.Roy.Astron.Soc., 144, 255

Davis, L., Greenstein, J.L., 1951, Astrophys.J., 114, 206

Doroshkevich, A.G., Lukash, V.N. and Novikov, I.D., 1975, Sov.Astron.AJ, 18, 554

Doroshkevich, A.G. and Zel'dovich, Ya.B., 1963, Astron.Zh., 40, 807

Drapatz, S., Michel, K.W., 1974, Astron.Astrophys., 36, 211

Dyck, H.M., Lockwood, G.W., Capps, R.W., 1974, Astrophys.J., 189, 89

Dyck, H.M., Capps, R.W., Beichmann, C.A., 1974, Astrophys.J., 188, L103

Dyck, H.M., Simon, T., 1976, P.A.S.P., 88, 738

Ennis, D., Backlin, E.E., Backwith, S., Elias, J., Gatley, I., Mathews, K., Neugebauer, G., Willner, S.P., 1977, Astrophys.J., 214, 478

Fabbri, R., Melchiorri, F., Natale, V., 1978, Astrophys.Space Sci, 59, 223

Fabbri, R., Melchiorri, F., Mencaraglia, F., Natale, V., 1979 Astron.Astrophys. (in press)

Fabbri, R., Melchiorri, B., Melchiorri, F., Natale, V., 1979 submitted to Phys.Rev.D

Fabbri, R., Melchiorri, F., 1979,, submitted to Astron.Astro-phys.

Fazio, G.G., Kleinmann, D.E., Noyes, R.W., Wright, E.L., Zeilik, M.,

Fazio, G.G., Stecker, F.W., 1976, Astrophys.J., Lett., 207, L49

Felli, M., Panagia, N., 1976

Field, G.B., Somerville, W.B., Dressler, K., 1966, Ann.Rev.
        Astron.Astrophys., 4, 207

Field, G.B., 1972, Ann.Rev.Astron.Astrophys., 10, 227

Field, G.B. and Perrenod, S.C., 1976, Astrophys.J., 215, 717

Forrest, W.J., Gillett, F.C., Stein, W.A., 1975, Astrophys.
        J., 195, 423

Forrest, W.J., Soifer, B.T., 1976

Frogel, J.A., Hyland, A.R., 1972, Mem.Soc.Roy.Science Liege,
        3, 111

Frogel, J.A., Persson, S.E., 1974, Astrophys.J., 192, 351

Gallagher, J.S., Ney, E.P., 1976, Astrophys.J., 204, L39

Gammon, R.H., Gaustad, J.E., Treffers, R.R., 1972, Astrophys.
        J., 154, 687

Gamow, G., 1952, Phys.Rev.86, 251

Gatley, I., Becklin, E.E., Mathews, K., Neugebauer, G.,
        Penston, M.V., Scoville, N., 1974, Astrophys.J., 191,
        L121

Gatley, I., Becklin, E.E., Werner, M.W., Wynn.Williams, C.G.,
        1977, Astrophys.J., 216, 277

Gaustad, J.E., Gillett, F.C., Knake, R.F., Stein, W.A., 1969,
        Astrophys.J., 158, 613

Gauthier, T.N., Fink, U., Treffers, R.R., Larson, H.P., 1976,
        Astrophys.J., 207, L129

Geballe, T.R., Rank, D.M., 1973, Astrophys.J., 182, L113

Gehrz, R.D., Hackwell, J.A., Smith, J.R., 1975, Astrophys.J.,
        202, 133

Gehrz, R.D., Hackwell, J.A., 1976, Astrophys.J., 206, L161

Geisel, S.L., Kleinmann, D.E., Low, F.J., 1970, Astrophys.J.,
        161, L101

Gezari, D.Y., Joyce, R.R., Righini, G., Simon, M., 1974,
        Astrophys.J., 191, L33

Giacconi, R., 1974, Astrophys.J., Suppl., 27, 37

Gillett, F.C., Low, F.J., Stein, W.A., 1967, Astrophys.J.,
        149, L97

Gillett, F.C., Stein, W.A., 1969, Astrophys.J., 155, L

Gillett, F.C., Stein, W.A., 1970, Astrophys.J., 159,817

Gillett, F.C., Forrest, W.J., 1973, Astrophys.J., 179, 483

Gillett, F.C., Forrest, W.J., Merrill, K.M., 1973, Astrophys.
        J., 183, 87

Gillett, F.C., Forrest, W.J., Merrill, K.M., Capps, R.W.,
        Gilman, R.C., 1969, Astrophys.J., 155, L185

Gilman, R.C., 1969, Astrophys.J., 155, L185

Gilra, D.P., 1972, Nasa-SP-210

Glass, I.S., Webster, B.L., 1973, MNRAS, 165, 77

Gold, T., Pacini, F., 1968, Astrophys.J., 152, L115

Goldreich, P., Kwan, J., 1974, Astrophys.J., 187, 243

Grasdalen, G.L., Joyce, R.R., 1976, Nature, 259, 187

Grasdalen, G.L., Gaustad, J.E., 1971, Astrophys.J., 76, 231

Greenberg, L.T., Dyal, P., Geballe, T.R., 1976

Grishchuck, L.P., Doroshkevich, A.G. and Lukash, V.N., 1972,
        Sov.Phys. IETP, 34, 1

Grishchuck, L.P., Doroshkevich, A.G. and Novikov, I.D., 1969,
        Sov.Phys. JETP, 28, 1210

Gull, S.F. and Northover, K.J.E., 1976, Nature, 263, 572

Gustincic, J.J., de Graauw, Th., Hodges, D.T., Luhmann, N.C.,
        1977, IEE Meeting (San Diego)

Gunn, J.E., 1978, in Gunn, J.E., Longair, M.S., Rees, M.J.,
        Observational Cosmology, Saas-Fee Proceedings

Hansen, O.L., Blanco, V.M., 1973, Astrophys.J., 78, 669

Hansen, O.L., Blanco, V.M., 1975, Astrophys.J., 80, 1011

Harper, D.A., 1974, Astrophys.J., 192, 557

Harper, D.A., Low, F.J., Ricke, G.H., Thronson jr., H.A.,
        1976, Astrophys.J., 205, 130

Hawey, P.M., Gatley, I., Werner, M.H., Elias, J.H., Evans,
        N.J., Zuckermann, B., Morris, G., Sato, T., Lituak,
        M.H., 1974, Astrophys.J., 189, L87

Hawey, P.M., Campbell, M.F., Hoffmann, W.F., 1976, Astrophys.
        J., 205, L69

Hawking, S.W., 1969, Mon.Not.Roy.Astron.Soc., 142, 129

Heckmann, O. and Schucking, E., 1962, in Gravitation: an In-
        troduction to Current Research, ed. L.Witten (Acade-
        mic Press, New York)

Hills, R., Jannsenn, M.A., Thornthorn, D.D., Welch, W.J.,
        1972, Astrophys.J., 175, L59

Hoyle, F., Wickramasinghe, N.C., MNRAS, 1962, 124, 417

Hyland, A.R., Neugebauer, G., 1970, Astrophys.J., 160, L177

Hyland, A.R., Becklin, E.E., Frogel, J.A., Neugebauer, G.,
        1972, Astron.Astrophys., 16, 204

Jacobs, K.C., 1968, Astrophys.J., 153, 661

Jameson, R.F., Longmore, A.J., McLinn, J.A., Woolf, N.J.,
        1974, Astrophys.J., 190, 353

Johnson, H.L., 1966a, Ann.Rev.Astron.Astrophys., 4, 193

Johnson, H.L., 1966b, Astrophys.J., 143, 187

Kaler, J.B., Aller, L.H., Czysak, S.J., Epps, H.W., Astro-
        phys.J., suppl., 31, 169

Kavara, K., Maihara, T., Noguchi, K., Oda, N., Sato, N.,

1976, P.A.S., Japan, 28, 163

Kislyakov, A.G., Chernishev, V.I., Lebskii, Yu.V., Maltsev, V.A. and Serov, N.V., 1971, Astron.Zh., 48, 39

Kleinmann, D.E., Low, F.J., 1967, Astrophys.J., 149, L1

Kleinmann, D.E., Gillett, F.C., Wright, E.L., 1976, Astrophys.J., 208, 42

Kleinmann, S.G., Sargent, D.G., Gillett, F.C., Grasdolen, G.L., Joyce, R.R., 1977, Astrophys.J., 215, L79

Knake, R.F., Capps, R.W., 1977, Astrophys.J., 216, 271

Kurskov, A.A., Ozernoi, L.M., 1978, Astrophys.Space Sci. 56, 67

Lake, G., Partridge, R.B., 1977, Nature 270, 502

Layzer, D., 1968, Astrophys.J.Lett., 1, L99

Layzer, D. and Hively, R., 1973, Astrophys.J., 179, 361

Lebofosky, M.J., Kleinmann, S.G., Rieke, G.H., Low, F.J., 1976, Astrophys.J., 206, L160

Lemke, D., Low, F.J., 1972, Astrophys.J., 177, L53

Lemke, D., Low, F.J., Thum, C., 1974, Astrophys., 32, 231

Lifshitz, E.M., Lifshitz,I.M. and Khalatnikov, I.M., 1971 Sov.Phys.JETP, 32, 173

Longair, M.S., 1978, in Observational Cosmology, Saas-Fee Proceedings

Low, F.J., 1974, Astrophys.J., 192, L23

Low, F.J., Kleinmann, D.E., Forbes, F.F., Aumann, H.H., 1969, Astrophys.J., 157, L97

Low, F.J., Kurtz, R.F., Vrba, F.J., Rieke, G.H., 1976, Astrophys.J., 206, L153

Low, F.J., Kurtz, R.F., Potreet, W.M., Nishimura, T., Astrophys.J., 214, L115

Low, F.J., Kurtz, R.F., Potreet, W.M., Nishimura, T., 1977, Astrophys.J., 214, L115

Low, F.J., Tucker, W., 1968, Phys.Rev.Lett., 21, 1538

Lukash, V.N., Novikov, I.D., Starobinsky, A.A. and Zel'dovich, Ya.B., 1976, Nuovo Cimento, 35B, 293

Lukash, V.N. and Starobinsky, A.A., 1974, Sov.Phys.JETP, 39 742

Mac Callum, M. and Ellis, G., 1970, Comm.Math.Phys.,19, 31

Martin, A.H.M., Gull, S.F., Lecture Notes in Physics, 42, Ed.Wilson, Downes

Mather, J.C., 1977, Cosmic Background Explorer, NASA Report

Matzner, R.A., 1969, Astrophys.J., 157, 1085

Matzner, R.A., 1971a, Ann.Phys., 65, 438

Matzner, R.A., 1971b, Ann.Phys., 65, 482

Matzner, R.A. and Misner, C.W., 1972, Astrophys.J., 171, 415

Mc Cammon, D., Münch, G., Neugebauer, G., 1967, Astrophys.J.,
        147, 575

Merrill, K.M., Stein, W.A., 1976a, P.A.S.R., 88, 285

Merrill, K.M., Stein, W.A., 1976b, P.A.S.R., 88, 294

Merrill, K.M., Stein, W.A., 1976c, P.A.S.R., 88, 874

Metzger,   1977

Misner, C.W., 1967, Phys.Rev.lett., 19, 533

Misner, C.W., 1968, Astrophys.J., 158, 431

Mitchell, R.J., Culman, J.L., Davison, P.J.N. and Ives, J.C.
        1976, Mon.Not.Roy.Astron.Soc., 175, 29P

Muehlner, D., Weiss, R., 1976, Infrared and submillimetre
        Astronomy, Reidel

Mufson, S.L., Lyon, J., Marionni, P.A., 1975, Astrophys.J.,
        201, L85

Nanos, G.P., 1973, Ph.D.Thesis, Princeton University (un -
        published)

Nariai, H., 1956, Sci.Rept.Tohoku Univ., 39, 213

Narlikar, J.V., Edmunds, M.G. and Wickramasinghe, N.C.,1975
        in Far Infrared Astronomy, ed. M.Rowan-Robinson
        (Pergamon Press, Oxford)

Natta, A., Panagia, N., 1976, Astron.Astrophys.

Neugebauer, G., Leighton, R.B., 1969, NASA SP-3047

Neugebauer, G., Garmire, G., 1970, Astrophys.J., 161, L91

Neugebauer, G., Becklin, E.E., Hyland, A.R., 1971, Ann.Rev.
        Astr.Astrophys., 9, 67

Neugebauer, G., Becklin, E.E., Beckwith, S., Mathews, K.,
        Wynn-Williams, C.G., 1976, 205, L139

Ney, E.P., Allen, D.A., 1969, Astrophys.J., 155, L193

Ney, E.P., Merrill, K.M., Becklin, E.E., Neugebauer, G.,
        Wynn-Williams, C.G., 1975, Astrophys.J., 198, L129

Novikov, I.D., 1968, Sov.Astron.AJ, 12, 427

Novikov, I.D., 1974, in Confrontation of Cosmological Theories
        with Observational Data, Ed. M.S. Longair (Reidel,
        Dordrecht, Holland), p.273

O'Dell, C.R., Krishna-Swamy, K.S., 1968, Astrophys.J.,151, L61

Ozernoi, L.M. and Chernin, A.D., 1968, Sov.Astron.AJ, 11,907

Ozernoi, L.M. and Chibisov, G.V., 1971a, Sov.Astron.AJ, 14,
        615

Ozernoi, L.M. and Chibisov, G.V., 1971b, Astrophys.Lett., 7,
        201

Panagia, N., 1974, Astrophys.J., 192, 221

Parijskij, Y.N., 1968, Sov.Astron.AJ, 12, 219

Parijskij, Y.N., 1973, Astron.Zh., 50, 453

Parijskij, Y.N., 1977, IAU Symposium, Tallinn, USSR

Partridge, R.B., 1967, Astrophys.J., 148, 377

Partridge, R.B., 1978

Partridge, R.B., Peebles, P.J.E., 1967, Astrophys.J., 148,377

Pauls, D., Downes, D., Mezger, P.G., Churchwell, E., 1976, Astron.Ap., 46, 407

Payne-Gaposchkin, C., 1957, "The Galactic Novae", North-Holland

Peebles, P.J.E., 1965, Astrophys.J., 142, 1317

Peebles, P.J.E., 1968, Astrophys.J., 153, 1

Peebles, P.J.E., 1971, Physical Cosmology (Princeton University Press, Princeton)

Peebles, P.J.E., and Yu, J.T., 1970, Astrophys.J., 162, 815

Penzias, A.A. and Wilson, R.W., 1965, Astrophys.J., 142, 419

Persson, S.E., Frogel, J.A., Aaronson, M., 1976

Pehosian, V., 1970, Astrophys.J., 159, 833

Pipher, J., 1973, IAU Symp., 52, 559

Puget, J.L., Stecker, F.W., Bredekamp, J.H., 1976, Astrophys. J.. 205, 638

Pottasch, S.R., 1968, Bull.Astr.Instr.Neth., 19, 469

Puget, J.L., Stecker, F.W., Bredekamp, J.H., 1976, Astrophys. J., 205, 638

Rees, M.J., 1978, Nature, 275, 35

Ricke, G.H., Low, F.J., 1973, Astrophys.J., 184, 415

Ricke, G.H., Low, F.J., Kleinmann, D.E., 1973, Astrophys.J., 186, L7

Ricke, G.H., Telesco, C.M., Harper, D.A., 1977, Astrophys.J.,

Righini, G., Simon, M., Joyce, R.R., 1976, Astrophys.J., 207, 119

Robertson, H.P. and Noonan, T.W., 1968, Relativity and Cosmology (Saunders, Philadelphia)

Robson, E.I., Clegg, P.E., 1977, in Radio Astronomy and Cosmology, Ed. D.Jauncey (Reidel)

Robson, E.I., Vickers, D., Huizinga, J.S., Beckman, J. and Clegg, P.E., 1974, Nature, 251, 591

Rouan, D., Lena, P., Puget, J.L., de Boer, K.S., Wijnbergen, J.J., 1977, Astrophys.J., 213, L35

Rowan-Robinson, M., 1974, Mon.Not.Roy.Astron.Soc., 168, 45P

Rubin, V.G., Ford, W.K., Tonnard, N., Roberts, M.S. and Gordon, J.A., 1976, Astrophys., 81, 687

Russell, R.W., Soifer, B.T., Puetter, B.C., 1977, Astr. Astrophys., 54, 959

Ryter, C., Puget, J.L., 1977, Astrophys.J., 215, 775

Sachs, R.K., Wolfe, A.M., 1967, Astrophys.J., 147, 73

Salpeter, E.E., 1974a, Astrophys.J., 193, 579

Savage, B.D., 1975, Astrophys.J., 199, 92

Schiffer III, F.H., Mathis, J.S., 1974, Astrophys.J., 194, 597

Sciama, D.A., 1966, Nature, 211, 277

Sciama, D.W., 1971, Modern Cosmology (Cambridge University Press, Cambridge, Eng.)

Scoville, N.Z., Solomon, P.M., Jefferts, K.B., 1974, Astrophys.J., 187, L63

Scoville, N.Z. and Solomon, P.M., 1975, Astrophys.J.Lett., 199, L105

Scoville, N.Z., Kwan, J., 1976, Astrophys.J., 206, 718

Serra, G., Puget, J.L., Ryter, C.E., Wijnbergen, J.J., 1978 Astrophys.J., 222, L21

Silk, J., 1974, Astrophys.J., 194, 215

Silk, J., White, S., 1978, Astrophys.J.Letters, 226, L103

Simpson, J., 1975, Astron.Astrophys., 39, 43

Smith, M.G. and Partridge, R.B., 1970, Astrophys.J., 159, 737

Smith, M.G., 1973, Astrophys.J., 179, L11

Smoot, G.F., Gorenstein, M.V. and Muller, R.A., 1977, Phys. Rev.Lett., 39, 892

Soifer, B.T., Russell, R.W., Merrill, K.M., 1975, Astrophys. J., 207, L83

Solomon, P.M.,Scoville, N.Z., Penzias, A.A., Wilson, R.W., Jefferts, K.B., 1972, Astrophys.J., 178, 125

Starobinsky, A.A. and Zel'dovich, Ya.B., 1972, Sov.Phys. JETP, 34, 1159

Stein, W., 1966, Astrophys.J., 144, 318

Sternn, F.W., Puget, J.L., Fazio, G.G., Goddard, S.C., X602-76-274

Strecker, D.W., Ney, E.P., 1974, Astronom.J., 79, 797

Sunyaev, R.A., 1974, in Confrontation of Cosmological Theories with Observational Data, ed. M.S. Longair (Reidel, Dordrecht, Holland), p.167

Sunyaev, R.A. and Zel'dovich, Ya.B., 1980a, Astrophys.Space Sci., 7, 20

Sunyaev, R.A. and Zel'dovich, Ya.B., 1970b, Astrophys.Space Sci., 7, 3

Sunyaev, R.A. and Zel'dovich, Ya.B., 1972, Astronom.Astrophys. 20, 189

Telesco, C.N., Harper, D.A., Locevenstein, R.F., 1976, Astrophys.J., 203, L53

Terzian, Y., Sanders, D., 1972, Astronom.J., 77, 350

Thorstensen, J.R., Partridge, R.B., 1975, Astrophys.J., 200, 527

Thompson, R.I., Schopper, H.W., Mitchell, R.I., Johnson, 1969, Astrophys.J., 158, L55

Thorne, K.S., 1967, Astrophys.J., 148, 51

Tinsley, B., 1973, Astron.Astrophys., 24, 89

Treffers, R.R., Cohen, M.,1974, Astrophys.J., 188, 545

Treffers, R.R., Fink, U., Larson, H.P., Gautier III, T.N., 1976, Astrophys.J., 209, 793

Treffers, R.R., Fink, U., Larson, H.P., Gautier III, T.N., 1976, Astrophys.J., 209, L115

Visvanathan, N., Sandage, A., 1977, Astrophys.J., 37

Vogt, S.S., 1973, Astrophys.J., 78, 389

Wagoner, R.V., 1974, "Confrontation of Cosmological Theories with Observational Data", ed. Longair, M.S., Reidel

Walker, R.G., Price, S.D., AFCRL-72-0588

Walker, R.G., Price, S.D., AFCRL-TR-75-0373

Walker, R.G., Price, S.D., AFGL-TR-76-0208

Ward, D.B., Dennison, B., Gull, G.E., Harwit, M., 1976, Astrophys.J., 205, L75

Weinberg, S., 1971, Astrophys.J., 168, 175

Weinberg, S., 1972, Gravitation and Cosmology (Wiley, New York)

Weiss, R., 1977, private communication

Werner, M.W., Gatley, I., Harfer, D.A., Becklin, E.E., Lowenstein, R.F., Telesco, C.M., Thronson, H.A.,1976, Astrophys.J., 204, 420

Westbrook, W.E., Becklin, E.E., Merrill, K.M., Neugebauer,G., Schmidt, H., Willner, S.P., Wynn-Williams, C.G., 1975, Astrophys.J., 202, 407

Westbrook, W.E., Werner, M.W., Elias, J.H., Gezari, D.Y., Hauser, M.G., Lo, K.Y., Neugebauer, G., 1976, Astrophys.J., 209, 94

Wink, J.E., Altenhoff, W.J., Webster jr., W.J., 1975, Astron. Astrophys., 38, 109

Wolfe, A.M. and Burbidge, G.R., 1969, Astrophys.J., 156, 345

Wollman, E.R., Gebolle, T.R., Lacy, J.H., Townes, C.H., 1976, Astrophys.J., 205, L5

Woody, D.P., Mather, J.C., Nishioka, N.S. and Richards, P.L., 1975, Phys.Rev.Lett., 34, 1036

Woody, D.P., Richards, P.L., 1978, preprint

Woolf, N.J., Stein, W.A., Strittmayer, P.A., 1970, Astrophys. Lett., 9, 252

Wynn-Williams, C.G., Becklin, E.E., 1974, P.A.S.P., <u>86</u>, 5

Zel'dovich, Ya.B., Illarionov, A.F. and Sunyaev, R.A., 1972,
        Sov.Phys.JETP, <u>35</u>, 643

Zel'dovich, Ya.B., Kurt, V.G. and Sunyaev, R.A., 1968, Zh.
        Eksp.Teor.Fiz., <u>55</u>, 278

Zel'dovich, Ya.B. and Novikov, I.D., 1974, Relativistic
        Astrophysics (University of Chicago Press, Chicago),
        vol. II

Zel'dovich, Ya.B. and Sunyaev, R.A., 1969, Astrophys.Space
        Sci., <u>4</u>, 301

RECENT PLANETARY PHYSICS AND CHEMISTRY

R. Smoluchowski[*]
Princeton University
Princeton, New Jersey 08540
**Present address: University of Texas,
Austin, Texas.**

ABSTRACT

Within the last few years considerable progress has been made in our knowledge and understanding of the origin of planets and of the structure of their interiors and atmospheres. Some of these advances including Venera and Viking results are summarized for all the planets (except Earth) with accent on those facts which seem amenable to theoretical analysis. Only the minimum background information is added for completeness. (The results of the 1978-79 Mariner-Venus Orbiter, Pioneer 11 and Voyager 1 and 2 missions as well as other observations are very briefly summarized in appropriate addenda).

I.   Introduction

Three factors contributed to the recent rapid development of our understanding of planets: a) progress in the theory of the formation of planets and of their satellites from the primitive solar nebula, b) the large number of new data obtained from terrestial and spacecraft observations and c) availability of much better equations of state for the planetary interiors based on experiment and on theory. Only highlights of these developments are mentioned in this survey and the choice is admittedly very subjective. The stress is placed on physics and chemistry of the planets rather than on descriptive morphological and geological aspects of their surfaces. The inner planets are rather briefly treated but recent data obtained from Venera and Viking missions are included. Jupiter occupies most of the space since many of its basic properties can be derived from first principles so that, from this point of view, it is the best known of all planets. (Many new observations and theoretical results made within the last year or two necessitate corrections and additions. These were introduced in a few places in the text and at the end of the appropriate sections.)

[*] Supported in part by NASA Grants # NSG-7283 and 7505.

P. L. Bernacca and R. Ruffini (eds.), Astrophysics from Spacelab, 493-525.
Copyright © 1980 by D. Reidel Publishing Company.

2.   Origin of Planets

        A study of the formation of planets from the primitive solar
nebula requires taking into account numerous processes: accretion of
matter, various gravitational instabilities which can lead to the formation of the sun or of proto-planets in the midplane, the removal of the
angular momentum of the collapsing cloud by meridional circulation currents, central heating and the radial temperature gradient. The presence
or absence of a central object prior to the formation of proto-planets
raises questions concerning the influence of solar wind during the solar
flare-up phase and the presence of strong tidal effects. A variety of
assumptions concerning these phenomena led to a variety of theories of
the early stages of the development of the solar system, and as yet no
unique rigorous answer is available. In his theory of the origin of the
giant planets, Cameron assumed that in the midplane of the rotating
nebula, the temperature had a relatively low value so that the nebula
consisted not only of gases but also of dust. The latter was made of
metals, metal oxides, sulfides, silicates and perhaps ice; the former
of volatile elements and compounds such as hydrogen, helium, methane,
ammonia and water. Through accretion and amalgamation, these several-
micron-size grains grew to larger particles (1-10mm) and eventually
formed a midplane layer of small solids that became gravitationally unstable and formed more massive bodies. Through collision the largest of
these bodies grew sufficiently to form planetary cores which could
attract their own gaseous clouds co-rotating with them. A hydrodynamic
collapse of the clouds released then a large amount of heat which dis-
sociated and ionized H_2 and partially ionized He.
        In the course of time these proto-planets, especially the outer
ones, acquired atmospheres made primarily of hydrogen and helium with
any in-falling dust probably evaporating. The subsequent development of
such planets was accompanied by a slow evolution of heat due to gravitational shrinkage and by progressive cooling. The essential feature
of this model is that, while the rocky cores themselves incorporate the
various nebular compounds in solar proportions and the external mantles
also retain the characteristic solar ratio of hydrogen to helium
(3.5 ± 1 by mass), there is no fixed relation between the two so that
the planets as a whole are not solar. In particular, there is a large
excess of water in the form of ice or vapor.
        A different model of the formation of planets has been applied to
Jupiter by Bodenheimer. He assumed that the planet was formed by a local
selfgravitational condensation of the nebular matter of initial density
$10^{-11} g \ cm^{-3}$ at T=40K in a volume with r=4600 R_J. If the sun was already
present then the initial density has to be about 500 times higher to
overcome disruptive tidal effects. The subsequent evolutionary contraction first proceeded rapidly in quasi-hydrostatic equilibrium until
the hydrodynamic collapse associated with the dissociation of H_2 had
occurred and a denser central region had formed. This was followed by a
further slow contraction, in hydrostatic equilibrium, from a radius of
about 5R_J until the present state was reached. Throughout this process
the composition was solar and homogeneous (Fig. 1).

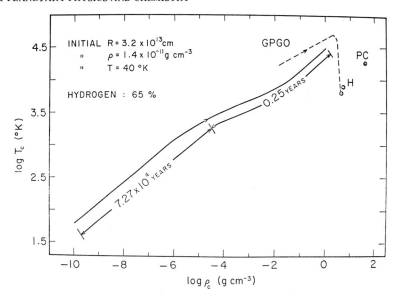

Fig. 1. Evolution of Jupiter from the solar nebula (according to P.
Bodenheimer). GPGO, H and PC indicate results of calculations
by H.C. Graboske, jr. et al., by W.B. Hubbard and by M. Podalak
and A.G.W. Cameron.

The striking difference between the size and densities of the
inner and the outer planets is the result of several factors: the
great difference between the volatility of the rocky materials (oxides,
silicates, etc.) on one hand and of the lighter gases and icy constituents on the other, the radial gradient of temperature within the
solar nebula, the effect of the solar T-Tauri wind which expelled, at
an early epoch, large amounts of gas from the sun and from its proximity
and finally the more efficient accumulation of matter by the more distant proto-planets because of their longer orbits and the weaker competing gravitational attraction of the sun. In fact, the giant planets
contain an order of magnitude more rocky material such as oxides,
silicates, etc. than all the other planets combined.

3. Mercury

Nearly all new information we have about Mercury comes from the
Mariner-Venus-Mercury fly-by in 1974-1975. Since the space craft
acquired a heliocentric orbit with a period twice as long as that of
the planet one was able to get useful data about Mercury on three
separate encounters referred to as I, II and III and spaced about 170
days apart. Probably the most unexpected result is the discovery that
Mercury has a permanent magnetic field (about 1500 times weaker than
that of the Earth). Other planets for which the existence of a field is
established are Earth, Mars, Jupiter and Saturn although strong indications are that Uranus has one also. Fig. 2 and 3 show the paths of
the spacecraft which identified the planet's magnetopause (1). The stag-

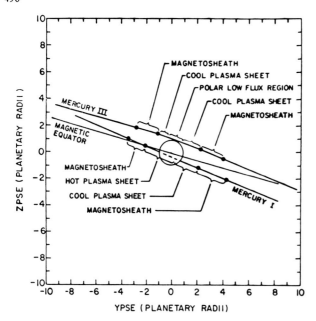

Fig. 2.
View from the sun
of Mariner 10
trajectories
near Mercury
(according to N.F. Ness).

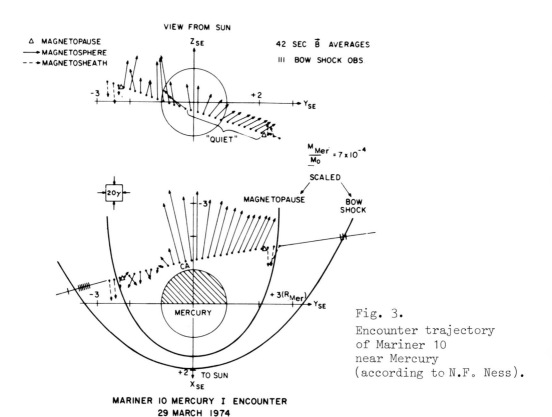

MARINER 10 MERCURY I ENCOUNTER
29 MARCH 1974

Fig. 3.
Encounter trajectory
of Mariner 10
near Mercury
(according to N.F. Ness).

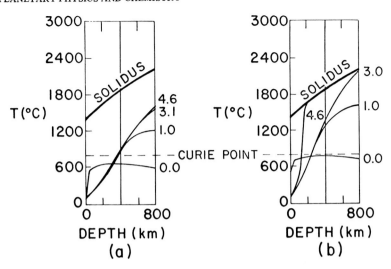

Fig. 4. Near-surface temperature profiles for two extreme models of heat
sources distribution in Mercury (according to R.W. Siegfried and
S.C. Solomon).

nation point towards the sun occurs at 1.6 planetary radii as compared
to about 11 for Earth in accord with the much stronger field of the
latter. Bursts of high energy electrons have been observed within the
magnetosphere. Similarly to the terrestrial and the Jovian fields the
field of Mercury is tilted about 12° away from the rotational axis. The
origin of this field is not well understood. The small size of the
planet (R=2440 Km) and high density (5.44 g cm-3) indicate that the external rocky mantle extends to a depth of a quarter of the radius, the
remainder of the planet being iron and nickel in some form. As Fig. 4
shows depending upon the assumed distribution of radioactive elements
the planet could have had, and perhaps still has, a liquid metallic core
which could permit the operation of the usual hydromagnetic dynamo. Also
the maximum depth which would now be below the Curie temperature is 400
Km if no core was formed and 200 Km if a core exists. Fossil magnetization
of this thin shell is not sufficient to account for the present magnetic
moment. The existence of a hydromagnetic dynamo is also supported by
consideration of the history of the planet's rotational and orbital
periods which are in a 3:2 resonance but earlier, when the planet
rotated faster, they were probably in a 2:1 resonance. It turns out that
in order to escape from the latter the viscosity of the liquid core of
the planet had to be roughly that of water (2). This value satisfies
the necessary requirements for the operation of a hydromagnetic dynamo.
On the other hand the rotation period of about 60 days is rather slow so
that the Reynolds magnetic number may be too low. Another possibility
is that the Malkus mechanism of magnetic field generation based on the
planet's precession is responsible for the observed field. This problem
remains to be clarified.

498   R. SMOLUCHOWSKI

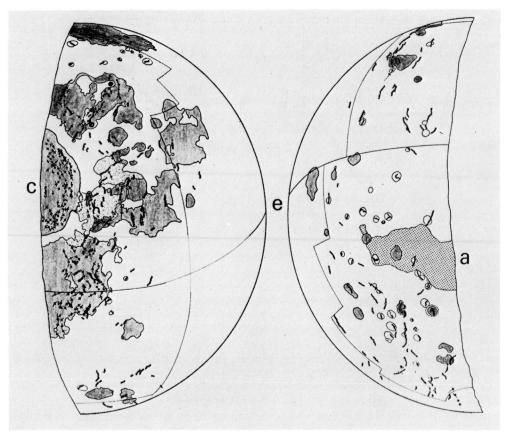

Fig. 5. Sections of a map of Mercury showing the Caloris basin (c), the
       equator (e) and the hilly and lineated terrain on the antipode
       (a) (according to R.G. Strom et al.).

     The surface of the planet observed with a resolution up to 450 m
looks very much like that of our Moon with interesting consequences for
the history of the spacial distribution of meteoroids. A striking feature is the huge (1400 Km diameter) Caloris basin, which resembles
somewhat Mare Imbrium on the Moon and is shown schematically (3) in Fig.
5. On its antipode one observes a comparable huge area of peculiar rough
terrain quite distinct from its neighborhood and thus one speculates
that both of them were formed by an impact of an enormous body which
produced a huge shock wave across the whole planet. Unlike the Moon the
surface of Mercury is not saturated with craters 20-50 Km in diameter.
The crust of the planet shows good evidence that the planet contracted
during cooling of its interior. Craters superimposed on the various
faults and hundreds of kilometers long cracks permit the establishment
of an approximate time sequence of these events.
     The planet has a very thin He atmosphere which is due to the
captured He component of the solar wind (hydrogen could escape right
away). As a curiosity one may add that because of the above mentioned

resonance between rotation and synodic year and the orbital eccentricity, which is highest among all planets, Mercury gets $2\frac{1}{2}$ times as much solar flux at 0° and 180° longitude as at 90° and 270°. Infrared observations indicate that the subsurface temperature at the poles is below 0°C while it is above 0°C at the equator.

4.   Venus

Enormous progress has been made in our understanding of Venus. Fly-by observations from various Mariners, orbiter and lander observations made by the Veneras as well as terrestrial radar studies enriched greatly our knowledge of this planet. Only some of these results will be here mentioned. Photographs of the planet taken in the ultraviolet by Mariner indicate a rotation period of the clouds of about 4 days. This period does not represent the actual rotation period of the planet itself which, from radar data, is known to be about 243 days. The rapid longitudinal prograde motion of the clouds is associated with a rapid vertical translation implying a gravitational wave as the source of energy. The velocity of these clouds relative to the surface of the planet is about 100 m sec^{-1} in agreement with velocities measured on certain Venera flights near 60 Km altitude discussed below. The pattern of the clouds agrees with the theoretically predicted Hadley circulation. Recent very high precision infrared CO_2 measurements (4) indicate that at still higher altitudes (near 120 Km) the main motion of the atmosphere is a flow (a few m sec^{-1}) away from the subsolar point the average velocity being close to zero.

Information about the atmospheric and surface conditions on Venus comes primarily from the Venera missions and especially from orbiters and landers Venera 9 and 10. Both landers were able to function on the planet's surface for about an hour and transmitted various data including images of the landscape. Fig. 6 shows the pressure and the temperature profiles in the lower atmosphere confirming the earlier results: T $\sim$ 750 K, P $\sim$ 90 bars. Fig. 7 shows the penetration of the solar radiative flux of various wave lengths and the rapid change of its attenuation at the lower boundary of the solid cloud cover. Actually about one per cent of sun light reaches the surface which is similar to conditions on Earth on a cloudy day. The atmosphere consists primarily of CO_2 and some CO. The clouds, at altitudes between 50 and 60 Km, consist of 1-3 micron size aerosols with a density of 100-500 cm^{-3}. Various optical measurements indicate that they are droplets of hydrated H_2SO_4, which originates from photolysis of COS formed on the surface (5). The fact that in this thick layer their size is constant in a narrow range in spite of a rapid change in pressure and temperature indicates the presence of an interesting balance between partial pressures of H_2SO_4, H_2O and SO_3, surface tension, composition and rates of reaction with CO at lower altitudes. At still lower altitudes the particles (which show strong back scattering) are about 2.5μ in size with a density of about 2 cm^{-3}. Their composition is not known (M.Ya. Marov, private communication).

Planetological arguments suggest that the total amount of CO_2 on Venus is comparable to that on Earth but here it is mostly bound

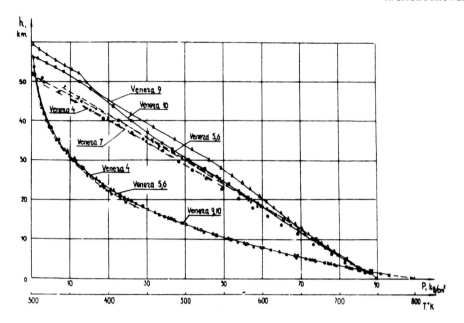

Fig. 6. Altitude dependence of pressure (exponential) and temperature
(linear) on Venus as obtained by various Venera missions (according to M.V. Keldysh).

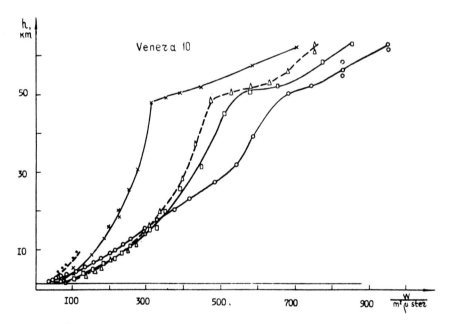

Fig. 7. Altitude profiles of the radiative energy fluxes on Venus for
wavelength varying from 0.75μ (right curve) to 2.75μ (left
curve) (according to M.V. Keldysh).

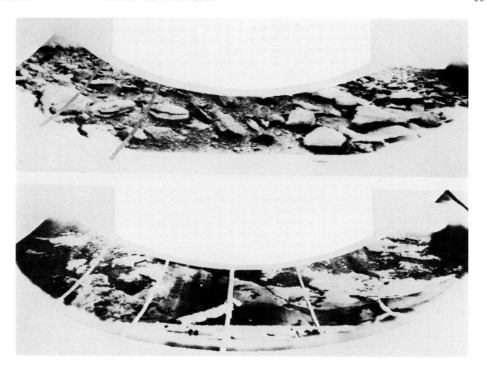

Fig. 8. Images of the surface of Venus obtained by Veneras 9 (top) and
        10 (bottom) landers. Note the sharp edges of the several deci-
        meter size rocks in the top picture. (according to M.V. Keldysh).

chemically in the rocks while on Venus, partly because of the greenhouse
effect, it has been released into the atmosphere. It has been speculated
that a similar greenhouse effect will develop on Earth because of increased solar heat and of CO_2 content in the atmosphere. A compensating
cooling by infrared emission of dust may however change the course of
events which otherwise would lead to an end of all terrestrial life.
        The landers of Venera 9 and 10 came down some 2000 Km apart, the
former in an area covered with sharp-edged granite rocks (10 to 30 cm
in size) the latter in an area covered by more eroded, smoother,
basaltic rocks Fig. 8. The presence of sharp rocks in the highly
erosive atmosphere suggests that they are of recent volcanic origin.
The nature of the rocks was determined from the measured content of
radioactive elements as shown in Table I. Fig. 9 shows the decrease of
winds from 60-100 m sec^{-1} near 50 Km altitude to about 1 m sec^{-1} or less
at the surface. The mass transport at 1 m sec^{-1} at 90 bar   .ponds
to a wind of about 300 Km hour^{-1} at 1 bar.
        Large scale information about the planet's surface has been ob-
tained using 12.6 cm radar echoes. These data are obtainable only in a
subterrestrial band that is near the planet's equator (6). Also, since
the rotation of Venus is coupled to the position of the Earth (the
effective center of gravity of Venus is some 2.5 Km away from its geometrical center) the planet shows always the same side at the inferior

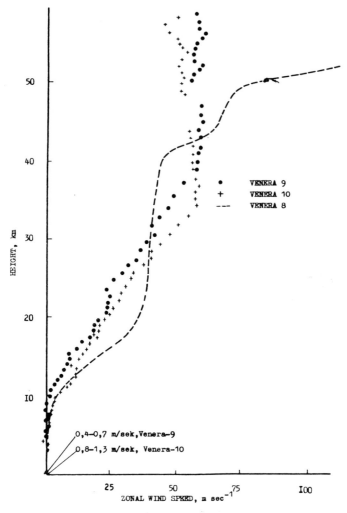

Fig. 9. Wind velocities on Venus as a function of altitude (according to
       M.V. Keldysh).

conjunction. These two factors limit considerably the radar coverage of
the planet. The echoes indicate differences in reflectivity (i.e. in
roughness, slope or dielectric constant) and in elevation with respect
to a mean surface. Many craters and canyons have been discovered in
this manner as shown for instance in Fig. 9. Each of these pictures
represents an area of about 750 Km diameter with the central stripe
blacked out because of inevitable distortions. The elevation contours
are 500 cm apart with white indicating higher altitudes. The channel in
Fig. 10 is about 120 Km wide and about 1.5 Km deep. Other pictures show
craters many tens of kilometers in diameter and up to one kilometer in
depth. A complete coverage of the planet's surface will be obtained by

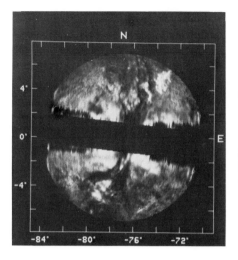 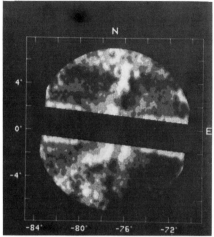

Fig. 10. Radar echoes indicating reflectivity (a) and altitude (b) in proximity of a canyon on Venus (according to R.M. Goldstein et al.).

Table 1

	U $(10^{-4}\%)$	Th $(10^{-4}\%)$	K $(\%)$
Venera 8	2.2 ± 0.7	6.5 ± 2.0	4.0 ± 1.3
Venera 9	0.5 ± 0.2	4.0 ± 0.25	0.9 ± 0.2
Venera 10	0.7 ± 0.2	1.1 ± 0.2	0.3 ± 0.1
Granite	9.04	21.9	3.24
Basalt	0.86	2.1	0.76

Table 1: Radioactive elements on the landing sites on Venus and in terrestrial rocks (according to M.V. Keldysh).

a radar mapping orbiter. From earlier measurements one knows that the magnetic field of Venus, if any, is less than 10^{-5} of the terrestrial field and that the planet's ionosphere interacts strongly with the solar wind forming a tail but there is no magnetosphere.

Note added in 1979: Much new information about the atmosphere of Venus, so far only partly analysed, has been obtained by the 1978 Mariner Venus Orbiter which also dropped one big and three smaller probes. The clouds appear to consist of three layers, an upper one with 1μ droplets of

H_2SO_4, a middle layer producing strong extinction of the solar light
with particles up to 2μ and a lower layer with 4-5μ droplets probably
of liquid sulphur at a concentration of 10 to 100 per cc. Below the
cloud layers there is a thin haze with about one droplet per cc but no
dust. The often observed Y-feature visible in the ultraviolet in the
high altitude clouds shows the strongest contrast at the evening ter-
minator. In a few cases huge convection cells seem to have been observed.
The ionosphere reaches 300 Km at the subsolar point and 3000 Km on the
nightside. There is a collar of dense clouds (215K) at 60-70°N latitude
and two hot spots (260K) near 80 N. A few percent of the solar energy
reaches the surface but less than one percent of it is radiated into
space. Winds are easterly: 1 m s^{-1} near the surface, 50 m s^{-1} at the
bottom of clouds and perhaps 200 m s^{-1} in the dense middle cloud layer.
The altitude of 90% of the surface is known from radar data with an
accuracy of ±200 m. There is a great rift valley, some 30 impact craters
12-260 Km diameter and four major plateaus with mountain ranges several
Km high. The age of the surface is estimated at 6×10^8 years. The upper
limit to the magnetic moment is about 10^{-4} gauss R_V^3. The gravity
anomalies resemble those of Mars and the Moon rather than the Earth.
ref. Science, 205, July 6, 1979

5.   Mars

     Our knowledge of the surface of Mars before 1976 was based on a
series of Mariner missions of which the last one, the orbiter Mariner
9, provided wide surface coverage. The simultaneous occurrence of a dust
storm (originating as usual at the summer solstice), which in a few
weeks covered nearly the whole planet, and its subsequent subsidence
provided a unique opportunity for the study of seasonal winds, dust
transport and atmospheric dynamics. The surface has many prominent
features the most spectacular being the largest known volcano Olympus
Mons (about 600 Km diameter at the base, 25 Km high with a 70 Km diameter crater) and the Valles Marineris canyon complex (about 4000 Km
long, 75 to 500 Km wide and several kilometers deep). There is much
evidence, such as sinuous dry river beds, of water erosion at an earlier
epoch and of volcanic and impact activity. The latter two are in many
ways analogous to those observed on the lunar surface, but the density
of craters is much lower. The bottom of Valles Marineris has particularly few craters indicating its fairly recent origin. It has been suggested that because of the highly eccentric orbit of the planet there
may have been periods of high intensity of bombardment by meteorites
originating in the asteroidal belt. The evidence for this interesting
effect may not be observable, however, because of surface erosion and
because of the atmosphere which was probably much denser at an earlier
epoch. It is interesting that, in contrast with the Moon, there are few
ballistic ejecta around craters but there is evidence for surface flow
carried by water originating from permafrost melted upon impact. The
interior of the planet has an iron rich core up to about half the radius
of the planet, a silicate mantle and a 100 Km thick crust rich in Al and
Si. The planet's magnetic moment is 3000 times smaller than that of the
Earth.

The very successful Viking 1 and 2 missions of 1976 are still
supplying new facts: the orbiter gives a detailed coverage of large
areas of the planet and the lander provides unique data on the surface
conditions. It should be pointed out that the one millimeter resolution
obtainable in some of the images made by the Viking lander is a factor
of 10^8 better than the 100 Km resolution available some 11 years ago.
High resolution photographs of the extremely steep banks of the Tithanius
Chasm (part of the Valles Marineris) give a several kilometers deep cross
section through the layers of the surrounding plane which is of great
interest for the history of the planet's crust. Similarly details of the
wind-carried sand dunes and of various features formed by water are now
available for the study of the erosion mechanisms. One of the major
findings is the large abundance of water (about 1 per cent bound water
in the soil near the Viking 1 lander) and the predominance of water,
rather than CO_2, in the polar caps. This has of course important bearing
upon the biological questions. The presence of considerable amounts of
H_2O is in agreement with the observation of clouds on the Martian
horizon by the lander and of the early morning mist and haze by the
orbiter.

The soil is rich in iron (14 per cent by weight), with Ca, Si, Ti,
Al and S also present, confirming the early suggestion that the red color
of the planet may be due to ferric oxide or oxyhydroxide. There is also
a considerable amount of silicates. Besides CO_2 the atmosphere contains
2-3 per cent N_2, 1-2 per cent Ar ($Ar^{40}/Ar^{36} = 2750$ as compared to about
300 on Earth), 0.1 - 0.4 per cent O_2 and some Xe and Kr. The presence of
O_2 and N_2 ($N^{15}/N^{14} = 6 \times 10^{-3}$ as compared to 3.7×10^{-3} on Earth) is of
biological interest while the amount of noble gases (Ar^{40} coming from
the decay of K^{40} and Ar^{36} being of primodial origin) indicates that in
its earlier history the planet was covered by perhaps 20 meters of
liquid water. This conclusion is based on the assumption that at an
earlier epoch the noble gases were in the same ratio to other, chemically
active, gases as they are now on Earth. It is possible that this epoch
was related to climatic changes caused by the variation of the solar
constant as well as to changes in the position of the highly eccentric
orbit of the planet. The present very low concentration of free carbon,
nitrogen and oxygen is not well understood. The isotopic ratio for
nitrogen indicates that there may have been as much as 30 mb of N_2 in
the past. The meteorological station of Viking 1, indicates rather low
winds (average 2.5 m sec^{-1}), a considerable daily variation of temperature (Fig. 11), humidity and pressure whose gradual drop is probably
associated with the process of condensation of CO_2 on the south polar
cap (Fig. 12). The atmospheric temperature is lower than that of the
soil by some 20°K which is similar to the situation on terrestrial
deserts. The images (Fig. 13) obtained by the landers added further
evidence of an earlier water erosion, of small local dust dunes formed
and transported by wind and of a surface crust of a strength of 10^3-10^4
dynes cm^{-2} covering a darker soil. The difference between the crust and
the soil under it is still an open question. The size of particles most
easily moved on the surface by wind is 100 - 200μ. The study of the
reflectivity of the sky, which is higher in the red than in the blue at
large angles from the sun and redder at high elevations than at low,

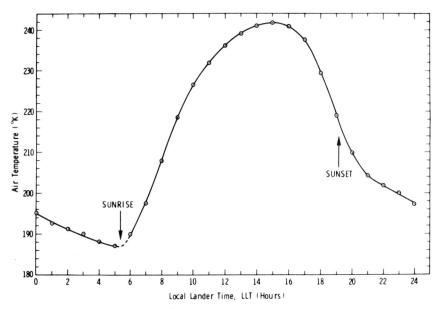

Fig. 11. Atmospheric temperature at Viking 1 landing site (according to
S.L. Hess et al.).

Fig. 12. Images of the surface of Mars obtained by Vikings 1 (top and
left bottom) and 2 (right bottom). The large rock in the top
picture is about 10 m away and 2 m across. Note the sand dunes.

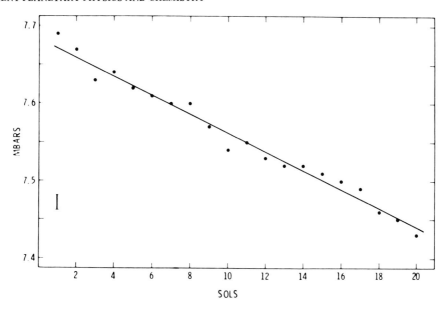

Fig. 13. Drop of atmospheric pressure at Viking 1 landing site (according to S.L. Hess et al.). Sol is a Martian day.

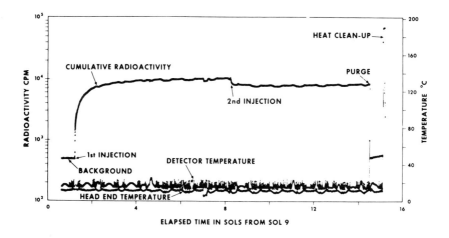

Fig. 14. Labelled gas release experiment made by the Viking 1 lander (according to H.P. Klein et al.). Sol is a Martian day.

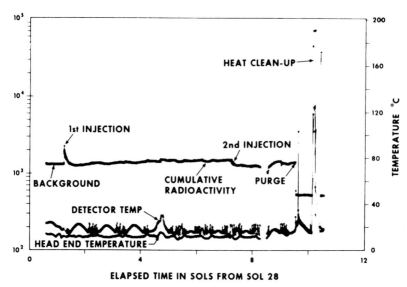

Fig. 15. Same as Fig. 12 except that the experiment was made on a sample preheated to 170°C for 3 hours.

indicates that there is much 1 micron dust in the atmosphere.

The interpretation of the various biological experiments performed by Viking 1 and 2 on the surface of Mars is not yet completed. These experiments are based on the assumption that the biological processes are carbonaceous, that the microorganisms biodegrade organic material and that the reactions are aqueous. It appears certain that the soil is chemically very active as if some simple biological processes analogous to terrestrial phenomena (such as photosynthesis of large carbon containing molecules, gas exchange and gas release from organic nutrients) were taking place. For instance the gas release experiment (7) detects metabolism by studying the release of $C^{14}O_2$ from soil injected with labelled nutrients such as formates. Fig. 14 shows the very rapid gas release at about 20°C after the first injection of the nutrient and a small drop of it after the second injection. The same experiment made, as a control, on part of the same sample of the soil heated to 170°C for 3 hours (which would sterilize a terrestrial soil) indicates, see Fig. 15, a radical drop in the amount of the released gas. This is analogous to effects observed with soil samples from certain regions in the Antarctic. In contrast to this seemingly positive result, the absence (less than 10^{-8}) of organic compounds containing more than three carbon atoms on both landing sites seems to argue against the presence of organic life as known on Earth. The presence of a substantial amount of water obstructs the mass-spectroscopic determination of molecules containing fewer than four carbon atoms such as propane or methanol. It may be that the activity and the high concentration of radicals in the soil have an inorganic explanation. Also it is probable that the intense ultraviolet flux incident on the surface in the presence of oxygen and

such oxidizing agents as nitrates and metal oxides, acting independently
or synergistically, destroys all larger organic molecules. Thus biological activity could be limited to deeper layers of the soil which are
protected from the radiation. Liquid water is very likely present only
in those pores in the soil in which the permafrost undergoes periodic
melting. A buoyancy effect (8) caused by the difference in density between the CO_2 and H_2O vapors tends to enhance the escape of water from
such traps but it prevents it in capillary pores which are closed from
above (9).

It should be added that preliminary results from the orbiter and
lander of Viking 2 are in good agreement with the data provided by
Viking 1 in spite of a different appearance of the two landing sites and
a distance of some 7000 Km between them. If confirmed by further landings
this result would indicate a surprising physical and chemical uniformity
of the surface of the planet in contrast to its very diversified morphological structure.

6.   Jupiter

More is known about the origin, evolution and present structure
of Jupiter than about other planet including Earth. The reason for this
somewhat unusual situation is the fact that, as mentioned below, Jupiter
is undoubtedly all fluid (i.e. has no true surface) and has a rather
simple overall chemistry. It should be stressed also that Jupiter
represents about 3/4 of the mass of all planets and, together with
Saturn, all of the mass but a few per cent.

The early evolution of planets, as described in the first part
of this paper, led to the formation of proto-Jupiter. The subsequent
evolution which followed the collapse of the central region has been
analyzed (10) starting with a body with a radius 35 R_J. One concludes
that the internal energy transport is primarily convective, while the
heat flux to the outside is controlled by radiative transfer in a
relatively thin atmospheric surface layer. The early luminosity of
Jupiter was $10^6 - 10^7$ times higher than it is now, which had an important effect on the densities of the Galilean satellites which decrease with the distance from the planet. At an epoch of about 10^5
years, Jupiter's central temperature reached 50,000K which may have
vaporized the central core. Interestingly enough much uncertainty concerning the later stages of this evolutionary process is related to
details of the equation of state of the condensed matter which even now
have not been completely clarified. In particular, it does not seem
possible to give a definite answer to the question whether the internal
energy built up in the past is sufficient to account for the present
Jovian luminosity. Bodenheimer's calculations of the very early history
of Jupiter, (Fig. 1) fit the results of Graboske et al. very well. It is
encouraging that varying -- sometimes quite drastically -- the initial
conditions of the early processes has only a minor effect on later
stages and of the final result. Even the presence or absence of a rocky
core, provided it is not too big a fraction of the total mass of the
planet, does not affect significantly the evolutionary outcome. From
the point of view of planetary models discussed below, it is important

to note that, all these evolutionary histories lead invariably to a
solar ratio of hydrogen to helium.

The impressive recent progress in our understanding of Jupiter
stems from the evolutionary studies described above, from the large
amount of data obtained by the fly-by Pioneer 10 and 11 missions and
from theoretical and experimental studies of the hydrogen-helium
system. Pioneer data of importance for the planetary interior are
oblateness, gravitational coefficients, internal heat emission and the
structure of the magnetic field. The agreement between oblateness calculated from the new values of the gravitational coefficients J_2 and J_4
and the observed oblateness indicates that, with good approximation,
the planet is in hydrostatic equilibrium. The fact that, within the
limits of error, J_3, J_6 and the tesseral coefficients vanish further
confirms this conclusion and points to a totally fluid planet. Inasmuch
as the gravitational coefficients depend critically upon the radial
density profile of the outer 3000 Km or so of the planet it is possible
to characterize accurately that region using gravitational sounding
provided a reliable equation of state is available. The two central
assumptions concern the hydrogen-helium mass ratio which, as mentioned
before, for a solar composition should be (11) 3.4 ± 0.1 (or (12)
3.8 ± 0.15) and the temperature at 1 bar, for which the best data indicate the value $165 \pm 5K$. In some models of Jupiter, the correct values
of J_2 and J_4 can be accounted for only if an excess of a heavy component
such as water or of another heavier solar constituent is present in the
outer layers. Other models based on somewhat different equations of
state, especially of the H_2 layer, appear to avoid this problem.

Various studies show that, at not too high pressures and temperatures, helium is not completely soluble in the metallic and in the
molecular hydrogen. This can lead to a separation (13) between He-rich
and He-poor layers in both phases in a planet as shown in Fig. 16. It
should be stressed, however, that the actual limits of these miscibility
gaps are not well established. If such boundaries are indeed present,
then adiabatic convection, heat transport and homogenization are possible only within each layer but not across the boundaries. This situation
may affect significantly the calculated temperature gradient and the
central temperature. Presence of the boundaries implies that, as the
planet cools, a progressive precipitation or separation of helium droplets takes place. These droplets drift towards the center of the planet
converting gravitational energy into internal heat (14, 15). This
mechanism can account for the observed emission of internal energy by
Jupiter (10^4 erg cm^{-2} sec^{-1}) and by Saturn which is nearly twice as
high as the energy which they receive from the sun. The mechanism is
also self-regulating. There are of course other possible sources of this
energy such as primordial heat, shrinkage of about 0.1 cm per year without a phase change, etc. It seems certain that on Saturn, and perhaps
also on Jupiter, such an internal source of energy is essential in order
to account for the present luminosity of both planets which appears to
be higher than that deduced from their evolutionary histories. In this
connection one should point out the known conclusion that if Jupiter were
some 70-100 times larger it would have become a white dwarf. Thus in this
sense it is almost a star. Fig. 17 illustrates a model (15) calculated

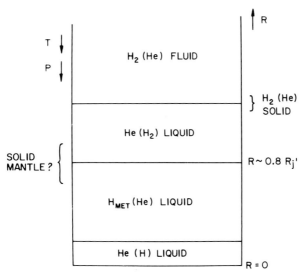

Fig. 16. Schematic sequence of possible hydrogen-helium layers on Jupiter. The range where islands of solid H_2 could exist is indicated (according to R. Smoluchowski).

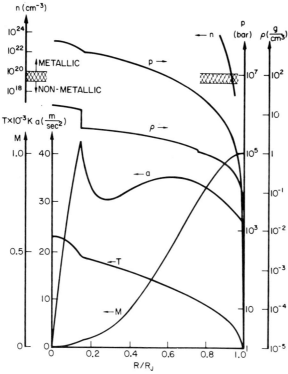

Fig. 17. Radial profiles of mass M, Temperature T, pressure P, density ρ and gravitational acceleration a on Jupiter (according to V.N. Zharkov et al.). Curve n indicates the concentration of conducting electrons in the H_2 layer (according to R. Smoluchowski).

assuming a fully adiabatic temperature gradient, a temperature of 140K
at 1 bar, a central liquid core made of SiO_2, MgO, Fe, Ni, NH_3, and H_2O,
no miscibility gaps and the presence of some heavier constituents added
to the solar mixture of H_2 and He in the outer mantle. The discontinu-
ities of the various quantities at the core boundary ($R/R_J \sim 0.14$) and
at the metallic-molecular boundary ($R/R_J \sim 0.75$) are clearly visible.
The mass of the central core is usually of the order of several per cent
of the mass of the planet, central density is around 20-30 g cm^{-3}, central pressure 20-100-M bar and central temperature 15-25,000K.

     Independent of the details of the models discussed above, it is
clear that within the molecular H_2 layer the pressures and the temperatures are very high. In fact, as shown in the upper part of Fig. 17,
below a radius of about 0.93 R_J the concentration of conduction electrons
(16) is high enough to produce metallic conduction. The main consequence of this result is that the usual dynamo mechanism for the
generation of a magnetic field in the planet does not need to be confined to the actual metallic phase of hydrogen and that the inclusion
of the highly conductive part of the H_2 region nearly doubles the
volume within which the dynamo can operate bringing it much closer to
the surface. This is important because a recent evaluation (17) of this
mechanism for Jupiter suggests that the generation of the magnetic field
occurs predominantly in the outer parts of that region where the conduction is metallic. This effect may explain the presence of intense
quadrupolar and octupolar components of the basically dipolar Jovian
magnetic field as observed by both Pioneers. The effective dipole
moment, which has a value of 4.24 gauss R_J^3 is tilted by about 10° away
from the rotational axis and is somewhat off-center. The extrapolated
values of the fields at the north and south poles are respectively 14
and 10.7 $\pm$ 0.3 gauss. The offset expressed as a fraction of the planetary
radius and the tilt are comparable to those found on Earth.

     One expects the Jovian magnetosphere to have a shape somewhat
like that shown in Fig. 18. Depending on the strength of the fluctuations
of the solar field the magnetosphere terminates on the sunward side at
50 to 100 R_J. It is worth pointing out that the diameter of the magnetosphere is so huge that from Earth it subtends an angle of about $\frac{1}{2}$ degree
that is the same as our Moon or Sun. On the lee side of Jupiter the solar
wind produces an enormous magnetotail which, according to very recent
Pioneer 10 data (18), reaches all the way to the orbit of Saturn which
is 5 AU away. Up to 15 R_J or so the field is essentially dipolar but
further out it extends into a magnetodisc in which the field lines are
stretched out by interaction with the plasma co-rotating with the planet.
The boundaries of the magnetosphere are displaced by the buffeting solar
wind at a velocity up to 1/10 of the velocity of light which enhances
the escape of many electrons into the space. These energetic electrons
(3 to 30 MeV) are emitted in the polar direction and since the axis of
the dipolar field (and of the magnetosphere) is tilted to the rotational
axis, they show an approximately 10 hour period. They have been observed
as far as Earth and Mercury.

     The well known but rather mysterious bursts of decametric (and
hectometric) radiation appear to be generated in specific several hundred kilometers wide areas in the planetary ionosphere. Some but not all

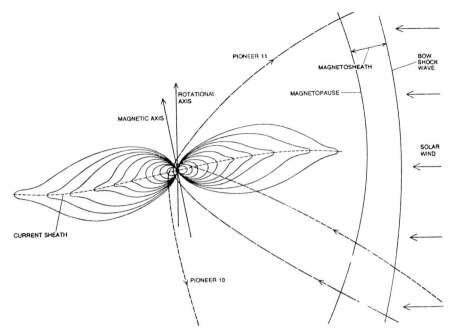

Fig. 18. Jovian magnetosphere showing the region of the dipolar field, the magnetodisc and the magnetopause (according to E.J. Smith).

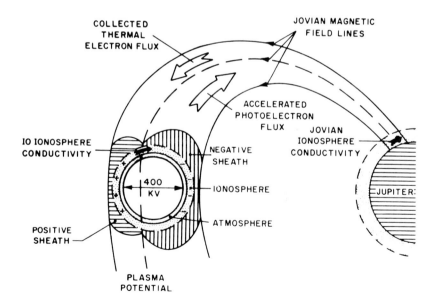

Fig. 19. Schematic configuration of the plasma sheath around Io and the flux tube (according to S.D. Shawhan et al.).

of these emissions are well correlated with the position of the satellite
Io. This correlation has been at least partially explained by a verification by Pioneer 11 of the presence of the theoretically anticipated
magnetic flux tube connecting a plasma sheath formed around Io with the
planet's ionosphere (Fig. 19). One expects the foot of this tube to
enhance instabilities in the ionosphere and thus affect the decametric
bursts. Io is subjected to such intense bombardment by the energetic
particles in the magnetosphere that 10^7 Na and 10^{11} H atoms per cm^2 are
sputtered off from its surface. These atoms form an atmosphere around
the satellite and spread into a torus along a large part of its path.
Decimetric radiation is generated by a synchrotron motion of electrons
trapped in a toroidal region contained between 1.5 and 6 R_J which is
analogous to the terrestrial Van Allen belt. The energy of these electrons is in the neighborhood of 17 MeV and their highest flux occurs at
about 2.5 R_J in accord with measurements *in situ*.

The composition and structure of the Jovian atmosphere and ionosphere is rather well known. Fig. 20 and Fig. 21 summarize the present
data for the upper atmosphere and for the lower atmosphere respectively
as obtained from an analysis of terrestrial and spacecraft observations.
The upper ionosphere is some 3000 Km thick, where at pressures lower
than 10^{-9} bar, the average proton and electron temperature is around
1000K. Pioneer radio occultation data suggest that there may be a hot
thermosphere heated by inertia gravity waves as if Jupiter sustained a
corona. At the upper boundary of the mesosphere, that is at 10^{-6} bar, the
temperature drops to around 170·K. Just above the mesosphere, heat may
be generated by the difference between the rotational velocities of the
atmosphere and of the magnetic field. The latter rotates faster and drags
with it ions and electrons which in turn accelerate neutral gas. The
mesosphere itself is isothermal down to a level at which the pressure
reaches 10^{-3} bars. In this region there is a radiative balance between
the heat originating in the planet and in the sun and that lost through
molecular infrared cooling.

At a pressure of 10^{-3} bar begins the stratosphere within which
the temperature drops rapidly down to a minimum of about 110 K at 0.1
bar. Below, in the troposphere, the temperature rises adiabatically,
reaching around 165 ± 5 K at 1 bar. This gradient extends down to pressures around 20 bars. Fig. 21 shows also the probable location and composition of the clouds. Besides H_2, CH_4 and NH_3 Pioneer 10 observations
have verified the presence of helium by measuring the resonant scattering
of the 584Å line. The abundance of H_2 is very close to solar but the
error in the determination of the hydrogen to helium mass ratio is so
high that the best one can say is that the assumption of a solar ratio
of 3.4 ± 0.1 is not inconsistent with the atmospheric data. Other species
besides those mentioned above are CO, PH_3, GeH_4, CH_3D and HCN. The
various brown, reddish and yellow colors of Jupiter may be associated
with the presence of amorphous red phosphorus, hydrogen polysulfides
H_2S_x, ammonium polysulfides or sulfur.

High resolution images of the clouds and radiation measurements
taken by Pioneer 10 and 11 have shown a wealth of data much of which is
not yet understood. In the bright zones (including the Great Red Spot)
the clouds are high, the velocity is upward, the temperature is high

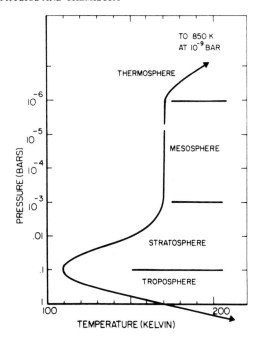

Fig. 20. Structure of the upper atmosphere of Jupiter (according to
D.M. Hunten).

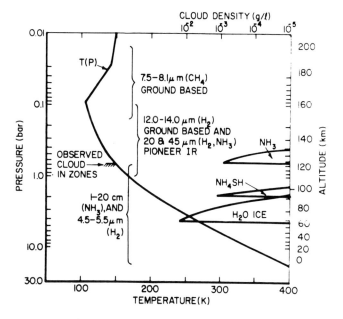

Fig. 21. Structure of the troposphere and the location of the principal
infrared absorbers and clouds (according to A.P. Ingersoll).

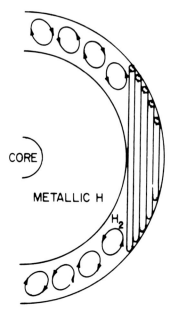

Fig. 22. Odd and even convention cells in the Jovian H_2 mantle (according to F.H. Busse).

while the infrared brightness is low. The opposite is true of the dark belts. The geostrophic balance between pressure and Coriolis forces results in prograde and retrograde motion of various cloud systems. Pioneer images indicate the presence of barotropic instabilities along poleward edges of belts and equatorward edges of zones as expected on theoretical grounds. The longevity of distinct Jovian cloud features appears to be related to the length of the infrared radiative time constant of the atmosphere which is about one year. Many new features such as several small red spots, white ovals, plumes, and blue spots in the equatorial zone as well as some detail within the Great Red Spot have been discovered. There is still uncertainty whether the Great Red Spot and the smaller ones are purely meteorological phenomena or solitons or external manifestations of Taylor columns which may be attached to solid H_2 (He-poor) islands floating at an appropriate level (perhaps 10^4 Km below the cloud level) in liquid H_2 (He-rich). None of these hypotheses seems to be entirely satisfactory and quantitative conclusions are rather elusive.

Pioneer data indicate that the temperature difference between the poles and the equator is only about 2 K in contrast to a 20 K difference expected on the basis of much higher insolation at the equator. A possible explanation of this phenomenon is a decrease of the outward radial thermal gradient near the equator produced by solar radiation with the result that the flow of the internally generated heat towards the poles is more than a factor of two higher than towards the equator. This difference in the heat flow may be related to the fact that the well known banded structure of Jovian clouds ceases at about 45° latitude and is

replaced by a more random cellular motion near the poles. The presence
of this limiting latitude also agrees with the conclusion (19) that, at
low latitudes, the internal heat is convected by even mode cells which
are parallel to the rotation axis of the planet and cannot penetrate
into the metallic phase (Fig. 22). At higher latitudes the usual odd
mode convection cells set in.

Note added in 1979: The wealth of new information about Jupiter and its
satellites obtained by both Voyager missions in 1979 is enormous. Briefly: the velocities observed in the atmosphere suggest mass rather than
wave motion but the motion is very intricate and convoluted; the well-
known belt zone pattern of clouds extends into the polar regions in
contrast to earlier conclusions; various cloud spots seem to be spaced
more or less regularly around the planet at one or two latitudes suggesting a wave interaction; the Great Red Spot and other red and white
spots are meteorologically similar, interact among themselves and show
internal detail in anticyclonic and cyclonic motions; the atmosphere
above the GRS is colder than in the nearby regions. Auroral emissions
in the polar regions and cloud-top lightnings in the ultraviolet have
been observed. Undoubtedly, the most unexpected result is the discovery
of a tenuous ring at 1.83 Jovian radii (width $\sim$6000 Km, opacity $\sim$10^{-4})
confirming an earlier conclusion based on anomalies in the proton and
electron belts observed by Pioneer spacecrafts. The ring is probably the
result of a break-up of an incident body and was severely eroded by
protonic and micrometeoritic bombardment. In fact a small satellite
($\sim$40 Km) which is present at the edge of the ring (1.8 R_S) may be the
remnant of this body. Another new feature is a disc of very fine material
which extends from the ring to the planet itself. Amalthea appears
faceted in shape, roughly 265 Km long and 150 Km wide and it is dark red
which probably results from the expected intense proton bombardment.
Eruptions of at least eight volcanoes with plumes extending up to 320
Km have been observed on Io; the satellite's interior is heated by tidal
effects and probably also by interaction with the powerful rotating
magnetosphere. Io's surface has reddish-yellowish (sulphur) and white
(SO_2frost) areas and is very smooth indicating obliteration of craters
by volcanic ejecta. These ejecta account probably for the presence of
Na and S in the atmosphere and for the ionized sulphur torus along the
orbit of the satellite. Particles from the torus can follow the magnetic
lines to the planetary poles and may account for the ultraviolet and
visible auroras mentioned above. The current flowing in Io's flux tube
is about 5×10^6 Å which may play a role in keeping its interior liquid;
the plasma contains S^{2+}, S^{3+} and O^{2+} ions as well as about 4500 electrons
per cc; the kilometric radio emission of Jupiter may be due to oscillations in Io's plasma. Europa is bright presumably because of an ice
cover and is crisscrossed by striking linear features over 100 Km long
and 200-300 Km wide caused probably by tectonic activity. The scarcity
of impact craters on Europa may be due to the softness of the ice (50
Km ice layer on water). Ganymede shows bright fresh craters, remnants
of ancient impact basins and grooved terrains also indicating tectonic
motions in its icy and rocky interior. Callisto is more heavily cratered
than any of the other three satellites but its easily deformable crust

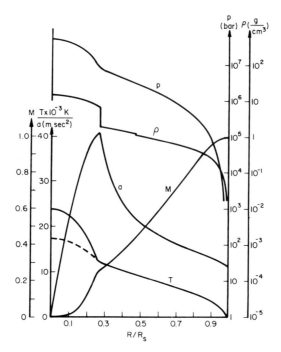

Fig. 23. Same as Fig. 15 but for Saturn.

made the craters rather indistinct although enormous, a few thousand
Km diameter, concentric rings indicate ancient impacts. It is probably
the oldest Galilean satellite.

ref. Science 204, June 1, 1979 and 206, November 23, 1979
     M.H. Acuña and N.F. Ness, J. Geoph. Res. 81, 2917 (1976)
     S.J. Peale, P. Cassen and R.T. Reynolds, Science 203, 894 (1979)
     R. Smoluchowski, Nature 280, 377 (1979)

7.   Saturn

      Saturn in many ways is very similar to Jupiter, and most of what
was said above about Jupiter is applicable to it although Saturn is
much less understood. One reason for this is that 50 to 70 per cent of
Saturn's interior is condensed H_2 (density 0.1 - 1 g cm^{-3}) for which the
equation of state is rather poorly known while on Jupiter H_2 accounts
for only about 25% of the mass. Nevertheless many models of the interior
of Saturn (mean density 0.7 g cm^{-3}) have been proposed. One of them (15)
is shown in Fig. 23 which indicates the internal temperature, pressure,
density, acceleration and mass distribution. It appears that the overall
composition of Saturn deviates further from solar than it does for
Jupiter mainly because the rocky core is much bigger and accounts for
about 25% of its mass.

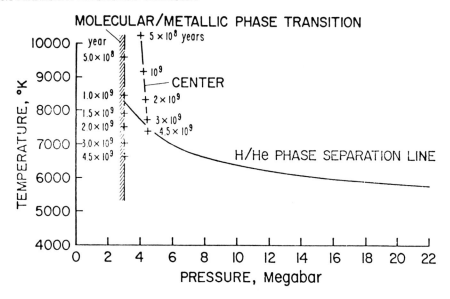

Fig. 24. Molecular-metallic phase transition, the limit of the hydrogen-
helium miscibility gap in the H-He system and the evolutionary
track of the center of Saturn (according to J.B. Pollack et al.).

The metallic layer reaches to 0.40 – 0.45 of the planets radius,
the rest being fluid H_2 – He with some enhancement of H_2O above the
solar value. Theory shows that separation of the He-rich and He-poor
phases in metallic hydrogen started after 2×10^9 years and reached the
center at about 3.7×10^9 years as shown in Fig. 24. Since the metallic
core is small and is deep in the planet, one would expect the magnetic
field, if any, to be weaker than that of Jupiter and to have a more
purely dipolar character. This conclusion is valid even if, in analogy
with Jupiter (16), one considers the presence of metallic conductivity
in the H_2 layer up to a radius 0.6 R_S (which more than doubles the
volume within which the field can be generated) as contrasted with
0.93 R_J for Jupiter. Actually for a long time it was suspected that
Saturn had no magnetic field, but recent observations (20) of, admittedly
weak, hectrametric and decametric radiation indicate the presence of a
magnetic moment of about one tenth of that of Jupiter. The absence of a
measurable decimetric radiation is accounted for by the fact that the
Van Allen belt is suppressed by the planet's rings which easily capture
all electrons spiraling along the magnetic lines (Fig. 25). Electrons
moving outside of the rings produce about 3 per cent of the intensity
of the Jovian decametric radiation and this would not be measurable
from Earth. Saturn, just like Jupiter, has an internal source of powerful thermal energy which could be accounted for by the same progressive
segregation of helium from the hydrogen-helium solutions. It appears
that this process may have been active for a billion years and may explain the present luminosity of Saturn which is appreciably higher than
that indicated by evolutionary calculations (21).

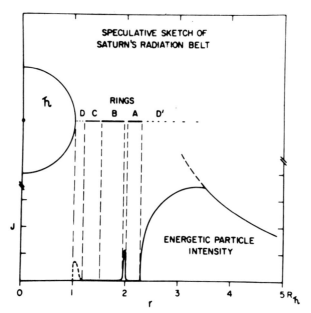

Fig. 25. Capture of electrons of the Saturnian Van Allen belt by the
        rings (according to J.A. Van Allen).

     Saturn's atmosphere is qualitatively similar to that of Jupiter
and contains H_2, CH_4 and NH_3. Recently ethane, C_2H_6, has been also
identified (22). The rather diffuse light and dark bands of clouds have
higher zonal velocities than on Jupiter and their motion is much more
differentiated. Various spots appear and disappear after a few days or
weeks. The visible clouds are ammonia ice; further below one expects to
find water ice clouds. Since the radiative relaxation time (23) of the
atmosphere is 10 years, which is shorter than the sideral period, there
may be seasonal effects.
     Considerable progress can be noted in our understanding of the
origin of Saturnian rings (24) whose chemistry, i.e. "dirty" ice, seems
to have been rather well confirmed. Their uniqueness appears to be a
consequence of the fact that at the time when the temperature in the co-
rotating proto-Saturnian nebula dropped low enough to condense ices,
other gases were already expelled so that the ice could form large
clusters without being slowed down by the drag of the gas. All this
happened of course within the Roche limit which prevented the formation
of a satellite. Of particular interest are the very thin rings; the
innermost ring D and the reasonably well established outermost ring D'.
Although these rings are not easily visible they may contain a large
amount of very finely divided material as contrasted to the other bright
rings in which one expects clusters up to 100 - 1000 m in diameter.
Polarization data and brightnes of the ansae indicate that at least in
some rings the particles are elongated with their major axes pointing
towards the planet. Much information about the rings has been obtained
from the study of the rate of increase of their temperature when the
rings emerge from the planet's shadow.

One has to mention also Titan the fascinating satellite of Saturn. Of all the known satellites it is the only one to have a dense atmosphere (about 4 times that of Mars) and, as very recent microwave measurements (25) have shown, its surface temperature is raised by a greenhouse effect from the expected equilibrium temperature of about 100K to nearly 180K.

Note added in 1979: New data on the molecular-metallic phase transition in hydrogen indicate a lower pressure than previously estimated and this increases by a factor cf 2 to 3 the volume within which a magnetic field can be generated on Saturn probably by a hydromagnetic dynamo. The measured magnetic dipole moment is 0.2 gauss R_s^3, which is less than previous estimates, and the magnetosphere is intermediate between those of Earth and Jupiter, but without the pronounced magnetodisc of the latter. The angle between the axis of the field and the planet's rotation axis is less than 1° which makes magnetic measurement of its rotation rate impossible. The magnetosphere reaches all the way to Titan and distinct absorption signatures in the distribution of entrapped particles are produced by the known satellites Dione, Mimas, Tethys and Enceladus but none by Janus at 2.66 R_S. New satellites (>170 Km) appear to be present at 2.534 R_S and at 2.343 R_S and quite possibly also at 2.82 R_S. A new F-ring between 2.336 and 2.371 R_S is separated from the outer edge of the A-ring, 2.292 R_S, by the Pioneer gap. The existence of the previously suspected E-ring is not confirmed. As expected there are essentially no radiation belt particles below 2.292 R_S. The temperature of the planet is about 100K which indicates an excess internal heat 2.5 larger than that due to insolation. The temperature of the rings is 65-75K which is lower than previously believed. The planet and the rings appear to be surrounded by a cloud of H gas which may come from ultraviolet dissociation of H_2O sputtered from the rings. Saturn has a large H^+ ionosphere reaching 1250K.

ref. Science 207, January 18, 1980.
     R. Smoluchowski, Phys. Earth Plan. Inter. October 1979.

8.   Uranus and Neptune

As one progresses further away from the sun, the amount of information and reliability of models of the planetary interiors decreases exponentially. Nevertheless, it is clear that both Uranus and Neptune deviate from the solar composition more than Jupiter and Saturn and thus their average densities give less information about their chemistry than they do for the other two. In that sense the two planets are intermediate between the giant hydrogen-helium planets and the terrestrial group. Neither planet is likely to have metallic hydrogen; their cores are large and contain perhaps as much as 30 per cent of the mass (15,26). It is possible that there is a thick layer of ice (H_2O and NH_3) between the core and the surrounding mantle (H_2, He and CH_4) which accounts for about a quarter of the mass.

Various one, two and three-layer models of the planets show only a fair agreement with the observed oblateness. This is partially to be blamed on the poorly known velocity of rotation ω. The gravitational

coefficients J_2 and J_4, on which nearly all detailed information about
a planet's interior is based, are respectively proportional to ω^2 and
ω^4, so that the resulting errors are huge. A model of Uranus which
assumes a rather high enrichment in CH_4 appears to be more satisfactory.
It should be pointed out, however, that very recent observations of
Uranus in the methane wavelength region suggested a radius 7 percent
larger than the value previously used. If confirmed, this would bring
the planet's density to below unity and would change the models of the
interior drastically.

It is not impossible that both planets have a weak magnetic field
because the ionic conductivity of the rocky core and of water at pressures and temperatures characteristic of the interior may be high enough
to satisfy the necessary conditions for a hydromagnetic dynamo. Tentative
evidence exists of bursts of radio noise from Uranus near 0.5 M H_z. The
planets may also possess sources of internal energy (27) similar to
those on Jupiter and Saturn but based on segregation within the rocky
core and the ices rather than in the H_2 layer.

The atmospheres of both Uranus and Neptune are composed of H_2 and
CH_4, the last one being much enhanced over its abundance in Jupiter and
Saturn. The clouds are most likely icy NH_3 with the gaseous form at
lower levels but no definite identification of this compound has as yet
been made. On Uranus there is considerable limb darkening and the
presence or absence of a reflecting layer of clouds below the visible
ones has been widely discussed. In a recent model there is an upper
layer of aerosol (28) haze (CH_4) separated by clear gas from a lower
haze layer (NH_4). The calculated (23) radiative relaxation time for
Uranus is 630 years which suggests that, at least at the level of the
visible clouds, there are no significant seasonal variations in spite
of the abnormal insolation mentioned below.

A fascinating aspect of Uranus is the origin of the 98° inclination
of its equatorial plane, including the satellites, to the plane of the
ecliptic. While it is possible to suggest a mechanism whereby some encounter or other perturbation resulted in a tilted axis of the planet,
it is difficult to see why the satellites should be similarly affected.
Perhaps a more satisfactory explanation assumes that the planet had (29)
originally a small inclination but captured a satellite of a mass of a
few per cent of the planet's mass in an orbit which had an inclination
of 140° and a mean radius of about $25R_u$. Calculation shows that, in the
course of time, through tidal evolution the planet's axis moved to its
present position, the satellite reached the Roche limit at about $6R_u$ and
broke up into the present three retrograde satellites. The unusual
orientation of the planet has the interesting consequence that the location of the hottest part of the planet moves during one full revolution
around the sun from one pole, across the equator, to the other pole and
back again. As a result each pole reaches a higher and a lower temperature than the equator.

Note added in 1979: New values of the radius and oblateness of Uranus
are 25,700 ± 200 Km and 0.035 ± 0.008 respectively. Nearly all rotation
periods of this planet suggested recently are longer than the old one
of about 10^h, the most likely correct one being $23^h 55^m$. The differences

in the solar irradiation appear to explain the fact that Neptune emits
three times as much energy as it gets from the Sun and Uranus essentially
none. A three layer model of Uranus has a 24% rocky core, 65% H_2O and
11% H_2-He atmosphere; for Neptune the respective values are 25, 68 and
6%. Both planets have a central temperature of about 7000K. The core of
Uranus appears to be partly solid, the H_2O layer liquid. The conductivity of the latter is too low, however, to explain the presence of a magnetic field which has to be thus generated (precessionally by Miranda
or thermally) in a very high conductivity H_2 layer. The thinness of the
H_2 layer on Neptune may account for the total absence, or weakness, of
a magnetic field on that planet. Occultation studies of Uranus led to
the discovery of several narrow (10-20 Km), roughly concentric, rings
and of one eccentric and wider ring. Their albedo is at most 0.03 and
they probably are remnants of a broken up body which are kept in
position by resonances with satellites.

ref: J.L. Elliot, E. Dunham and D. Mink, Nature 267, 328 (1977)
     Bull. A.A.S. 11, #3 (1979) several abstracts
     W.B. Hubbard and J.J. MacFarlane, Icarus, in press
     M. Torbett and R. Smoluchowski, Geoph. Res. Lett. 6, 675 (1979)
     R. Smoluchowski, Comments Astroph. 8, 69 (1979)

9.   Pluto

     Essentially nothing can be said about the interior of Pluto since
even its density of about 5 is known at best with an error of 20 per
cent because of uncertainties both in the mass and in the radius. This
fact and the high inclination of its orbit, 17°, makes it almost certain
that this planet did not originate during the process of condensation
of the solar nebula. Its origin and structure remain a great mystery.
Recent observations indicate that there is CH_4 and water ice on its
surface.

Note added in 1979: The new important development concerning Pluto is
the discovery of a presumably synchronous satellite with a period of
6.39 days and a mass (5 to 10) x 10^{23}g as compared to the planet's mass
of about 10^{25}g which, with the planet's radius of about 1500 Km, gives
a density of 0.7. The distance between the two bodies is 17,400 Km which
is comparable to the distances of the two satellites of Mars from the
parent planet. The atmosphere of Pluto must have a heavy constituent
such as argon to retain the observed CH_4 at 55K. The low density of
Pluto seems to support the suggestion that it is an escaped satellite
of Neptune.

ref: J.W. Christy and R.S. Harrington, Astron. J. 83, 1005 (1978)
     Bull. A.A.S. 11, #3 (1979) several abstracts

10. Conclusions

Recent advances in our knowledge of the planets have opened new
possibilities for the application of theoretical and experimental physics
and chemistry to the problems of the origin and structure of the solar
system. Geological and biological investigations have answered some important questions but also raised a host of others which will require
systematic investigations by future planetary missions.

11. Acknowledgements

The author is indebted to Drs.: J.L. Anderson, T. Gehrels,
R.M. Goldstein, M.Ya. Marov, G.H. Pettengill, I.S. Rasool, C. Sagan and
B.A. Smith for supplying him on very short notic with recent data and
slides.

General References (contain much of the material used and quoted in this
paper)

"An Introduction to Planetary Physics: The Terrestrial Planets",
W.M. Kaula, J. Wiley, New York, 1968.
"A Survey of The Outer Planets", R.L. Newburn, jr. and S. Gulkis, Space
Sc. Res. 3, 179 (1973).
"Jupiter", T. Gehrels ed., U. of Arizona Press, Tucson, Ariz. (1976).
"Proceedings IAU 16th General Assembly", August 1976, Grenoble, France.
(all planets including Venera and Viking missions).
"Proceedings Intern. Symp. on Solar and Terrestrial Physics", June 1976,
Boulder, Colo. (all planets).
"Proceedings NATO Advanced Study Institute", April 1976, Newcastle-Upon-
Tyne. (Solar system).
"Proceedings 19th COSPAR Meeting", June 1976. Philadelphia. (all planets,
Venera missions).
Science, 183, Mar 29 (1974) Mariner 10 mission to Venus
        185, July 12 (1974) Mariner 10 mission to Mercury
        193, Aug. 27 (1976) Viking mission to Mars
        194, Oct. 1 (1976) Viking mission to Mars
J. Geophys. Res. June 10 (1975). Mariner 10 mission to Mercury.
J. Geophys. Res. September 30 (1977) Viking Mission
Icarus, 28, June 1976, Pre-Viking
        28, Aug. 1976, Mercury
        29, Oct. 1976, Jupiter
        30, 605, 1977, Venera Mission (M.V. Keldysh)
Scientific American, Sept. 1975 (all planets – popular).
American Scientist, Nov.-Dec. 1976 (Jupiter – popular).

References (Specific topics not necessarily covered by the general
references)

1. N.F. Ness, Icarus, 28, 479 (1976).
2. S.J. Peale and A.P. Boas, J. Geophys. Res. 82, 743 (1977).

3. R.G. Strom, N.J. Trask and J.E. Guest, J. Geophys. Res. 80, 2475 (1975).
4. A.L. Betz, M.A. Johnson and E.C. Sutton, Astroph. J. 208, L141 (1976).
5. M.B. McElroy in "Chemical Kinetics", Ed. D.R. Herschbach, Butterworths, London 1976.
6. R.M. Goldstein, R.R. Green and H.C. Rumsey, J. Geophys. Res. 81, 487 (1976).
7. G.V. Levin, Icarus 16, 153 (1972).
8. A.P. Ingersoll, Science 168, 972 (1970).
9. R. Smoluchowski (to be published).
10. H.C. Graboske, jr., J.B. Pollack, A.S. Grossman and R.J. Olness, Astroph. J. 199, 265 (1975).
11. A.G.W. Cameron, Space Sci. Rev. 15, 121 (1973).
12. A.B. Makalkin, Sov. Astr. 18, 243 (1974).
13. R. Smoluchowski, Astroph. J. 185, L95 (1973).
14. E.E. Salpeter, Astroph. J. 181, L83 (1973).
15. V.N. Zharkov, V.P. Trubitsyn, I.A. Tsarevsky and A.B. Makalkin, Physics of Solid Earth, 10, 610 (1974).
16. R. Smoluchowski, Astroph. J. 200, L119 (1975).
17. F.H. Busse, Phys. Earth, Plan. Int., 12, 350 (1976).
18. J.H. Wolfe, D.H. Mihalov, H.R. Collard, D.D. McKibbin, L.A. Frank and D.S. Intriligator, Science January 18, 1980.
19. F.H. Busse, Icarus 29, 255 (1976).
20. L.W. Brown, Astroph. J. 198, L89 (1975).
21. J.B. Pollack, A.S. Grossman, R. Moore and H.C. Graboske, jr., Icarus 29, 35 (1976).
22. A. Tokunaga, R.F. Knacke and T. Owen, Astroph. J., 197, L77 (1975).
23. P.H. Stone, Space Sci. Rev. 14, 444 (1973).
24. J.B. Pollack, Space Sci. Rev. 18, 3 (1975).
25. J.R. Dickel, April 1976 Meeting Am. Astr. Soc. Div. Pl. Science.
26. M. Podolak and A.G.W. Cameron, Icarus 22, 123 (1974).
27. L.M. Trafton, Astroph. J. 207, 1007 (1976).
28. R.I. Murphy and L.M. Trafton, Astroph. J. 193, 253 (1974).
29. R. Greenberg, Icarus 23, 51 (1974).

REMOTE SENSING OF EARTH RESOURCES AND ENVIRONMENT FROM SPACELAB

JOHN PLEVIN
Application Programmes Department,
European Space Agency

ABSTRACT

Although Man's knowledge of the Earth and his environment has
increased spectacularly in recent years, the demand on natural resources and the pressures applied by mankind on the environment seem to
increase at an even faster rate, creating an urgent need for new and
improved information sources. In this chapter the value of remote
sensing satellites as a provider of earth resources information is discussed and, in particular, an assessment made of the role of Spacelab as
a remote sensing platform. Spacelab is seen as a platform best suited
to conduct geoscientific and technological experiments using a wide range
of advanced sensors (e.g. optical images, active and passive microwave
sensors, cameras). In summary, Spacelab is seen as a bridge linking the
remote sensing aircraft of today with the operational and long lifetime
automatic satellites of the future.

NEED FOR EARTH RESOURCES SURVEYS

The world we live in faces many problems related to the management
and utilisation of its natural resources. The demands on these resources
are intense and presently little controlled in any global or large area
sense. It is generally recognised that improvements are necessary in the
way decisions are made that relate to the utilisation of natural resources.
In order to achieve these improvements, new and more effective sources of
information are required that are both relevant to earth resources requirements and objectives, and available in a suitable form and timescale.
Remote sensing techniques appear today the most likely candidates to fill
this information-gathering role. This need for information exists in both
the developed and developing areas of the world and is required on local,
regional and global scales and in differing levels of detail. Often the
clearest common denominator in this spectrum of requirements is the technology of the sensing process itself.

Although the importance of new data sources is recognised, we need
to remember that remote sensing provides information, nothing more and
nothing less. The prime aim of many applied research programmes today

P. L. Bernacca and R. Ruffini (eds.), Astrophysics from Spacelab, 527-562.

is to attempt to understand what this information means and how it can
be used to solve the complex problems we are facing. These problems are
diverse and often appear contradictory; for example, remote sensing
data can be used to increase growth in developing areas, whereas similar
information may be used to limit or control growth in a developed area.
In attempting to categorise the many different applications, a regional/
global distinction is possible :

a)   Regional applications :

     -   to monitor dynamic features in developed regions such as Europe,
         and possibly in selected areas within the developing regions;

     -   to provide basic resources information to be used within the
         framework of development aid programmes.

b)   Global applications :

     -   to monitor the global oceans;

     -   to monitor the global atmosphere, in particular with respect to
         the climate and climate changes, and air quality measurements;

     -   to map global resources such as basic food supplies, soils and
         forests.

     Studies have shown that,in the development of spaceborne remote
sensing methods, high priority must be given to the implementation of
regional space systems providing data that can be applied both in Europe
and within the framework of development aid programmes. The key missions
for such a satellite system are :

Monitoring of Developed Regions

     The problems of developed regions are usually associated with the
management and conservation of known resources and monitoring an overstrained environment rather than on such aspects as resources exploration and exploitation, which are often the primary goals of developing
countries. Exceptions exist such as Greenland which needs to be
considered very much as a developing area, and the Continental Shelf
where considerable basic exploration is still required. Similarly, the
information requirements of the Canadian Northern Provinces will differ
considerably from those in the southern regions of the country. Not-
withstanding these exceptions, the main applications of satellite remote
sensing in developed regions will be related to the critical problems
facing a modern society such as food production including fisheries,
monitoring how the use of land evolves, determining levels of environmental pollution and like topics.

     An analysis of objectives for a European remote sensing programme(1)
conducted by the European Space Agency (ESA) has shown that priority
should be given to :

- Statistical information on agricultural products (e.g. crop inventory, yield prediction for key crops, forest inventories);

- Land use classification and mapping;

- Water resources management (e.g. snow melt, soil moisture, storm run-off);

- Coastal zone surveys (e.g. Continental shelf operations, sea-ice, oil pollution, fisheries);

- Monitoring of the Northern polar regions (e.g. Greenland, ice surveys).

In the application of remote sensing data acquired over the European land and coastal regions we need to bear in mind that there are factors that are peculiarly, but not uniquely, European, and these will dictate the directions for future progress. The spectral information we receive about the surface of the European landscape bears little relation to its natural condition; modifications of the natural state by the hand of man over hundreds and sometimes thousands of years have made the interpretation of remote sensing data difficult. Further factors are the small size and complex patterns of agricultural units which present formidable problems in the interpretation of remote sensing data, particularly when obtained from satellite altitudes. Climatic conditions are an additional factor, not only as a result of often near continuous cloud cover, but also due to the dynamic character of the small scale climatic variations.

In addition to these landscape and climatic factors, we need to remember that an enormous amount of information is already available in developed regions; detailed maps exist and organisational structures are available that use statistically derived earth resources data as essential inputs to planning decisions. There is no need to duplicate these existing methods, but rather to propose improvements, particularly in monitoring the dynamic features of our landscape, where these improvements can be seen to show clear economic or social benefits.

Development Aid

The problems of a developed and sometimes overstrained society as found in Europe and elsewhere are relatively well understood; it has been easier to be articulate on the problems we face than on the solutions we should adopt. There is however, a growing appreciation at all levels that developed regions such as Europe are only a component of a global system, and that as a part can only expect to work as well as the whole if any long term global stability is to be achieved. This growing concern with global problems finds an outlet where remote sensing methods can help, namely aid programmes to developing countries. It should be clear even to the most insular of us that imbalances between societies create stresses that are usually disruptive as far as human progress is concerned. Development aid is aimed, in theory, at reducing these imbalances and hopefully their attendant stresses. Many forms of aid have been tried, some more successfully than others, but the net result often appears to be a widening in the gap between the developed and developing nations. Clearly there is a need for more information on the way a development aid programme works. Remote sensing methods are a potential source of this information, providing both the receiver and giver of aid with data on the status of their programmes and early ideas on whether overall objectives are likely to be met.

Key applications of spaceborne remote sensing data in developing regions are :

- Sustained production of biomass (food, fibres, animals);
- Mineral exploration aimed mainly at providing sources of foreign currency to fund development programmes;
- Disaster warning and damage assessment;
- Small to medium scale cartographic and thematic mapping.

For many of the developing regions, basic data of this type is inadequately known or inaccurate. The major programme objectives are to acquire detailed data bases that would permit optimised development schemes to be planned and implemented.

It is likely that satellites designed to meet these requirements would be similar in design and performance to the present-day Landsat system(2). The main sensors of interest are multispectral scanners providing high resolution (typically 80 m) imagery in the visible and near - IR bands of the electromagnetic spectrum. The need for all-weather sensors in areas such as tropical rain forests where near continuous cloud cover prevails is anticipated.

The proper use of this monitoring capability administered through a global agency or by bilateral agreements could provide a key to help reduce the gap separating the rich from the poor. In addition to this overall objective aimed at improving conditions in developing countries, it must be remembered that a developed area such as Europe relies on the developing countries to supply basic products such as energy, many vital mineral resources, and food. It is in our own interest to reduce the existing imbalances and through this equality eliminate one of the basic causes of stress and conflict. Finally, and perhaps more obscurely, the use of remote sensing by developing countries is not dependent on the existance of an aid programme. Data and information can be used as required by the developing countries themselves to accelerate and guide their own development. In this way, access to information is provided rather than the more material forms of aid which seem sometimes more connected to the objectives of the giver than the receiver.

Monitoring the Oceans

The study of the oceans is today a relatively new science, perhaps still in its early phases of collecting basic data as individual pieces of a jigsaw puzzle, and not yet in a position to piece these together to form an integral picture of the ocean system. Although the science may be new, its importance both from the regional and global view-points is clearly recognised. Today the demands made on the ocean are increasing rapidly as a source of food, expanding in the future through the use of fish farming methods; as a transportation medium; as an energy supplier either directly from tides, currents or temperature gradients or indirectly as with offshore drilling; and as a potential source of minerals, initially by the dredging of manganese modules from the deep ocean floor, but perhaps eventually by direct extraction from sea water. The demands we place on the ocean are increasing at a faster rate than our ability to understand the effects these may eventually have on the oceans as a whole. Again, this is a clear case where add-itional information is needed and where remote sensing can help.

The global scale of the oceans (approximately 70% of the earth's surface) and their dynamic characteristics normally result in reduced requirements for ground resolution and increased requirements for areal coverage and measurement repetition rate. For many applications (e.g. surface wind field, sea surface temperature), resolutions in the kilometre range are satisfactory. The character of the majority of these missions results in the selection of high inclination orbits that provide the necessary global coverage.

The applications of spaceborne remote sensing to monitoring of the sea surface can be divided into two parts having largely different measurement requirements and objectives; these are :

- monitoring of the coastal and inland seas;
- monitoring of the global oceans.

Man's use of the seas adjacent to the coastal states has been somewhat blasé in the past, even though these areas have historically provided important sources of food and transport. The recognition, however, of the increasing economic significance of the coastal regions has changed this situation dramatically at least at a political level. The import-ance of fisheries as sources of food and employment, the recognition of the damage and bitterness caused by fishing disputes, the enormous economic impact of the oil and gas fields located on the continental shelves, the potential of using tides and waves as energy sources, the value of the coast as an area of natural beauty in an otherwise over-crowded society, all lead to the need to understand the dynamic pro-cesses of this complex environment and to conserve and husband the ocean's resources for future generations. This preoccupation with the value and fragile character of the coastal and inland seas has been manifested in the decisions and recommendations emanating from the recent conferences on the law-of-the-sea. The introduction of the 200-

nautical-mile Exclusive Economic Zone (EEZ) has presented the coastal
nations with additional responsibilities for areas where knowledge is
scant and existing information sources largely inadequate. A first
systematic assessment on the use of spaceborne sensors for monitoring
the ocean's surface and near-surface features was initiated in 1978
using remote sensing instruments on the NASA Nimbus-7 and Seasat satellites.

Monitoring of the global oceans has to be considered in conjunction
with the atmosphere. The interrelationships between ocean, weather and
climate plus the global state of the measurements to be made lead to
the need to establish international ocean monitoring programmes along
lines similar to those adopted for the Global Atmospheric Research
Programme (GARP).

Monitoring of Environment

Environmental subjects are attracting increasing attention in both
developed and developing nations. As far as the European countries are
concerned, there are major problems forming which are already causing
concern at all levels of society. Words such as environmental pollution,
limits to economic growth, quality of life, ecology, etc., are beginning to appear as household words, even before their real meaning is
fully understood. The dynamic character of many of these phenomena
clearly suggests that this is an area where remote sensing methods can
help. Many environmental problems are local in character and will not
require monitoring by satellites. Others, such as the measurement of
upper atmosphere trace gases on a global basis, will benefit from a
satellite approach.

The limited absolute accuracy of many of the imaging sensors proposed for an operational role, particularly in the area of environmental monitoring, has led to an increased interest in ground based
automatic platforms fitted with a wide range of precision sensors
that would be interrogated and if necessary located by data collection
satellites passing overhead. Typical platforms are buoys, balloons,
air and water quality sensor packages, flood and earthquake warning
systems and many others. Collection of data from ground based platforms
is likely to form an important element or sub-system in all the remote
sensing satellite systems described above as a support to the imaging
sensors.

REMOTE SENSING FROM SPACE TODAY

In the previous section, some general ideas were presented on the
value and on the wide range of potential applications of remote sensing
methods. In this section, a closer look is taken at remote sensing from
satellite altitudes, the sensors used and the results obtained to date.
This understanding of the particular advantages and capabilities of
satellite methods is necessary in order that the preferred role for
Spacelab as one possible space platform can be identified.

The Bird's-eye View

Remote sensing techniques for earth resources investigations can be applied from different types of observation platforms. Each possible platform, mobile or static, has its own significance especially in relation to the distance from the target. Although there is a notable overlap between the different modes of observation, there is no argument supporting the view that one platform can be completely substituted by others. As an example, local and infrequent surveys such as large-scale mapping (e.g. 1:10,000 scale) are best performed using aircraft. On the other hand, if the survey is required at a continental or even global scale, as is the case in many environmental or oceanographic applications and where repetitive measurements are necessary, clearly satellite methods are to be preferred. The final choice of platform will depend on many factors, including the size and character of the area to be surveyed, requirements for repetition, ground resolution, the sensors to be carried, operational or experimental aspects, and of course costs. The development of spaceborne sensing methods will, in addition, be based on a series of carefully controlled ground and airborne experiments aimed at perfecting the measurement methods and optimising sensor performance prior to committing the programme to an expensive space evaluation phase. This staged development towards the final satellite sensing techniques is illustrated in figure 1.

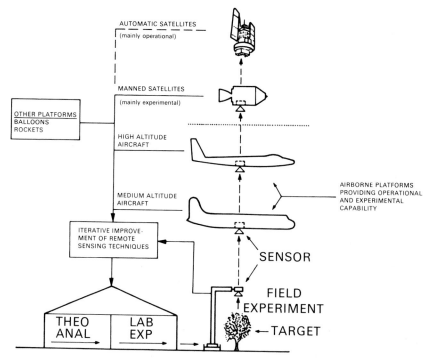

Fig. 1   Multi-platform Remote Sensing Operations

In practice, many different sources will be used in any typical survey. A good land use planner will, for example, often combine data from old aerial photographs, ground-collected statistics and sophisticated satellite imagery to produce the quality of result he requires knowing that the resulting combination is likely to be better than each single data source taken alone. In addition, the small scale (large area) satellite image will often be used to identify smaller localised areas where more detailed airborne or ground surveys are indicated; the economic benefits of not having to conduct these detailed and therefore expensive - surveys for the total area are self-evident.

Selection of System Elements

The main characteristics of space systems that are of particular interest for earth resources survey area :

- synoptic overview of regional relationships in small scale (e.g. 35,000 km^2 for each Landsat image);

- almost complete coverage of the earth (e.g. I8 days for Landsat);

- repeated observation of changing phenomena of the earth's surface;

- observation under comparable or identical conditions of illumination (solar or spacecraft produced);

- long duration observations (several years for an automatic satellite);

- simultaneous observation of large areas to determine regional interrelationships;

- observation of worldwide relationships in different earth science disciplines.

Different technical options are available for the choice of satellite system for both experimental and operational missions that are based on :

- the spacecraft (e.g. manned or automatic);

- the sensors (e.g. active or passive);

- the orbit (e.g. low altitude or geostationary orbit).

These technical options need to be balanced against the measurement requirements imposed by the geographical regions to be imaged and the applications themselves.

As far as the choice of spacecraft is concerned, the two fundamental options that need to be considered are the use of manned systems (e.g. Spacelab) or automatic satellites (e.g. Landsat). Present studies show that operational systems will primarily comprise automatic satellites, that can provide high measurement repetition rates over long periods of time. The role of manned spacecraft such as Spacelab is seen to be almost exclusively in the experimental phases of the programme. Exceptions are one-off missions such as cartographic or geological mapping, or limited area surveys (e.g. coastal erosion, urban growth) that do not need frequent repetition. Typical system parameters derived from European remote sensing regional experiments point to the need for high spatial resolution images and weather independence; the latter being considered essential for missions where measurements of dynamic phenomena are involved and where certain guarantees on data availability are required.

The choice of sensor depends primarily on the requirements for :

-   ground resolution;

-   spectral characteristics;

-   day and night operations ;

-   all-weather operations.

The performance characteristics of some of the main sensors that are presently under evaluation within the framework of experimental programmes are given in Table 1; further details are available in the literature (3,4).

The precise choice of sensor will depend largely on the requirements of the application. For example, agricultural surveys in relatively cloud-free areas are best performed using multispectral scanners; sea-ice surveys in the upper latitudes require high resolution all-weather data as provided by synthetic aperture radar systems; global measurements of sea surface temperature requires high measurement repetition but reduced spatial resolution and are best performed using passive infrared or microwave radiometers.

Choice of Orbit

The main orbital parameters under our control are altitude and inclination. Of the many combinations of orbital altitude and inclination that are possible, only a limited number are of direct interest for earth resources satellites; these are :

(a)   Sun-synchronous orbits

A combination of altitudes and inclinations exists that results in sun-synchronous orbits where the angle between the orbital plane and the earth-sun line remains constant. The main characteristic of this type of orbit is that measurements made along the sub-orbital track are always at the same local time. Small variations in altitude result in a drift of the orbital track to the East or West, permitting global coverage to be achieved. Landsat-1 was launched into this type of orbit

SENSOR	Passive / Active	Spectral Bands	Ground Resolution	Day & Night Operation	All-weather Operation	Examples of Applications	REMARKS
Cameras (Metric and multispectral)	Passive	Visible to near-IR (0.4-0.9 microns)	10 m	no	no	Cartography, Basic land-use maps, Geometric calibration	Basic sensors used in aerial photography. Many space photographs available from manned satellites, particularly Apollo 9 and Skylab.
Line scanning Instruments (single- and multi-channel)	Passive	UV to thermal IR (0.4-14 microns)	good (30-100 m)	Day only (except thermal-IR band)	no (IR band has some haze penetration)	Agriculture (including land use, crop and forest surveys), Hydrological surveys	Up to 24 channels for airborne systems. Satellite multi-spectral scanner with high resolution (50-80 m) flown on Landsat and Skylab.
Radiometers	Passive	Thermal-IR (10-14 microns)	average	yes	no (some haze penetration)	Thermal pollution, Surface temperature	Non-imaging sensor providing accurate measure of surface temperature.
		Microwave (mm + cm)	poor	yes	yes (short wavelengths affected by cloud and rain)	Surface temperature, Soil-moisture surveys, Polar-ice surveys, Surface wind field	All-weather sensor measures thermal emission at microwave frequencies. Flown on Nimbus 7 and Seasat.
Scatterometer	Active	Microwave (cm wavelengths)	poor	yes	yes (short wavelengths affected by cloud and rain)	Sea state, Surface wind field	Research instrument aimed at increasing understanding of material properties. Microwave wavelengths. Flown on Seasat.
Synthetic Aperture Side Looking Radar (SAR)	Active	Microwave (cm wavelenths)	good (10-50 m)	yes	yes (short wavelengths affected by cloud and rain)	General reconnaissance, Wave measurements, Sea-ice, Land use sunrays	High resolution all-weather sensor. Normally single frequency but multi-frequency, multi-polarisation systems needed for future. Flown on Seasat.

Table 1 Typical Characteristics of the Main Remote Sensing Instrumentation

(altitude 910 km, inclination 99.1°) providing complete global coverage every I8 days.

(b)   Low-inclination orbits

Where complete global coverage is not required, as is the case for resources surveys of near-equatorial areas, then lower orbital inclinations can be chosen. This results in higher measurement rates with data obtained at different local times. The 55° inclination Skylab orbit was of this type, but lower inclinations are possible and perhaps desirable in surveys of near tropical regions.

(c)   Geosynchronous orbits

The most popular of the geosynchronous orbits is the 24-hour geostationary orbit (altitude 37,000 km) which has to date been used mainly by communication satellites. Interest in this orbit for earth observation is increasing as indicated by the present range of geosta- tionary meteorological satellites (e.g. GOES, Meteosat). The use of this orbit for earth survey where high resolution is required has been studied by ESA(5) and NASA where it appears that ground resolutions of the same order as Landsat are possible using large aperture (1-1.5 m) telescopes. The main interest in this type of orbit is in monitoring rapidly changing and transient phenomena (e.g. ocean upwelling, change in soil moisture in arid regions) particularly at the low and middle latitudes where the earth's sphericity does not excessively distort the images obtained. Different types of geosynchronous orbits are possible (e.g. I2-hour, inclined 24-hour).

Historical Background

The idea of looking at the earth from space in a practical sense goes back as far as the ballistic launches of the Aerobee, Viking and Atlas rockets in the late 1940s and early 1950s. The techniques and methods developed from these early sub-orbital flights prepared the way for the later manned USA and USSR flights culminating in the recent Skylab and Salyut missions.

The objectives of these early missions - in particular Gemini and Apollo - were generally not directed towards the development of earth observation techniques, but rather to the ideas of lunar landings and establishing manned space stations.Most of the flights carried simple photographic cameras, usually handheld by the astronauts and obliquely directed towards targets of opportunity on the ground. Similar missions were carried out by the USSR Soyuz and Salyut series of manned satell- ites.

In parallel to the early Gemini and Apollo efforts, more sophisti- cated space systems were under development by NASA. These were the Earth Resources Technology Satellite (now named Landsat) and the Skylab

The importance of the Skylab EREP flights was not simply due to the increased time in orbit compared to the earlier Gemini and Apollo flights, but also and perhaps more importantly to the increased sophistication of the remote sensing instruments and the extension of the range of spectral measurements from the visible and near-IR to include the thermal-IR and microwave regions. Data from EREP have enlarged the experimental possibilities for the use of satellite remote sensing. The concept of Skylab in its broadest sense shows the significant importance of manned space flights for predominantly experimental missions and conversely its restricted use for operational monitoring where mission duration of several years are usually mandatory.

Current Programmes and Future Trends

New imaging sensors are under development that provide finer spatial or spectral resolution and which place increasing emphasis on measurements in the microwave region of the electromagnetic spectrum either simply to avoid constraints introduced by cloud cover, or to take advantage of the unique features offered by long wavelength (cm) sensing methods. The main thrust in the USA has been in the definition of a range of automatic satellites that, although experimental in concept, are clearly aimed at later operational roles where a continuous data service is needed. The key characteristics of these NASA automatic satellites scheduled for launch in the late 1970s are listed in Table 4.

In addition to these approved NASA programmes, European plans for remote sensing satellites are maturing rapidly and a first satellite, the French national earth observation satellite SPOT (Satellite probatoire d'observation de la Terre) is in the final stages of design by CNES (Centre National d'Etudes Spatiales) and is scheduled for launch in early 1984. At the wider European level studies are underway for a range of satellites for different land and coastal ocean survey applications (see Figure 2) that are based on the use of a standard platform (6).

The different missions listed in Figure 2 will necessitate the development of a range of sensors and measurement methods in order that the required quality of information can be derived from the acquired data. As a first step the ESA Remote Sensing Preparatory Programme has been approved and will initiate the study and early development of key critical areas associated with the optical and microwave sensors and with the interfaces between the remote sensing satellite payloads and the common bus platform. The final development of these sensors and measurement methods will take advantage of the feasibility offered by Spacelab to test in space sophisticated instrumentation and to perfect methods to handle the vast quantities of data aquired. This role of Spacelab as a test platform is developed further in the next section.

Earth Resources Experiment Package (EREP) programmes. Landsat is an
unmanned (automatic) satellite and to date three have been launched
(Landsat-1 on 23 July 1972 and Landsat-2 on 22 January 1975), the
third and last in the series (Landsat-3) was launched on 5 March
I978. Summary details of Landsats 1 and 2 are given in Table 2,
Landsat-3 is similar but with a fifth thermal IR channel on the MSS
and a panchromatic high resolution 2 camera RBV system.

Table 2   Main Characteristics of Sensors on Landsat 1 and 2

Data / Sensor	No. of Channels	Spectral Range (microns)	Spatial Resolution (m)	Swath Width (km)
Multispectral Scanner (MSS)	4	0.5-1.1	80	185
Return Beam Videcon (RBV)	3	0.47-0.83	80	185

The first of the four Skylab missions started in May 1973 with the
completion of the final mission in February 1974. Key features of the
sensors carried by Skylab (EREP) are given in Table 3.

Table 3   Skylab Earth Resources Experiment Package   (EREP)

Data / Sensor	Type of Instrument	No. of Channels	Spectral Bands	Ground Resolution (m)	Swath Width (km)
S 190 A	Multispectral Cameras	6	0.4-0.9 microns	30	163
S 190 B	Earth Terrain Camera	1	0.4-0.9 microns	10	109
S 191	Infrared Spectrometer	2	0.4-2.4 6.2-15.5	430	-
S 192	Multispectral Scanner	10	0.4-12.5 microns	80 200 (IR)	68
S 193	Radiometer/Alt/ Scatterometer	1	2.2 cm	11 km	-
S 194	Microwave Radiometer	1	21 cm	110 km	-

Table 4   Summary Characteristics of NASA Remote Sensing Satellites for Earth Survey, Marine and Environmental Monitoring Applications (Approved Programmes Only)

Satellite	Orbit Altitude /km	Orbit Inclination	Orbit Repetition Rate	Launch Date	Main Imaging Sensors	Ground Resolution /m	No. of Spectral Bands	Swath/Fov /km	Other Sensors
Landsat-1 and -2	920	99	18 days	7/72 and 1/75	MSS RBV	80 / 80	4 / 3	105 / 185 × 185	Data collection
Landsat-3	920	99	18 days	3/78	MSS RBV	80 / 40	5 / 1 (stereo)	185 / 130 × 130	Data collection
HCMM (Heat capacity mapping mission)	600	98	12 h	4/78	Heat capacity mapping radiometer	500	2	700	—
Seasat	800	108		7/78	SAR SMMR	25 / 16–144 km (function freq.)	1 / 5	100 (250–350 off nadir)	1. Radar altimeter 2. Vis/IR Scanner 3. Microwave scatterometer
Nimbus-7	925	99	2–3 days	10/78	CZCS SMMR	800 / Similar to Seasat	6 / 4	600	

MSS   Multispectral Scanner
RBV   Return Beam Videcon
SAR   Synthetic Aperture Radar
SMMR Scanning Multichannel Microwave Radiometer
CZCS Coastal Zone Colour Scanner

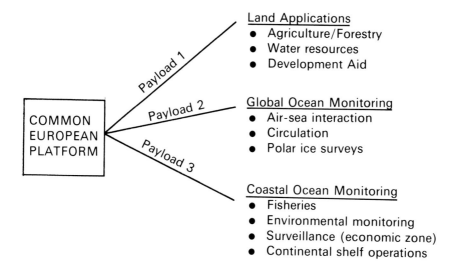

Fig 2 - Payloads for European Remote Sensing Satellite Programme

Table 5 - Spacelab Capabilities for Remote Sensing Mission

	MAIN PAYLOAD PERFORMANCE PARAMETERS	SUMMARY USER (EARTH RESOURCES) VIEWPOINTS
CONFIGURATION	Pressurised Volume Pallet Length Payload Weight	1. Preferred configuration for earth observation missions is the short module and 3 element pallet. 2. Pallet is volume rather than weight limited (requirement for large deployable antennae).
POWER	Average Power Energy (from Orbiter) Voltages available	1. Power levels are adequate for user; main limitation is total energy available in one power kit. 2. Earth survey missions require peak energy over some test areas (e.g. in Europe), but also a continuous power drain throughout mission as a result of long term background monitoring.
DATA	Digital Data Storage Rate Real Time Transmission On-board Data Storage	1. High data rates expected for multispectral scanner (50-100 Mbps) and SLR. 2. There is no urgent requirement for real time data transmission. Partial (quick-look) data transfer and communications to ground teams are necessary.
MISCELLANEOUS	Crew Number Optical Window Stable and Pointing Platforms	1. Available crew total is adequate, probably remote sensing technicians during early missions rather than earth scientists. 2. Metric quality required for optical window. 3. Pointing platforms (for spectrometers and radiometers) and stable platform (for SLR) are required.

SPACELAB AS A REMOTE SENSING PLATFORM

Capabilities of the Platform

        Prior to discussing the role of Spacelab, we need first to under-
stand the concept of the system within the framework of the different
platform types described in the previous section. Details of the Spacelab design are presented elsewhere in this book; in this section,
system aspects that are particularly relevant to the earth survey
disciplines are examined, namely orbits and the role of man as a crew
member. The main characteristics of Spacelab, together with summary
assessments of their impact on earth resources survey missions, are
given in Table 5.

Orbital Characteristics

        The ultimate use the earth resources survey community will make of
Spacelab is in many ways dependent on the performance capabilities of
the Shuttle orbiter. The main orbital elements that have a major impact
on the mission are :

- altitude;
- orbital inclination;
- local time at nodal passage.

The choice of altitude is determined by the requirements for ground
resolution and areal coverage. The operating altitudes presently considered for the Shuttle orbiter lie in the range 185 km to 550 km. The
possible orbital inclinations lie between 28° and 104° with respect to
the equatorial plane. The inclination does, however, depend on the
launch range used; early Spacelab flights will be made from the Eastern
Test Range (ETR-maximum inclination 57°), but these should be followed
two or more years later by Western Test Range launches (WTR-maximum
inclination 104°) with the high inclination required by many European
users.

        Orbits that are suited for earth survey missions can be rationally
selected depending on the mission requirements for :

- ground resolution;
- geographical locations of areas to be overflown;
- measurement repetition required during the mission;
- local time of measurement;
- total areal coverage.

The requirements listed above will vary considerably between different
missions. However, a certain number of orbit types exist that are of
clear interest for many earth observation missions; these are listed
in Table 6.

Table 6 - Preferred Orbits for Earth Resources Survey Missions (7)

Inclination (deg)	Altitude (km)	Revolution (min)	Characteristics
55	187	88.2	High resolution, small areal coverage, optimum repetition/day between 50-55° latitude Drift between 1. and 17. orbits ∿ 1° longtitude Aperiodical repetition during mission
55	370	91.9	Synoptic coverage, less high resolution drift between 1. and 17. orbit Repetition as above
96.6	275	90.0	Exact sunsynchronous orbit Maximum repetition/mission (every 24 h, twice at cross points) Repetition at same local time-sun-angle Low areal coverage, one major ground track only Preffered for experimental measurement mission
96.6	262/288	89.7/90.3	Drifting sunsynchronous orbit Lowest repetition rate Greater areal coverage (global coverage) at 1° per day
97.4	567	96	Exact sunsynchronous orbit for operational applications (drifting orbit ∿ 550 km if required) - SPACELAB Sensor Testbed Missions - Satellite Deployment Missions

Two of the orbits listed in Table 6 are of particular interst and deserve a more detailed description; these are:

- an orbit of 55° inclination, 187 km altitude out of ETR;
- an orbit of 96.6° inclination, 275 km altitude out of WTR.

The first of these, the near earth 55° inclination orbit, is the best we can expect from ETR, but only provides coverage up to the middle European latitudes, the northern parts of Europe and Canada will not be within the range of this orbit. The ground track for the first three days (41 orbits) is shown in Fig. 3(a) for this orbit on a global scale. This orbit does offer certain advantages for areas between 50° and 55° latitude due to the flattening of the sub-satellite trajectory within this range. This effect can be seen in Fig 3(b) where the detailed ground track for three consecutive orbit passes over Western Europe is shown.

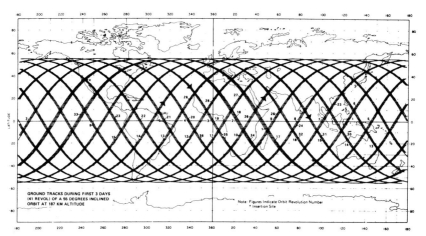

Fig 3(a) - Ground Track (first 3 days) for 55° Inclination, 187 km Altitude Orbit

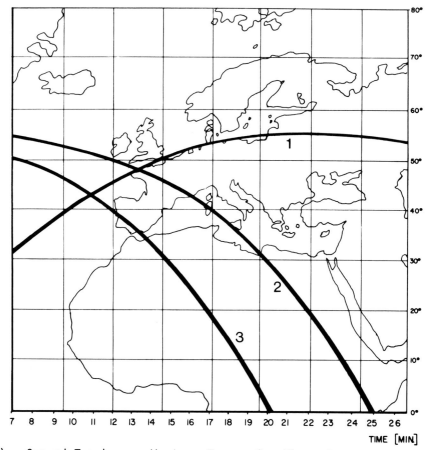

Fig 3(b) - Ground Track over Western Europe for Three Consecutive Passes (55°, 187 km)

The second orbit of particular interest is a sun-synchronous orbit
(96.6° inclination, 275 km altitude). In this case, the angle between
the orbital plane and the earth-sun line remains constant. The exact
sun-synchronous orbit is a special case with an integer number of orbits/
24 hours (16 for 275 km case). The sub-orbital track for this type of
orbit is shown in Fig 4 where it can be seen that the ground track is
identical for each day (orbit 17 exactly superimposed over orbit 1,
etc.), and that the displacement between two consecutive orbits is
22.5° or 2500 km (at the equator).

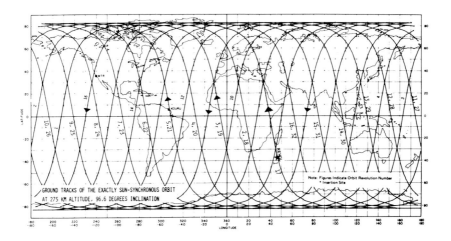

Fig 4 - Ground Track Sun-synchronous Orbit with Integer Number of
Orbits/24 Hours

Fig 4 shows, that, although the total areal coverage is low, the
repetition rate along the ground track is once/day and twice/day at
the crossover points. The main advantage for experimental missions are:

-    repeated measurements along subsatellite track (important for
     experimental missions);

-    observations made under constant illumination conditions
     (same local time of measurement).

The disadvantages are that the areal coverage is low and that
within the framework of an experimental programme it will be necessary in some cases to establish test sites along the sub-orbital track.
A detailed example of such a ground track over Western Europe and
North Africa is given in Fig 5. Although it can be seen from this
figure that ground test areas would have to be located along the satellite track, the areas covered permit a wide range of experiments from
the Arctic regions (e.g. sea-ice studies), through the mid-latitude
temperate regions (e.g. vegetation, surface water studies) to the Med-
iterranean (sea pollution) and North African areas (arid climate
studies).

A small variation in the altitude of the exact sun-synchronous orbit causes a gradual sideways drift of the ground track. An altitude reduction of 13 km to 262 km results in a 1°/day drift eastwards; an increase leads to a westward drift. This 1° of longitude means at the equator a distance of 111 km, reducing to 72 km at 50° latitude. The advantage of such a drifting orbit compared to the exact sun-synchronous orbit is the gain in geographical coverage. This means, of course, that repeated coverage over specific points is not possible during a 7-day mission, other than in the limited areas of overlap between adjacent ground swaths or at the crossover points between ascending and descending tracks.

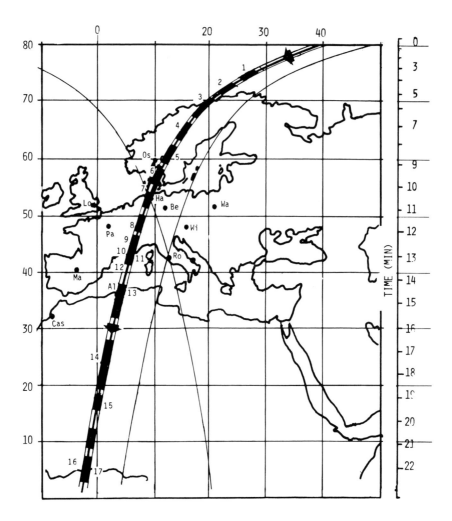

Fig 5 - Time Sequence of a Typical Descending Ground Track of a Sun-Synchronous Orbit

Role of Man

In considering the role of man in Spacelab missions, two approaches are normally taken; these are :

- to consider man as a technician whose main tasks are to operate, maintain and, if necessary, repair the equipment in his charge;

- to consider man as an experimenter on-board Spacelab performing experiments aimed at providing new perceptions to the subject under study.

Both these approaches are an improvement over previous manned missions where, with the possible exceptions of Skylab; the tasks and capabilities of the astronaut crew were not normally directed towards the scientific objectives of the mission or the technical requirements of the payload, but rather on getting there and back.

The main advantages of the first "technician" approach are that automatic equipment can be in some cases eliminated and that a low-cost approach to system testing and reliability is possible if on-board repairs can be carried out without endangering crew safety. The benefits of adopting the "scientist-in-orbit" approach for earth observation missions are clear in relation to the study of transient phenomena on an unscheduled basis (e.g. cloud formation, volcanic eruptions) and in rejecting non-valuable data. His value is presently less clear as far as the scheduled mission operations are concerned due to the limited duration of the sortie missions and the need to conduct the detailed data interpretation on the ground after the missions are completed.

Based on the above, it is likely that the technician crew member approach will be adopted at least until we have a better understanding of both the value and capability of man to perform scientific research in space. The determination of the optimum role for man is in itself an important experiment, and one that must be studied and resolved before we can seriously start to think of permanent space stations.

ROLE OF SPACELAB

In the previous sections, information has been presented on the characteristics of Spacelab that are of importance to earth observation missions. This discussion provides us with a starting point to identify the optimum role of Spacelab for remote sensing missions and eventually to design and produce payloads. We need first, however, to make certain assumptions on the position of Spacelab within the more general frame-work of remote sensing programmes comprising different platform types, covering local as well as global survey areas and consisting of both experimental and operational programmes. Based on these assumptions, ideas on the way the earth scientist will use and operate Spacelab begin to emerge.

Basic Assumptions

Any discussion on an earth resources payload for Spacelab must at this time be based on certain assumptions. Some of the more important of these are :

(a)   The duration of a sortie mission will be initially 7 days, with the possibility later on of extended missions up to 30 days.

(b)   The first Spacelab mission is scheduled for launch in 1981. In other words, the time period we are primarily concerned with is the 1980s. This point is important because it is necessary to recognise that many of the experimental programmes of today are expected to become operational by the mid-1980s (e.g. sea surface temperature, sea-ice mapping, basic land-use mapping, plus many others). New and in many cases more complex problems will be posed in the coming years and it is towards these that Spacelab has to be aimed.

(c)   Orbital inclinations from near equatorial to sun-synchronous will be possible, although inclinations higher than 55° are not possible during the first years of Shuttle operations.

(d)   Most data collected by Spacelab will be returned to the ground for detailed interpretation with a minimum of on-board processing.

(e)   Many different sensor platforms will be available (balloons, rockets, aircraft and automatic satellites) for both experimental and operational programmes. It is necessary to consider the advantages and disadvantages of each of these different platform types in order that the optimum role of Spacelab be realised.

(f)   The main emphasis on future space programmes (1980-1990 time period) will be in understanding more fully the data obtained from existing sensor types rather than in developing radically new sensor systems.

(g)   European interests will concentrate on sensors having a high resolution and an all-weather capability. This is expected to place the main emphasis on sensors operating in the microwave spectral bands.

Advantages and Limitations of 7-day Missions

Missions that require repetitive coverage over long periods part- icularly where large or normally inaccessible areas are involved are probably best achieved with automatic satellites. Conversely, aircraft are more suited for local missions covering limited areas and by necessity of short duration. Similar arguments and logic can be applied to Spacelab based on the assumptions listed previously and considering

the advantages and disadvantages offered by the Space Shuttle. Some of
the more obvious of these are listed below :

(a)   Advantages :

- large payload weight and volume ;
- presence of scientist/technician crew member enabling
  increased flexibility in mission planning, sensor control,
  on-board data processing, etc. ;
- return of equipment (e.g. films) to earth;
- easy access for experiments and quick turn-around time
  between flights compared to automatic satellites.

(b)   Disadvantages :

- short mission duration (normally 7 days with later extension
  up to 30 days);
- need to share resources with other experiments (e.g. non-
  dedicated mission);
- high launch costs compared to aircraft, although lower cost/
  kg than automatic satellites.

From this background, it is possible to see that the operational
capabilities of Spacelab will be limited. The main interest for operational satellite systems will be for monitoring dynamic features over
long time periods. Clearly automatic satellites are best suited to
this operational role. Conversely the inherent flexibility resulting
from the large volume and weight capability and the presence of on-board
specialists points towards an important experimental role for Spacelab.
This concept of Spacelab as an experimental platform acting as a bridge
between the aircraft programmes of today and the operational automatic
satellite of the future is illustrated in Figure 6, and enlarged in the
next section.

Experimental programmes using Spacelab

There are two main and interrelated objectives associated with any
experimental earth observation programme using advanced sensing methods;
these are :

- to understand the capabilities and limitations of the various
  sensor systems proposed and to develop measurement techniques
  that can be applied to the different application areas under
  investigation;
- using the proven remote sensing instruments and methods derived
  above, conduct experiments in the earth observation disciplines
  aimed either at purely scientific objectives or at perfecting
  measurement methods for later applications-orientated missions.

Many of the important contributions that are required to achieve
these two objectives will come from careful exhaustive ground based
and airborne measurement programmes. These measurements are, however,
not enough in themselves to define and specify completely the later
automatic satellite systems; an intermediate phase is required using
experimental space systems such as Spacelab.

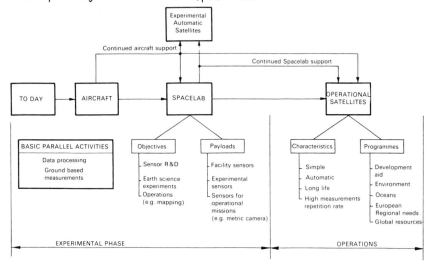

Fig 6 - Role of SPACELAB in the development of remote sensing
instrumentation and measurement methods (8)

If we examine the requirements for this intermediate experimental
orbital phase, there are several ways these two objectives can be met
using either manned or automatic satellites or a combination of both.
It is generally recognised that the first "technological" objective is
best met using manned systems such as Spacelab. This is because extended periods of observation and high measurement repetition rates are
not essential and considerable advantages can accrue from the flexibility provided by a manned platform. It is more difficult to decide
between manned and automatic satellites to meet the second "geoscientific" objective where extended measurements and high repetition rates
are normally required. In this second case, the final choice must
depend on particular applications and perhaps above all on programme
costs.

The role of Spacelab in meeting the two principal objectives
outlined above has been studied (7,8) and this had led to a clearer
understanding of the types of missions that could be flown on Spacelab.
In summary, the main uses foreseen for Spacelab are in meeting the
following objectives :

Technological objectives

a)   Development of on-board data processing, handling, compression
       and storage techniques.

b)   Investigations into the role of man in earth survey missions, e.g. targets of opportunity, sensor control, data processing, mission flexibility, etc.

c)   Space testing of sensors and other equipment (e.g. deployment of large antennae) for later use in automatic satellites.

Scientific objectives

(a) Experimental missions aimed at understanding the physical processes involved between the full spectral range of electromagnetic radiation and the earth's surface, and determining and correcting the influence of the intervening atmosphere. These programmes should eventually lead to the definition of physical and mathematical models describing the theoretical basis for the understanding and interpretation of data from automatic satellites. This function has to date been performed using well-equipped aircraft,but will eventually require a flexible well-equipped orbital laboratory such as Spacelab.

(b) Development of experiment methodologies and techniques, e.g. comparison in space of different sensor types, combinations of sensors or spectral bands, for use in future Spacelab missions or in automatic satellites.

(c) Experiments in the different earth observation disciplines using established remote sensing methods for scientific purposes or to prepare operational systems.

(d) Operations in conjuncture with automatic satellites and aircraft as part of a multi-platform, multi-level mission.

(e) Operations in conjunction with automatic satellites and aircraft support of operational automatic satellites and providing them with an absolute calibration.

Operational objectives

     In addition to the primarily experimental objectives outlined above, other missions are possible of an operational character where the limited stay in space is not an insurmountable constraint; these are typically :

(a) Medium to small scale cartographic mapping (9) where the need for repeat flights over the same area is limited and where the return to earth capability of Spacelab is of immense value in the retrieval of photographic films.

(b) Geological mapping missions (6) where a wide range of sensors can be flown but again where repeated measurements are largely unnecessary

(c)   Special purpose operational missions such as disaster assessment, where data is required rapidly to support relief or rescue operations.

Certain key words emerge from the above preliminary list of mission types; first and foremost is the word experimental; others are flexibility and support. These words, or rather the ideas they imply, do much to state in general terms the approach adopted to define possible payloads for Spacelab.

As a final but important point, it is clear that there are many possible earth resources payloads for the Spacelab incorporating known or experimental sensors in different combinations. It should also be clear that optimum payloads will not appear overnight; the initial systems and objectives will be relatively simple, increasing in complexity and capability as experience grows. The secret of a well-designed Spacelab will be its ability to keep pace with this phased growth and its ability to provide an adequate degree of flexibility.

SPACELAB PAYLOADS

Parametric Presentation of User Requirements

Proposals for experimental missions to be conducted from Spacelab need to be translated into parametric terms that can be used to define clearly the payload and mission requirements. The main parameters to be identified are listed in Figure 7.

MEASUREMENT TASK	MEASUREMENT CONDITIONS	MEASUREMENT DEVICES	ROLE OF MAN
– DISCIPLINES	– REQUIRED COVERAGE	– PHYSICAL QUANTITIES	– CREW SKILL
– SPECIAL APPLICATION	– TARGET SIZE	– SPECTRAL BAND	– GROUND COMMUNICAT
– PRESENT METHODS	– REPETITION RATE	– PROVED AND POTENTIAL SENSORS	– LINK TO OTHER PLATFORMS
– MEASUREMENT INDICATOR	– TIME OF YEAR		
	– LOCAL TIME		
	– TEST SITE		

Fig. 7   Listing of Required Mission Parameters (11)

Sensor Types

An interesting difference between the earth and space scientists
as far as experiments in satellites are concerned is that where the
space scientist often designs and produces his own sensors, the earth
scientist generally does not. Instead, he will normally rely on receiving data from a few facility sensors that are common to many investigators and disciplines. A good example of this is the Landsat-1 satellite where approximately 300 principal investigators used in common
two imaging sensors (MSS and RBV) and one data collection system.

If we examine this difference against the background of a flexible
manned laboratory such as Spacelab, a typical earth resources survey
payload is likely to consist of three main groups of sensors, namely :

-    facility sensors such as synthetic aperture radar (SAR), passive
     microwave radiometry (MR), multispectral scanners (MSS);

-    standard support sensors and equipment such as cameras, precis-
     ion radiometers, tracking telescope, sensor control and display
     consoles, etc.;

-    experimental sensors to be flown on later missions; examples
     are laser devices, second generation solid state optical
     imagers, and multifrequency radars.

The quantities of data these sensors are capable of producing are
in excess of what we can presently handle; this is true even when we
do not look too deeply into developments that may become available from
the current military sensor programmes. Based on this it seems reasonable to assume that sensors initially incorporated into Spacelab will
be similar to, and derived from, sensors at present installed in aircraft facilities, or flown in the Landsat, Seasat, Nimbus or the Skylab
EREP programmes. Sensors falling into this category include multispectral scanners, synthetic aperture radar, passive microwave radiometers, and probably active optical sensors. The main emphasis will be
placed on achieving a deeper and less empirical understanding of the
data produced by these sensors rather than on radical new technical
developments in data acquisition.

It is likely that various combinations of facility and support
sensors will be selected to form basic modules that could be flown
separately or in combination depending on the requirements of specific
missions. Each module would comprise one facility sensor, e.g. multispectral scanner, radar or microwave radiometers plus the necessary
supporting sensors, e.g. cameras, spectrometers, etc. Additional modules can be envisaged, e.g. active optics or an atmospheric sensor (air
quality) package. These modules can be combined and flown on Spacelab
as an integrated remote sensing payload as illustrated in Figure 8.

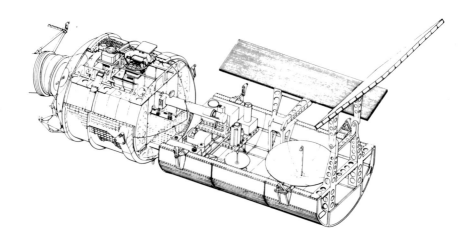

Fig 8 - Spacelab Earth Resources Survey Payload (12)

The Spacelab configuration shown above represents a general purpose remote sensing laboratory that could be used by many earth scientists to develop an understanding of the capabilities of space-borne remote sensing, and to perfect measurement methods for a number of applications over different test areas. In practice the facility would be extremely expensive to construct and it is more likely that individual modules would be flown as separate units on mixed discipline missions, particularly those where the imposed constraints on the Spacelab orbit and attitude are not critical as is often the case for life sciences and material processing missions. In the limit, of course, individual sensors can be flown such as the Large Format Camera shown in Figure 9(a) which is presently under study by NASA, or the metric camera under development by DFVLR in Germany for the first flight. (See Fig 9(b).

Perhaps the most important task for Spacelab will be the testing and operation in space of experimental sensors. These could be extensions or modifications to the basic facility sensors, or new sensors specially developed for manned and automatic satellite programmes. Possible experimental sensors include :

- synthetic aperture radar systems for soil moisture measurements;
- laser radar (scatterometer and scanner);
- imaging optical wave length spectrometer;
- derivative spectrometer: to reduce slowly varying spectral functions, e.g. atmospheric scattering;
- polypanchromatic side looking radar.

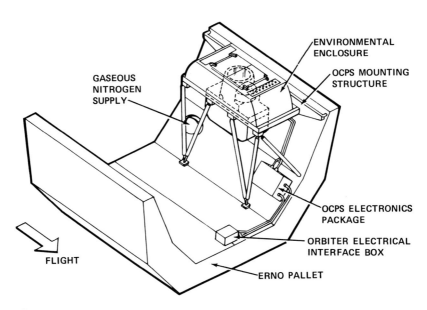

Fig. 9(a)   Large Format Camera mounted on Spacelab Pallet (12)

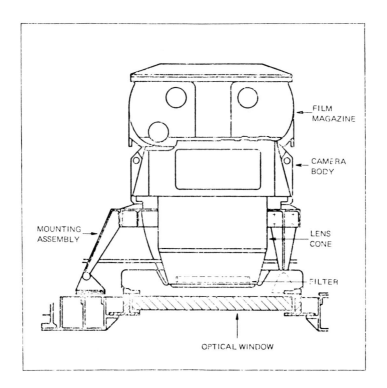

Fig. 9(b)   Metric Camera to be flown on First Spacelab Flight

Apart from the development of the new sensors, additional programmes would be required to develop and test in space large deployable
antennas for imaging microwave radiometers, various types of antennas
for radar systems, instrumentation for automatic earth observation
satellites, precision pointing platforms, on-board data processing
methods, etc. In conclusion it is expected that the complex remote
sensing instruments developed for future automatic satellites would as
a matter of routine be tested in Spacelab before committing the programme to a launch of the automatic satellite itself.

REMOTE SENSING EXPERIMENTS ON THE FIRST SPACELAB PAYLOAD

Two European remote sensing experiments are to be flown in the
first Spacelab payload to be launched in 1981 into a 250 km, 57^{0}
inclination orbit; they are :

- metric camera;

- microwave experiment.

The two experiments form a part of a multidisciplinary payload
comprising some 36 experiments (24 European and 12 American) from the
space and life sciences, earth observation and materials sciences. The
different experiments share the Shuttle and Spacelab resources during
the seven-day mission based on an orbital operations profile that
allocates resources such as attitude, crew time, power, energy, data
handling capacity in an equitable way.

Metric Camera

The camera (see Fig.9(b)) to be flown is a modified version of the
Zeiss RMK A30/23 305 mm focal length aerial survey camera (14). The
objectives of the experiment are to test the mapping capabilities (map
generation and map revision) of high resolution (better than 20 m)
space photography taken at the resolution limit of image motion on
large film format (23 x 23 cm). The photographic films obtained during
the flight over known test areas will be used to prepare topographic
and thematic maps at scales in the range 1:50,000 to 1:100,000 or
smaller. Different types of film will be flown (black and white panchromatic, colour and IR films), and a range of analogue and digital
data processing methods will be used to evaluate the results obtained.
Key features of the camera are summarised in table 7.

The camera, which weighs about 180 kg, will be mounted inside
the pressurised module and will view the earth's surface through the
high quality optical window.

Microwave Remote Sensing Experiment (MRSE)

The microwave insrument to be flown will be the first European
microwave remote sensing experiment to be placed in orbit (15, 16).

Table 7 - Key features of metric camera experiment

CAMERA TYPE	CARL ZEISS RMW A30/23
Lens	Topar
Focal length	305 mm
Total weight	178.5 kg
No. of cassettes	3
Film size	23 x 23 cm
Film length	120 - 150 m
Photos/cassette	550

The design of the instrument is such that different sensor operating modes will be possible, enabling a range of experiments to be performed at a single operating microwave (X band) frequency. Three operating modes are scheduled :

- a main mode as a Two Frequency Scatterometer (2FS)
- a high-resolution mode as a Synthetic Aperture Radar (SAR)
- a passive mode as a Passive Microwave Radiometer (PMR).

The key features of the experiment are listed in Table 8, and a sketch of the 2m x 1m pallet mounted antenna is shown in Fig 10.

a) 2FS Measurements

The main objective is to measure ocean surface wave spectra and directions for long ocean waves between the wavelength range 10 m and 500 m using a dual-frequency scatterometer technique (17).

The instrument will measure the complex backscattering of the ocean surface at two adjacent microwave (X band) frequencies. Processing of the backscattered signal received at the experiment will produce a cross correlation term proportional to a component of the sea-wave spectrum depending on look angle, depression angle and difference frequency; rotation of the antenna provides information on wave direction. Data gathering will be done in ocean areas parallel to the Space-lab ground-track and calibration of the system is obtained by the comparison of space measurements with sea truth data collected from test areas.

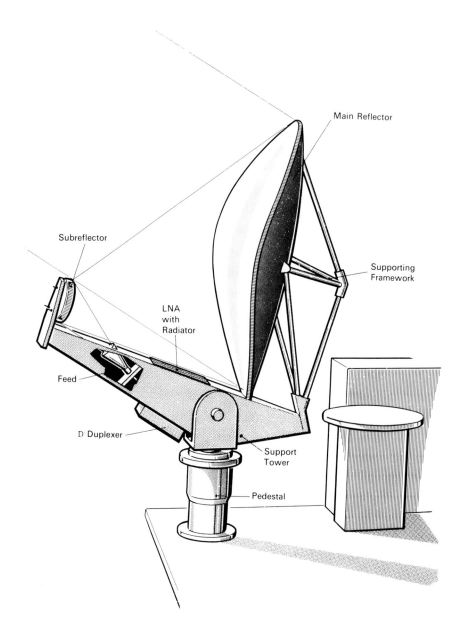

Fig. 10 Conceptional View of the Microwave Experiment Antenna System

Table 8   Key Features of Microwave Experiment

MICROWAVE EXPERIMENT	FEATURE
Ground Resolution	
2FS	9 km (1dB footprint)
SAR	25 m (9 km image swath)
PMR	$\pm$ 1°C (temperature resolution)
Antenna	Paraboloid section with Casse-grain feed
Size	1 m x 2 m (see fig. 10)
Frequency	8.5 GHz
Antenna Pedestal	
Elevation angle	40° $\pm$ 15°
Azimuth	$\pm$ 40° to -35°
Angular rate (max)	$\pm$ 0.5°/sec

b) SAR Operating Mode

Areas of the earth's surface will be imaged with the instrument operating in a synthetic aperture radar mode. The backscattered data are coherently recorded and off-line processing will provide imagery with a ground resolution of 25 x 25 m over a limited swath width of 9 km. Motion compensation is carried out during the ground processing using data from the Spacelab attitude sensors.

The SAR will be used over both land and ocean surfaces for a range of experiments related to measurements of detailed surface features (land use, surface roughness, sea ice, oil slicks, etc.).

c)   Radiometer_Mode

This mode is an add-on mode to support the 2FS study;   the PMR
mode will be used in time multiplex with the other modes and will
measure the microwave emission from the earth (surface and atmosphere)
providing coarse spatial resolution data of interest principally to
studies of sea surface temperature.

CONCLUSIONS

Remote sensing of the earth's resources and environment cannot
be classed as a space science such as astronomy, nor as a clear area
of applications such as telecommunications satellites. Our understanding
of the potential of spaceborne remote sensing is largely based on a
number of experiments conducted by a geographically and organisationally
dispersed community of earth scientists. The transfer of these experimental activities into operational programmes will require considerable
technical and scientific developments on the one hand, and political
and organisational decisions of a far reaching character on the other.

In its simplest terms, the main role of Spacelab is seen as an
orbital laboratory acting as a bridge between the experimental airborne
activities of today and the automatic operational satellites needed
for tomorrow; this role includes :

-   the development and optimisation of measurement methods
    using a wide range of sensor types;

-   the testing in orbit of sensors and critical components prior
    to their use in automatic satellites.

Secondary roles of a more operational character have also been
identified, in particular where synoptic data is required at infrequent
intervals such as the generation of high resolution photographs for
mapping purposes.

The importance of the experimental role of Spacelab must not be
under-emphasised. Many of the problems we are facing in Europe today
will not be solved by simple means. The requirements to monitor
accurately the dynamic features of a complex surface and to construct
and refine environmental models are of major importance to the
European community. The eventual success of these programmes will
depend to a large extent on the thoroughness and the care taken
during the experimental preparatory phase. It is clear that much of
this experimental activity can be performed using well-equipped aircraft. However, it is also believed that an intermediate orbital
experimental phase is required prior to the final definition of the
operational system. It is during this orbital experimental phase that
the main benefits from using Spacelab are likely to materialise.

REFERENCES

1. European Remote Sensing Programme : Mission Objectives
    and Measurement Requirements
    ESA/EXEC (77)3,April, 1977

2. Landsat Data Users Handbook
    NASA/GSFC document No. 76SDS4258

3. Manual of Remote Sensing
    American Society of Photogrammetry, 1975

4. Active Microwave Workshop Report
    NASA, SP 376, 1975

5. Geosynchronous Multidisciplinary Earth Observation
    Satellite (MEOS) Study
    ESA study carried out by Dornier System, CR(P)-898, 1976

6. The ESA Remote-Sensing Programme:Present Activities and
    Future Plans
    J.Plevin & I.Pryke, ESA Bulletin No. 17, 1979

7. Definition of the Technical Requirements for an Earth
    Resources Payload for Spacelab
    ESA study carried out by MBB, CR(P)-374, 1973

8. Utilisation of Spacelab for Remote Sensing of Earth
    Resources: European contribution to Applications Summer
    Study, National Academy of Engineering, Snowmass,
    Colorado, ESA, 1974

9. Use of Metric Camera in Spacelab
    M.Schroeder and G.Konecny, ESA SP 134, May 1978

10. Practical Applications of Space Systems : Supporting Paper
    No. 6 - Extractable Resources
    National Academy of Sciences, 1975

11. Remote Sensing from Spacelab : A Case for International
    Cooperation
    J.Plevin, Environmental Remote Sensing 2, Edward Arnold 1977

12. European Earth Observation Payload for Spacelab
    B.Kunkel, IAF XXV Congress, Amsterdam, 1974

13. Large Format Camera
    NASA Information Brochure, 1977

14. Experiment Definition of metric Camera for First Spacelab
    Mission
    DFVLR study carried out by MBB, 1977

15. Microwave Scatterometer Experiment Final Report
    DFVLR study carried out by Dornier System, 1977

16. Sea-State Measurements and Radar Imaging from Spacelab 1
    F.Schlude et al, ESA SP 134, May 1978

17. Mission Definition Group Report on Microwave Experiment
    for First Spacelab Payload
    ESA, 1976

MATERIAL SCIENCES IN SPACE

G. Seibert
European Space Agency, Paris

Abstract

In less than a decade, space processing has developed from an idea, to a discipline of material science. The results from space experiments performed during several Apollo flights, the Skylab mission and the Apollo-Soyuz Test Project are summarised. The results show that the benefits of performing material science experiments under near-weightlessness in space can be applied to numerous materials of high technological importance ranging from crystals for electronic components, glasses, metals and alloys, to biological specimens.

In addition the paper provides a survey of material sciences experiments and instrumentation to be flown as part of the First Spacelab Payload in 1981.

1.   GENERAL GOALS AND OBJECTIVES OF MATERIAL SCIENCES IN SPACE

The weightless environment of an orbiting space system opens up new possibilities for the utilisation of space both to develop new industrial processes and to advance research in various aspects of material science. The absence of strong gravitational fields and the presence of a high vacuum of unlimited pumping capacity in space is of great interest for fundamental research and technology, with direct applications on Earth.

The basic objective of material sciences in space - often called "Space Processing" - is to expand our knowledge of physical and chemical processes in materials and provide a sound basis for new process inventions, which will include new or improved processes for use on Earth and processes that will work only under free-fall conditions.

The specific goals of material sciences are the development of

563

P. L. Bernacca and R. Ruffini (eds.), Astrophysics from Spacelab, 563-608.

existing materials or products, where the physical properties can be
improved, and the development of new materials with exceptional properties.

2.   THE DIFFERENT ENVIRONMENTAL CHARACTERISTICS OF EARTH AND SPACE

        The Earth environment through its two fundamental characteristics
of gravity and atmosphere determines the basic conditions of materials
on Earth. More precisely, the specific environment of Earth dictates
not only the chemical and physical states of materials but also their
possible methods of fabrication.

        These conditions imply :

- firstly, that materials exist on Earth only in the specific chemical
  and physical states determined by natural thermodynamical equilibrium.
  Accordingly, materials on Earth tend to return towards their stable
  states of oxides, sulphides, carbonates, etc.

- secondly, that among the number of methods we can conceive to manufacture interesting and useful products, only a few can be fully
  utilised in practice due to the limiting effect of Earth's environment.

        The basic characteristics of the space environment are the ab-
sence of both gravity and a gaseous atmosphere. Gravity-induced heat
convection in liquids is a very significant factor in processes involving the behaviour and handling of molten materials on Earth.

        In the absence of gravity under nearly zero-g conditions, tech-
niques are being developed to facilitate the manipulation or handling
of materials. In particular, the shape of fluids will be determined
mainly by molecular and surface tension forces. Consequently, handling
fluids can present problems, since they might not remain in an open
container, and, similarly, solids tend to float.

        The mutual interactions between the solid, liquid and gaseous
states is different from those to which we are accustomed on Earth.
For example, materials of different density always form physically
stable mixtures in the liquid state and liquids and gases should mix
perfectly.

        Further, the absence of pressure and reactive atmosphere, as
well as weightlessness, changes the rate of attainment and nature of
thermodynamic equilibrium.

        It is evident that the only property of the space environment
which cannot be simulated on Earth for extended periods is the absence
of gravity. It ought to be noted, however, that the vacuum on Earth
is limited by pumping capacity and may contain residual reactive constituents with which some materials could combine.

In the near-zero gravity environment, it is possible to exploit :

- the levitation of liquids and solids,

- the absence of fluid motion induced by gravity ,

- molecular forces, which in space become of significant importance
  for the processing and manufacturing of materials.

In addition, the absence of a reactive atmosphere enables developments to take place in the utilisation of pressure-sensitive phenomena or processes, such as evaporation, etc., and the change in position
of thermodynamic equilibrium.

Consequently, the extraction and processing of highly purified
materials represents an attractive form of space utilisation.

On the basis of the above considerations, it is apparent that
the general processes and experiments in the materials field which
will exploit space conditions to the optimum are the following :

- the melting and solidification, including fluid motion

- the preparation of large, defect-free, single crystals

- the preparation of materials of high purity

- the preparation of new heterogeneous materials of significant
  technological importance (for example, fibre-reinforced materials).

However, gravity is the strongest, but not the only possible
motive force for fluid motion. Surface forces generated by surface
tension gradients on free liquid surfaces, solid liquid interfacial
tension forces, thermal volume expansions, and some not too well
understood "interfacial tensions" due to differences in physical
properties between the hot and colder portions of a fluid, are also
possibilities.

3. SUMMARY OF PAST MATERIAL SCIENCE EXPERIMENTS IN SPACE

In less than a decade, space processing has developed from an
idea to a discipline of material science. The results from space
experiments and the supporting research have repeatedly shown the
advantages of weightless processes for many diverse materials.

This chapter will give a survey of material science/space processing experiments since its inception in the late 1960s. Demon-
strations of composite solidification, electrophoresis and fluid
flow were performed during several Apollo missions. Fourteen different
experiments in metal processing and single crystal growth were done

in Skylab, and on the Apollo-Soyuz Test Project (ASTP) four metallurgy
experiments, three crystal growth experiments and two electrophoresis
investigations were carried out. In addition, two rocket programmes
(SPAR : Space Processing Applications Rockets, in the USA, and TEXUS :
Technologische Experimente unter Schwerelosigkeit, in Germany) are
at present running, in preparation for the Shuttle-Spacelab era.

3.1   Apollo demonstrations

Eleven composite casting experiments were done during Apollo-14
in a small hand-held furnace (1). These were designed to determine
the potential advantages of space processing for this class of materials
and to assist in the planning of future experiments.

Dispersions of fibres, microspheres or gas in an indium-bismuth
matrix were melted, shaken (to disperse the solid phase) and then solidified for comparison with ground samples. Samples containing paraffin
and sodium acetate trihydrate were similarly processed, showing more
homogeneous mixtures than were achievable on Earth.

Surface-tension-driven convection was demonstrated on both
Apollo-14 and -17 by heating from one side a thin layer of oil containing aluminium flakes. The cell sizes and shapes agreed well with predictions, although a lower temperature gradient initiated the observed
cellular flow.

An electrophoresis apparatus was tested on Apollo-14, and the
electrophoresis of polystyrene latex was photographed on Apollo-16[2].
Two sizes of the spherical, submicron particles were used as single
species and combined in a three-cell geometry, in order to model the
electrophoresis of particulate materials. The periodic photograph
showed that eliminating gravity-induced thermal convection and sedimentation allowed cohesive bands with sharply defined particle fronts
to be formed, although   electro-osmosis caused the bands to have a
bullet shape. Research into coatings to control electro-osmosis became
a research goal before electro-phoresis was done again on the ASTP
mission.

3.2   Skylab experiments

Table 1 lists the experiments carried out on Skylab. The
following section will give only the highlights of a few experiments,
since the complete results are described in the Proceedings of the
Third Space Processing Symposium[3] and subsequent publications by the
Principal Investigators. In addition - in later paragraphs - on
"Separation and Purification Processes", "Crystal Growth" and "Metallurgical Processes", further information on Skylab experiments is provided.

TABLE 1
Skylab Space Processing Experiments

SKYLAB EXPERIMENTS	SKYLAB MISSION		
	II	III	IV
MATERIAL PROCESSING FACILITY			
M551: Metals-Melting Experiment, R.M. Poorman, MSFC Astronautics Lab		X	
M552: Exothermic Brazing Experiment,J.R. Williams, MSFC Product Eng. Lab		X	
M553: Sphere-Forming Experiment,E.A. Hasemeyer, MSFC Product Eng. Lab		X	
MULTIPURPOSE FURNACE SYSTEM			
M556: Vapor Growth of IV-VI Compounds, H. Wiedemeier, Rensselaer Polytechnic Institute		X	X
M557: Immiscible Alloy Compositions, J.L. Reger, TRW Systems		X	X
M558: Radioactive-Tracer Diffusion, A.O. Ukanwa, MSFC Space Sciences Lab		X	
M559: Microsecration in Germanium, F.A. Padovani, Texas Instruments		X	
M560: Growth of Spherical Crystals, H.U. Walter, Univ. of Alabama		X	X
M561: Whisker-Reinforced Composites, T. Kawada, National Institute for Metal Research		X	X
M562: Indium Antimonide Crystals, H.C. Gatos, MIT		X	X
M563: Mixed III-V Crystal Growth, W.R. Wilcox, USC		X	X
M564: Alkali Halide Eutectics, A.S. Yue, UCLA		X	
M565: Silver Grids Melted in Space, A. Deruytherre, Katholieke Univ., Leuven, Belgium		X	
M566: Copper-Aluminum Eutectic, E.A. Hasemeyer, MSFC Product Eng. Lab		X	X
EXPERIMENTS PERFORMED ON EACH MISSION:	3	11	7

The experiments now discussed were performed in the Multipurpose Electric Furnace System, which provided controlled heat-up, soak and cool-down, using less than 200 W to attain a maxium temperature of 1000°C.

In Experiment M560, one end of a cylindrical single crystal sample of indium antimonide was melted and resolidified in space. The crystal shape was formed by expansion of the indium antimonide during solidification plus intrinsic forces of surface tension and adhesion. In addition to the unusual shape, extremely well developed facets were observed on the crystal surface: for example, the overall flatness of some facets was of the order of 10^{-8}m. It was also found that the

number of imperfections in each crystal decreased steadily as the crystal
was regrown. In the middle of the space-grown crystal, the dislocation
density decreased by a factor of six in a distance of 1 cm. Dr Walter
has concluded that highly perfect single crystals can be prepared by
seeded, containerless solidification, and this technique offers major
advantages to processing high temperature and reactive materials.

Dr. A. Ukanwa, now at Howard University, designed Experiment M558:
Radioactive Tracer Diffusion, to determine the self-diffusion coefficients
of liquid zinc in a gravity-free environment. Rods of pure zinc had
pellets containing radioactive zinc-65 bonded to the end. The rods,
about 5 cm long and 6 mm in diameter, were encased in cartridges and
exposed to a temperature gradient of 45°C/cm and midway temperature of
550°C. After melting and solidification in space, the returned rods
were sliced into thin sections and the radioactivity of each section
measured to assess the migration of the radioactive atoms in space.
Whereas the ground-processed sample showed a uniform distribution of
radio-active zinc, the space sample showed good agreement with the
theoretically-calculated diffusion.

The final Skylab experiment discussed here is Experiment M564, which
produced a fibre-like NaCl-NaF eutectic mixture by directional solidification in the furnace. It was found that NaF fibres were aligned
and continuous along the axis for a major portion of the ingot. Optical
transmittance measurements of transverse and longitudinal sections show
that the ingots grown on Skylab have a significantly improved transmittance over material grown on Earth.

3.3 ASTP experiments

- ### Metallurgy
         Four metallurgy experiments were carried out on the ASTP
    mission (4):

    (i)   Surface-Tension-Induced Convection by Diffusion in a
             Liquid Pb and Pb - 0.05 at % Au couple (Reed and Adair);

    (ii)   Oriented Growth in LiF-NaCl Eutectic (Yue, Yue and Lee);

    (iii)   Monotectic Alloy Pb-Zn and Syntectic Alloy AlSb (Ang and
             Lacy); and

    (iv)   Processing of Permanent-Magnet Materials MnBi and (Cu,
             Co)$_5$Ce (Larson).

    (i)   The first was designed to investigate the phenomenon of
    surface-tension-induced convection by setting up a liquid diffu-
    sion couple of pure Pb and Pb-0.05 at % Au alloy in a micro-
    gravity environment. This system was chosen because the Au could
    subsequently be activated in a reactor and its concentration

determined by autoradiography. The couple was contained in two
types of ampoules, one wetted by the liquid (1015 steel) and the
other not (LTJ graphite). The couples were in a molten state
for about 2 hours at temperature of 650° and 470° C, during
which time the gold diffused about 2 cm. The concept was that
if no convective stirring effects due to the presence of surface-
tension differences between the Pb and the Pb-0.05 at % Au occurred, a normal concentration/distance profile of the Au would be
found in the diffusion couples. The presence of surface-tension-
driven convection would be expected to be observable in the
autoradiographic patterns, and the two types of containers would
demonstrate the effect of wetting on the convection. The two
temperatures would permit the diffusion parameters to be obtained
for the Pb-Au system. This carefully designed experiment was
thus closely related to the experiment on the self-diffusion of
zinc carried out by Ukanwa on Skylab. Analysis of the results
(by radiography) showed that, in the space sample, the solidification interface is distinctly curved, the gold apprently diffusing
more rapidly in the centre of the specimen than at the edges.
However, this does not necessarily mean that the Au did, in fact,
diffuse more rapidly along the centre of the ingot; such curvature
could be caused by a curved interface during solidification.
Since the equilibrium distribution coefficient (from the Pb-Au
phase diagram) for Au during solidification is < 1, the results
would imply a concave solidification interface, which is consistent with the observed coring.

The analogous specimen grown on the ground exhibited a
surprising feature. It was expected that the Au would be evenly
distributed throughout the ingot, but the Au concentration was
greater near the bottom than near the top. Since crystallisation
proceeded from the bottom to the top, and the distribution coefficient is < 1, the opposite concentration gradient would be
expected. A full interpretation of these results is still out-
standing. Further experiments of this type foreseen on Spacelab might be required in order to arrive at an explanation.

(II)   The phenomenon studied with the Halide Eutectic Growth
experiment is that of directional solidification, which can
lead to unique metallurgical microstructures and materials of
potentially great commercial importance. These materials are
composed of long fibres or platelets of one phase in the other
and are thus in-situ composite materials with unique properties.

The aim of the experiment was to grow LiF fibres in a
NaCl matrix, which is not a typical metallurgy experiment, but
in halide eutectics, the emphasis is on obtaining unique optical
properties, whereas in metal eutectics, the emphasis is on mechanical properties. In near-zero gravity, in both cases, a more
uniform microstructure with fewer flaws and longer, more continuous fibres is expected. Rods of NaCl-28.8 wt % LiF eutectic

mixture were melted back in the furnace both on the ground and
in space and then solidified unidirectionally in a steep temperature gradient. The light-transmittance data obtained from
both samples clearly demonstrate that the eutectic fibres in
the space-grown samples have a nearly perfect structure, whereas
the Earth-based sample shows that such a perfection in the structure cannot be achieved on Earth. As a conclusion from this
experiment and the Skylab experiment of Hasemeyer, it is re-
commended to grow superalloys for engine components (e.g.,
turbine blades) in the near-zero gravity environment.

(iii)   The Monotectic and Syntectic Alloys experiment consists
of two parts, the overall intent of which was to exploit the
near-zero gravity conditions of space, to produce alloys with
a more homogeneous microstructure than is possible on Earth.
The first part of the experiment (monotectic alloys) was concerned with producing homogeneous alloys in systems where there
is a miscibility gap in the liquid. The system chosen for study
was Pb-Zn. The density difference between lead (11.7 g/cm^3) and
Zn (7.14 g/cm^3)would clearly make production of a homogeneous
alloy difficult at 1 g for almost any composition. This experiment is, therefore, similar to Skylab experiment M557 (immiscible
alloy compositions), conducted by J.L. Reger. The temperature
for complete miscibility is, according to the phase diagram for
Pb-Zn, 795°C. The temperature reached in the space experiment
was 835°C, and the soak time at this temperature was 1.6 hours.

The results of this experiment clearly indicate a
separation into two macroscopic regions. There was one zinc
region and one lead region, with a zinc bubble in the lead region.

The experiment offered three hypotheses that would
explain these results: a) the consolute temperature is actually
higher than that published in the open literature (Pb-Zn phase
diagram of Hansen), b) a layer of oxide prevented complete inter-
diffusion in the liquid or c) the temperature was not high enough
to cause diffusional mixing in the time allotted.

Recent work shows explanation a) to be the most reasonable of these hypotheses. Further experiments on immiscible
alloys must be performed.

The second part of this experiment was concerned with
the synthetic growth of the Al-Sb system. Al-Sb is expected to
have electronic properties that would lead to several important
device applications, but preparation difficulties have inhibited
its exploitation. This experiment was more successful. The
problem of Earth-grown Al-Sb is the intergranular phase in an
Al-rich impurity phase, as often demonstrated by X-ray dispersion
investigations using scanning-electron microscopes. The space-
grown sample shows a much smaller amount of intergranular phase.

(iv)   The experiment "Zero-g Processing of Magnets" had several aspects. One concerned the fluid shape for various crucible fill factors. This greatly influences the heat flux and thermal profile, which are critical to solidification experiments. A further aspect concerned the solidification behaviour of the important Mn-Bi and Cu-(CuCo)$_5$ eutectics systems. The magnet analysis of the space-grown samples showed that the coercive force is so high that a field of 150 k Oe could not saturate them, i.e., the space-grown samples are much better than the best samples grown on the ground, which show appreciable gravitationally dependent segregation.

The general conclusion drawn from the ASTP metallurgy experiments is the following: the near zero-g environment has permitted the production of metallurgical materials with properties improved over those produced on the ground. The reasons for these improvements have not been established in all cases, but they would appear to be the absence of convection and the minimisation of buoyancy forces. It can be expected that with the use of more sophisticated equipment, substantial improvements might be forth-coming in the Shuttle/Spacelab era.

- Crystal Growth

The objectives of the various experiments differ in detail; however, they all have in common the exploration of growth conditions in a near-zero gravity environment. All experiments were conducted with a dual purpose: a) to gain fundamental information and data which could help bridge the still-existing gap between theory and experiment, and thus be instrumental in advancing the inherently deficient materials-processing technology on Earth, and b) to explore the potential advantages of space processing.

The following crystal growth experiments were performed during the ASTP mission:

(i)   Growth from Solution of Insoluble Crystals (M.D. Lind);

(ii)   Germanium Crystal Growth (H.C. Gatos, H.F. Witt);

(iii)   Crystal Growth from the Vapour Phase (H. Wiedemeyer).

(i)   The objective of the solution growth experiment was to investigate the growth of single crystals of insoluble substances by a process in which two or more reactant solutions are allowed to diffuse towards each other through a region of pure solvent. This novel method of crystal growth was conceived for orbital space flights to take advantage of the absence of gravity-driven convection which, on Earth, predominates over diffusion as a mechanism of material transport. The experiment performed during

the ASTP mission was designed both as a first attempt to determine the feasibility of the method and as a starting point for
further development.

On Earth, a gel growth method is used, where two or
more reactant solutions - separated by a gel - diffuse slowly
together and react. The reasons for choosing the three systems
given below where mainly because they have been studied extensively by the gel growth technique on the ground, and because of the
technological importance of Ca-CO3 and Pb-S. The primary function of the gel are suppresion of gravity-driven convection and
support of the growing crystal. Thus, the crystal grows in a
cell of solution surrounded by a flexible gel structure and has
no contact with container surfaces.

The ASTP experiments were conducted under the assumption
that, in an environment of sufficiently low gravity, convection
and sedimentation become negligible, with diffusion being the
predominant mixing mechanism even in the absence of gel. It
was, furthermore, assumed that such a diffusion-controlled
solution growth process (in the absence of gel) may lead to
crystals of higher quality, and the results obtained could
ultimately contribute to a better understanding of solution and
gel growth processes on Earth.

The three reactions chosen as growth experiments on
the ASTP are:

Calcium tartrate:
$CaCl_2 + NaHC_4H_4O_6 + 4H_2O =$
$CaC_4H_4O_6 . 4H_2O + NaCl + HCl$

Calcium carbonate:
$CaCl_2 + (NH_4)_2CO_3 = CaCO_3 + 2NH_4Cl$

Lead sulphide:
$PbCl_2 + CH_3CSNH_2 + H_2O = PbS + CH_3CONH_2 + 2HCl$

The post-flight analysis of the space-grown crystals
confirmed that the experimental approach taken under near-zero
gravity conditions yielded macroscopic single crystals of high
quality, while on Earth, only material of microscopic dimensions
can thus be obtained. Plate-like crystals of calcium tartrate
with a maximum length of 10 mm exceed in size, for comparable
growth periods (116 hours), even crystals produced by optimised
gel growth techniques on Earth. The calcium-carbonate experiments produced numerous well-formed clear rhombohedral up to
0.5 mm on an edge. The largest of these are large enough for
easy observation of their birefringence. The lead-sulphide
experiments produced crystals up to 0.1 mm in size.

While the preliminary results indicate conspicuous improvements in growth conditions for calcium tartrate and calcium carbonate, no major effect of zero-gravity environment could be observed for growth of lead sulphide.

(ii) The scientific objective of the Germanium Crystal Growth experiment was to investigate :

- the microscopic growth behaviour for directional solifification in a gradient furnace

- the dopant segregation behaviour and its functional dependence on the microscopic growth rate

- the absence or presence of convection phenomena

- the wetting characteristics if the melt under near-zero gravity conditions

- the effects of near-zero gravity conditions on the heat-transfer characteristics of the solidification system.

Gallium-doped germanium single crystals were partially melted and successfully grown from the melt with simultaneous interface demarcation during the ASTP mission. The interface demarcation was applied by transmitting current pulses at 4 sec intervals across the advancing crystal/melt interface in order to permit a quantitative post flight analysis of the growth conditions.

The results of the post flight analysis can be summarized as follows :

- Contrary to observations on earth, the germanium melt was not in wetting contact with the quartz ampoules, what means a significantly reduced contamination interaction.

- All regrown crystal portions were single crystalline over a length exceeding 3.5 cm; eventual breakdown of the matrices was expected and could be attributed to constitutional supercooling. Crystals grown on earth in the same configuration exhibited matrix breakdown after regrowth of approximately 2 cm or less.

- The microscopic rate of growth, determined from the spacings of interface demarcations, exhibited an initial steep transient and ultimately reached a value of about 9 μm/s. Ground-based tests in the same system indicated identical growth behaviour and it is therefore concluded that laminar convection, unavoidable on earth, does not contribute significantly to

heat transfer in directional solidification systems. It is
also concluded that the observed initial rate transients are
unavoidable in any real system and, as a result, existing
segregation theories cannot be expected to yield agreement
with the experiment.

- Compositional (dopant) fluctuations on the microscale are
  absent in the portions of the space-grown materials that
  were not subject to constitutional supercooling.

- The initial dopant concentration in the space-grown material
  is significantly larger than that predicted from presently
  available data.

- There is no evidence for the existence of surface-tension
  driven convection effects in the bulk of the germanium.

        The data acquired with this experiment contribute sign-
ficantly to the understanding of the crystal growth and segregation on earth and demonstrate the existance of favourable
growth conditions in the near zero-gravity environment.

(iii)   The primary objectives of the vapour growth experiments
were to investigate the effects of micro-gravity on the morphology of single crystals of mixed systems and to determine the
mass transport rates of these growth systems using the chemical-
transport technique. For this purpose, three vapour transport
experiments were performed during the Apollo-Soyuz mission
employing three chemical compounds, two transport agents and
one inert gas in one common temperature gradient. The systems
employed were those shown below.

$$GeSe_{0.99} \ Te_{0.01} - GeCl_4 \ (A)$$

$$GeS_{0.98} \ Se_{0.02} - \ GeCl_4 \ (B)$$

$$and \ GeS - GeCl_4 - Ar \ (C).$$

The materials were enclosed in evacuated sealed ampoules of
fused silica and transported in the temperature gradient 604-
507°C of the multipurpose furnace on board the Apollo-Soyuz
vehicle.

        The analysis and evaluation of the space experiments
is based on a direct comparision of crystals and of mass transport rates obtained under ground-based (prototype) and near-zero
gravity conditions. During ground-based studies it was observed
that the crystal morphology and mass fluxes of all the above
systems are affected by transport conditions, in particular by
gravity-driven convective interference; it was thus shown that
the crystal quality decreases with increasing convective contribution to the overall transport process.

The experiments conducted in space demonstrate a measureable improvement of the quality of the grown crystals in terms of surface perfection, crystallographic and chemical microhomogeneity, and density of defects. It was also found that the mass transport rates in the micro-gravity environment are in fact significantly greater than expected according to established transport relationships. This finding is of fundamental scientific and technological significance. The data obtained cannot be explained by present models for diffusive or convective vapour transport and suggest the existence of other transport components in reactive solid/gas-phase systems. These transport components are presumed to be related to thermo-chemical effects of gas-phase reactions.

In summary it may be stated that the vapour transport experiments conduced under near-zero gravity conditions yielded high-quality single crystalline materials which, according to every characterisation procedure applied, are markedly superior to materials grown by the same technique on earth. A study of the mass flux achieved in space showed that the presently available theoretical concepts are inadequate to explain the results obtained.

As a general conclusion from the ASTP crystal growth experiment it can be said all three growth techniques yielded material which are of significantly better quality than that achieved by the same growth technique on earth.

Adverse effects due to Marangoui type convection could not be observed. Several unexpected phenomena of potentially far-reaching consequences were indeed reported and require further and deeper study during the coming flight opportunities of the Shuttle/Spacelab era.

From the scientific point of view, the experiments are of extreme value. The data obtained not only give strong indications that some presently accepted theoretical concepts are deficient and based on erroneous assumptions, but they more importantly provide the basis for new approaches to be taken.

Electrophoresis

The scientific requirement for the two electrophoresis experiments on the ASTP flight arose from a study concerning the application of electrophoresis in future biological experiments on board Spacelab. It is expected that separation capability will be increased at zero gravity and that this will lead to new applications which depend on improved resolution. The experiments were devised to test electrophoresis hardware and to verify electrophoresis experiments by operation in space.

The American MA-011 entitled "ASTP Electrophoresis Technology Experiment" (G.V.F. Seaman, R.E. Allen, G.H. Barlow and M. Bier) and the German MA-014, "Flow Electrophoresis" (K. Hannig and H. Wirth) experiment were very different systems intended for very different purposes. The American apparatus was based on models already tested on board Apollos-14 and 16 (1) and was a static system with limited sample volume, developed for high resolution and for testing various electrophoretical methods, e.g. iso-focussing or iso-tachophoresis techniques.

The German MA-014 apparatus was designed for free-flow electrophoresis with special provision for high sample volume processing at high resolution (5). An advanced data-collection and storage system was to deliver data on the influence of heat generation.

The major objectives of experiment MA-011 were to demonstrate static free-fluid electrophoresis and iso-tachophoresis of biological cells in a space environment and to verify concepts by operation in this environment. The overall objective was oriented towards the testing of electrophoretic hardware sample handling, maintaining viability of samples through all phases of the experiment and the demonstration of electrophoresis technology under near zero gravity conditions.

The following samples were processed :

- Fixed red blood cells from man, horse and rabbit used as model mixtures with well known properties. A total of about 1.5×10^7 cells were used.

- Human kidney cells, a fraction of which produces the enzyme urokinase, which can dissolve blood clots. 2×10^6 cells were electrophoresed.

- Human lymphocytes, a class of white blood cells involved in the body's immunity against disease. 1.5×10^7 cells were used.

- Fixed human and rabbit erythrocytes. This mixture was used to demonstrate iso-tachophoresis.

The test equipment consisted of four major elements :

- an electrophoresis unit with camera

- a cryogenic freezer

- 8 experiment columns, i.e. 6 electrophores and 2 iso-tachophoresis.

The following results of experiment MA-011 were obtained:

a) Separation of red blood cell model mixture with high quality, showing sharp zone boundaries.

b) Successful separation of kidney cells, widened by a six-fold increase in urokinase production in the respective cell population after growth in culture back on earth.

c) No separation of lymphocytes could be achieved, but preservation techniques could be successfully demonstrated.

d) Iso-tachophoresis separation was limited by lysis of red blood cells, the Cl-ion working as leader ion. The time allowed for separation was therefore too short and only the initial state could be observed.

The experiment was not totally successful because the fluid lines in some of the columns were blocked. Pre-flight testing was not possible because of contamination problems of the enclosed sterile buffer.

Electro-osmosis which has been a major obstacle to successful electrophoretic separations in a closed cylindrical tube apparatus, was eliminated in the ASTP electrophoresis experiment. The coating techniques developed for this purpose significantly increased the resolution capability of the electrophoresis unit and will be valuable for ground-based and future space electrophoretic applications.

Experiment MA-014 was based on the so-called "free flow" or "continuous flow electrophoresis" concept in which the sample to be separated is introduced continuously in a fine stream into a flowing buffer curtain, and a strong electric field is applied perpendicular to the direction of flow. Particles with different surface charge densities are deflected from the buffer flow direction at an angle determined by the flow rate and the electro-phoretic mobility of the particles.

At the end of the buffer curtain, the separated zones can either be collected continuously or - as in the MA-014 experiment - analysed by monitoring the optical absorption pattern across the buffer curtain.

Since this electrophoresis system is a continuous one, there are no theoretical limits in respect of the amount of cells to be separated, in contrast to a static system. A second advantage is the possibility of changing samples relatively frequently without interrupting the experimental conditions, i.e. without disturbing established equilibrium. In combination with an advanced system providing for automatic operation, this principle of electrophoresis can be used for analytical, preparative and routine applications.

The following samples were separated during the ASTP mission :

- rat bone-marrow cells

- rat spleen cell

- mixture of human and rabbit erythrocytes

- rat lymph-node cells and human erythrocytes

Inspite of a partial malfunction (excessively bright) of the lamps of the absorption measurement system, it was possible to evaluate and analyse the data and the following conclusions can be drawn :

a) The bone-marrow sample showed an excellent separation pattern, sharper than that measured with comparable ground equipment.

b) Spleen cells showed the best result of all. The volume of recorded data was sufficient to identify details of the separation quality in spite of high-rate sample processing, only possible under zero gravity.

c) Less information was stored in the case of the erythrocyte mixture, precluding the extraction of well-founded results.

d) Lymph node cells showed a somewhat poorer distribution, due to the lack of information stored.

Both electrophoresis experiments of ASTP proved the applicability of electrophoresis under near zero gravity conditions and confirmed that successful adaption of hardware to the space vehicles can be achieved.

High resolution electrophoris as well as up to a tenfold higher sample processing rate than on earth are possible in space.

4.   INTEREST IN AND EXPECTATIONS FOR MATERIAL SCIENCES IN SPACE

The benefits of performing material science experiments under a near-weightless conditions in space is projected to have applicability to numerous and various types of materials. These materials of high technological importance range from electronic materials, crystals and glasses, metals, alloys and compounds, to biological specimens.

Scientists and engineers have suggested a wide variety of ideas

for which there is reasonable technical justification in the sense
that novel or improved results should be achievable through the use
of methods made possible by space flight.

The basic characteristics of the space environment are the absence of both gravity and a gaseous atmosphere, of which the dominant
criterion for the selection of processes and manufacturing techniques
suitable for application in orbit will be the absence of gravity.
Vacuum can be simulated on Earth, but it is limited by pumping capacity
and may contain residual reactive constituents with which certain
materials may combine.

Space processing requires, besides heavy voluminous instrumentation, high power, large energy provisions and, of course, recoverability. Since conventional satellites do not fulfil these prere-
quisites, space-processing experiments have only become possible within the Skylab programme, after some very preliminary experiments conducted during the Apollo flights.

As well as recoverability and short lead time, the main advantage of Spacelab, the European element of the Space Shuttle Programme,
is its large payload capacity, which will also substantially relax
space-qualification criteria by allowing more conventional equipment
to be used. In addition, the involvement of man will allow simplification of laboratory equipment. Limitations on storage space in which to
return processed samples to Earth are not expected to arise with the
Shuttle/Spacelab system.

A number of surveys and studies to identify and evaluate the
research and development topics of Material Sciences/Space Processing
have been performed in the United States and in Europe. From these
activities it can be concluded that the first phase of the utilisation
of the near zero-gravity environment should relate to applied research.
The results and conclusions of the first phase should stimulate industry's interest in more elaborate experiments before moving on in
a subsequent phase of exploitation proper.

In the following chapter, the interest and expectations in the
fields of Electrophoretic Separation, Metallurgy, Crystal Growth and
Fluid Physics/Fluid Dynamics shall be discussed.

4.1   Electrophoresis

In biochemistry the isolation of a molecular class with specific
qualities opens the way to the understandinf of biological mechanisms
at the molecular level. Therefore, separation techniques applying
sophisticated methods based on tenuous differences in biophysical
and biochemical parameters of induvidual molecules are considered as
an essential tool of progress.

It is evident that separation on earth always involves gravity, even if this parameter is not selected for its own sake as it is in the centrifuge for example. From this it follows that zero gravity could have a genuine significance for these separation techniques where gravity acts as a nuisance as in most electrophoretic separation methods.

Electrophoresis is the movement of charged particles, macro-molecules or ions in a fluid medium under the influence of an applied electrical field. The technique may be used for analytical purposes and/or for the preparation or isolation of materials differing in their electrophoretic mobilities. For biological cells in media of the ionic strength used in these studies, the electrophoretic mobility is independent of the size, shape or orientation of the particle.

The separation and isolation of biological and chemical materials by electrophoresis is limited due to gravity sedimentation of the com-ponents or stirring by thermal convection in the fluid phase.

During the past space experiments (ASTP) two methods of electro-phoretic separations have been applied :

- the "free flow electrophoresis", which is a dynamic method

- the "static zone electrophoresis"

The principle of both methods of which the first is mainly preparative whereas the second is mainly analytical, shall be described in the following sections :

4.1.1 Free Flow Electrophoresis

The principle of free-flow electrophoresis is to separate charged particles in a streaming liquid curtain by the application of an electric field perpendicular to the curtain flow. A thin sample stream is in-jected through the one wall into the liquid curtain. According to the surface charge of particles in the sample, the sample stream splits as a result of the electric field applied and beyond the curtain a number of particle streams or bands are observable.

As the charges and the electrical forces acting upon them are very small, all parameters influencing the migration path of the charged particles must be stabilised in order to achieve a high-quality separat-ion. Besides the electric field, one of the most significant para-meters is the liquid curtain velocity and its profile within the separat-ion gap.

Since most of the liquids used in electrophoresis are aqueous solutions and show electrolytic conductivity, the electric field gene-rates electric current and Joule heat. Temperature gradients across the separation gap are the unavoidable result of the finite thermal conductance of the buffer liquids. In consequence, buoyancy and thermal convection occur, and degrades the resolution of quality of the separation process.

On earth, application to the preparative separation of fractions from a mixture of proteins or particles such as living cells, has not been particularly successful. The scaling-up of equipment to accomodate industrial capacities has produced a series of problems associated with natural convections due to heat dissipation, forced convections due to density differences, diffusion effects and particle interactions.

Terrestrial electrophoresis has been hampered especially by the effect of Joule heat on the liquid medium and the electrical streaming effects, termed electro-osmosis, so that a wide variety of systems have been developed to meet particular separatory needs.

4.1.2 Static Zone Electrophoresis

For the moment the scope of static electrophoresis in space is not at all rigidly determined. The methods have multiple variants such as phoresis, focussing and iso-tachophoresis (used in ASTP), most of which show high resolution but low yields. The static electrophoresis is oriented towards batch-type equipment performed in a tube (like a column) useful for analytical work.

Zone electrophoresis on earth is subject to two major difficulties which make it unsuitable for the separation of biological cells or larger particles :

(a) cell or particle sedimentation

(b) convection arising from Joule heating during electrophoresis.

Although various techniques have been developed to overcome these problems on earth, their elimination can be accomplished best in the near-zero-gravity environment of space.

Zone electrophoresis is a potentially powerful separation technique for charged materials. Under terrestrial conditions, the method has been successfully applied to the separation of macromolecules and small particles, but its use for biological cells or larger particles is limited by gravity-driven processes such as sedimentation and convection.

A micro-gravity environment offers the elimination of sedimentation and flotation problems as well as any density-driven convection arising from either Joule heating or concentration differences. The advantages of such an environment were recognised, and resulted in initiation of the original 'space electrophoresis' programme.

Electrophoresis in a free fluid was first demonstrated in space during
the Apollo-14 mission, and subsequently during the Apollo-16 mission
(15). The Apollo-16 electrophoresis demonstration was performed using
two different sizes of polystyrene latex particles. The sample bands
were severely distorted by the effects of electro-osmosis, but a certain
degree of separation was achieved.

The ASTP Electrophoresis Technology Experiment (11) was performed
to demonstrate static free-fluid electrophoresis and iso-tachophoresis
of biological cells in a space environment and to verify experiment
concepts by operation in this environment.

The method "Iso-tachophoresis" consists of placing a mixture
of charged particles (sample) of the same sign (co-ions) in a column
or on some solid support medium, containing in addition two electro-
lytes, one of which has all of its co-ions more mobile than any sample
co-ions, while the other electrolyte has all of its co-ions less mobile
than any sample co-ion, the entire electrolyte having a common counter-
ion. Upon application of a constant direct current, the sample co-ions
initially move at different velocities until a steady state is reached
in which the sample ions have separated into contiguous zones with
sharp interface in order of their mobilities. The zones migrate at
the same velocity and adjust themselves into compartments or various
lengths, concentrations and conductivities such that the product of the
co-ion mobility and the field gradient is the same within every zone.
The polarity of the electric field must be such that the leading ion
migrates towards the electrode that is placed on the same side of the
sample as the leading electrolyte.

4.1.3   Specific Aspect of Electrophoresis at Near Zero Gravity

Electrophoresis is an electrokinetic phenomenon based on the
transport of charged particles in an electric field. It is used predominantly in biology and medicine, but also in engineering and chemistry, to separate particles which possess different charges, such as
proteins, viruses and even cells.

Electrophoresis can be carried out on earth with high resolution, but its processing capacity is limited. Gravity effects such as
heat convection and sedimentation then limit its efficiency, and the
cross-sectional dimensions of the electrophoresis chambers must be
kept small. Under zero-gravity conditions, however, there should be
no limits in these respects, but some other limiting factors do emerge,
such as heat transfer from the liquid medium in the chamber. A rough
calculation shows that effectiveness can be increased by a factor of
ten with equal or better resolution than similar ground equipment.

Electrophoresis may be of assistance in studying many phenomena
such as growth, metabolism, genetics and immunity response which function
differently under low-gravity conditions.

On earth, the problem of convection has been solved or rather kept within given limits through several technical tricks such as by the use of paper or gels on unidimensional strips and of paper or a buffer stream with adequate cooling on two-dimensional curtains. In space - as a result of the absence of gravity and in - as far as there are no acceleration gradients in the spacecraft during the experiment - the absence of weight involves absence of heat and concentration induced convection currents.

Is this absence of convection solely advantageous? When the convection currents, which favour heat exchange towards the cell-wall disappear all heat transfer must be achieved through conduction and clearly this mechanism has its proper limitation.

On the other hand, absence of convection opens the way to higher electrical fields and to a thicker cell with higher throughout. Slower flow rates could also be applied to increase dwelling time between the electrodes and this increases separation distance or resolution.

4.2   Metallurgy

Weightlessness, together with the vacuum available in space offers great opportunities for various types of applied research or manufacturing processes in the field of metallurgy. This is particularly true for the purification of metals or alloys and the processing of levitated materials. Other areas of potential benefit are the possibilities of controlling the nucleation in molten metals are forming liquid metals without mould or crucibles.

First the advantage of the space environment for metallurgical processes, of which the strongly reduced gravity will have the greatest effect, shall be summarized.

The absence of buoyancy and convection results in the stability of liquid-solid, liquid-liquid and liquid-gas mixtures so that metal matrix composites, dispersion strengthened alloys, metal foams and new alloys from immiscible liquids become possible. Complex ordered structures such as directionally solidified eutectics or composites can be made if fluid flow in the melt is reduced. Investigations into the nature of the microstructure of eutectics obtained by various solidification techniques revealed that many defects are induced by variable solidification rates. Thus, an increased control over the solidifaction process is necessary before the advantage of space processing can be clearly shown. However, once we attain this knowledge, the order of thousands of eutectic materials consisting of metals, ceramics, intermetallics and organics become possible candidates for space processing.

Processing in an extreme vacuum (around 10^{-14} torr) by use of a wake shield associated with the Shuttle may lead to materials with impurity content orders of magnitude lower than the best now

obtainable. Since the history of materials development contains frequent reference to the sensitivity of physical properties to the nature
and quantity of impurities, theoretical property values may be approached
by space processing.

The reduced gravity conditions allow :

A) Preparation of gravity-segregation-free alloys due to
   lack of buoyancy

   - composite materials by composite casting (no sedimentation
     of the reinforcing material)

   - alloys with super-fine and homogeneous distribution of a
     second phase precipitated in liquid state, i.e. from melts
     with miscibility gaps

   - foam metal by incorporating gas bubbles in melts.

B) Preparation of alloys in absence of convection currents
   (due to ineffectiveness of density differences)

   - growth of defect-free single crystals

   - preparation of improved aligned eutectics

C) Levitation melting of metals and alloys (without induction
   fields)

   - preparation of ultrapure metals and alloys by crucible-
     free melting

   - improvement of floating-zone refining by removing the
     constraints due to the weight of the liquid

   - homogeneous solidification of metallic materials resulting
     in finer grain structures.

D) Study of material and heat-transport phenomena like diffusion
   processes in liquid and gaseous materials normally concealed
   by gravity.

It will be possible to prepare alloys without gravity segregation, which normally occurs on Earth. This results in a stabilisation
of mixtures in the liquid state, e.g. of molten metals and alloys,
which should allow the preparation of two or multi-phase alloys unknown on Earth to-day.

4.2.1   Composite Materials

Space conditions offer the possibilities for preparing isotropic composite materials by homogeneous dispersion of short fibres
and particles in a molten metal matrix. Uniform dispersion of the
fibres as well as good bonding between fibres and matrix, requires
besides near-zero gravity conditions both good wetting between the fibres
and matrix and outgassing of the molten metal matrix.

In space, it will be possible to produce composite materials
by composite casting, the simplest method for shaping these materials.
This would be a great advantage, because at the moment it is only possible to produce composites with continuous fibres. Short fibres or
particles incorporated in a melt will segregate due to gravity. Anisotropy of properties and difficult joining properties - as they cannot
be welded without damage - are the most serious drawbacks of earth
produced composites with discontinuous short fibres and particles. In
space the gravity induced segregation it expected to disappear completely, i.e. one should be able to achieve perfectly stable mixture. Composite materials strengthened by short fibres or particles are of great
technical interest, because they have a random distribution of the
reinforcing materials. They therefore will show no anisotropy in their
properties, but do show an improvement in strength, especially warm
strength. However, these kinds of materials are very difficult or
impossible to prepare on earth, because the reinforcing components
immediately segregate in the molten bath as they always differ widely
in density from the matrix. For these materials, the weightlessness
of space represents an ideal and unique environment for processing
without promoting any gravity-segregation.

The influence of gravity on the sedimentation rates of the
possible reinforcing components has been thoroughly studied in the
simulation experiments of Steuer & Kay (6), who estimate that for
certain common differences in density between fibre and matrix (e.g.
for carbon, Al_2O_3 or SiC fibres in aluminium), the sedimentation rate
may be as high as a few centimetres per second.

To return to the considerable interest of the space environment
for the preparation of these kinds of composites, it should be mentioned
that one finds in this category the most promising composite systems,
namely those possessing the best combinations of fibres and matrices,
e.g. the very high modulus, high-strength carbon, boron Al_2O_3, SiC
fibres or whiskers with metal matrices. All of these high-strength
fibres are very light compared with the heavy metal matrices in which
they should be embedded and hence the weightlessness of space is here
of still more significant value.

4.2.2   Immiscible Alloys

Another possibility offered by space processing is the pro-

duction of two-phase alloys by solidification with a miscibility gap
in the liquid state. If such an alloy is heated above the critical
point, it will be homogeneous. By passing through the miscibility gap
during cooling, separation into two different liquids occurs, resulting
in the precipitation of droplets of the second phase, which are considered to have a fine and homogeneous distribution.

        The possibility that the processing of immiscible materials
consisting of phases of different specific densities, in a micro-gravity
environment can be used to make products with unusual structures and
properties. The advantages of a space processing would be that segregation effects which arise as a result of density differences can be
avoided or reduced to very low levels.

        There are approximately 500 binary metallic alloy systems
which demonstrate a miscibility and a very large number of systems
based on glasses. Amongst the metallic systems are such alloys as
Al-Pb, Cu-Pb, Ga-Bi, Pb-Zn, Al-In etc., all of which show miscibility
gaps.

        As an added complication, convection currents on earth can
occur in single-phase fluids and in multiphase (liquid-liquid and
liquid-solid) systems because of density variations in the liquid phase
produced either by temperature or compositional gradients within the
liquid phase.

        The overall compositional heterogeneity induced by these effects
leads to unsatisfactory properties, which are well recognised by metal-
lurgists.

        There is another feature of these gravity-driven convection
processes which leads to coarse microstructures and this concerns the
occurence of various agglomeration processes (7). Coalescence can be
produced by any of the following phenomena: direct particle-particle
contact as a result of growth in particle size; random collisions of
the Brownian type; collisions produced by Stokes flow which is controlled by particle size; particle contact produced by velocity gradients
within the matrix fluid; contact brought about by variations in interfacial energy which result from temperature gradients (this is termed
the Marangoni effect); particle growth as a result of overall reduction
in interfacial energy (known also as Ostwald ripening)(8). Presuming
that embryos lead to stable nuclei with radii greater than a minimum
critical value, then growth will continue to occur as the overall free
energy of the system decreases. The coalescence processes briefly
outlined above are brought about by a reduction in total interfacial
energy and the point of interest for a micro-gravity environment is
to establish which of these processes will be rate-controlling if
gravitational forces are substantially lowered.

        There are thus two features of immiscible-alloy systems which
merit attention, namely segregation and particle-size distribution.

In a microgravitational environment, segregation will be significantly
reduced and particle sizes will be controlled by diffusional growth
rather than by convection-driven collisions. Given that these two features can be realised in practice, the question then arises as to the
potential advantages of materials with a homogeneous composition and a
uniform distribution of fine stable particles. Immiscible-alloy systems
can produce such structures and present understanding suggests that
they can result in such desirable properties as high strength, tough-
ness, wear resistance, superconductivity, superplasticity and good
bearing characteristics. Potential applications for such microstructures
prepared from immiscible alloys have been identified (9) and even though
the list is short it includes applications as diverse as diverse as
catalysts and nuclear control systems.

    The first results of the Skylab experiments were irregular,
and the mixtures showed sometimes unstability, probably because of
coarsening of the droplets.

    New space experiments are urgently required to answer a number
of questions :

- Does the absence of gravity associated with space processing enable alloy systems which contain an immiscible liquid
  region to solidify from the single-liquid state with structures different from those obtained on earth?

- In particular, does the absence of gravity enable such
  alloys to be produced with fine stable dispersions of one
  phase in a solid matrix, and do such materials possess special
  or unusual properties in such systems?

- Can values for parameters such as diffusion coefficient and
  interfacial energy be determined?

- Are current theories of particle-size stability valid for
  liquid-liquid immiscible systems?

4.2.3   Foam Metals

    Another possibility is the preparation of a foamed metal in
space by incorporating gas bubbles in a melt. Such metals can already
be produced on Earth, but by another, very complicated process. Foamed
metals have special properties and offer some advantages for particular
applications, e.g. controlled thermal conductivity, improved shock resistance, etc.

    On the ground, collision effects appear dominant in determining
large effective dimensions of the dispersed particles or bubbles, due
to aggregation. In zero-gravity diffusion effects should be dominant
instead. To avoid the effects of convection due to surface tension,

it is proposed to take advantage of the role of oxide films on the metals as containers.

Bubbles may form by boiling of the liquid, by ebullition of dissolved gases, or they may be introduced artificially into the melt.

4.2.4   Defect-Free Crystals and Aligned Eutectics

Besides the absence of segregation the second major advantage of the space environment as regards the preparation of alloys is the absence of convection currents. These currents are normally beneficial during the preparation of alloys on Earth, but are sometimes disturbing. They are induced in a molten metal by the density differences resulting from temperature differences. Their absence will probably allow the production of single crystals with greater perfection and an improvement in the perfection of unidirectionally solidified eutectics.

Levitation melting will be feasible in a near zero-gravity environment and it will not require any support, and melting under space conditions will be a very effective purification process, it would seem that during solidification the nucleation process should only be homogeneous. It may be assumed, therefore, that a very great degree of supercooling will be observed and that during solidification a single crystal will be formed, especially if solifification is induced just by contact with the solid material. Because of the surface cleanliness of the melt, a seed crystal touching the supercooled liquid will transmit its orientation through the molten material, and it should be possible to prepare single crystals under these conditions free from any defects.

Eutectic solidification is the mechanism leading from a single liquid phase to two distinct solid phases of different composition during cooling. It is thus a natural composite. In contrast to artificial composites they are called "in-situ composites". These eutectic materials, for which there are already numerous projected applications (magnetism, optics, mechanics, electronics) have differences in morphological aspects.

The Skylab results (10) and the ASTP results (11 and 12) were encouraging for fibrous morphologies like alkali halides, and were relatively disappointing for lamellar metallic eutectics. It is necessary to bear in mind the complexity of the oriented eutectic growth process which in space, as on the ground, involves numerous mechanisms: homogeneous and heterogeneous germination, heat and mass transport (chemical and thermal diffusion), couples growth of the phases, the influence of boundaries or external surfaces, equilibria of surface tensions, liquid decantation phenomena, relationship between phase spacing and growth rate, agitation of the bath, and the consequent mechanical or solute redistribution effects.

The co-participation of these various mechanisms leads to oriented structures which usually contain many defects. Some of the above-mentioned mechanisms will be modified to various degrees by the space environment, but they are not necessarily the ones which will improve the morphology. Consequently, though convection was originally considered to be the main reason for structural defects, it is no longer possible to assign it sole responsibility : we now know the influence of the growth parameters (thermal gradient and speed) on interphase crystallographic orientation relationships. In some systems one of the phases can develop in the liquid ahead of the principle interface. Under these conditions hydrodynamic forces created by convection currents will vary from one system to another. Their effects will be greater for small eutectic fibres than for more voluminous lamellae, or for more fragile alkali halide fibres than for metallic fibres. In fact, in order to study the influence of the lack of convection (due to weightlessness) it will be more useful to study the modifications occurring for the same system, obtained with fibres or lamellae of sufficiently different dimensions by modifying the solidification conditions.

A large number of space experiments will be necessary in these areas. However, the results will depend critically on the quality of experiment preparation. The space experiments give scientists the opportunity to achieve improvements of the microstructure, which might have enormous economic impact like for example the improvement of super alloys of turbine blades might have.

4.2.5   Levitation Melting

The most efficient method of purification for metals on earth is the zone-melting method first devised by Pfann. Under space conditions, the diameters of zone-refined ingots are limited by the capabilities of the Spacelab only, because weightlessness will remove the constraints imposed on floating molten zones by the weight of the liquid on earth. Moreover, the absence of convection in the melt will probably improve the refining process. Additional advantages also seem to be possible from refinements in the control over heat and mass transport in melting.

The possibility of levitation melting, which provides a melting process without crucibles or walls, will further favour the purification of a metal or alloy.

The process of levitation melting in orbit offers the advantage of handling molten metals without containers or other physical restraints, thus avoiding contamination and other disturbances due to contact with foreign materials. On earth this process in restricted to small quantities and conducting materials. In space, the technique can be applied to both conducting and non-conducting materials without any restrictions in the quantity of material.

A further application of the handling of molten metals in space is the possibility of forming them into particular shapes and sizes. This can be done by using magnetic and electric fields, ultrasonics, surface tension pulling, adhension casting, or blow casting. In conclusion it may be pointed out that most of the advantages of the purification and levitation melting processes outlined above are not limited to metals and alloys but that these processes can also be used for the preparation of electronic and ceramic materials. This will give rise to a wide variety of space applications in the field of materials sciences.

4.3   Crystal Growth

The main requirements of single crystals of electronic materials are crystalline perfection, high purity and homogeneity, controlled doping, large dimensions and surface perfection.

Crystal defects (dislocations, stacking, faults, swirls) cause malfunctioning of devices, rapid ageing and low reliability, as well as yield problems in manufacture. Today quite a lot is known about their influence on device properties as well as about methods for avoiding such defects. However, this knowledge has been found empirically and a true understanding of the underlying physics and chemistry seldom exists. Little is known about the relation between lattice defects and the shape of the solid/liquid interface.

The performance of electronic materials in use depends critically on the methods of their preparation, which in many cases involve processes that can be affected by gravity. Therefore, one of the main advantages of a zero-gravity environment foreseen for achieving the above requirements is that gravity-driven convection is eliminated.

The increasing development of solid-state components in the electronics industry has induced a search for ever larger and more perfect single crystals. Whatever the growth method both the perfection and size of the crystal are limited by the occurence of quite stringent factors, one of which is gravitational field.

Generally, this field causes free convection in the liquid or vapour, thus generating so-called 'growth striations' which are detrimental to crystal homogeneity. Furthermore, it limits the size of the crystals grown because of:

(i)   convection, which becomes more important as the liquid volume increases,

(ii)   vibrations, which greatly enhance spurious nucleation in solution and bulk growth and

(iii)   weight, in the case of vertical withdrawal.

Finally, gravity indirectly results in a source of contamination, because of the need for a container in most cases and the occurence of
an atmosphere.

A new and attractive aspect of crystal growth is the possibility
of preparing crystals in the near zero gravity environment of Spacelab.
Some of the common methods of growing electronic materials from the melt
are (a) Czochralski, (b) Bridgman, (c) zone melting, (d) Verneuil, (e)
flux, (f) hydrothermal, and (g) solution growth. Techniques (a) -
(d) are the most used commercially. They are generally suitable for
space-growth experiments and details are given in the literature (13-
14). Methods (e) - (f) can yield good optical-quality single crystals,
but they are generally very small.

The selection of growth techniques involves heating methods
as well as materials. The latter include technically and economically
important crystal like silicon, and model substances like InSb for
basic studies of crystal growth.

The Skylab and ASTP experiments have shown that crystal growth
in space may bring advantages. In the following, some fields are outlined where problems arise in terrestrial crystal growth and where
advantages might be expected of space-grown crystals. In each particular
case, however, it will be necessary to analyse in detail the origins of
the problems of crystal growth and the effect of gravity on these origins.

4.3.1   Elementary semiconductors

Silicon and - to a minor degree - germanium single crystals
are the most important base materials of the present-day electronic
industry. They are used for a large variety of applications in discrete
and integrated electronic components. As a result of their great importance, these crystals also are of exceptionally high quality. It is
possible to grow crystals without dislocations and with extremely small
impurity concentrations (10^{10}/cm^3, i.e., 12 N in the chemical notation
of purity) which have diameters of several inches. Because of their
high quality, silicon and germanium crystals would be particularly
suitable to study the effect of zero gravity on the crystal quality
and to point out differences and improvements. Only in special cases
may Si-crystal growth in space be expected to bring significant advantages. Examples include the growth of silicon crystals with diameters
larger than 10 cm for thyristor applications or as an extremely homogeneous base material for highly integrated electronic components, such
as charge-coupled imaging or data storage devices.

For elementary semiconductors other than silicon, an improvement of the crystal properties is expected, in particular for tellurium,
since tellurium crystals have high dislocation densities, probably because of their high atomic weight and the low energy required for the
formation of dislocations. These dislocations give rise to infrared

absorption and limit the lifetime of photoelectrically generated charge carriers. Tellurium is physically and technologically of interest because of its exceptionally high nonlinear optical coefficient, and may be used as a base material for electronic components operating in the infrared.

4.3.2   Compound semiconductors

High-quality single crystals are required today and in the future in the following fields:

- Optoelectronics and optical communications,

- Computer memories,

- Microwave engineering,

- Piezoelectric and ultrasonic components,

- Surface wave engineering.

For some of the above fields, III-V semiconductors (in particular GaAs) are suitable. The industrial production of GaAs can be controlled today, but there are still problems relating to the growth of high-quality single crystals of large diameters (several inches).

These diffuculties could be overcome by crystal growth in space. The space-grown crystals could then be used as substrates for light-emitting diodes, for lasers, for high-frequency components, such as Gunn diodes, and also for piezoelectric and surface wave components.

Light-emitting diodes are made today primarily of Ga(As,P) with red emission and of GaP with green and yellow emissions. Ga(As, P) can be prepared today only in the form of thin epitaxial layers deposited on GaAs or GaP substrates; bulk crystals are not sufficiently homogeneous and have too many non-radiating centres which reduce the light yield substantially. Space-grown Ga(As,P) crystals will bring advantages if they are of adequate luminescence quality and no additional epitaxy is necessary for the fabrication of light-emitting diodes. Large homogeneous Ga(As,P) crystals of luminescence quality could be grown, for instance, by the travelling-solvent technique.

Improvements of the crystal quality may also be expected for high-melting compounds and for compounds with melts which tend to attack the container walls. Skylab experiments have shown that the contact between container and melt is reduced in the space environment. This phenomenon should be studied by an additional number of orienting experiments, in order to gain a better insight. Suitable semiconductors would be BN, AlP, BP and SiC. Improved crystal growth of these materials could provide new details concerning the implementation of high-effi-

ciency blue luminescence diodes, high-temperature devices and cold
cathodes.

4.3.3   Oxide single crystals

The new oxidic materials, e.g., mixed niobates or tantalates,
are the basic constituents for advanced components of electro-optic
and electronic technology. Their use, however, is in most cases limited
because of their low structural perfection.

The results of the Skylab experiments justify the assumption
that, under micro-gravity conditions, crystals of higher structural
perfection and with more uniform doping can be grown. Such crystals
represent a high value per unit weight, and, for some applications,
e.g., volume-holographic memories, relatively large quantities are
required.

The space processing facilities of Spacelab should be used to
deepen our knowledge on fundamental crystal growth processes, to improve the performance of existing devices, and, possibly, to develop
new crystals with unique properties not attainable on Earth.

4.4   Fluid Physics/Fluid Dynamics

The near-zero-gravity environment of Spacelab offers wide-
ranging possibilities for performing basic studies in connection with
physical and physico-chemical phenomena in fluids and for developing
processes of materials both on Earth and in Space. Often these phenomena are fairly complicated and masked on Earth by the effects of
gravity.

The most sophisticated potential source of process control
based on weightlessness arises from the possibility of maintaining
relatively larges masses of fluids in an almost quiescent state. In
this new environment heat and mass transport in liquids and gases should
be governed solely by the well-known partial differential equations
that describe diffusion, heat conduction and radiation. These equations
have unique solutions determined by initial and boundary conditions
which can be controlled by appropriate apparatus design. It seems
possible that any physical or chemical process that is controlled by
heat and mass transport can be made to follow a predetermined course
in a properly designed space experiment, even in heterogeneous material
systems that include liquid and vapour phases. This type of process
control has potential applications in practically every type of solidification and crystal-growth technique.

Since weightless fluids can be kept quiescent in the presence
of intense temperature and composition gradients, it should be possible,
for example, to predict and control the effects of constituional super-

cooling in most solidification processes, or to grow highly perfect
crystals on unsupported seeds by diffusion-controlled transport in
solutions. Similarly, weightlessness is expected to make a wide range
of idealised transport and temperature conditions available for crystal
growth from the vapour.

Also, the study of the floating zone must be considered a fundamental experiment for purification and crystal growth. The linear
stability of a rotating liquid rod and Marangoni convection, as well
as the influence of viscosity on the dimensions of the fluid zone,
should be studied. Theoretical models of near-zero gravity zones (16)
as well as useful simulative experiments on earth using Plateau techniques (17), e.g. Carruthers and Grassi (18) and more recently aboard
Skylab (19) have been performed.

The diffusion of the flames may be pertinent to the study of
the inflammability of material in Spacelab. Fluid convection under
reduced-gravity conditions should be of particular interest for studying
the performance of phase change materials used for spacecraft thermal
control. Under heating conditions these materials melt, freezing when
the temperature falls again. The convection provides high heat transfer
rates that are desirable in most cases, since known phase change mate-
rails are fairly poor heat conductors.

Gravity may have important effects in chemical reactions.
Although it is fairly well established that chemical reactions in a
homogeneous reactor are independent of gravity, because of the small
mass of the molecules and atoms involved, in most cases the rate and
effectiveness of the chemical reactions are influenced by the availability of the reactants. As the reaction advances, reactants are
depleted and must be replenished by some diffusion process. When the
the species react rapidly once they are mixed, their arrival at the
reaction zone is the controlling mechanism. If the diffusion process
is induced by buoyancy, as in the ordinary candle, the net attainable
speed of the reaction depends on gravity.

Surface tension and interfacial tension (Marangoni convection)
appear in liquid-gas or miscible liquid-liquid interfaces. The surface
tension varies with temperature, generally decreasing when the temperature increases. Surface tension gradients induce surface tractions.
This mechanism is dominant under normal gravity conditions only if the
fluid-layer thickness is small enough. In space, it is nearly always
dominant if an interface exists.

Surface tension is extremely sensitive to small gradients in
the chemical composition of a fluid. Once the surface tension gradients
exceeds a certain threshold value (to be studied), a convection process
starts.

All the above mentioned non-gravity driving forces, which are
usually suppressed by gravity include :

- surface or interfacial tensions

- thermal volume expansions

- g-jiter (as it could be produced by spacecraft manoeuvers)

- magnetic and electric forces

may cause natural convection also in the space environment (20).

The insert of unstable motion is highly dependent on the confining boundaries like (21)

- the magnitude and direction of residual accelerations

- geometric configuration

- material properties

- imposed boundary conditions.

As far as convection due to magnetic and electric fields is concerned the following should be noted:

Electric and magnetic fields induce body forces (like gravity) so that they can generate convection and phase separation similar to those in a gravitational field. Whereas in a gravitational field convection is generated by density differences, in an electrical field the flow is generated by the differences in electrical conductivity and by differences in susceptibility in a magnetic field. The electrical conductivity and magnetic susceptibility are temperature-dependent, so that temperature differences are usually required to obtain the flows. Both convectional and unstable types of convection are possible with such fields. The phenomena have not been fully explored or exploited in the space programme.

Theoretical studies of static capillary stability and of the dynamic stability of fluid interfaces have, because of their intrinsic scientific interest, a long history in the development of classical physics. Their relevance to space-processing activities has generated renewed interest in this field, and at the same time the opportunity to conduct experiments in genuine micro-gravity seems likely to lead to fresh advances. Theoretical studies of certain problems are already well developed, especially of the static stability of axisymmetric surfaces subject to various end-constraints, the dynamic behaviour of axisymmetric surfaces in steady rotation about the axis of symmetry, and the static stability of thin liquid films on solid surfaces. The last-named topic offers a possible route to the accurate measurement of long-range intermolecular forces.

Experimental studies are also in progress, some being in true micro-gravity, whilst others make use of Plateau's neutral buoyancy technique. Further use of the Plateau technique seems likely to be beneficial, both as a simple simulation tool in studying the practical feasibility of proposed space experiments and (in cases where side-effects due to the viscosities and other properties of the two fluids are unimportant) as a direct route to intrinsically useful results.

5.   MATERIAL SCIENCE EXPERIMENTAL FACILITIES AND EXPERIMENTS FOR THE FIRST SPACELAB MISSION

The first Spacelab flight is scheduled for late - 1981 on the first operational flight of the Space Shuttle.

ESA considers material sciences, besides atmospheric physics, the most important payload element of the first Spacelab mission. 39 experiments from a total of 61 European experiments selected for the first mission are material sciences experiments.

Electrophoresis experiments have not been included because of the special first mission constraints of "no continuous power during the last 9 days prior to launch", limited access during pre-launch period, etc.

The material science part of the first Spacelab payload (FSLP) includes :

- General-purpose facilities like furnaces and process chambers, which are used to perform experiments.

- A number of autonomous experiments, which do not need any general purpose facility.

The technical characteristics of the general purpose facilities and the scientific objectives of the experiments are summarized in the following chapters :

5.1   General-purpose facilities

a) Isothermal Heating Facility

The Isothermal Heating Facility is a multi-purpose facility for different types of experiments, including solidification studies, diffusion fundamental, casting of metals and composites, preparation of new and/or improved glasses and ceramics.

A multi-foil multi-purpose furnace capable of providing 1500° C (later 2400° C) from tungsten heating elements clamped to water-cooled electrodes forms part of this facility. The heating elements are enclosed by a tungsten tube

which is enveloped by 1 cm of multi-foil insulation. The
supporting and protective envelope of the furnace is a
steel shell (15 cm diameter) with copper cooling coils
attached. Accoustic positioning and mixing devices, mechanical mixing devices, transportation pistons, etc., can
be mounted to the structure via quick-coupling locks.
Vacuum and noble gas provisions will be available (including a vacuum bell jar); cooling will be provided via Spacelab's experiment heat exchanger, and peripheral equipment
may be connected to the Spacelab avionics air loop.

To simplify operations for the FSLP, experiments relying
on the Isothermal Heating Facility should, whenever possible, use cartridges in which the material components can
be dispersed ultrasonically prior to solidification.

TABLE 1   Isothermal Heating Facility Preliminary Performance
           Characteristics

Temperature range	200° - 1500°C
Temperature accuracy	$+$ 5°C
Temperature constancy	$\mp$ 2°C at 1500°C
Axial temperature distribution	$\mp$ 0.5°C/cm
Heat-up rate (max.)	500°C/min
cool-down rate (max.)	150°C/min
Pressure range	10^{-5}Torr to 1 atm noble gas
Max. size of cartridges	50 mm diameter, 100 mm length

b) Acoustic Stirring and Positioning

Since the accelerations expected in Spacelab are not zero,
but of the order of 10^{-4} to 10^{-5} g, contactless positioning and manipulation of molten materials is required for
metallurgical processes, glass preparation, crystal growth
and physical/chemical processes in fluids. In addition,
many potential experiments require one material (liquid or
solid) to be mixed or dispersed in another liquid which
will subsequently be solidified.

The most relevant parameters for acoustic levitation and
stirring within a tube furnace have been identified theoretically and experimentally (22). Multi-axial positioning
with ultrasonic standing waves have been shown to be feasible
with a single transducer, but is a rather complex technique
in a tube furnace with changing temperatures.

Ultrasonic stirring (dispersing, emulsifying, homogenising,
etc.) of liquid melts with and without particle inclusion
is feasible when the radiating transducer is in direct contact with the melt.

c) Gradient Heating Facility - Low temperature

This is a multi-purpose facility for crystal growth, unidirectional solidification of eutectic and other experiments (Table 2). It contains a furnace with three heating
elements which may be controlled independently, so that a
variety of temperature profiles (isothermal, three-zone)
can be achieved. Thermal insulation is provided by axial
heat shields, a low-conductivity radiation shield, multi-
foil insulation and an outer protective shield. The top
cover of the furnace allows parallel injection of three
cartridges. The furnace is capable of achieving 1600°C,
but, for power reasons, it will be operated only up to
1200°C.

TABLE 2   Gradient Heating Facility Preliminary Performance
          Characteristics

Volume	: for three 300 mm long, 25 mm diameter cartridges
Heating rate	: 2 h to reach 1200°C (500 W), operating power 400 W
Cooling rate	: 10-15°C/min (700 - 1200°C), faster with He flow
Gradient	: up to 150°C/cm

d) Mirror Heating Facility

The mirror is designed for a variety of experiments: e.g.,
floating zone/travelling heater techniques for crystal
growth, metallurgy and glasses; two different versions are
under consideration. One facility uses a single lamp and
the other two lamps as heat source. The two-lamp facility
provides a more homogeneous temperature distribution (Table
3). Temperatures in excess of 2300 K, depending on sample
size and absorption properties, have been achieved.

The heat source of the single-lamp facility is a tungsten
filament lamp, located at one focus of an ellipsoidal gold-
plated quartz mirror, whereas the crystal rod is moved
through the focus by pushing and pulling rods. The integrated sample holders provide for feed and turning motion,
and vacuum and noble gas supply facilities are provided.
A water-cooled jacket is connected to the Spacelab experiment heat exchanger.

TABLE 3   Mirror Heating Facility Preliminary Performance
           Characteristics

Max. sample diameter	: 20 mm
Zone length	: 10 - 20 mm
Max. growth length	: 110 mm
Max. feed length	: 150 mm
Minimum turn velocity	: 0.1 per minute
Max. turn velocity	: 10 per minute
Minimum feed velocity	: 10^{-5} mm/min.
Maximum feed velocity	: 50 mm/min.
Max. variation for turn and feed for 1 run	: 1 : 20
Operating temperature range	: 200° - 2100°C depending on sample diameter and material

typical values :

a) Al_2O_3 : 10 mm dia, 15 mm zone length : 2045°C

b) silicon : 10 mm dia, 20 mm zone length : 1420°C

c) Al   : 12 mm dia, 15 mm zone length : 600°C

Power : for case a) 800 W, for case b) 650 W and for case
         c) 150 W are required.

Temperature setting accuracy : + 5°C

Temperature constancy   : $\pm$ 0.5°C

Atmosphere : noble gas (max. 1,2 bar) or vacuum

Heat-up/cool down rates : very fast rates possible.

Power/energy consumption is extremely low as compared to
conventional heating methods, especially if the very high
heating and cooling rates are taken into account.

Using the laboratory model of the mirror furnace for preparation of growth experiments on Earth, among others, the
following crystals could be obtained (23, 24).

- dislocation-free silicon size: 15 mm dia. length: 60 mm

- GaSb single crystals (with very few dislocations) from
  a Ga-solution in a quartz ampoule, size 10 mm inner dia.
  length : 90 mm.

e) Fluid Physics Module

This facility will be used to investigate certain stability
limits of rotating fluid masses in zero-gravity. The instabilities of interest are relevant to crystal growth experiments of various kinds and are of fundamental scientific
value in providing an understanding of the basic processes
in a liquid melt. These investigations involve the growth
of axisymmetric and nonaxisymmetric perturbations of fluid
masses constrained between slowly rotating circular boundaries and the study of mass transport.

The present study is aimed at developing an apparatus (25,
26) which allows study of the static and dynamic characteristics of liquids by spinning, oscillating or vibrating
a liquid zone. The complete process can be recorded on
film for evaluation on Earth.

Table 4   Fluid Physics Module Preliminary Performance
           Characteristics

Test volume	: 120 mm max. height, 100 mm diam. (max.)
Selectable disc diameters	: 15, 30, 40, 60, 80 and 100 mm $\pm$ 0.5 mm
Rotation speeds	: 5 - 100 rotations per minute $\pm$ 0.1 rpm or 1%
Applied vibration	: 0.1 - 5 Hz, 0.05 - 0.5 mm amplitude, pre-setting $\pm$ 5%, measurement accuracy: 0.1%
Lateral movements	: Pre-set steps of 0.1 mm to a total of 2 mm
Heat-up capabilities	: 60°C of 1 disc
Electrical potential	: 100 V DC between 2 end plates

The concept of the Fluid Physics Module is demonstrated in
(Figure 1).

5.2   Experiments of the First Spacelab Payload

5.2.1   Crystal Growth Experiments

Crystal growth in space is characterised by the micro-gravity
environment which means, in the ideal case, the absence of gravity induced convection. It is expected that semiconductors crystals can be
grown with a degree of perfection and chemical homogeneity not otherwise
obtainable. Model substances like InSb for fundamental studies and

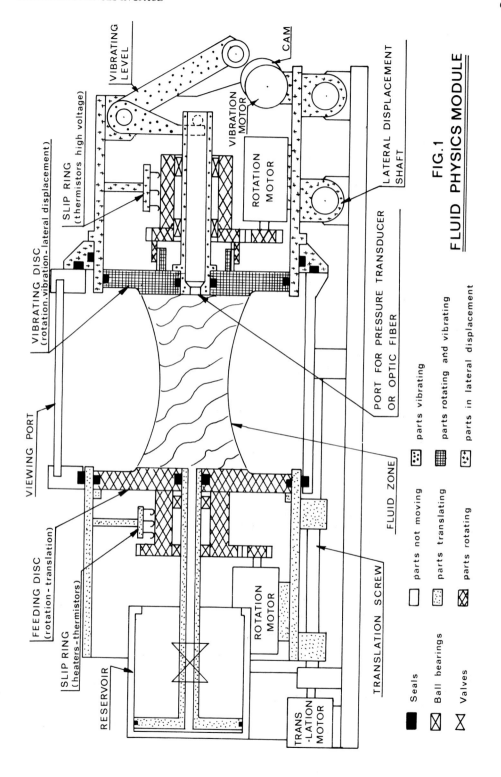

FIG. 1
FLUID PHYSICS MODULE

technically important crystals like silicon have been chosen to clarify
the nature and effect of convection currents. Use is also made of the
possibilities that the near gravity-free environment allows crucible-
free zone melting of materials which melt without decomposition and
negligible vapour pressure.

   Experiment F 10 uses the Bridgeman technique for growing
single crystals of lead telluride. It is intended to determine quantitatively the influence of the parameters : nucleation conditions, presence
of a crucible and convection on the growth of a crystal. The crystallographic, electrical and optical characteristics of single crystals
grown on earth and in space with and without seeds (oriented) and also
of space grown crystal obtained by levitation growth shall be evaluated.

   Experiment D 47 uses the travelling solvent method to grow
Cd Te. It is expected that the space grown crystals show better homogeneity and a lower density of intrinsic defects of the crystals.

   Experiment D 48 intends to use the crucible-free floating-
zone technique for the growth of InSb from the melt. Form the space
experiment improvement in the homogeneity of dopant distribution and
lower dislocation densities are expected.

   D-44 is an experiment which intends to study the influence
of convection and of the Marangoni effect on the formation of striations
(dopant inhomogeneities) in silicon single crystals.

   Experiment D 45 is a liquid and gas-zone crystallisation of
Si-spheres using a mirror furnace. Under micro-gravity it is expected
to distinguish the influence of diffusion and convection of crystal
growth with controlled dopant distribution.

   The objective of experiment DK-58 is to grow crystals of various
organic charge-transfer complexes (like TTF-TCNQ), which show quasi one-
dimensional conductivity and which will be grown from solution by diffusion processes. It is believed that weightless conditions provide a
unique opportunity to achieve highest crystal perfection, which is a
prerequisite for desired high conductivity in a narrow temperature
interval (60 K).

   The study of the magnetic domain structure and the study of
the generation of growth defects on space-grown $MnCO_3$ crystals is the
objective of experiment F-62.

   It is envisaged that in micro-gravity the amount of spurious
nucleation will be mostly eliminated and that steady thermal conditions
can be obtained. Experiment F-71, vapour growth of HgI_2, intends to
study the improvements of space grown crystals regarding :

   - non-regular and heterogeneous incorporation of impurities

- deviations from chemical stoichiometry and

- large strain fields.

Experiment D-49 aims at the diffusion growth of large single crystals of proteins like 2β Hemoglobin (M = 32 000) and β - Galactosidase (M = 520 000). Such large single crystals are necessary for the X-ray structure analysis and neutron diffraction analysis which allow the determination of the relation between function and structure of proteins.

5.2.2   Metallurgy Experiments

Since the number of metallurgy experiments selected for the first Spacelab is rather large (21) it is not the objective of this section to discuss experimental details of each proposal but to deal with the reasons for the proposals and the information or insight that might be gained by them.

One category concerns the solidification of immiscible alloys, i.e. systems containing a liquid-phase miscibility gap. They consist of two or more liquid phases which form a homogeneous single-phase liquid at higher temperature. The possibility that the processing of immiscible materials, consisting of phases of different specific masses in a micro-gravity environment, can be used to make products with unusual structures and properties. Reasons for these new properties are that the constituents of these materials normally segregate and cannot combine, or the production of fine homogeneous dispersion of one minor phase in a metallic matrix are not possible on earth. On earth segregation and agglomeration can occur. Typical immiscible alloy systems chosen for the first mission are AlIn, AlPb, ZnPb.

Another category of interest for space processing are isotropic composite materials, produced by homogeneous dispersion of short fibres and/or particles in a molten matrix metal. On earth composite materials with a hard second phase finely distributed for dispersion hardening in a matrix material can only be produced by powder metallurgical methods, which result in a material and undesired pores and particle clusters. In space it is expected to achieve homogeneous distributions and large numbers of hard particles in the molten matrix of, for example, high conductivity material for electrical contacts. Examples of composite materials experiments selected for the first Spacelab flight are Al_2O_3 powder and fibres (SiC) in Al - and in a Cu - matrix material.

In the field of growth of eutectic alloys advantages may be expected from the space environment, because under ground conditions eutectic materials include a great number of defects, which have been attributed to convection movements in the liquid phase. The results obtained by past unidirectional growth in space have been irregular, because strongly reduced convection is an important but not the only parameter to be considered. On the ground as well as in space numerous mechanisms are effecting the eutectic growth : homogeneous and hetero- genous germination, chemical and thermal diffusion, coupled growth of phases, the influence of boundaries or external surfaces, equilibria of surfaces, equilibria of surface tensions etc. These mechanisms will be modified to various degrees by the space environment.

In general the solidification process in space should be more regular leading to a better structure which determines the properties of the solidified alloys. First experiments will be made with Al-Al$_2$Cu and AgGe eutectic alloys.

It is also intended to study the possibility of a controlled unidirectional solidification of a AlZn vapour emulsion having a fine regular dispersion of voids. Under 1 g conditions there is a very low stability of the gas-liquid emulsion caused by the gravity field.

Another experiment concerns the unidirectional solidification of Te-doped In Sb-Ni Sb eutectics, in which the NiSb needles grow per- pendicular to the phase boundary liquid solid. Since both length and direction of the Ni-Sb needles and the distribution of the dopant are influenced by gravity it is expected that the space-grown material shows superior properties.

An experiment of practical interest is the gas turbine blade investigation, with the objective to improve the mechanical properties at high temperatures. The blades based on directionally solidified eutectics of chromium alloys are coated with a thin refractory layer by plasma spraying, chemical vapour deposition and cementation. In space the turbine plate will be remelted and solidified without chang- ing their geometry.

Also two diffusion experiments have been selected for the first mission of Spacelab. In microgravity in the absence of convection the thermodiffusion will become very important due to larger temperature gradients and will result in composition changes (called Soret Effect). In space it will therefore be much easier to measure the thermodiffusion parameters and their temperature dependance. The study will be per- formed on Sn + $\mathcal{E}$ Sb, Sn + $\mathcal{E}$ In and Sn + $\mathcal{E}$ Ag systems.

Another diffusion experiment concerns the self - and inter- diffusion in liquid metals. Measurement of the diffusion of two iso- topes (Sn 112 and Sn 118) and its temperature dependance and of two Sn-In alloys near the eutectic mixture will be made using La$_2$O$_3$ - markers. The selfdiffusion experiment shall help to find out which of the current theories of diffusion in liquids (quasi crystalline fluctuation and critical volume) is adequate.

On earth one is unable to decide which of the current theories of diffusion in liquids is the most probable one, because the accuracy of measurements is limited by the unavoidable convection.

It is the objective of another experiment to investigate how the characteristics of mating surfaces affect adhesion and friction, as well as to discriminate between the relevant mechanical and chemical mechanisms. Examinations of the surface morphology on microscopic (electron microscopy) and submicroscopic (SIMS) level will be performed. Small metallic spheres of controlled variable momenta will be thrown against a flat very clean metallic surface and the conditions for sticking will be investigated. In microgravity environment there is no need for counterweights which cause disturbing friction in the equipment. This experiment aims at information about the field of force of atoms at an interface.

5.2.3 Glass Investigations in Micro-Gravity

The only glass experiment of the first Spacelab mission concerns the investigation of reaction kinetics in glass melts.

The reactions between solids and glass melts or between glass melts of different compositions is controlled by the reaction through the interface. At normal terrestrial gravity conditions and high temperatures, i.e. at low viscosity conditions the mass transport due to diffusion is superimposed by convective transport due to density gradients (gradients in composition and/or temperature). Therefore, on earth a separation of mass transport due to diffusion and due to convection is not possible. Micro-gravity conditions will be used to enlarge the range of temperature within which undisturbed diffusion experiments can be realised.

Since silicate systems (Na_2 0.3 SiO_2 / Rb_2 0.3 SiO_2) are important for both glass and ceramic technology, binary alkali-silica systems have been choosen for the experiments. Applying controlled heating and cooling procedures it is expected that a reaction depth of 100 μ m during 1 hour experimental time at 1100°C will be achieved, which will allow the evaluation of the diffusion coefficients.

5.2.4 Fluid Physics Experiments

A Fluid Physics Module, one of the multipurpose experimental facilities to be flown as part of the First Spacelab Payload, will be used to study several phenomena connected with the hydrodynamics of floating liquid zones. Floating zone techniques are widely used to avoid container contamination in crystal growth in terrestrial laboratories. The maximum stable length of a vertically suspended liquid zone is determined by the balance between the surface tension forces and the hydrostatic pressure on earth. In the microgravity environment of space the constraints imposed on the length of a zone and the buoyancy-induced convection are strongly reduced, which renders the zone more accessible than on earth.

Several methods of controlled disturbances of the equilibrium of the zone are considered like iso and counter-rotating flows, non-isothermal conditions and application of electric fields. Also the study of oscillatory injection and removal of small volumes of liquid through an orifice in the centre of one of the end discs. Here the occurance of standing waves at the liquid surface shall be studied as function of the radius to length ratio of the zone.

Another disturbing method is the axial vibration of one of the end discs with low frequencies and low amplitudes. In all these stability studies the influence of several fluid-mechanical parameters, such as density, surface tension, viscosity, etc. on the behaviour of the floating zone shall be investigated. These parameters will be especially important in the non-isothermal experiments. It is expected that surface-tension gradients induced by temperature gradients may cause convection through the well-known Marangoni effect. This convection will stirr the liquid, modifying the temperature field of non-rotating zones.

Another investigation foreseen is the study of adhesion between phases (liquid/solid) in the absence of electric and magnetic forces. In many capillary systems the work of adhesion is opposed by the work of cohesion of the liquid so that spreading and non-spreading situations are found according to which of the two is the greater. Since the forces (van der Waal's forces) involved in this process fall off very rapidly with distance, it is found in terrestrial experiments that gravitational forces mask all attempts to study these forces in larger macroscopic systems. In the Fluid Physics Module (FPM) experiment it is intended to study the properties of a liquid bridge zone between the two solid discs (end plates) of the FPM. From the shape of the liquid bridge the interaction forces will be derived and the critical point at which the bridge becomes unstable will be determined.

Another related experiment will study the kinetetics of spreading of liquids on solids. The objective is to study the flow of a liquid in response to forces generated at the non-equilibrium line of intersection of a liquid/vapour interface with a plane solid surface. In micro-gravity it is feasible to study large systems in which the curvature effects are small and in which fluid flow is predominantly driven by movements of the contact line.

Using the FPM a spherical liquid drop of known volume is formed at one of the end plates with its contact line constrained by an anti-spreading barrier. Then contact is made with the other smooth plane end plate and the spreading behaviour and internal fluid motion will be visualised with the aid of solid tracer particles.

Another FPM experiment intends to study coupled motions of liquid-solid systems in micro-gravity. Coupled motions represent an unsolved classical problem in analytical mechanics. The experiment uses a number of differently shaped containers partially or completely filled with liquid. The containers will be fixed to one end plate and the FPM system will be used to excite the liquid motion by spin-up from rest, and by forced vibration sweeps. The internal liquid motions and the free surface behaviour will be examined.

REFERENCES

1.   Yates, I.C. Jr., Apollo-14 Composite Casting Demonstration, NASA MSFC Report TM X-64641, March 1972.

2.   Snyder, R.S. et. al., Separation and Purification Methods, 2,259 (1974).

3.   Space Processing and Manufacturing, NASA MSFC Report MS-69-1, 21 October 1969.

4.   ASTP Science Report, Report N° NASA-TM X-58173, February 1976.

5.   Hannig, K., Separation of Cells and Particles by Continuous Free Flow Electrophoresis, in "Techniques of Biochem. and Biophys. Morphology", John Wiley and Sons, Inc., New York, Vol. 1 pp. 191-232, 1972.

6.   Steuer, W.H. & Kay, S., Preparation of Composite Materials in Space, Convair Aerospace Division of General Dynamics, Report GDCA-DGB 73-0014, January 1973.

7.   Markworth, A.J., Gelles, S.H., Duga, J.J. & Oldfield W., "Immiscible Materials and Alloys", Proc. Third Space Processing Symposium, II, 1003-1029, 1974.

8.   Ostwald,W., Z.Phys. Chem. 37, 785 (1901).

9.   Reger J.L. "Study on Processing Immiscible Materials at Zero Gravity", NASA contract NA 58-2867, 1973.

10.   Proceedings of the 3rd Space Processing Symposium "Skylab Results", Vol. I and II, Marshall Space Flight Center, Huntsville, Alabama 30.4 to 1.5. 1975.

11.   Apollo-Soyus Test Project, Preliminary Science Report, Report N° NASA TM-X 58175, February 1976.

12.   ESA SP-114 "Material Science in Space" Proceedings of 2nd European Symposium on Material Science in Space, Frascati (Italy), 6-8 April 1976.

13.   Brice, J.C. The Growth of Crystals from Liquids, North Holland Publishing Company, Amsterdam, 1973.

14.   Laudise, R.A. "The Growth of single Crystals" Prentice-Hall,
        Inc., New Jersey, 1970.

15.   Snyder, R.S. Bier, M. Griffin, R.N. Johnson, A.L. Leidheiser,
        H. Micale, F.J. Vanderhoff, J.W. Ross, S. & van Oss, C.J. Free
        Fluid Particle Electrophoresis on Apollo 16 "Sep. Purif. Methods"
        2.2. 259-282 (1973).

16.   Fowle, A.A., Haggerty, J.J. & Strong, P.F., NASA CR 143876,
        October 1974.

17.   Plateau, J.A.F., Smithsonian Inst. Ann. Rep. 1863, p. 250.

18.   Carruthers, J.R. & Grasso, M., J. Appl. Phys. 43, 2, 436-445
        (1972).

19.   Carruthers, J.R., NASA Report M 74,5, V2, 837-856.

20.   Grodzka, P.G., Types of Natural Convection in Space Processes:
        Summary and Report, Lockheed Missiles and Space Co., Inc.,
        Huntsville Research & Engineering Center. HREC-5577-4, CMSC-
        HREC TR D30650, January 1973.

21.   Ostrach, S., Natural Convection in Enclosures, Advances in
        Heat Transfer, Vol. 8, Chapt. 3, Academic Press, 1972.

22.   ESA Contracts SC/39/HQ (1974), SC/67/HQ (1975) and 2678/76/F/
        WMT (1976) to Battelle (Frankfurt): "Study on Positioning and
        Stirring of Molten Materials in Space by Ultrasonic Methods"

23.   A. Eyer, R. Nitsche and H. Zimmermann "Zonenzüchtung von
        Kristallen mittels Optischer Heizung", Status Seminar, Werk-
        stoff und Verfahrenstechnik, 6-8.10. 1976, Bad Kissingen,
        issued by Bundesministerium für Forschung und Technologie,
        Bonn.

24.   Private Communication, H. Zimmermann, Kristallograph. Institut,
        University of Freiburg, W. Germany.

25.   ESA Contract SC/95/HQ (1976) to CASA (Madrid): "Fluid Physics
        Module".

26.   ESA Contract 3122/77/F/HGE (SC) (1976) to Fiat (Turin):
        "Fluid Physics Module".

ABUNDANCE DETERMINATIONS FOR INTERSTELLAR GAS

Donald G. York
Princeton University Observatory

Presented in Trieste, Italy, September 8, 1976
at the conference on "Physics and Astrophysics from Spacelab"

Abstract

The results of general abundance studies, using the Copernicus satellite are reviewed. The general pattern of depletion of some elements such as Si, Al, Mn, Fe, and Mg, near the Sun is described. Techniques for obtaining abundances for various species are discussed, as are the aspects of the analysis which are uncertain. Variations in gas phase abundance from region to region are found for Fe and Si, similar to variations now known to exist for Ca and Ti based on visible spectra. A new generation of space instruments, with higher efficiency and resolution, in particular for 920 Å $< \lambda <$ 1400 Å, can provide detailed interstellar line profiles, thus putting abundance determinations based on equivalent widths on a more firm footing, and yield more detailed information reflecting the make up and binding energies of interstellar dust particles as well as regional changes in the products of nucleonsynthesis. The most important wavelengths and techniques relevant to these future studies are tabulated and discussed.

I. INTRODUCTION

Interstellar absorption measurements made with Copernicus have provided us over the last four years with a large amount of material on equivalent widths for the elements H, D, B, C, N, O, F, Mg, Al, Si, P, S, Cl, Ar, Cr, Mn, Fe, Ni, Cu, and Zn, as well as for several molecules. All of these species were previously not observed at sufficient resolution to guarantee freedom from stellar line blends for the equivalent width measures. Interstellar features with $\lambda >$ 3000 Å, accessible from the ground, have very low abundances (Ti) or are trace ionization

P. L. Bernacca and R. Ruffini (eds.), Astrophysics from Spacelab, 609-644.

stages, requiring additional knowledge of the electron density, n_e, to convert the
equivalent width measures to element abundances (Na I, Ca II, K I). Abundances
determined from equivalent widths of very weak lines are now known for several lines-
of-sight to an accuracy of better than 50%, and the corresponding H I column densities are known to an accuracy of 20% to 50% for over 100 stars. However, several
species suffer from saturation problems, so that various assumptions have to be made
in interpreting the equivalent widths and in evaluating the true uncertainties of
the determinations. I shall describe attempts to evaluate what assumptions should
be made, using in particular, observations of stars near the sun and of very distant
unreddened stars (§II), then summarize our present evaluation of the abundance pattern,
compared to the sun, and the interpretation of that pattern (§III). In §IV, I shall
summarize results on the abundances of interstellar species which might be forthcoming using a new generation of space-borne equipment, using higher sensitivity and/or
resolution than that presently available. These measurements need to be as free as
possible from the ambiguities of saturation and line blends: the specific wavelengths are tabulated which will probably be the most useful. General conclusions
are stated in §V. Abundances of molecules, including H_2, and of isotope ratios
(D/H, C^{13}/C^{12}) are not covered in detail.

II.   PRESENT PROBLEMS

Interstellar gas exists in a wide variety of physical conditions, with tem-
peratures ranging from 20°K to $\gtrsim 10^6$°K, and densities from 10^{-3} cm^{-3} to $>10^5$ cm^{-3}.
Most aggregates of gas are probably affected by mechanical energy input, in the
form of shock waves, as well as by radiation input, and by internal heating due
to chemical processes. Thus, the gas phase abundances of both atoms and molecules,
which we attempt to measure using well known laboratory transitions, may be expected
to undergo modification from time-to-time. There may be some "typical" grain type,
which is formed under conditions which can be reproduced at many points on a galactic scale, such as atmospheres of M stars. The existence of such grains would
produce some characteristic absence of atomic species from the gas phase. Additional depletion may occur under selected conditions and be detectable in only

some of the absorption measures. Grains may be broken down by sputtering or spallation. The destruction will occur to different degrees in various aggregates of gas,
thus introducing cosmic noise on any attempt to find a typical pattern.

The satellite Copernicus, consisting of a one meter all-reflecting spectrometer
mounted in an 80 cm telescope [described by Rogerson, Spitzer et al. (1973)], is
effectively limited to studying gas within 2000 pc of the Sun. On this scale, it
seems reasonable to neglect gradients in abundances, produced by different stellar
populations, noticeable across the extent of an entire galaxy. Our main idea, then,
in trying to interpret abundance determinations is to try to understand the budget-
ing of atoms, presumed to be in the same relative abundances as we find in the Sun,
between gas and solid (grain) phase. There may be some selection effect in our interpretation of surface solar abundances as representative of the pre-solar gas, but
this reference provides a fairly well determined comparison, and with respect to this
standard we may speak of an element as being depleted if the ratio of column densities $N(X)/N(H)$, where "X" is the species in question, is low. It is possible that
spectroscopic details of solid material might be observable, but, based on the various
discussions of a possible set of lines of this class, the unidentified visible diffuse
interstellar lines (e.g. Snow, York, and Welty 1976), there is little overlap between
lines of sight where gas phase abundances can be studied at present and those where
the diffuse bands readily appear (Herbig 1975). The prospects for independently determining gas phase and solid phase abundances are dim.

An attempt to determine a baseline depletion pattern, a pattern characterizing
gas processed through possible grain formation sites but not yet subject to further
accretion of gas, has been made by studying stars near the sun (Rogerson, York, et
al. 1973; York 1975). These stars, in principle, should provide us with simple lines-
of-sight for study. In practice, simple, approximately triangular, profiles can
result, as demonstrated by the NI lines in Figure 1. However, for species that represent the dominant stage of ionization in both H I and H II regions, the lines of sight
are clearly complex. This fact is directly visible for the nearly unreddened star, ζ
Cen. The silicon profiles in Figure 2 show insipient resolution of the stronger H I
region from the weaker H II region (long wavelength) features. At greater distances,
the unreddened stars provide even more complex situations as is amply demonstrated by

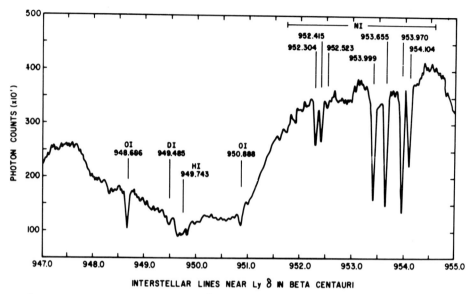

Figure 1:   Interstellar lines near 950 Å in the spectrum of β Cen. Most of the

features are due to N I.

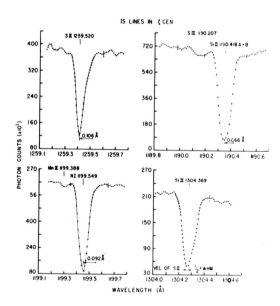

Figure 2:   Profiles for ζ Cen, a B2.5 IV star about 85 pc from the Sun. While the

N I and strong Si II features are symmetrical, the S II, S III, and

Si II 1304 features show incipient resolution of two profiles: one due

to H I gas, the other due to H II gas.

the spectra of μ Col (Shull and York 1977), to be discussed later. For some lines of
sight, the H I and H II gas overlap, as for α Virginus. The velocities plotted in
Figure 3 show no separation into components, even though ionized and neutral gas are
clearly present. The ionized and neutral gas are present and separated in the spectrum
of λ Sco, as shown in Figure 4 (York 1975). The pattern of ionized gas (N II, Si III,
N II*) at more positive velocities than H I gas also appears in β Cen and υ Sco.

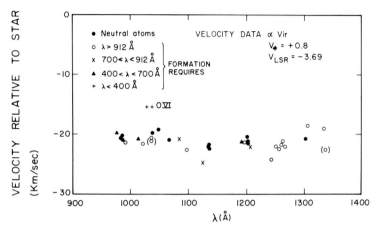

Figure 3: A plot of velocities versus wavelength. The observations cover a period
           of about four days (λ increased with time), but show no systematic drifts
           and no systematic difference between H I and H II region species.

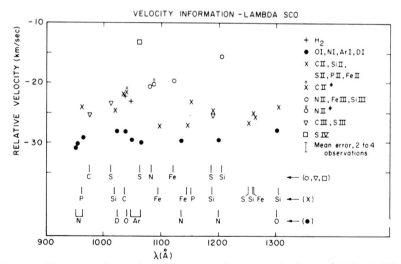

Figure 4: Same as Figure 3, for λ Sco. Features due to ionized gas (N II, N II*,
           C II*) are separated from neutral ions.

More distant and reddened lines of sight are clearly more complex, as demonstrated by the 1 km/sec spectra published by Hobbs (1969) of neutral sodium. Some
stars, such as ρ Leo, show up to 5 components within a 10 km/sec range, a situation
which will cause confusion in the analysis of lower resolution data. In principle,
one can model the velocity distribution of absorbers, based on observations of
trace ionization stages, but these species may be unreliable for predicting distributions of the dominant ionization stages.

It is usually clear from ground based observations of sodium and calcium that
there are multiple components, unresolvable at Copernicus resolution, and that the
densest gas has a velocity dispersion on the order of 1 km/sec in Na I and K I.
Since this gas is at 120°K or less, based on the relative populations of the two
lowest levels of molecular hydrogen (Spitzer, Drake, et al. 1973), the thermal
broadening is less than 1.3 km/sec for all species, and 0.3 km/sec for Na I. The
typical value of 1 km/sec thus indicates a value which should apply to all species.
However, preliminary results by Morton et al. (1973) for ζ Oph showed that for most
species, curves-of-growth with characteristic b-values of 8.6, 10.4, 8.3, 6.3, and 8
km/sec were needed for, respectively, ξ Per, α Cam, λ OriA, ζ Oph, and γ Ara, all of
which have column densities in hydrogen $N(HI) + N(H_2) \gtrsim 3 \times 10^{20}$. Such large b values
are certainly due to confusion caused by unresolvable components.

Abundances derived using curves of growth may be regarded at the outset as
being uncertain. Consequences of some rather simplified cases are summarized in
Figures 5 and 6. As discussed in the figure captions, the curves of growth for
multiple but unresolved components differ substantially from single component curves
of growth. If a few weak lines are available, for a given species, this difference
is of little consequence. If components with different column densities are involved, this situation may be detected, if several strong lines are present. However,
in general the available lines-of-sight are more complex than represented for Figures
5 and 6. If two components have similar relative abundances in all elements, a curve
of growth determined from weak and strong lines of one element can be used to derive
abundances for other species for which only stronger lines exist. However, as shown
later, some lines of sight contain two components with similar iron abundances, but
quite different abundances of P II, S II, and Ar I, so that the technique of using

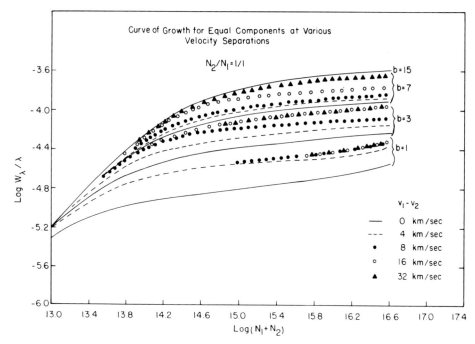

Figure 5:   Theoretical two component curves of growth.   The two components have

identical b-values and column densities.   For each b, the velocity sep-

aration is different.

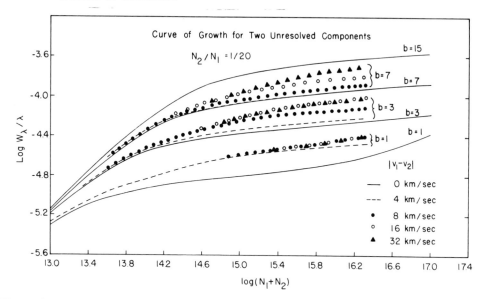

Figure 6:   Similar to Figure 5, but the column densities in the two components differ

by the ratio 20/1.

an empirical curve of growth must be justified independently for each line of sight.
The results derived later suggest that one may derive general abundance patterns in
this way, at the expense of masking the full range of variation for elements such as
Fe and Si from component to component.

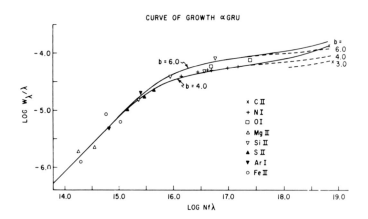

Figure 7: A curve of growth for α Gru, a B7 IV at a distance of about 29 pc. The
           b-values were determined independently for each species, but agree well
           with each other. Even for this short line of sight, b-values are higher
           than the 1 km/sec inferred in denser clouds for Na I.

III.   PRESENT TRENDS

     With these preliminary remarks, we can proceed cautiously to a discussion of
the inferred abundances for various types of interstellar gas.

     Abundances for unreddened stars have been determined from weak lines or curves
of growth. Figure 7 contains a curve of growth for α Gru, similar to one derived
for α Vir by York and Kinahan (1979). The curves derived independently for several
elements agree quite well in yielding b values of 3-5 km/sec. For λ Sco (Figure 8),
a somewhat different picture results, in which H I and H II region species define
different curves, consistent with the separation in velocity noted earlier, and
the theoretical two component curves already shown. The neutrals still show b ~
4 km/sec.

     That the curves-of-growth for neutrals are not dominated by multicomponent
effects, thereby compromising the results, is suggested by the composite curves for

D I and H I, for five nearby stars by York and Rogerson (1976). They find that $\underline{b}$(DI) = 6.8 km/sec and b(HI) = 10 km/sec. The lighter mass of H relative to D, and of D relative to N and O, together with the typical Doppler widths, suggests a thermal dependence in the $\underline{b}$-values, a result generally confirmed by observations of other nearby stars. A small amount of apparent turbulence is needed for a perfect fit in a plot of $\underline{b}$ vs. $1/\sqrt{m}$, perhaps due to very weak components at $v_{LSR} \leq 10$ km/sec. However, this general dependence lends some credibility to the derived $\underline{b}$ values and hence to the derived abundances.

In deriving relative abundances, one need not worry about ionization stages formed by radiation of higher energies than stellar photons. Figure 9 shows for α Vir that only the expected stages of ionization are important. The main uncertainties in the abundances summarized below are therefore in the curve-of-growth effects.

For reddened stars, the situation is more complex, as indicated for the previously quoted (large) $\underline{b}$ values for several stars with E(B-V) > 0.10. However, the

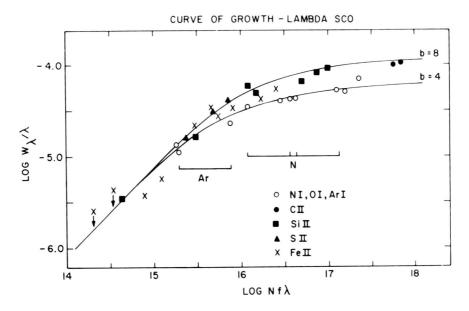

Figure 8: A curve of growth for λ Sco (B1.5 IV, d ∼ 102 pc), showing a neutral species curve similar to α Gru. The species which occur in both H I and H II regions form a different curve, as expected for the observed velocity shift and the theoretical curves in Figure 5.

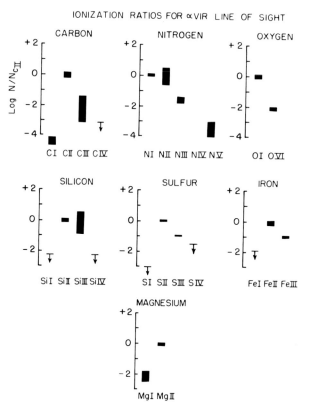

Figure 9: Relative abundances of different ionization stages in α Vir. The figure
            demonstrates that one need not worry about ionization stages except those
            expected based on starlight ionization in determining total abundances.

clouds seen in these lines of sight are dense enough to form detectable column

densities of H_2 and of trace ionization stages such as S I and K I. These species

generally form a different empirical curve of growth than do the dominant stages,

as demonstrated by Morton (1975) and de Boer and Morton (1974). The lower b-values

of b ∿ 1 km/sec implied for neutrals in ζ Oph and in o Per (Snow 1976) imply that

we may be selecting out a single cloud in the trace elements, as opposed to the 2

or more components present in H_2 and in dominant stages of ionization.

The trace ionization stages are useful for relative abundance determinations

because, assuming radiative recombination and ionization by stellar photons are the

only processes which need to be considered in the ionization balance, the electron

densities can be eliminated in the ionization balance equation and the ratio of the dominant stages can be directly related to ratios of trace ionization stages. For

instance, if α and Γ are the radiative recombination and photoionization rates,

respectively,

$$\frac{N(Na\ II)}{N(S\ II)} = \frac{N(Na\ I)}{N(S\ I)} \frac{(\alpha/\Gamma)_S}{(\alpha/\Gamma)_{Na}}$$

This technique is relatively insensitive to locations of the clouds with respect

to stars, since the two values of Γ, the photodissociation rate, tend to scale, pro-

viding the energy distribution approximates that of an O9-B1 star. Table 1 summar-

izes the application of this technique to several elements in ζ Oph (Morton 1975)

and in o Per (Snow 1976), both of which are quite reddened (A_V > 1). In both cases,

the column densities still depend on curves of growth (the $\underline{b}$ values are low), but

since the derived $\underline{b}$ values are near the physically reasonable value of 1 km/sec, the

abundances are fairly reliable. The observations are normalized to S I for purposes

of this table.

ζ OPHIUCHI INTERSTELLAR LINES

COMPOSITION OF HI CLOUDS RELATIVE TO SUN

$LOG \left(\frac{N}{N_H}\right) - LOG \left(\frac{N}{N_H}\right)_\odot$

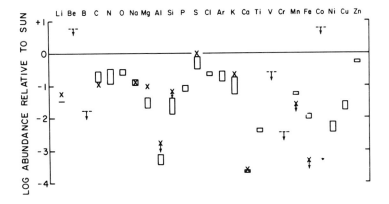

Figure 10: Depletion in the interstellar gas as observed in ζ Oph (Morton 1975).

     X's represent depletions with respect to sulfur based on recombination

     species only. Rectangles are Morton's values based on an empirical

     curve of growth. The two techniques agree very well, except for Fe and

     Mn (see text).

Table 1: Depletions Based on Trace Element Abundances

	α/Γ[1]	log N[2]		N/N(S)[5]		log N/N(S)			D	
		ζ Oph[3]	o Per[4]	ζ Oph	o Per	ζ Oph	o Per	Sun[6]	ζ Oph	o Per
Li I	0.048	9.4	--	5.7(-5)	--	-5.2	--	-4.02	-1.2	--
C I	0.15	15.6	15.7:	2.8	2.1	+0.5	0.3	1.36	-0.9	-1.0
Na I	0.47	13.9	14.0:	1.8(-2)	1.5(-2)	-1.7	-1.8	-0.97	-0.8	-0.9
Mg I	0.085	14.2		2.0(-1)	--	-0.7	--	+0.33	-1.0	--
Al I	0.0057	<10.2		3.0(-4)	--	-3.5	--	-0.79	<-2.7	--
Si I	0.0036	<12.6		1.2(-1)	--	-0.9	--	+0.34	<-1.2	--
S I	0.0085	13.9	14.1:	1.0	1.0	0.0	--	0.0	0.0	0.0
Cl I[7]	0.10	14.0		--	--	--	--	-1.56	--	--
K I	0.15	12.2:	12.1	1.1(-3)	5.6(-4)	-3.0	-3.3	-2.16	-0.7	-1.1
Ca I	0.016	9.7	9.1	3.3(-5)	5.3(-6)	-4.5	-5.3	-0.88	-3.6	-4.4
Ca II	33	11.8	12.3	--	--	--	--	-0.88	(-3.65)[8]	(-3.2:)[9]
Mn I	0.11	<11.6	12.0	3.9(-4)	--	-3.4	--	-1.79	<-1.6[10]	--
Fe I	0.04	<11.5		8.5(-4)	2.7(-3)	-3.1	-2.6	+0.19	<-3.3[10]	-2.8

1   Values of α/Γ are taken from Morton (1975). Γ is based on the WJ1 field of deBoer and Pottasch (1974). α is from Aldrovandi and Pequignot (1973) for C, Mg, S, and Si; from Seaton (1951) for Al, Cl, and Mn; from Herbig (1968) for Li, Na, K, Ca, and Fe.

2   Uncertainties quoted in the original sources are ±0.1 unless the numbers are followed by a colon: in that case, the errors are ±0.2.

3   Morton 1975.

4   Snow 1976.

5   Numbers in parentheses are powers of ten:
    $5.7(-5) = 5.7 \times 10^{-5}$.

6   Solar values adopted from Morton 1974.

7   Jura and York (unpublished) show that, as predicted by Jura (1975), Cl I virtually contains all the chlorine. The ionized chlorine is neutralized through reaction with H_2.

8   Based on observed Ca II ($N(CaIII)/N(CaII) < 5\%$) Morton 1975.

9   Based on observed Ca II, Snow 1976. The disagreement between depletion evaluated using trace element abundances and those using direct observations of the dominant stage of ionization (Ca II) arises from the fact that Ca II from other clouds contributes to the observed equivalent width.

10 Based on measuring the dominant stage of ionization (Mn II, Fe II), Morton (1974) gives depletions of -1.3 (manganese) and -2 (iron). Apparently, iron and manganese from other regions contribute to the observed equivalent widths. Elements with variable abundances (Ca II, Ti II) are quite strong in a small cloud at -28 km/sec in ζ Oph: Mn II and Fe II may also be strong (per unit hydrogen atom in this low column density cloud). The main dense cloud is at -15 km/sec. The postulated effect also explains the very low electron densities derived by Morton (1975) using manganese and iron.

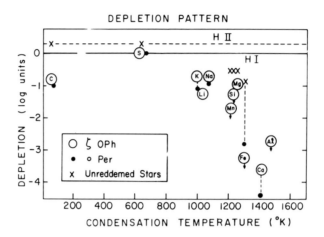

Figure 11: Depletions in unreddened stars (X), ζ Oph (0) and o Per (0), plotted
          versus condensation temperature, defined in the text. Elements with
          higher condensation temperatures are more depleted than those with lower
          condensation temperatures. Depletions are greater in dense clouds than
          in the tenuous H I regions near the Sun.

For ζ Oph, Morton used an empirical curve of growth to determine the abundances
for elements up to an atomic number of 30. These results are given in Figure 10.
The "X's" show abundances derived from recombination species, assuming $N(S)/N(H)$ is
solar. The two techniques are in good agreement, and we can be sure of the reality
of the depletions in order of magnitude. The recombination species show more severe
depletion of some species such as Fe and Mn in ζ Oph and Ca in o Per. This is quite
understandable in ζ Oph, where a low column density cloud at -28 km/sec is known to
be anomalous in having strong Ti II and Ca II lines. One may suppose that the Fe II
lines are also strong and blended with those at -15 km/sec. (Note the separation of
most of the Fe II features in Morton's figure 3.) This slide emphasizes our earlier
statement that the contrast manifested in varying depletions from one cloud to another
is potentially lost in empirical curve-of-growth analyses, even though the qualitative
statement that depletion does or does not exist may be made with some confidence.

The results of abundance analyses near the sun and in dense clouds can be summarized in a single plot, which we give in Figure 11. The abscissa is the condensation temperature tabulated by Field (1974) as the temperature at which half the atoms

of a given species will condense out of gas phase under equilibrium conditions at

some assumed pressure. The temperatures come mainly from studies of meteorites and

from considerations of possible conditions in the pre-solar nebula. The X's repre-

sent the results for a number of unreddened stars near the sun, including λ Sco, β

Cen, and α Vir. With respect to hydrogen, S II and (less reliably) C II are over-

abundant, probably reflecting the fact that the H II regions of the stars observed

are comparable in column density to the intervening H I gas. The results using only

the recombination species, normalized to sulfur, are shown for ζ Oph and o Per. While

the same qualitative pattern is observed in the reddened and unreddened stars, but

the dense regions show the effect to a much greater degree. The observations are

consistent with condensation of gas into grains under some equilibrium conditions,

with T > 1000°K. Based on Figure 11, the process could be rather incomplete, per-

haps terminating at somewhat higher temperatures to explain the unreddened star ob-

servations. Further accretion onto the grain cores must then occur in the dense

clouds. Alternatively, the condensation process may be complete down to T = 1000°K,

as indicated by the reddened stars. The data for unreddened stars would then be

explained by some process which selectively destroys about 10%-30% of the grains in

low density intercloud gas, or in low column density clouds which make up the tenuous

H I regions observed. Several authors (Barlow and Silk, 1976; Jura 1976; Spitzer

1976) have discussed the possible importance of shock waves in destroying the grains.

The former authors in particular have pointed out that the lower density regions may

experience more shocks than the dense clouds.

The depletion of carbon, apparently real in both ζ Oph and o Per (see discussion

below) does not fit into the picture of condensation in an equilibrium environment.

Carbon must be accreted onto the grain surfaces in the interstellar clouds themselves

to be consistent with the above picture. However, the extent of such possible

accretion depends on several unknown factors, including the charge on the grains and

their total surface area (Snow 1975).

The fact that different density regions of the O km/sec gas show quantitatively

different depletions seems well established, at least for the few cases discussed.

Recent work on more distant stars has shown that clouds of comparable column density

in the same line of sight show different depletions depending on the velocity of the clouds. Spitzer (1976) and Barlow and Silk (1976) have conveniently summarized the

observations of ζ Ori (Drake and Pottasch 1976); μ Col and HD28497 (Shull and York

1977); HD50896 (Shull 1977); and HD75821 (Jenkins et al. 1976). Depletions range

from 30 in 0 km/sec components for Si and Fe to nearly 0 in components with veloci-

ties between 40 and 80 km/sec. The depletions measured in the 0 velocity gas could

be more severe, as noted for ζ Oph, when individual components near 0 km/sec can be

studied in detail. More observations of several species at comparable resolution in

the same clouds are necessary to define the behavior of depletion with velocity. The

depletion of carbon in low column density, but dense clouds at 0 km/sec might be

instructive. Jura (1975) has suggested that Na and C are undepleted in such gas

(where $2N(H_2)/N(H) + 2N(H_2) < 10^{-4}$), but more observations are necessary. The deple-

tion of N I and O I also needs to be measured, particularly in dense clouds, to define

observationally the site of most of the heavy atoms (C, N, and O). Ground based ob-

servations of Ca, Ti, K, and Na exist in some detail at high enough resolution, but

determination of hydrogen column densities in individual components is lacking.

(Determinations based on the Lyman lines refer to the integrated column densities of

all clouds). This gap might be filled by observations of S I at high resolution.

Abundances for most species can be determined in principle for H I regions using

very weak lines of dominant ionization stages or the lines of recombination species,

as described above and detailed in §IV. Three special cases where different tech-

niques are useful should be mentioned.

York and Rogerson (1976) and Vidal Madjar et al. (1977) have demonstrated the

use of damping profiles of the Lyman lines with n = 2 and greater. Such determina-

tions are uncertain mainly due to problems in defining the underlying stellar con-

tinuum, but N(HI) can be obtained to ±30% in most cases. Bohlin (1975) has completed

a large survey of H I column densities using Ly α, and he quotes similar uncertain-

ties. It is desirable, in any case, to have as many profiles of different Lyman

lines as possible, to ensure that high velocity gas is not affecting the shape of

the damping wings at Ly α or Ly β in an adverse manner.

The high abundance of carbon, and the intrinsically strong damping in the 1036Å

line in particular, can be used to provide severe constraints on the carbon abundance

(Morton 1975, Snow 1976). We show here preliminary results for two stars, 29 CMa

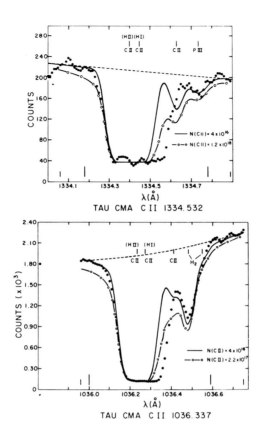

Figure 12: Carbon profiles in τ CMa (09Ib, d ∼ 1000). Theoretical profiles for
four components are represented by lines (see legend). The damping
wings of the 1036 Å line are a sensitive way of determining C II abundances. Column densities for the component labelled H I differ for the
two theoretical lines, as shown in the legend. The other components
have (from left to right): $N(CII) = 4 \times 10^{16}$ cm^{-2}, $N(CII) = 2 \times 10^{13}$ cm^{-2},
and $N(PIII) = 4 \times 10^{13}$ for the 1334 Å plot. The column densities are
the same for 1036 Å, except P III is not present whereas $N(H_2) = 2.5 \times$
10^{14} and $N(H_2) = 2.5 \times 10^{13}$. The column density of the strongest H_2
component is from Spitzer et al. 1974; the weaker one was added to fit
the present data. b = 10 km/sec for the H II region, b = 6 km/sec for
other components. Another C II component is clearly needed longward of
the "H I" component.

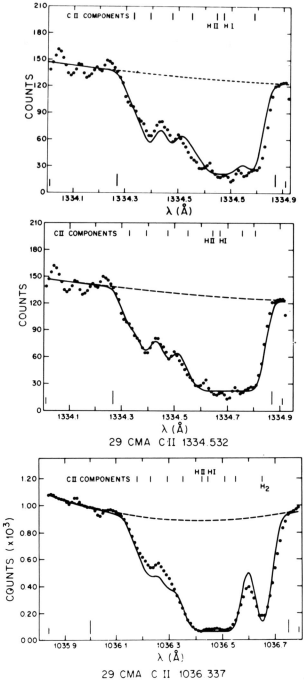

29 CMA C II 1334.532

29 CMA C II 1036.337

Figure 13

and τ CMa, stars with $N(HI) = 5 \times 10^{20}$ (Bohlin 1975). Figure 12 shows profiles for
τ CMa. Solid dots represent the actual observations. The solid line is a theoretical
profile for the three C II components shown and two H_2 components (for 1036Å) or a
P III component (for 1334Å). The velocities of the central C II components were
chosen to correspond to the velocities of H I gas (manifested in C I and H_2) and to
the H II gas (manifested in P III and S III). The H II column density was fixed at
4×10^{16} cm^{-2}, expected for an H II region with n_H = 1 around an O8 star (Spitzer
1968). τ CMa and 29 CMa have spectral types of O7f and O9, respectively. Another
component is clearly needed to fit the longward wing in both 1036 and 1334. The
dashed lines are least square fits where only the column density of the H I component is allowed to vary. The need of an extra component in the longward wing
has allowed the column density to rise to values clearly inconsistent with the shortward wing. Two points are noteworthy: 1) the 1036 Å component constrains the H I
component of C II more severely than does the 1334Å component, due to its larger

Figure 13:   Carbon profiles for 29 CMa (O7f, d $\sim$ 1000 pc). Different combinations
             of components were tried in fitting the 1334 line. The parameters for
             the middle solution were used as input to the 1036 profile (bottom).
             In this profile, N(HI) was then varied to get a best least squares fit
             to the whole profile. For N(C)/N(H) to be equal to the solar value,
             $N(CII) = 2 \times 10^{17}$ cm^{-2}.

             For the top frame, the column densities of C II are, from left to right,
             1.5×10^{13}, 4.8×10^{13}, 4.8×10^{13}, 4.3×10^{13}, 4.0×10^{15} (labelled
             "H II"), 4.0×10^{16} (labelled "H I"), and 1.0×10^{15} cm^{-2}. For the middle
             frame N = 1.5×10^{13}, 3.5×10^{13}, 4.5×10^{13}, 4.5×10^{13}, 4.0×10^{16}
             (labelled "H II"), 4.0×10^{16} (labelled "H I"), 1.0×10^{15}, and 1.5×10^{15}.
             For the 1036 Å feature, the column densities are the same as for the middle
             frame, except that the component labelled "H I" has $N(CII) = 3.9 \times 10^{16}$
             (based on least square fitting), and $N(H_2) = 2.3 \times 10^{15}$ (Spitzer et al. 1974).
             The continuum fitting regions used are delineated by vertical marks at the
             bottom of the page. For all components but the last C II component, $\underline{b}$ =
             6 km/sec. For the last C II component, $\underline{b}$ = 4 km/sec.

damping constant and   2) it is clear that for the 1036Å feature, $N(CII) = 4 \times 10^{16}$
cm^{-2} is a better fit to the profile than $N(CII) = 2.2 \times 10^{17}$. For $N(HI) = 5 \times 10^{20}$,
$N(CII) = 2.5 \times 10^{17}$. Thus, carbon appears to be depleted along this line of sight,
by a factor of 5.

Figure 13 shows similar plots for 29 CMa. The H I and H II regions were similarly defined. Though 29 CMa is only 1° from τ CMa, the interstellar profile is much
more complex, requiring at least 8 components for even an approximate fit, with a
maximum separation of 110 km/sec. However, since the 1334Å feature is not too affected
by damping, the many high velocity components can be at least formally found in that
profile. Using these values as input to the 1036 profile, and allowing the main H I
profile to vary in obtaining a least squares fit, we find again $N(CII)$ equal to $4 \times$
10^{16}, representing a depletion of about 5. While further work is needed to define
the profiles for these two stars, and to finalize the apparent depletion of carbon,
it is clear that the technique is quite powerful. A comparison of abundances determined in this way with those obtained from the ratio $N(CI)/N(SI)$ would provide an
important internal consistency check on our understanding of the gas phase abundance
of carbon. If carbon is generally depleted, it probably accounts for most of the
mass tied up in grains.

Finally, we should mention that abundances in H II regions are difficult to
determine empirically due to blending of the main features with those of H I regions.
(The presence of H II region features, of course, compromises determinations for
some H I species as well.) In general, higher resolution is required in order to
separate the two types of gas. However, as pointed out by Spitzer and Jenkins (1975),
the fine structure levels of silicon, nitrogen, and carbon are populated in a well
determined manner (quantitatively), i.e. by electron collisions. The ratio $N(SiII^*)/$
$N(CII^*)$ is then related to $N(SiII)/N(CII)$ by a fairly well defined constant:

$$\frac{N(SiII^*)}{N(CII^*)} = \frac{N(SiII)}{N(CII)} \frac{<\sigma v>_{CII}}{<\sigma v>_{SiII}} \frac{A_{SiII}}{A_{CII}}$$

where $<\sigma v>$ is the rate coefficient for collisional excitation with electrons and
A is the rate of radiative de-excitation from the excited level.

A similar equation holds for N II and C II or Si II and N II. In practice,
the column densities of Si II and N II* are easy to determine, due to the presence
of weak lines for these species, whereas the C II* lines are generally saturated.
York et al. (1977) have shown spectra for these species in the H II regions around
ζ Oph and V Pup. For V Pup, the relative populations may be affected by optical
pumping, but for ζ Oph and α Vir (York and Kinahan 1979) the observations indicate
that silicon is depleted by at least a factor of 20.

IV. WORK FOR THE FUTURE

Based on broad band extinction measurements and polarization measurements
several specific formulae for interstellar grains have been evolved (Greenberg 1974).
Abundance measurements have confirmed that Fe, Si, Ca, Al, Mn, Mg, and other species
are part of the grains, but confirmation of the main constituents by mass has not
been forthcoming. Thus, higher resolution observations are necessary, as indicated
several times already, to verify our tentative conclusions and to further our search
for the knowledge of grain constituents.

The next generation of space spectrometers promises to allow a question of
equal or surpassing importance, to be attacked: namely, how extensive is the distribution of elements with A > 8 in the universe? How extensive is the apparent
uniform distribution of deuterium, an important consideration in cosmology? The
abundance of species accessible in the UV, the large oscillator strengths, and the
detectability of ions which are dominant for temperatures up to 4×10^5 °K contribute to the fact that absorption measurements using extragalactic sources as background sources will produce an unprecedented sensitivity to material near and around
galaxies, and within the galaxies of the local group. Detection of metals in intergalactic space or within clusters of galaxies could provide severe constraints on
models for element production within galaxies. Column densities of hydrogen of
10^{14} cm^{-2} are readily detectable directly, and heavy elements are detectable at a
level corresponding to 10^{16} cm^{-2} H atoms, at all temperatures from 0° to 4×10^5 °K,
assuming solar abundances.

Both the high resolution observations for purposes of grain studies and the
somewhat lower resolutions required for cosmological problems represent studies

that can be done with relatively small aperture, high efficiency instruments which

orbit for periods of 1 week to a year. The high and low resolution programs can

be performed by a grating change in the same instrument, or separate instruments

could be designed — one for high resolution work, particularly at the shortest

wavelengths, and one for lower resolution work which could perform useful experi-

ments at $\lambda > 1200$ Å, if that restriction proved necessary.

The main instrumental requirements for more detailed observations of inter-

stellar gas are use of coatings with high UV reflectivity, minimization of reflection

losses, maximum photocathode efficiency (possibly using opaque photocathodes), and

use of multi-channel photon limited detectors. If it appears likely that the sta-

bility requirements of the anticipated long focal length spectrographs can be met

or worked around, relatively small aperture space instruments will add significantly

to our knowledge and interpretation of interstellar gas abundances.

We summarize here the particular lines of interest for high resolution abun-

dance studies. Except for H I and C II, where the damping wings can probably pro-

vide secure abundance measurements, work on the weakest lines seems most useful.

Hence, we show in Figures 13 and 14, the strengths of the lines of various elements,

in the plot of log $Nf\lambda$ versus $1/\lambda(\mu^{-1})$. The column densities of the dominant stage

of ionization for each element were calculated assuming that $N(H) = 3.2 \times 10^{19}$ and

that abundances of B, C, N, O, P, S, Cl, and Ar are solar. Other elements were

taken as being depleted by a factor of 10. The plotted ordinate is scaled so that

log $Nf\lambda = 15$ corresponds to $10(\lambda/10^3)$mÅ, a level at which, even with $\underline{b}$-values as low

as 1 km/sec, saturation effects will only begin to be important. The optical depth

at the line center $(\tau_0 \alpha Nf\lambda/b)$, $\tau_0 \sim 1.5$, then varies inversely with $\underline{b}$, so that log $Nf\lambda$

$= 15.6$, $\underline{b} = 4$ yields comparable saturation. Features lying below log $Nf\lambda = 15$ are

mostly on the linear portion of the curve so that reliable column densities may be

derived if several lines of the same species with different oscillator strengths

are available. The weaker lines become detectable [log $Nf\lambda = 10^{14}$ for $W_\lambda = 1(\lambda/10^3)$

mÅ] as the column density increases, and the stronger lines get weaker as the deple-

tion increases. The features noted in Figure 14 also appear in Table 2, where the

wavelengths and oscillator strengths are indicated. While reliable abundances can

in principle be obtained using $\lambda/\Delta\lambda = 2 \times 10^4$, comparable to Copernicus, individual

Table 2: SELECTED WEAK INTERSTELLAR FEATURES

(DOMINANT STAGES OF IONIZATION-HI)

Sp	λ	$0.1/\lambda(\mu)$	$\log Nf\lambda(N_H=19.5)$	$\log f\lambda(\overset{\circ}{A})$	$\log N$
H I	----				
Li I	----				
D I	937.548	1.07	15.4	0.864	14.5
	949.485	1.05	15.6	1.122	14.5
B II	1362.460	0.73	14.3	3.05	11.2
C II	----				
N I	951.080	1.05	14.7	-0.9^*	15.6
	952.523	1.05	15.1	-0.5^*	
	952.789	1.05	15.4	-0.2^*	
	955.265	1.05	14.2	-1.4^*	
	955.529	1.05	<13.9	$<-1.7^*$	
	955.881	1.05	<13.9	$<-1.7^*$	
	959.540	1.04	14.9	-0.7^*	
	1159.817	0.86	<13.9	$<-1.7^*$	
	1160.937	0.86	<13.9	$<-1.7^*$	
O I	988.581	1.01	16.0	-0.3	16.3
	1355.598	0.74	13.0	$-3.3^\dagger$	
F I	954.825	1.05	?	----	
	951.871	1.05	?	----	
Mg II	946.769	1.06	14.6	$< 0.6^*$	14.0
Mg II	946.703	1.06	14.6	$< 0.6^*$	
	1025.968	0.97	14.2	$+0.18$	
	1026.113	0.97	13.9	-0.12	
	1239.925	0.81	14.1	0.079	
	1240.395	0.81	13.8	-0.222	
Al II	1670.786	0.60	16.4	3.498	12.9
	935.20	1.07	?		

Table 2(Continued)

Sp	λ	0.1/λ(μ)	log Nfλ(N=19.5)	log fλ(Å)	log N
Si II	1808.003	0.55	14.9	0.83	14.1
	1020.699	0.98	15.8	1.69	
P II	1152.81	0.87	15.2	2.3	12.9
	1301.87	0.77	14.6	1.72	
	1532.51	0.65	14.1	1.16	
S II	1250.586	.80	14.5	0.825	14.7
Cl I	1096	.91	15.4	1.2**	13.2
	1347.238	.74	14.4	2.18	
Cl II	1063.83	.94	13.9	0.75	13.2
	1071.05	.93	14.4	1.23	
Ar I	1066.660	.94	16.0	1.802	14.2
K	----				
Ca II	1226.73	.82	12.7	-0.123	12.8
	1227.03	.81	12.4	-0.424	
	1341.890	.75	12.8	0.048	
	1342.554	.75	12.5	-0.253	
	1649.858	.61	12.4	-0.369	
	1651.991	.61	12.1	-0.670	
	3933.663	.25	16.2	3.432	
	3968.468	.25	15.9	3.131	
Sc II	4020.399	.25	--	3.040	--
	3996.607	.25	--	2.086	
	3907.476	.26	--	3.206	
	3642.785	.27	--	3.293	
	3580.927	.28	--	2.964	
	3567.701	.28	--	2.487	
Ti II	3057.395	.33	12.3	1.113	11.2

Table 2 (Continued)

Sp	λ	$0.1/\lambda(\mu)$	log $Nf\lambda$ (N=19.5)	log $f\lambda(\text{Å})$	log N
Ti II	3066.354	.33	13.3	2.115	
	3072.971	.33	13.7	2.525	
	3121.599	.32	13.9	2.722	
	3148.033	.32	13.4	2.209	
	3203.435	.31	14.0	2.819	
	3229.193	.31	13.5	2.297	
	3241.984	.31	15.0	2.819	
	3383.761	.30	14.2	3.037	
V II	2683.09	.37	12.8	2.038	10.5
	2739.715	.37	12.5	2.279	
Cr II	2011.13	.50	?	--	12.2
	2013.65	.50	?	--	
	2016.90	.50	?	--	
	2055.59	.49	14.7	2.54[†]	
	2061.54	.49	?	--	
	2065.46	.48	?	--	
Mn II	1197.172	.84	14.0	2.06[*]	11.9
	1199.388	.83	14.0	2.08[*]	
	1201.124	.83	13.7	1.84[*]	
	2576.107	.39	14.8	2.871	
	2593.731	.39	14.7	2.762	
	2605.697	.38	14.5	2.615	
Fe II	1055.269	.95	14.9	1.02[*]	13.9
	1063.982	.94	14.5	0.61[*]	
	1096.886	.91	15.5	1.61[*]	
	1106.362	.90	14.7	0.79[*]	
	1121.987	.89	15.3	1.35[*]	
	1133.678	.88	14.8	0.85[*]	
	1142.334	.88	14.8	0.90[*]	

Table 2(Continued)

Sp	λ	$0.1/\lambda(\mu)$	log Nfλ(N=19.5)	log fλ($\overset{\circ}{A}$)	log N
Fe II	1143.235	.87	15.1	1.23[*]	
	1144.946	.87	16.1	2.23[*]	
	1260.542	.79	15.3	1.39[*]	
	1608.456	.62	--	--	
	2343.495	.43	16.3	2.403	
	2366.864	.42	14.7	0.751	
	2373.733	.42	15.9	1.972	
Co II[**]	1984.86	.50	10.9	-0.1[†]	11.0
	1998.55	.50	10.7	-0.3[†]	
	2011.546	.50	13.0	2.0[†]	
	2025.76	.49	12.6	1.6[†]	
	2058.818	.49	12.2	1.2[†]	
Ni II[**]	1317.22	.76	15.0	2.20[†]	
	1345.878	.74	13.5	0.76[†]	12.8
	1370.20	.73	14.9	2.14[†]	
	1381.685	.72	12.8	-0.23[†]	
	1393.324	.72	14.2	1.40[†]	
	1454.96	.69	14.7	1.87[†]	
	1467.85	.68	13.6	0.78	
	1703.41	.59	13.9	1.13	
	1741.56	.57	14.9	2.07	
Cu II	1358.764	.74	13.9	2.86[†]	11.0
Zn II	2025.512	.49	13.8	2.92	10.9
	2062.016	.48	13.5	2.62	

[*]   Determined from <u>Copernicus</u> data by deBoer, Pottasch, Morton and York (1974)
      or York (unpublished).

[†]   Theoretical f-values, mostly from Kurucz (1974).

[**]   Line lists quite incomplete. Selected lines are representative of region where
      species may be detected, and of line strengths.

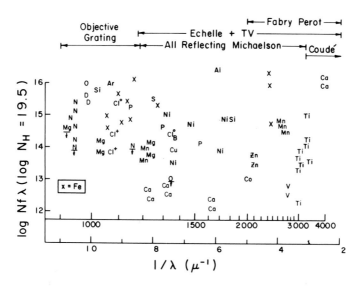

Figure 14: Strengths of lines from dominant stages of ionization in interstellar H I
regions, as a function of wavelength. Only weak lines are plotted. The
figure is discussed in the text. Instruments useful in different regions
are noted at the top.

regions can only be studied at much higher resolution, as noted previously.

Several facts are apparent from this figure. First, the most productive region
for study is the region from 920Å to 1200Å, though wavelengths as high as 2800Å could
be useful. The only hope for reliable nitrogen column density lies in this region,
as do the features for direct determination of abundance for H, D, and C. Second,
the line density per unit frequency interval is high enough that even at the highest
resolutions, several interesting features could be studied based on a single exposure,
using small (<12 cm) linear detector arrays. For special purpose, short flights,
it might even be feasible to operate an instrument with no moving parts, observing
for one region of 10-50Å in several stars, then using the same instrument at a
different grating angle in subsequent flights to study other spectral regions.

The large number of features listed shows that a properly designed instrument
can in fact be used to determine abundances for C, N, and O, so important for understanding the main constituents by mass of the grains. Furthermore, by properly
choosing features, a wide variety of chemical elements can be studied, hopefully

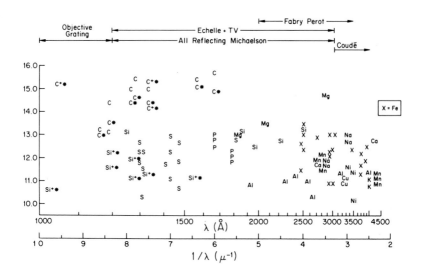

Figure 15: Same as Figure 14, for trace ionization stages.

representing a large enough range of binding energies and condensation temperatures

that detailed models of gas depletion can be constructed.

Figure 15 shows a similar plot for trace elements such as C I, S I, etc.   In

this plot log $Nf\lambda$ increases as N_H and as n_e (and as a function of n_H and T for C I*).

Excited levels of Si II, and C II, due to collisions of Si^+ or C^+ and electrons,

and of C I, due to collisions of C° and neutral hydrogen, are included.   There is

a small dependence on $T(N(X)\alpha T^{-0.7}$ for recombination species, $T^{-0.5}e^{-\Delta E/kT}$ for

collisionally excited species).   The numbers in the table refer to T = 50°K, n_e =

10^{-2} except as noted.   Based on the previous discussion, the densest region, where

accretion of grain mantles may most efficiently occur, can best be studied using the

ratios of recombination species.   A survey of S I at high resolution could be immed-

iately useful in terms of depletions of K I, for which high resolution profiles already

exist.   Table 3, containing a list of the features in Figure 15, also includes data

for the strong potassium doublet at 7681Å.   They are comparable in $Nf\lambda$ for solar

abundances, and hence refer to the same regions (see earlier discussion on the ad-

vantages of using trace elements).   Therefore, even lower resolution work on S I

could be useful, using K I profiles to construct appropriate curves of growth.

Table 3: SELECTED FEATURES DUE TO TRACE ELEMENTS

$(HI, n_H = 10^2, T = 50\text{-}100°K)$

Sp	λ	$0.1/\lambda(\mu)$	$\log N f\lambda$	$\log f\lambda(\text{Å})$	$\log N$	$\log n_e \ \alpha/\Gamma$
C I†	1157.186	.86	13.1	−0.2*	13.3	−2.8
	1193.996	.84	14.4	1.1*		
	1260.736	.79	15.0	1.68		
	1277.245	.78	15.4	2.089		
	1328.833	.75	15.0	1.717		
	1560.310	.64	15.4	2.102		
	1656.928	.60	15.7	2.352		
C I*$\dagger$	1157.405	.86	13.1	0.0*	13.1	
	1192.451	.84	13.5	+0.4*	$(<\sigma_{H+C°}v> = 5 \times 10^{-10}, T = 50°,$	
	1261.122	.79	14.4	1.299	$(n_H = 10^2; n_H \ \gamma/A = 0.63)$	
	1277.513	.78	14.6	1.457		
	1329.100	.75	14.4	1.337		
	1560.683	.64	15.1	1.977		
	1657.380	.60	14.9	1.750		
C II*	1037.018	.96	15.2	2.113	13.1	$n_e\gamma/A \sim 10^{-3}$
	1335.662	.75	14.3	1.196		
	1335.708	.75	15.3	2.150		
Na I	2852.811	.35	12.0	0.621	11.4	−2.3
	2853.013	.35	11.7	0.320		
	3302.369	.30	13.0	1.646		
	3302.979	.30	12.7	1.345		
	5889.950	.17	15.0	3.586		
	5895.924	.17	14.7	3.285		
Mg I	1827.94	.55	13.0	1.983	11.0	−3.1
	2025.824	.49	13.5	2.512		
	2852.127	.35	14.7	3.734		

Table 3(Continued)

Sp	λ	$0.1/\lambda(\mu)$	log Nfλ	log fλ(Å)	log N	log n_e α/Γ
Al I	1931.92	.52	10.9	2.179	8.7	-4.2
	2263.463	.44	11.0	2.345		
	2367.053	.42	11.2	2.452		
	2567.983	.39	11.0	2.272		
	2652.475	.38	10.3	1.571		
	3082.153	.32	11.4	2.738		
	3944.006	.25	11.4	2.664		
Si I	1255.276	.80	13.1	2.441	10.7	-4.4
	1841.152	.54	--	--		
	1845.520	.54	13.1	2.449		
	1977.579	.51	12.5	1.789		
	2207.978	.45	12.8	2.114		
	2514.316	.40	13.3	2.594		
Si II*	1023.693	.98	10.7	1.692	9.0	$n_e\gamma/A \sim 10^{-7}$ @ 50°
	1194.496	.84	12.3	3.287		$\sim 10^{-5}$ @ 100°
	1197.389	.84	11.6	2.588		
	1264.730	.79	12.0	3.037		
	1265.02	.79	11.1	2.082		
	1309.274	.76	11.3	2.284		
	1533.445	.65	11.1	2.067		
P I	1671.68	.60	12.5	2.590	9.9	(~-3)
	1674.61	.60	12.8	2.891		
	1679.71	.60	13.0	3.067		
	1774.99	.56	12.4	2.465		
	1782.87	.56	12.2	2.289		
	1787.68	.56	11.9	1.988		
S I	1295.661	.77	12.7	2.148	10.6	-4.1
	1296.174	.77	12.3	1.671		

Table 3(Continued)

Sp	λ	$0.1/\lambda(\mu)$	log Nfλ	log fλ(Å)	log N	log n_e α/Γ
S I	1303.42	.77	11.9	1.326		
	1316.57	.76	12.3	1.657		
	1316.61	.76	11.5	0.909		
	1316.62	.76	10.3	-0.267		
	1401.541	.71	11.9	1.346		
	1425.065	.70	13.0	2.410		
	1425.24	.70	11.1	0.487		
	1425.229	.70	12.3	1.662		
	1474.005	.68	12.7	2.062		
	1474.390	.68	11.9	1.314		
	1474.569	.68	10.7	0.138		
	1807.341	.55	12.9	2.307		
K I	4044.136	.25	11.1	1.391	9.7	-2.8
	4047.206	.25	10.8	1.090		
	7664.899	.13	13.4	3.718		
	7698.959	.13	13.1	3.417		
Ca I	2721.645	.37	11.7	2.736	9.0	-3.8
	4226.728	.24	12.8	3.832		
Mn I	4034.480	.25	11.3	2.392	8.9	-3
	4033.073	.25	11.2	2.267		
	4030.755	.25	11.0	2.091		
	2794.817	.36	12.1	3.243		
	2798.270	.36	12.0	3.118		
	2801.084	.36	11.8	2.942		
Fe I	2447.708	.41	11.4	0.896	10.5	-3.4
	2462.645	.41	12.6	2.072		
	2483.270	.40	13.4	2.937		
	2501.130	.40	12.4	1.946		
	2522.848	.40	13.2	2.645		

Table 3 (Continued)

Sp	λ	0.1/λ(μ)	log Nfλ	log fλ(Å)	log N	log n_e α/Γ
Fe I	2719.027	.37	12.9	2.405		
	2912.158	.34	10.9	0.413		
	2936.904	.34	12.1	1.588		
	2912.158	.34	10.9	0.413		
	2936.904	.34	12.1	1.588		
	2966.901	.34	13.0	2.454		
	2983.574	.34	12.3	1.772		
	3020.643	.33	12.0	2.471		
	3440.610	.29	12.4	1.862		
	3649.304	.27	11.3	0.798		
	3679.915	.27	11.7	1.202		
	3719.935	.27	12.7	2.186		
	3824.444	.26	11.9	1.388		
	3859.913	.26	12.5	2.042		
Ni I	2207.74	.45	--	--	9.8	(∿ -3)
	2310.952	.43	--	--		
	2207.736	.45	--	--		
	2255.873	.44	--	--		
	2320.026	.43	--	--		
	3369.573	.30	11.6	1.817		
	3391.050	.29	11.4	1.623		
	3409.578	.29	10.2	0.448		
Cu I	3247.540	.31	11.2	3.155	8.0	(-3.0)
	3273.957	.31	10.9	2.854		

*   See note to Table 2

†   See last footnote, Table 2.

Depending on depletions, a number of other abundance ratios could be determined, as already demonstrated for ζ Oph. The near UV Fe I lines are stronger than those already detected in the visible. Reliable carbon abundances are best determined using the features below 1200Å to minimize saturation problems.

The very strong lines useful for determining column densities of very tenuous extragalactic regions are included in the list by Spitzer and Zabriskie (1964) and are well known due to their appearance in the absorption and emission spectra of quasi stellar objects. For purposes of spectrometer design, it is important to emphasize that the temperature regimes near 80,000°K (which would be appropriate to a region of high star formation rate and the corresponding shocks from supernovae) and 4×10^5°K (possibly characteristic of halos of galaxies at the present epoch) are represented by features below 1200Å: C III 977Å in the first case, and O VI 1031Å in the second. Furthermore, the deuterium abundance, important for cosmological regions, can only be studied well using the higher Lyman lines $\lambda < 1026$ Å.

The spectral regions in which different instruments should be useful are noted in Figures 14 and 15. Below 1200Å, the low reflection coefficients of most feasible surfaces make single element spectrometers efficient by comparison with other types. A 25 cm grating in a Wadsworth mount is about a factor of four more efficient at 1050Å and a factor of 25 at 950Å than a telescope of the same aperture feeding a single reflection spectrometer, assuming Li F coatings for all reflecting surfaces. The resolving power is limited by the focal length achievable without introducing additional reflections. Various detectors (linear or area) might be used, but they must be windowless.

At wavelengths above 1400Å, interferometric techniques may be employed, using an all reflecting Michaelson interferometer, with three reflecting surfaces (Kruger, et al. 1971). Above 2000Å, Fabry-Perot interferometers may conceivably be used, perhaps with a grating as a pre-disperser. The entire region above 1200Å could be covered with an echelle spectrograph and a large area detector. For studies of a single small region at high resolution, the photoefficiency of the single photoelectric cell used as a detector for the interferometric devices may be appreciably higher than that achievable with area detectors. This fact may balance the chief

advantage of the latter scheme: the multichannel nature of the echelle/area detector
combination.

V. CONCLUSION

Observations of interstellar features in the ultraviolet have shown that
the study of velocity distributions and physical conditions in the interstellar
medium would benefit substantially from high resolution observations ($\lambda/\Delta\lambda \sim 10^5$).
Such studies are necessary to verify abundances determined at lower resolution,
and to allow abundance ratios in separate, closely spaced velocity components to
be ascertained. The variations in gas phase abundances from cloud-to-cloud provide
the best potential opportunity for developing a detailed model of interstellar
grains. Lower resolution work with more efficient collecting and detecting systems
will provide an opportunity for an increase in sensitivity of about 1000, over
present techniques, to extragalactic gas in very tenuous forms. Both types of
studies can be done with rather general spectroscopic equipment, though small,
special purpose payloads may offer the best chance for substantial increases in
our understanding of interstellar and intergalactic material in the near future.
Specific studies and lines requiring either very high resolution or very short
wavelength coverage ($\lambda < 1100\text{Å}$) have been suggested, and material for selecting
the best spectral features have been summarized.

Several of the considerations summarized here have evolved from discussions
with Drs. E. Jenkins, L. Spitzer, and A. Vidal-Madjar. I also wish to acknowledge
useful discussions with M. Jura, F. Roesler, and J. Silk. This work was supported
by NASA grant NAS 5-1810 to Princeton University. This profile fitting program
used for the C II features was developed by C. Laurent, A. Vidal-Madjar, Dan Davis,
and Dan Welty.

The reader may be interested in two papers which appeared after the present
paper was written. Lugger, York, Blanchard and Morton (Ap.J. 224, 1059, 1978) discuss N I f-values for far UV lines in detail. Jura and York (Ap.J. 219, 861, 1978)
discuss observations of chlorine and f-values of the required lines.

References

Aldrovandi, M.V., and Pequignot, A. 1973, Astr. and Ap., 25, 137.

Barlow, M.J., and Silk, J. 1977, Ap.J., 211, L83.

Bohlin, R. 1975, Ap.J., 200, 402.

deBoer, K.S., and Morton, D.C. 1974, Astr. and Astrophys., 37, 305.

deBoer, K.S., Pottasch, S.R., Morton, D.C., and York, D.G. 1974, Astr. and Ap.,

   31, 405.

Drake, J.F., and Pottasch, S.R. 1977, Astr. and Astrophys., 54, 425.

Field, G.B. 1974, Ap.J., 187, 453.

Greenberg, M. 1974, Ap.J., 189, L81.

Herbig, G. 1975, Ap.J., 196, 129.

Hobbs, L.M. 1969, Ap.J., 157, 135.

Jenkins, E.B., Silk, J., and Wallerstein, G. 1976, Ap.J. Suppl., 32, 681.

Jura, M. 1975, Ap.J., 200, 415.

Jura, M. 1976, Ap.J., 206, 691.

Kruger, R., Anderson, M., and Roesler, F. 1972, Applied Optics, 62, 942.

Kurucz, R.L., 1974, Smithsonian Astrophysical Observatory, Special Report No. 360.

Morton, D.C. 1974, Ap.J.(Letters), 193, L35.

Morton, D.C. 1975, Ap.J., 197, 85.

Morton, D.C., Drake, J.F., Jenkins, E.B., Rogerson, J.B., Spitzer, L., and York, D.G.,

   Ap.J.(Letters), 181, L103.

Morton, D.C., and Smith, W.H. 1973, Ap.J.Suppl., 26, 333.

Rogerson, J.B., York, D.G., Drake, J.F., Jenkins, E.B., Morton, D.C., and Spitzer, L.,

   Ap.J.(Letters), 181, L110.

Rogerson, J.B., Spitzer, L., Drake, J.F., Dressler, K., Jenkins, E.B., Morton, D.C.,

   and York, D.G. Ap.J.(Letters), 181, L97.

Seaton, M.J. 1951, MNRAS, 111, 390.

Shull, J.M. 1977, Ap.J., 212, 102.

Shull, J.M., and York, D.G. 1977, Ap.J., 211, 803.

Snow, T.P. 1975, Ap.J., 202, L87.

Snow, T.P. 1976, Ap.J., 204, 759.

Snow, T.P., York, D.G., and Welty, D. 1976, A.J., 82, 113.

Spitzer, L. 1968, Diffuse Matter in Space (New York: J. Wiley and Sons).

Spitzer, L. 1976, Comments on Astrophysics, 6, 177.

Spitzer, L., Cochran, W.D., and Hirshfeld, A. 1975, Ap.J.Suppl., 28, 373.

Spitzer, L., Drake, J.F., Jenkins, E.B., Morton, D.C., Rogerson, J.B., and York,

    D.G., 1973, Ap.J.(Letters), 181, L116.

Spitzer, L., and Jenkins, E.B. 1975, Ann. Rev. Astr. and Ap., 13, 133.

Spitzer, L., and Zabriskie, F.R. 1959, Pub. A.S.P., 71, 412.

Vidal-Madjar, A., Laurent, C., Bonnet, R., and York, D.G. 1977, Ap.J., 211, 91.

York, D.G. 1975, Ap.J.(Letters), 196, L103.

York, D.G., Flannery, B., and Bahcall, J. 1976, Ap.J., 210, 143.

York, D.G., and Kinahan, B. 1976, Ap.J., 228, 127.

York, D.G. and Rogerson, J.B. 1976, Ap.J., 203, 378.

ULTRAVIOLET OBSERVATIONS OF INTERSTELLAR MOLECULES
AND GRAINS FROM SPACELAB

Theodore P. Snow, Jr.
Laboratory for Atmospheric and Space Physics
University of Colorado

ABSTRACT

Observational results obtained to date on interstellar grains and molecules are briefly reviewed, and several promising areas for further research with Spacelab are suggested. Regarding grains, useful data can be expected on the shape of the ultraviolet extinction curve for new interstellar regions; the nature of UV extinction at short wavelengths, down to the Lyman limit; the presence or absence of structure in the UV extinction curve comparable to the visible-wavelength diffuse bands; the scattering properties of grains in new kinds of clouds and nebulae; and the polarization properties of grains in UV wavelengths. The principal advances which may be expected in observations of molecules will include the ability to probe more heavily-obscured regions, where molecular species are more abundant than in the diffuse clouds observed to date; coverage of wavelength regions (such as λλ 1400-3200) not well-studied with previous instruments such as Copernicus; and the capability of observing in optical absorption species detected in the same line of sight in radio emission, which provides unique information on cloud geometry and physical conditions.

I. INTRODUCTION

The advent of space-borne astronomical devices has, over the past decade, produced copious quantities of new data in many fields of research, but the greatest advances have come about in our understanding of the physics of the interstellar medium. A series of ultraviolet experiments, beginning with sounding rockets and culminating with Copernicus and the International Ultraviolet Explorer has led, through a progression of increasing sensitivities and/or spectral resolutions, to a reasonably clear understanding of the distribution, kinematics, physics, and chemistry of the diffuse gas in space in the local portion of the galaxy. General reviews of recent interstellar research results and techniques have been given by Spitzer and Jenkins (1975) and by Snow (1976b). York (1979; this volume) has summarized some particularly useful observations of the interstellar gas which are expected to be carried out in the next few years.

P. L. Bernacca and R. Ruffini (eds.), Astrophysics from Spacelab, 645-660.

This chapter will discuss observations of interstellar dust and molecular species which will be important additions to the store of information already available, and which will be possible with experiments expected to be orbited with future Spacelab missions. The emphasis will be on ultraviolet observations, where the greatest gains are still to be made. In section II, interstellar dust observations will be described, and in section III molecular problems will be treated. In each case a brief review will precede the discussion of suggested new observations.

II. THE INTERSTELLAR DUST

A.   Recent Results
Since the earliest observations of ultraviolet extinction (Stecher, 1965; Carruthers, 1969), it has been apparent that the observed $1/\lambda$-dependence of the extinction law in visible wavelengths does not extend uniformly throughout the ultraviolet. The pronounced hump in the curve at 2200 A (Stecher, 1965) and the far-UV rise (Bless and Savage, 1972; York, et al., 1973) both represent significant departures. Both features seem to be general throughout the local interstellar medium (Bless and Savage, 1972; Savage, 1975; Nandy and Thompson, 1976; Snow, et al., 1980). While many suggestions have been made regarding the origin of the 2200 A feature (e.g. Gilra, 1972, 1973; Wickramasinghe and Nandy, 1971; Huffman and Stapp, 1971; Duley and McCullough, 1977), none can yet be unambiguously chosen as the likely carrier. The strength of the bump correlates extremely well with visual extinction measures (Dorschner, 1972; Savage, 1975), indicating that it is probably produced by the same grains as those responsible for visual reddening and polarization.

The sharp increase of extinction towards short wavelengths below 1500 A has led to the conclusion, based on considerations of scattering theory, that a second population of grains with a much smaller size must exist in space (Witt, 1973). The poor correlation of this extinction rise, which seems to continue at least to 1000 A, with either visual extinction or the 2200 A bump (Bless and Savage, 1972), seems to confirm that it is caused by a second, distinct population of grains. Again, the detailed nature of the responsible particles is unknown, except that they are probably small (< 100 A). It is possible that the far-UV extinction rise is primarily due to scattering rather than absorption, since the albedo of the grains is inferred from diffuse galactic light measurements to be large at 1500 A and possibly increasing shortward of there (Witt and Lillie, 1972,1973; Lillie and Witt, 1976). The 2200 A bump, on the other hand, has been inferred from the same data to be primarily an absorption feature.

Savage (1975) has searched for structure in the UV extinction curve longward of 2200 A, concluding that any features like the unidentified visible-wavelength diffuse bands (e.g. Herbig, 1975; Smith, Snow and York, 1977) have central strengths $E(\lambda-V)/E(B-V) < 0.^m08$, a strength exceeded in the visible only by a few features too narrow to be seen with the 20 A resolution available to Savage. Copernicus data have been used to search for features of width 2-10 A in the far UV (Snow, York and Resnick, 1977), but the abundance below 2300 A of photospheric lines in O- and B- star spectra makes the detection of weak, broad features quite difficult, resulting in upper limits for features with FWHM > 5 A of $E(\lambda-V)/E(B-V) = 0.31$.

B. Future Observations

Upcoming space-borne ultraviolet experiments could add important data for five principal areas of research on grains: (1) determining the shape of the curve for more distant and/or heavily obscured stars than possible to date; (2) extending the wavelength coverage of extinction measures down to the Lyman limit at 912 A; (3) searching for structure in the extinction curve, for stars with significantly greater color excess than presently possible, enhancing the detectability of diffuse features over the noise introduced by mismatching of stellar lines between pairs of stars whose spectra are compared; (4) making more sensitive measurements of scattered light in reflection nebulae, to refine and extend existing data on grain albedos and scattering phase functions; and (5) making polarization measurements in ultraviolet wavelengths. The first three of these require high photometric sensitivity and accuracy as well as at least moderate spectral resolution, while the fourth and fifth require less spectral resolving power but perhaps even greater sensitivity, and may be accomplished with instruments having relatively large entrance apertures. The five areas of interest are described in more detail in the following.

1.   Extinction curves in faint objects

While the shape of the UV extinction curve is generally quite constant from one direction to another, for the present this statement applies only to the rather local portion of our galaxy which has been well-sampled. There are some indications that substantial variations may exist in interstellar regimes which cannot yet be adequately probed. Bless and Savage (1972), Snow and York (1975), and Wu, et al. (1980) have all reported cases of unusually low far-UV extinction which seem to be associated with dense clouds. It will be important both to sample more such regions to determine whether this is a general result, and to probe more heavily-obscured regions to see whether the ususual behavior becomes more extreme. For these observations a relatively small entrance aperture is needed, to reduce confusion due to scattered nebular light.

It is likely that low far-UV extinction indicates a relative under-abundance of the small grains; the association of this phenomenon with high-density regions may be a result of conversion of these small grains to larger ones by the growth of mantles, which would be expected to be density-dependent (e.g. Carrasco, Strom, and Strom, 1973; Snow, 1975b), or it may result from the selective expulsion of the small grains by radiation pressure due to newly-formed stars in dense aggregates of material.

Besides allowing the derivation of extinction curves for more heavily-obscured stars, the ability to observe faint stars will also allow more distant regions to be probed than possible at present. Borgman, Van Duinan, and Koorneef (1975) have used ANS data to show that the 2200 A bump seems to be absent in the 30 Doradus region of the Large

Magellanic Cloud (recently, however, Borgman (1978) has shown that the extinction could be normal); further evidence of unusual grain properties in other galaxies could have important implications for the chemical evolution of galaxies.

Finally, it will be of interest to determine the extinction properties of grains under various peculiar conditions such as in circumstellar material. Willis and Wilson (1975), using TD1 data, found an exceptionally strong 2200 A feature in the spectrum of a Wolf-Rayet star; it is probable that this anomaly resulted from some modification of the grains in the nebula surrounding the star. Many of the most interesting objects with circumstellar material are too faint to be well-observed with present instrumentation, so derivation of the UV extinction properties of such regions will be an important contribution of Spacelab. Caution must be exercised, however, since these objects are by definition peculiar and therefore difficult to match with comparison stars.

2.   Extinction below 1000 A

At least one of the suggested origins for the 2200 A extinction bump predicts the existence of a second feature centered near 1000 A (Gilra, 1973). If the far-UV rise is created by a distinct population of grains from those which produce the 2200 A feature, then the second bump near 1000 A would be superposed on the dominant far-UV rise, and might not be conspicuous. If the grains responsible for the far-UV rise are the ones which produce the bump at 2200 A, however, then possibly the observed rise towards 1000 A can be one wing of a second bump.

Present UV extinction curves terminate at 1000 A (York, et al., 1973; Snow, et al., 1980), and indicate that the steep rise continues at least that far with no sign of a turnover. With Copernicus data, it may be possible to derive isolated points in the curve down to 950 A, using only slightly-reddened stars. To improve the certainty of such results, as well as to extend the curve to the Lyman limit, it will be necessary to use an instrument with greater sensitivity below 1000 A than Copernicus and a resolving power of at least 2000, so that stellar continuum levels may be picked out between the many H_2 lines.

The question of matching stellar flux distributions in stars of similar spectral type will have to be carefully considered in deriving extinction curves below 1000 A, where the flux is dropping off rapidly in all but the early O stars. Small differences in chemical composition or temperature between reddened and comparison stars could result in substantial flux distribution differences at these short wavelengths, creating misleading extinction curves.

3.   Structure in the UV extinction curve

With the exception of the prediction by Duley (1974) that solid CO on grain mantles would produce interstellar absorption features in the 1400 to 1500 A region, there has been little specific reason to expect extinction structure in the far-UV. The work of Egan (1974) on optical properties of silicates indicates that the 2200 A bump might show structure if it is produced in silicate grains, however.

While no identification has yet been made of the absorbers which produce the diffuse interstellar bands in visible wavelengths, and therefore no specific predictions have been offered concerning the UV spectrum of these absorbers, the presence or absence of UV features could eventually play a role in identifying the source of the bands.

In order to unambiguously detect structure in the UV extinction in this portion of the spectra of hot stars. This could be the case if heavily-reddened stars were observed; hence high sensitivity and at least moderate spectral resolution (to avoid confusion with atomic or molecular lines) are needed.

4.   Scattered nebular light

Ultraviolet measurements of light scattered by interstellar grains have yielded valuable information on the nature of the scattering particles. In principle, both the grain albedo and the scattering phase function can be derived from such data, particuarly if some information can be deduced regarding the angle of incidence of the light prior to scattering. Witt and Lillie (1972, 1973), and Lillie and Witt (1976), using OAO-2 data on the diffuse galactic light, found that the grain albedo has a pronounced minimum near 2200 A, and then appears to rise steeply towards shorter wavelengths, down to about 1500 A. Hence it was deduced that the 2200 A extinction bump is an absorption feature, while the extinction shortward of there, at least as far as the beginning of the steep far-UV rise, is probably dominated by scattering.

It will be extremely interesting to extend the scattered light measurements to wavelengths shorter than 1500 A, where the dramatic rise in UV extinction begins. If the albedo continues to rise, and the extinction is therefore primarily due to scattering rather than absorption, then the challenge will be to find a substance which is nearly a perfect reflector of far-UV radiation (whatever such grains are made of, it might be a useful substance for coating the optics of future ultraviolet space instruments!).

Spacelab should provide opportunities to extend the scattered light measurements to a greater variety of regions and, with UV survey cameras, could provide dependable calibrations of field star contamination; a possible shortcoming of previous studies is that visible-wavelength star counts were used to determine the UV background. It will ultimately be of great interest to make scattered light measurements in other galaxies, but this may have to wait for the Space Telescope.

5.   Ultraviolet polarization measurements

Very little is known to date about the polarizing properties of the grains which produce the observed ultraviolet extinction. Gehrels (1974) used balloon data to show that the linear polarization decreases steadily with wavelength from visual wavelengths down to about 2200 A, implying that a single grain population dominates the observed polarization that far into the ultraviolet. This is consistent with the picture of UV extinction which has developed, in which the grains responsible for visual extinction dominate as far shortward as the 2200 A bump, but a population of smaller grains becomes more important in the far-UV. In this view, the polarization due to the larger grains should gradually disappear below 2000 A.

The polarizing properties of the small grains remain completely unknown. If the far-UV extinction is primarily due to scattering, then the potential for strong polarization is certainly present, if the small grains are nonspherical and tend to be aligned. The detection of such polarization and the determination of its wavelength dependence could provide additional constraints on the size and composition of the small grains.

Another interesting question is whether polarization structure exists within the 2200 A extinction feature, since for large classes of transitions in solids some polarization dependence on wavelength is expected (e.g. Martin and Angel, 1974). The single data point of Gehrels (1974) which was within the 2200 A bump cannot by itself be used to answer this question, since it was broad-band and contained no information on wavelength dependence on a sufficiently small scale. The fact that this data point fits the general trend starting in visual wavelength may, however, be indicative that no strong polarization structure exists within the extinction feature.

III.   MOLECULAR ABSORPTION LINES
A.   Recent Results

Until the advent of space astronomy, only the diatomic molecules CH, CH^+, and CN had been detected in interstellar absorption line measurements. Very recently two more species, OH and C_2, have been detected from the ground (Crutcher and Watson, 1976a; Lutz and Souza, 1977; Chaffee and Lutz, 1977). Both were also detected in UV absorption at about the same time with Copernicus (Snow, 1976a, 1978). Ultraviolet observations have added to the list H_2, HD, ^{12}CO, and ^{13}CO. In addition, upper limits have been determined for a number of species. Table 1 summarizes the observational results on molecular absorption lines, including a variety of upper limits. For species detected by several authors, the references given in the list are representative, and cite other relevant literature in their own references.

As Table 1 indicates, oscillator strengths are now unknown for a number of important species, which makes the results difficult to interpret and of only limited usefulness. Furthermore, for many simple and possibly abundant species, the absorption spectra have not yet been measured or analyzed, delaying any possibility of identifying them in space. It is imperative that the necessary laboratory work be done; for some of the molecular ions and very active radicals which are difficult to work with in normal laboratory conditions, perhaps it will be easiest to make the appropriate measurements in orbit, possibly on the Space Shuttle.

Two distinct mechanism have been discussed for molecular formation in interstellar clouds: formation on grain surfaces, followed by ejection into the gas phase; and direct formation by two-body interactions in the gas.

Formation on grains has been discussed in detail by Brecher and Arrhenius (1971), Watson and Salpeter (1972a,b) and Allen and Robinson (1975), who used available data on chemical processes on surfaces to discuss specific reactions which might take place on grains under interstellar conditions. Although the details of such calculations are rather uncertain, to a rough order of magnitude this approach results in predicted abundances for several simple species (Watson and Salpeter,

1972). These predicted abundances are only marginally in agreement with the observations, but one may speculate that a significantly greater total grain surface area, and hence greater molecule formation rates, might exist if a substantial portion of the depleted interstellar material is in the form of very small (< 100 A) grains such as those thought to be responsible for the far-UV extinction (Witt, 1973). Hence the grain-surface formation mechanism seems viable in the light of theoretical considerations.

Recently, however, intensive searches have failed to detect two species for which grain formation is the only feasible source. Crutcher and Watson (1976b) placed a restrictive upper limit on the abundance of NH, and Snow and Smith (1977) failed to detect NaH. For both species no significant concentration can build up through gas-phase reactions, either because key branching ratios strongly favor other species, or because the sequence leading to these products includes endothermic reactions. Since neither NH nor NaH was detected, however, at levels at least two orders of magnitude below the observed CH abundances, it now seems established that the abundant species such as CH (and probably the others detected so far, except for H_2) must be formed in gas phase, and that grain surface reactions are generally incapable of producing simple diatomics (other than hydrogen) in quantities greater than about $10^{-10} N_H$ (where N_H is the total column density of hydrogen nuclei).

H_2 probably is formed on grains, however, because the most favorable gas-phase reaction sequence (McDowell, 1961) fails to produce the observed quantities. Formation of this species on grains has been discussed by Hollenbach and Salpeter (1971), and more recently by de Jong (1972) and Dalgarno and McCray (1973), who made detailed comparisons of the relative efficiencies of H_2 formation in gas phase and on grains. York (1976) concluded that formation on grains dominates even in the low-mass clouds in front of unreddened stars, where the relatively high UV flux enhances the gas-phase formation of H_2 via associative detachment of H .

The importance of ion-molecule reaction sequences for forming molecules in gas phase in interstellar clouds was first pointed out by Watson (1973) and Black and Dalgarno (1973a), who showed that HD and OH should result from sequences beginning with charge-exchange between H^+ and either O or D, followed by $H_2 + O^+ \rightarrow OH + H$ or $H_2 + D^+ \rightarrow HD + H^+$. It soon became evident that reactions of this sort, which generally have very large cross sections, can produce substantial quantities of a number of diatomic (or even polyatomic) species in diffuse clouds (see, for example, Black and Dalgarno, 1977; Glassgold and Langer, 1976; and Mitchell, Ginsburg and Kuntz, 1977, and references cited therein). The abundances predicted from theoretical considerations of these reactions agree quite well in general with the observations, including the diatomics observed in dark clouds, although there are some discrepancies probably related to poorly-known bunching ratios or rate coefficients.

Besides providing information on chemical activity in space, observations of molecular species also yield data on physical conditions. For example, UV absorption measure of CO yield cloud density information from the J=1/J=0 ratio, assuming that the excited level is populated by

Table 1. Interstellar molecules observed in optical absorption

Species	Transition	Wavelength* (Å)	Star	N (cm^{-2})	N/N_H	Reference[†]
H$_2$	B$^1\Sigma_u^+$-X$^1\Sigma_g^+$	912-1108	ζ Oph	4.47 x 10^{20}	0.32	12, 19
HD	B$^1\Sigma_u^+$-X$^1\Sigma_g^+$	912-1108	ζ Oph	1.58 x 10^{14}	1.12 x 10^{-7}	19
C$_2$	D$^1\Sigma_u^+$-X$^1\Sigma_g^+$	2312.160	ζ Oph	1.2 x 10^{13}	8.57 x 10^{-9}	4, 8
C$_3$	A$^1\Pi_u$-X$^1\Pi_g$	4050.	ζ Oph	<2.7 x 10^{12}	<1.9 x 10^{-9}	5
CH	A$^2\Delta$-X$^2\Pi$	4300.321	ζ Oph	3.39 x 10^{13}	2.40 x 10^{-8}	1, 7
CH$^+$	A$^1\Pi$-X$^1\Sigma^+$	4232.539	ζ Oph	1.2 x 10^{13}	8.57 x 10^{-9}	21
^{13}CH$^+$	A$^1\Pi$-X$^1\Sigma^+$	4232.28	ζ Oph	1.32 x 10^{11}	9.36 x 10^{-11}	20, 21
CN	B$^2\Sigma^+$-X$^2\Sigma^+$	3874.608	ζ Oph	4.79 x 10^{12}	3.40 x 10^{-9}	7
CN$^+$	f$^1\Sigma$-a$^1\Sigma^+$	2181.10	ξ Per	<1 x 10^{12} f^{-1}	----	9
CO	C$^1\Sigma^+$-X$^1\Sigma^+$	1087.867	ζ Oph	1.2 x 10^{15}	8.6 x 10^{-7}	13
^{13}CO	A$^1\Pi$-X$^1\Sigma^+$	1395.3	ζ Oph	<5.62 x 10^{13}	<3.99 x 10^{-9}	9
CO$^+$	A$^2\Pi_i$-X$^2\Sigma^+$	4250.94	ζ Oph	<5.75 x 10^{12}	<4.08 x 10^{-9}	2, 7
CS	A$^1\Pi$-X$^1\Sigma^+$	2577.47	ζ Oph	<2.55 x 10^{13}	<1.81 x 10^{-8}	16
CH$_2$	$\tilde{D}^3\Pi_u$-X$^3\Sigma_g^-$	1396.8	ζ Oph	<1.56 x 10^{12} f^{-1}	----	16
CO$_2$	A^{1}B$_2$(Δu)-X$^1\Sigma_g^+$	1088.6	ζ Oph	<2.95 x 10^{13} f^{-1}	----	16
NH	A$^3\Pi_i$-X$^3\Sigma^-$	3358.06	ξ Per	<7 x 10^{11}	----	6
NH$^+$	C$^2\Sigma^+$-X$^2\Pi$	2889.76	ζ Oph	<1.65 x 10^{11} f^{-1}	----	16
N$_2$	p$^1\Sigma^+$-X$^1\Sigma^+$	958.51	δ Sco	<3.8 x 10^{12}	<2.62 x 10^{-9}	11
NO	A$^2\Sigma^+$-X$^2\Pi$	2262.28	ξ Per	<1.70 x 10^{15}	<1.21 x 10^{-6}	9
NO$^+$	A$^1\Pi$-X$^1\Sigma^+$	1312.94	ζ Oph	<1.07 x 10^{14}	<7.59 x 10^{-8}	9
OH	D$^2\Sigma$-X$^2\Pi$	1222.071	ζ Oph	5.24 x 10^{13}	3.72 x 10^{-8}	3, 16

Species	Transition	Wavelength* (Å)	Star	N (cm^{-2})	N/N$_H$	Reference†
O$_2$	$^3\Sigma_u^+ - X\,^3\Sigma_g^-$	1144.34	ξ Per	$<3.47 \times 10^{11}\,\underline{f}^{-1}$	----	15
H$_2$O	$3p\,^1\tilde{B} - X\,^1A_2$	1239.728	ζ Oph	$<1.7 \times 10^{13}$	$<1.21 \times 10^{-8}$	14, 18
NaH	$A\,^1\Sigma^+ - X\,^1\Sigma^+$	3990.88	ζ Oph	$<1.7 \times 10^{11}$	$<1.21 \times 10^{-9}$	17
MgH	$A\,^2\Pi - X\,^2\Sigma^+$	5187.06	ζ Oph	$<2.2 \times 10^{11}$	$<1.57 \times 10^{-10}$	7, 10
MgH$^+$	$A\,^1\Sigma - X\,^1\Sigma^+$	2806.02	ξ Per	$<2 \times 10^{11}\,\underline{f}^{-1}$	----	9
AlH	$C\,^1\Sigma^+ - X\,^1\Sigma^+$	2241.60	ζ Oph	$<2.56 \times 10^{11}\,\underline{f}^{-1}$	----	16
SiH	$A\,^2\Delta - X\,^2\Pi$	4119.48	ζ Oph	$<1.41 \times 10^{12}$	$<1.00 \times 10^{-9}$	7
SiO	$J\,^1\Pi - X\,^1\Sigma^+$	1310.01	ζ Oph	$<3.24 \times 10^{11}\,\underline{f}^{-1}$	----	16
SH	$G\,^2\Delta - X\,^2\Pi$	1257.34	ξ Per	$<1.45 \times 10^{11}\,\underline{f}^{-1}$	----	15
HCl	$C\,^1\Pi - X\,^1\Sigma^+$	1290.257	ζ Oph	$<1.29 \times 10^{12}$	$<9.21 \times 10^{-10}$	14, 22
CaH	$F\,^2\Sigma - X\,^2\Sigma$	2716.92	ζ Oph	$<9.34 \times 10^{10}\,\underline{f}^{-1}$	----	16

* Wavelengths shortward of 2000 Å are measured in vacuo; others are air wavelengths. All wavelengths refer to the line arising from the ground vibrational and rotational state. Most are from the compilation of Hsu and Smith (1977).

† References to Table 1:

1. Black and Dalgarno (1973b)
2. Bortolot and Thaddeus (1969)
3. Chaffee and Lutz (1977)
4. Chaffee, et al. (1979)
5. Clegg, Lambert, and Snell (1979)
6. Crutcher and Watson (1976b)
7. Herbig (1968)
8. Hobbs (1979)
9. Jenkins, et al. (1973)
10. Kirby, Saxon, and Liu (1979)
11. Lutz, Owen, and Snow (1979)
12. Savage, et al. (1977)
13. Smith, Kirshna Swamy, and Steche (1978)
14. Smith, Yoshino, and Parkinson (1979)
15. Snow (1975a)
16. Snow (1976a)
17. Snow and Smith (1977)
18. Snow and Smith (1980)
19. Spitzer, et al. (1973)
20. Vanden Bout (1972)
21. Vanden Bout (1979)
22. Wright and Morton (1979)

collisions with H_2, for which cross-sections are given by Green and Thaddeus (1976).

In addition, kinetic temperatures may be inferred from the observed rotational populations in homonuclear molecules, since these species have no allowed dipole transitions between rotational levels, so that the excitation is controlled by collisional equilibrium (Dalgarno, Black, and Weisheit, 1973). H_2 and recently C_2 (Chaffee, et al., 1980; Hobbs, 1979) have been analyzed in this way. A confusing effect, however, is provided by ultraviolet pumping, a process that populates high rotational levels due to photoexcitation to electronic states and subsequent downwards transitions through the vibration-rotation levels. For H_2, this is especially important for J=4 and higher levels so that the J=1 to J=0 ratio can still safely be used to derive kinetic temperatures. For C_2, the UV pumping mechanism has not yet been fully analyzed, so its effects on kinetic temperature determinations is not well understood.

The rotational level populations in the homonuclear species can be used to derive ultraviolet radiation field intensities, if the details of the pumping mechanism are known, and Jura (1975a,b) has carried out such analyses of H_2 for several lines of sight, identifying a few cases of high excitation, indicating close cloud-star proximity. Spitzer and Morton (1976) found a number of such clouds with negative velocities, perhaps indicating the presence of expanding circumstellar shells or bubbles, predicted (Caster, McCray, and Weaver, 1975; Weaver, et al., 1977; Hollenbach, Chu, and McCray, 1976) to arise due to stellar winds.

B.   New Observations
    1. H_2 and HD

While Copernicus has provided a great deal of data on H_2 and HD, some important additions are of interest, having to do with either extended sensitivity over that of Copernicus, so that more heavily-reddened lines of sight may be probed, or greater spectral resolution, allowing more detailed analyses of velocity structure.

An extension to fainter stars could provide H_2 column densities, hence overall hydrogen abundances (complementing Ly-α data obtained with the IUE), in dense clouds where depletions are of particular interest. Certainly in such regions nearly all the hydrogen is molecular, so that H_2 column densities are required in order to assess the ratios of other elements to hydrogen. For this work a large effective aperture, wavelength coverage at least as far shortward as 1108 A, and a spectral resolving power of 5000 or greater are required.

Useful velocity structure analyses, on the other hand, would require a resolving power of 10^5 or better, so that components separated by only a few km s^{-1} could be resolved. Again, coverage as far shortward of the (0-0) Lyman band at 1108 A would be needed, and it would be highly desirable to extend this to include some of the higher bands at shorter wavelengths to obtain a spread of oscillator strengths. The instrumental sensitivity for this purpose need not be so great, however; stars of similar brightness to those accessible to Copernicus could provide a great deal of new information. The primary scientific interest in this project would be the detection and analysis of cloud components that may be identifiable with circumstellar shells or bubbles, or with interstellar shocks due to supernovae. By resolving velocity structure, one could derive kinetic temperatures (in optically thin cases, where the J=0 and J=1 lines are not too strong) or radiation field intensities and densities (through the analysis of UV pumping of high rotational levels) for individual clouds.

Another argument for attempting to observe H_2 absorption towards faint stars is provided by the recent detection of forbidden infrared quadruple radiation from H_2 in the Orion nebula (Gautier, et al., 1976) which has led to theoretical work on mechanisms for producing hot H_2 clouds behing shock fronts (Kwan, 1977; London, McCray, and Chu, 1977; Hollenbach and Shull, 1977). Although the particular portion of the Orion nebula in which the detection of emission occurred is too heavily obscured to be observable in absorption against background stars, it may well be possible, using these theoretical models as a guide, to identify other regions where similar conditions exist, which can be observed in absorption. If so, a great deal more data may be obtained on the nature of such cloud-shock interactions, which have been invoked by some (e.g. Woodward, 1976) as triggers for star formation.

Finally, Black and Dalgarno (1973a) pointed out that the abundance of HD in clouds is a sensitive indicator of the cosmic ray ionization rate, since the production of HD is initiated by the reaction $D^+ + H_2 \rightarrow HD^+ + H$. Extension of HD observations into dense clouds could provide information relevant to the penetrating ability of cosmic rays into the clouds, and hence on the strength of their magnetic fields.

2.   Other Species

The primary contributions Spacelab can make to the study of molecules other than H_2 and HD are the search for new species in wavelengths not accessible to Copernicus, and the capability of probing somewhat denser clouds than presently possible, so that better data may be obtained even for species with transitions at wavelengths covered by Copernicus. There are three primary goals in the search for new molecules: (1) the detection of new species, providing new check-points for assessing the validity of models for interstellar cloud chemistry, and refining our knowledge of rate coefficients and branching ratios; (2) the detection in absorption of species observed in the same lines of sight in radio emission, yielding detailed information on physical conditions and star-cloud geometry; and (3) it may be hoped that with a moderate increase of sensitivity some species may be detected which are formed on grain surfaces, thereby providing specific information on grain surface chemistry.

In Table 2 are listed a variety of undetected species for which specific predictions have been made on the basis of gas-phase reaction schemes. Observational upper limits (from Table 1) are given for those species which have been sought. Most of those for which no limit is given have unknown spectra. In each case the prediction refers to a cloud with density $\sim 10^3$ cm^{-3} and T $\sim 50°$K; the cited references should be checked for details on the assumed conditions.

As pointed out by Snow (1975a), Crutcher (1976), and Knapp and Jura (1976), simultaneous detection of a molecular species in radio emission and optical absorption in the same cloud can yield unique data on the cloud conditions and the relative positions of the star and the cloud. The radio data have the advantage of having much better velocity resolution and accuracy than the optical data, thereby providing direct information on the internal velocity dispersion, whereas the optical data have much better spatial resolution, and furthermore place limits on the distance to observed clouds, which must be in front of their background stars. The combination of the two can yield: (a) improved column densities within a cloud, especially since the directly-measured velocity dispersions can be applied to the saturated optical lines, thereby reducing curve-of-growth uncertainties for species expected to exist primarily in the volume producing the radio emission; (b) more accurate knowledge of the cloud density and temperature, since the optical data yield rotational level population ratios, whereas radio observers normally have to assume an excitation temperature in order to derive the column density of a molecular species; and (c) improved knowledge of the relative positions of the star and the observed cloud. Furthermore, the radio mapping produces data on the extent of the cloud in the plane of the sky, something possible with absorption measures only if there is a homogeneous distribution of background stars.

At present, only a very few clouds have been investigated in this manner because most clouds optically thin enough for UV absorption measures to be feasible with Copernicus or other existing instruments are not sufficiently dense to produce detectable radio emission. A small increase in sensitivity would greatly increase the overlap between radio and UV optical astronomy and would enhance our knowledge of cloud conditions as well as of chemical proesses in space.

This work has been supported by National Aeronautics and Space Administration contract NAS5-23576 with Princeton University, and grant NSG-5355 with the University of Colorado.

Table 2.   Undetected molecules expected to be abundant in diffuse clouds*

Species	Predicted log N† (cm^{-2})	Predicted N/N_H (dex)	Reference††
CH_2	13.5 - 13.97	-7.5 - -7.2	2, 5
CH_3	12.85	-8.30	2
CH_3^+	11.0 - 12.58	-10.0 - -8.57	2, 5
C_2H	12.2 - 13.5	-8.8 - -7.5	1, 5
C_2H^+	10.5 - 10.79	-10.5 - -10.36	2, 5
$C_2H_2^+$	12.21	-8.94	2
NH	10.14 - 10.4	-11.01 - -10.6	1, 2
N_2	11.1 - 13.03	-9.9 - -8.12	1, 2
HCN	11.67	-9.48	2
OH^+	11.3 - 13.0	-9.7 - -8.0	1, 3
H_2O	12.03 - 13.3	-9.12 - -7.7	1, 2, 3
H_2O^+	11.3	-9.7	3
H_3O^+	11.7	-9.3	3
O_2	11.15 - 11.3	-10.00 - -9.7	1, 2
HCO	10.90 - 11.30	-10.10 - -9.85	2, 3
HCO^+	10.69	-10.46	2
H_2CO	10.93	-10.22	2
NO	11.2 - 11.54	-9.8 - -9.61	1, 2
HCl	13.46 - 13.63	-7.52 - -7.24	2, 4
HCl^+	12.94	-8.21	4
H_2Cl^+	12.41	-8.74	4

*   The predicted abundances given here are for clouds with $N_H \sim 10^{21}$ cm^{-2} and n $\sim$ 100-1000 cm^{-3}.   The cited references should be consulted for specific details of the models.

†   Molecules are included for which the column density in a diffuse cloud with $N_H \lesssim 10^{21}$ cm^{-2} is at least 10^{10} cm^{-2}.   When a wide range of values is given, this is an indication that some key parameter, such as a reaction rate or a branching ratio, is poorly known.

††   References for Table 2.
   1.   Barsuhn and Walmsley (1977)
   2.   Black and Dalgarno (1977)(specific model of ζ Oph cloud)
   3.   Glassgold and Langer (1974)
   4.   Jura (1974)
   5.   Mitchell, Ginsburg, and Kuntz (1977)

REFERENCES

Allen, M. and Robinson, G.W. 1975, Ap. J., 195, 81.
Barsuhn, J. and Walmsley, C.M. 1977, Astr. Ap., 54, 345.
Black, J.H. and Dalgarno, A. 1973a, Ap. J. (Letters), 184, L101.
Black, J.H. and Dalgarno, A. 1973b, Astrophys. Letters, 15, 79.
Black, J.H. and Dalgarno, A. 1977, Ap. J. (Suppl.), 34, 405.
Bless, R.C. and Savage, B.D. 1972, Ap. J., 171, 293.
Borgman, J. 1978, Astr. Ap., 69, 245.
Borgman, J., van Duinan, R.J., and Koorneef, J. 1975, Astr. Ap., 40, 461.
Bortolot, V.J. and Thaddeus, P. 1969, Ap. J. (Letters), 155, L17.
Brecher, A. and Arrhenius, G. 1971, Nat. Phys. Sci., 230, 107.
Carrasco, L., Strom, S.E., and Strom, K.M. 1973, Ap. J., 182, 95.
Carruthers, G.R. 1969, Ap. Space Sci., 5, 387.
Castor, J., McCray, R., and Weaver, R. 1975, Ap. J. (Letters), 155, L107.
Chaffee, F.H., Lutz, B.L., Black, J.H., Vanden Bout, P.A., and Snell, R.
    1979, AP. J., in press.
Chaffee, F.H. and Lutz, B.L. 1977, Ap. J., 213, 394.
Clegg, R., Lambert, D.L., and Snell, R. 1979, private communication.
Crutcher, R.M. 1976, Ap. J. (Letters), 206, L171.
Crutcher, R.M. and Watson, W.D. 1976a, Ap. J. (Letters), 203, L123.
Crutcher, R.M. and Watson, W.D. 1976b, Ap. J., 209, 778.
Dalgarno, A., Black, J.H., and Weisheit, J.C. 1973, Ap. J. (Letters),
    184, L77.
Dalgarno, A. and McCray, R. 1973, Ap. J., 181, 95.
de Jong, T. 1972, Astr. Ap., 20, 263.
Dorschner, J. 1973, Ap. Space Sci., 25, 405.
Duley, W.W. 1974, Ap. Space Sci., 26, 199.
Duley, W.W. and McCullough, J.D. 1977, Ap. J. (Letters), 211, L145.
Egan, W. 1974, private communication.
Gautier, T.N., Fink, U., Treffers, R.R., and Larson, H.P. 1976, Ap. J.
    (Letters), 207, L129.
Gehrels, T. 1974, Ap. J., 79, 590.
Gilra, D.P. 1972, Scientific Results from OAO-2, ed. A.D. Code (NASA
    SP-310), p. 395.
Gilra, D.P. 1973, Ph.D. Dissertation, University of Wisconsin.
Glassgold, A.E. and Langer, W.D. 1976, Ap. J., 206, 85.
Green, S. and Thaddeus, P. 1976, Ap. J., 205, 766.
Herbig, G.H. 1968, Z. f. Ap., 68, 243.
Herbig, G.H. 1975, Ap. J., 196, 129.
Hobbs, L.M. 1979, Ap. J. (Letters), 232, L175.
Hollenbach, D.J., Chu, S.-I., and McCray, R. 1976, Ap. J., 208, 458.
Hollenbach, D.J. and Salpeter, E.E. 1971, Ap. J., 163, 155.
Hollenbach, D.J. and Shull, J.M. 1977, Ap. J., 216, 419.
Huffman, D.R. and Stapp, J.L. 1971, Nature Phys. Sci., 229, 45.
Jenkins, E.B., Drake, J.F., Morton, D.C., Rogerson, J.B., Spitzer, L.,
    and York, D.G. 1973, Ap. J. (Letters), 181, L122.
Jura, M. 1974, Ap. J. (Letters), 190, L33.
Jura, M. 1975a, Ap. J., 197, 575.
Jura, M. 1975b, ibid., 197, 581.
Kirby, K., Saxon, R.P., and Liu, B. 1979, Ap. J., 231, 637.

Knapp, G.R. and Jura, M. 1976, Ap. J., 209, 782.
Kwan, J. 1977, Ap. J., 216, 713.
Lillie, C.F. and Witt, A.N. 1976, Ap. J., 208, 64.
London, R., McCray, R., and Chu, S.-I. 1977, Ap. J., 217, 442.
Lutz, B.L., Owen, T., and Snow, T.P. 1979, Ap. J., 227, 159.
Lutz, B.L. and Souza, S.P. 1977, Ap. J. (Letters), 213, L129.
Martin, P.G. and Angel, J.R. 1974, Ap. J., 188, 517.
McDowell, M.R.C. 1961, Observatory, 81, 240.
Mitchell, G.F., Ginsburg, J. L., and Kuntz, P.J. 1977, Ap. J., 212, 71.
Morton, D.C. 1975, Ap. J., 197, 85.
Nandy, K. and Thompson, G.I. 1976, Mon. Not. R. A. S., 173, 237.
Savage, B.D. 1975, Ap. J., 199, 92.
Savage, B.D., Bohlin, R.C., Drake, J.F., and Budich, W. 1977, Ap. J.,
    216, 291.
Smith, A.M., Krishna Swamy, K.S., and Stecher, T.P. 1978, Ap. J., 220,
    138.
Smith, P.L., Yoshino, K. and Parkinson, W.H. 1979, IAU Symp. 87, Inter-
    stellar Molecules, ed. B. Andrew (Dordrecht:Reidel), in press.
Snow, T.P. 1975a, Ap. J. (Letters), 201, L21.
Snow, T.P. 1975b, ibid., 202, L87.
Snow, T.P. 1976a, ibid., 204, L127.
Snow, T.P. 1976b, Earth and Extraterrestrial Sciences, 3, 1.
Snow, T.P. 1978, Ap. J. (Letters), 202, L93.
Snow, T.P. and Smith, W.H. 1977, Ap. J., 218, 124.
Snow, T.P. and Smith, W.H. 1980, in preparation.
Snow, T.P. and York, D.G. 1975, Ap. Space Sci., 34, 19.
Snow, T.P., York, D.G., Welty, D.E., and Hornack, P. 1980, in preparation.
Snow, T.P., York, D.G., and Resnick, M. 1977, Pub. A.S.P., 89, 758.
Spitzer, L., Drake, J.F., Jenkins, E.B., Morton, D.C., Rogerson, J.B.,
    and York, D.G. 1973, Ap. J. (Letters), 181, L116.
Spitzer, L. and Jenkins, E.B. 1975, Ann. Rev. Astr. Ap., 13, 133.
Spitzer, L. and Morton, W.A. 1976, Ap. J., 204, 731.
Stecher, T.P. 1965, Ap. J., 142, 1683.
Vanden Bout, P.A. 1972, Ap. J. (Letters), 176, L127.
Vanden Bout, P.A. 1979, preprint.
Watson, W.D. 1973, Ap. J. (Letters), 183, L17.
Watson, W.D. and Salpeter, E.E. 1977a, Ap. J., 174, 321.
Watson, W.D. and Salpeter, E.E. 1977b, ibid., 175, 659.
Weaver, R., McCray R., Castro, J., Shapiro, P. and Moore, R. 1977,
    Ap. J., 218, 377 (eratum: Ap. J., 220, 742).
Wickramasinghe, N.C., and Nandy, K. 1971, Nature Phys. Sci., 229, 81.
Willis, A.J. and Wilson, R. 1975, Astr. Ap., 44, 205.
Witt, A.N. 1973, IAU Symp. 52, Interstellar Dust and Related Topics, ed.
    G.M. Greenberg and H.C. van de Hulst (Boston: D. Reidel), p. 53.
Witt, A.N. and Lillie, C.F. 1972, Scientific Results from OAO-2, ed.
    A.D. Code (NASA SP-310), p. 199.
Witt, A.N. and Lillie, C.F. 1973, Astr. Ap., 25, 397.
Woodward, P.R. 1976, Ap. J., 207, 48.
Wright, E.L. and Morton, D.C. 1979, Ap. J., 227, 483.
Wu, C.-C., Snow, T.P., York, D.G., and van Duinan, R.J. 1980, in
    preparation.

York, D.G. 1976, Ap. J., 204, 750.
York, D.G. 1979, Proc. Conf. Ap. Space Sci. from Spacelab, ed. P.L.
    Bernacca and R. Ruffini (Dordrecht: D. Reidel), in press.
York, D.G., Drake, J.F., Jenkins, E.B., Morton, D.C., Rogerson, J.B.,
    and Spitzer, L. 1973, Ap. J. (Letters), 182, L1.

INDEX OF SUBJECTS

ASTROPHYSICS AND SPACE SCIENCE LIBRARY

Edited by

J. E. Blamont, R. L. F. Boyd, L. Goldberg, C. de Jager, Z. Kopal, G. H. Ludwig, R. Lüst,
B. M. McCormac, H. E. Newell, L. I. Sedov, Z. Švestka, and W. de Graaff

1. C. de Jager (ed.), *The Solar Spectrum, Proceedings of the Symposium held at the University of Utrecht, 26–31 August, 1963.* 1965, XIV + 417 pp.
2. J. Orthner and H. Maseland (eds.), *Introduction to Solar Terrestrial Relations, Proceedings of the Summer School in Space Physics held in Alpbach, Austria, July 15–August 10, 1963 and Organized by the European Preparatory Commission for Space Research.* 1965, IX + 506 pp.
3. C. C. Chang and S. S. Huang (eds.), *Proceedings of the Plasma Space Science Symposium, held at the Catholic University of America, Washington, D.C., June 11–14, 1963.* 1965, IX + 377 pp.
4. Zdeněk Kopal, *An Introduction to the Study of the Moon.* 1966, XII + 464 pp.
5. B. M. McCormac (ed.), *Radiation Trapped in the Earth's Magnetic Field. Proceedings of the Advanced Study Institute, held at the Chr. Michelsen Institute, Bergen, Norway, August 16–September 3, 1965.* 1966, XII + 901 pp.
6. A. B. Underhill, *The Early Type Stars.* 1966, XII + 282 pp.
7. Jean Kovalevsky, *Introduction to Celestial Mechanics.* 1967, VIII + 427 pp.
8. Zdeněk Kopal and Constantine L. Goudas (eds.), *Measure of the Moon. Proceedings of the 2nd International Conference on Selenodesy and Lunar Topography, held in the University of Manchester, England, May 30–June 4, 1966.* 1967, XVIII + 479 pp.
9. J. G. Emming (ed.), *Electromagnetic Radiation in Space. Proceedings of the 3rd ESRO Summer School in Space Physics, held in Alpbach, Austria, from 19 July to 13 August, 1965.* 1968, VIII + 307 pp.
10. R. L. Carovillano, John F. McClay, and Henry R. Radoski (eds.), *Physics of the Magnetosphere, Based upon the Proceedings of the Conference held at Boston College, June 19–28, 1967.* 1968, X + 686 pp.
11. Syun-Ichi Akasofu, *Polar and Magnetospheric Substorms.* 1968, XVIII + 280 pp.
12. Peter M. Millman (ed.), *Meteorite Research. Proceedings of a Symposium on Meteorite Research, held in Vienna, Austria, 7–13 August, 1968.* 1969, XV + 941 pp.
13. Margherita Hack (ed.), *Mass Loss from Stars. Proceedings of the 2nd Trieste Colloquium on Astrophysics, 12–17 September, 1968.* 1969, XII + 345 pp.
14. N. D'Angelo (ed.), *Low-Frequency Waves and Irregularities in the Ionosphere. Proceedings of the 2nd ESRIN-ESLAB Symposium, held in Frascati, Italy, 23–27 September, 1968.* 1969, VII + 218 pp.
15. G. A. Partel (ed.), *Space Engineering. Proceedings of the 2nd International Conference on Space Engineering, held at the Fondazione Giorgio Cini, Isola di San Giorgio, Venice, Italy, May 7–10, 1969.* 1970, XI + 728 pp.
16. S. Fred Singer (ed.), *Manned Laboratories in Space. Second International Orbital Laboratory Symposium.* 1969, XIII + 133 pp.
17. B. M. McCormac (ed.), *Particles and Fields in the Magnetosphere. Symposium Organized by the Summer Advanced Study Institute, held at the University of California, Santa Barbara, Calif., August 4–15, 1969.* 1970, XI + 450 pp.
18. Jean-Claude Pecker, *Experimental Astronomy.* 1970, X + 105 pp.
19. V. Manno and D. E. Page (eds.), *Intercorrelated Satellite Observations related to Solar Events. Proceedings of the 3rd ESLAB/ESRIN Symposium held in Noordwijk, The Netherlands, September 16–19, 1969.* 1970, XVI + 627 pp.
20. L. Mansinha, D. E. Smylie, and A. E. Beck, *Earthquake Displacement Fields and the Rotation of the Earth, A NATO Advanced Study Institute Conference Organized by the Department of Geophysics, University of Western Ontario, London, Canada, June 22–28, 1969.* 1970, XI + 308 pp.
21. Jean-Claude Pecker, *Space Observatories.* 1970, XI + 120 pp.
22. L. N. Mavridis (ed.), *Structure and Evolution of the Galaxy. Proceedings of the NATO Advanced Study Institute, held in Athens, September 8–19, 1969.* 1971, VII + 312 pp.

23. A. Muller (ed.), *The Magellanic Clouds. A European Southern Observatory Presentation: Principal Prospects, Current Observational and Theoretical Approaches, and Prospects for Future Research, Based on the Symposium on the Magellanic Clouds, held in Santiago de Chile, March 1969, on the Occasion of the Dedication of the European Southern Observatory.* 1971, XII + 189 pp.

24. B. M. McCormac (ed.), *The Radiating Atmosphere. Proceedings of a Symposium Organized by the Summer Advanced Study Institute, held at Queen's University, Kingston, Ontario, August 3–14, 1970.* 1971, XI + 455 pp.

25. G. Fiocco (ed.), *Mesospheric Models and Related Experiments. Proceedings of the 4th ESRIN-ESLAB Symposium, held at Frascati, Italy, July 6–10, 1970.* 1971, VIII + 298 pp.

26. I. Atanasijević, *Selected Exercises in Galactic Astronomy.* 1971, XII + 144 pp.

27. C. J. Macris (ed.), *Physics of the Solar Corona. Proceedings of the NATO Advanced Study Institute on Physics of the Solar Corona, held at Cavouri-Vouliagmeni, Athens, Greece, 6–17 September 1970.* 1971, XII + 345 pp.

28. F. Delobeau, *The Environment of the Earth.* 1971, IX + 113 pp.

29. E. R. Dyer (general ed.), *Solar-Terrestrial Physics/1970. Proceedings of the International Symposium on Solar-Terrestrial Physics, held in Leningrad, U.S.S.R., 12–19 May 1970.* 1972, VIII + 938 pp.

30. V. Manno and J. Ring (eds.), *Infrared Detection Techniques for Space Research. Proceedings of the 5th ESLAB-ESRIN Symposium, held in Noordwijk, The Netherlands, June 8–11, 1971.* 1972, XII + 344 pp.

31. M. Lecar (ed.), *Gravitational N-Body Problem. Proceedings of IAU Colloquium No. 10, held in Cambridge, England, August 12–15, 1970.* 1972, XI + 441 pp.

32. B. M. McCormac (ed.), *Earth's Magnetospheric Processes. Proceedings of a Symposium Organized by the Summer Advanced Study Institute and Ninth ESRO Summer School, held in Cortina, Italy, August 30–September 10, 1971.* 1972, VIII + 417 pp.

33. Antonin Rükl, *Maps of Lunar Hemispheres.* 1972, V + 24 pp.

34. V. Kourganoff, *Introduction to the Physics of Stellar Interiors.* 1973, XI + 115 pp.

35. B. M. McCormac (ed.), *Physics and Chemistry of Upper Atmospheres. Proceedings of a Symposium Organized by the Summer Advanced Study Institute, held at the University of Orléans, France, July 31–August 11, 1972.* 1973, VIII + 389 pp.

36. J. D. Fernie (ed.), *Variable Stars in Globular Clusters and in Related Systems. Proceedings of the IAU Colloquium No. 21, held at the University of Toronto, Toronto, Canada, August 29–31, 1972.* 1973, IX + 234 pp.

37. R. J. L. Grard (ed.), *Photon and Particle Interaction with Surfaces in Space. Proceedings of the 6th ESLAB Symposium, held at Noordwijk, The Netherlands, 26–29 September, 1972.* 1973, XV + 577 pp.

38. Werner Israel (ed.), *Relativity, Astrophysics and Cosmology. Proceedings of the Summer School, held 14–26 August, 1972, at the BANFF Centre, BANFF, Alberta, Canada.* 1973, IX + 323 pp.

39. B. D. Tapley and V. Szebehely (eds.), *Recent Advances in Dynamical Astronomy. Proceedings of the NATO Advanced Study Institute in Dynamical Astronomy, held in Cortina d'Ampezzo, Italy, August 9–12, 1972.* 1973, XIII + 468 pp.

40. A. G. W. Cameron (ed.), *Cosmochemistry. Proceedings of the Symposium on Cosmochemistry, held at the Smithsonian Astrophysical Observatory, Cambridge, Mass., August 14–16, 1972.* 1973, X + 173 pp.

41. M. Golay, *Introduction to Astronomical Photometry.* 1974, IX + 364 pp.

42. D. E. Page (ed.), *Correlated Interplanetary and Magnetospheric Observations. Proceedings of the 7th ESLAB Symposium, held at Saulgau, W. Germany, 22–25 May, 1973.* 1974, XIV + 662 pp.

43. Riccardo Giacconi and Herbert Gursky (eds.), *X-Ray Astronomy.* 1974, X + 450 pp.

44. B. M. McCormac (ed.), *Magnetospheric Physics. Proceedings of the Advanced Summer Institute, held in Sheffield, U.K., August 1973.* 1974, VII + 399 pp.

45. C. B. Cosmovici (ed.), *Supernovae and Supernova Remnants. Proceedings of the International Conference on Supernovae, held in Lecce, Italy, May 7–11, 1973.* 1974, XVII + 387 pp.

46. A. P. Mitra, *Ionospheric Effects of Solar Flares.* 1974, XI + 294 pp.

47. S.-I. Akasofu, *Physics of Magnetospheric Substorms.* 1977, XVIII + 599 pp.

48. H. Gursky and R. Ruffini (eds.), *Neutron Stars, Black Holes and Binary X-Ray Sources*. 1975, XII + 441 pp.

49. Z. Švestka and P. Simon (eds.), *Catalog of Solar Particle Events 1955–1969. Prepared under the Auspices of Working Group 2 of the Inter-Union Commission on Solar-Terrestrial Physics*. 1975, IX + 428 pp.

50. Zdeněk Kopal and Robert W. Carder, *Mapping of the Moon*. 1974, VIII + 237 pp.

51. B. M. McCormac (ed.), *Atmospheres of Earth and the Planets. Proceedings of the Summer Advanced Study Institute, held at the University of Liège, Belgium, July 29–August 8, 1974*. 1975, VII + 454 pp.

52. V. Formisano (ed.), *The Magnetospheres of the Earth and Jupiter. Proceedings of the Neil Brice Memorial Symposium, held in Frascati, May 28–June 1, 1974*. 1975, XI + 485 pp.

53. R. Grant Athay, *The Solar Chromosphere and Corona: Quiet Sun*. 1976, XI + 504 pp.

54. C. de Jager and H. Nieuwenhuijzen (eds.), *Image Processing Techniques in Astronomy. Proceedings of a Conference, held in Utrecht on March 25–27, 1975*. XI + 418 pp.

55. N. C. Wickramasinghe and D. J. Morgan (eds.), *Solid State Astrophysics. Proceedings of a Symposium, held at the University College, Cardiff, Wales, 9–12 July 1974*. 1976, XII + 314 pp.

56. John Meaburn, *Detection and Spectrometry of Faint Light*. 1976, IX + 270 pp.

57. K. Knott and B. Battrick (eds.), *The Scientific Satellite Programme during the International Magnetospheric Study. Proceedings of the 10th ESLAB Symposium, held at Vienna, Austria, 10–13 June 1975*. 1976, XV + 464 pp.

58. B. M. McCormac (ed.), *Magnetospheric Particles and Fields. Proceedings of the Summer Advanced Study School, held in Graz, Austria, August 4–15, 1975*. 1976, VII + 331 pp.

59. B. S. P. Shen and M. Merker (eds.), *Spallation Nuclear Reactions and Their Applications*. 1976, VIII + 235 pp.

60. Walter S. Fitch (ed.), *Multiple Periodic Variable Stars. Proceedings of the International Astronomical Union Colloquium No. 29, held at Budapest, Hungary, 1–5 September 1976*. 1976, XIV + 348 pp.

61. J. J. Burger, A. Pedersen, and B. Battrick (eds.), *Atmospheric Physics from Spacelab. Proceedings of the 11th ESLAB Symposium, Organized by the Space Science Department of the European Space Agency, held at Frascati, Italy, 11–14 May 1976*. 1976, XX + 409 pp.

62. J. Derral Mulholland (ed.), *Scientific Applications of Lunar Laser Ranging. Proceedings of a Symposium held in Austin, Tex., U.S.A., 8–10 June, 1976*. 1977, XVII + 302 pp.

63. Giovanni G. Fazio (ed.), *Infrared and Submillimeter Astronomy. Proceedings of a Symposium held in Philadelphia, Penn., U.S.A., 8–10 June, 1976*. 1977, X + 226 pp.

64. C. Jaschek and G. A. Wilkins (eds.), *Compilation, Critical Evaluation and Distribution of Stellar Data. Proceedings of the International Astronomical Union Colloquium No. 35, held at Strasbourg, France, 19–21 August, 1976*. 1977, XIV + 316 pp.

65. M. Friedjung (ed.), *Novae and Related Stars. Proceedings of an International Conference held by the Institut d'Astrophysique, Paris, France, 7–9 September, 1976*. 1977, XIV + 228 pp.

66. David N. Schramm (ed.), *Supernovae. Proceedings of a Special IAU-Session on Supernovae held in Grenoble, France, 1 September, 1976*. 1977, X + 192 pp.

67. Jean Audouze (ed.), *CNO Isotopes in Astrophysics. Proceedings of a Special IAU Session held in Grenoble, France, 30 August, 1976*. 1977, XIII + 195 pp.

68. Z. Kopal, *Dynamics of Close Binary Systems*, XIII + 510 pp.

69. A. Bruzek and C. J. Durrant (eds.), *Illustrated Glossary for Solar and Solar-Terrestrial Physics*. 1977, XVIII + 204 pp.

70. H. van Woerden (ed.), *Topics in Interstellar Matter*. 1977, VIII + 295 pp.

71. M. A. Shea, D. F. Smart, and T. S. Wu (eds.), *Study of Travelling Interplanetary Phenomena*. 1977, XII + 439 pp.

72. V. Szebehely (ed.), *Dynamics of Planets and Satellites and Theories of Their Motion. Proceedings of IAU Colloquium No. 41, held in Cambridge, England, 17–19 August 1976*. 1978, XII + 375 pp.

73. James R. Wertz (ed.), *Spacecraft Attitude Determination and Control*. 1978, XVI + 858 pp.

74. Peter J. Palmadesso and K. Papadopoulos (eds.), *Wave Instabilities in Space Plasmas. Proceedings of a Symposium Organized Within the XIX URSI General Assembly held in Helsinki, Finland, July 31–August 8, 1978*. 1979, VII + 309 pp.

75. Bengt E. Westerlund (ed.), *Stars and Star Systems. Proceedings of the Fourth European Regional Meeting in Astronomy held in Uppsala, Sweden, 7–12 August, 1978*. 1979, XVIII + 264 pp.

76. Cornelis van Schooneveld (ed.), *Image Formation from Coherence Functions in Astronomy. Proceedings of IAU Colloquium No. 49 on the Formation of Images from Spatial Coherence Functions in Astronomy, held at Groningen, The Netherlands, 10–12 August 1978*. 1979, XII + 338 pp.

77. Zdeněk Kopal, *Language of the Stars. A Discourse on the Theory of the Light Changes of Eclipsing Variables*. 1979, VIII + 280 pp.

78. S.-I. Akasofu (ed.), *Dynamics of the Magnetosphere. Proceedings of the A.G.U. Chapman Conference 'Magnetospheric Substorms and Related Plasma Processes' held at Los Alamos Scientific Laboratory, N.M., U.S.A., October 9–13, 1978*. 1980, XII + 658 pp.

79. Paul S. Wesson, *Gravity, Particles, and Astrophysics. A Review of Modern Theories of Gravity and G-variability, and their Relation to Elementary Particle Physics and Astrophysics*. 1980, VIII + 268 pp.

80. Peter A. Shaver (ed.), *Radio Recombination Lines. Proceedings of a Workshop held in Ottawa, Ontario, Canada, August 24–25, 1979*. 1980, X + 284 pp.